Saas-Fee Advanced Course 22
Lecture Notes 1992

S. N. Shore   M. Livio
E. P. J. van den Heuvel

Interacting Binaries

Saas-Fee Advanced Course 22
Lecture Notes 1992
Swiss Society for Astrophysics and Astronomy

Edited by H. Nussbaumer and A. Orr

With 201 Figures

Springer-Verlag
Berlin Heidelberg New York
London Paris Tokyo
Hong Kong Barcelona
Budapest

Professor S. N. Shore
GHRS Science Team, Computer Sciences Corporation, Code 681, Goddard Space Flight Center, Greenbelt, MD 20771, USA
and
DEMIRM, Observatoire de Meudon, F-92190 Meudon Principal Cedex, France

Professor M. Livio
Space Telescope Science Institute, Homewood Campus, 3700 San Martin Drive, Baltimore MD 21218, USA

Professor E. P. J. van den Heuvel
Astronomical Institute, University of Amsterdam, Kruislaan 403, 1098 SJ Amsterdam, The Netherlands

Volume Editors:

Professor H. Nussbaumer
Institut für Astronomie, ETH-Zentrum, CH-8092 Zürich, Switzerland

A. Orr
Observatoire de Genève, ch. des Maillettes 51, CH-1290 Sauverny, Switzerland

This series is edited on behalf of the Swiss Society for Astrophysics and Astronomy:

Société Suisse d'Astrophysique et d'Astronomie
Observatoire de Genève, ch. des Maillettes 51, CH-1290 Sauverny, Switzerland

ISBN 3-540-57014-4 Springer-Verlag Berlin Heidelberg New York
ISBN 0-387-57014-4 Springer-Verlag New York Berlin Heidelberg

CIP data applied for

Printed in Germany

Typesetting: Camera ready copy from the author/editor
55/3140 - 5 4 3 2 1 0 - Printed on acid-free paper

Foreword

Observational and Theoretical Issues of Interacting Binaries was the topic of the 22nd Advanced Course of the Swiss Society for Astrophysics and Astronomy. It was the first time that binary systems were the center of attention of our course. The established concept and organisation of the Advanced Course has been retained: three scientists, all acknowledged experts in their respective fields, were each invited to give nine one-hour lectures within the period of a week. The Advanced Course took place from April 6 to 11, 1992, at Les Diablerets, a charming resort in the Swiss alps. The high level of the lectures, the international background of the 65 participants, including many young students, and the beauty of the surroundings all contributed to the success of the course. The lecture notes of this course, the 22nd in our series, are also the third to be published by Springer-Verlag.

Well over half of all stars seem to exist in binary systems. The study of binary evolution is therefore essential for our understanding of stellar evolution in general. The evolution of interacting binaries contains in itself many of the problems met in other fields of modern astrophysics. This is very apparent in these lecture notes.

The lecturers were asked to cover different fields: Steven Shore puts the emphasis on observations and their connection to relevant physics. He also discusses symbiotic stars. Cataclysmic variables are the subject of Mario Livio's course, whereas Edward van den Heuvel concentrates on more massive binaries and X-ray binaries. This book covers a gap in the existing scientific literature. Its unique concentration of information and references to different aspects of interacting binaries will be invaluable to both students and researchers.

It is a pleasure to thank our three lecturers for their lively presentations which always provoked enriching discussions. We are particularly indebted to Steven Shore for preparing the index of this volume. We greatly appreciate the efficiency of Springer-Verlag in the publishing and distribution of this series. We also thank our secretary, Mrs Irène Scheffre, and other members of our institutes for helping in the organization of this course.

Zurich and Geneva,
May 1993

H. Nussbaumer
A. Orr

Contents

Observations and Physical Processes in Binary Stars

By Steven N. Shore (With 44 Figures)

Topics in the Theory of Cataclysmic Variables and X-Ray Binaries

By Mario Livio (With 65 Figures)

Interacting Binaries:
Topics in Close Binary Evolution

By Edward P.J. van den Heuvel (With 92 Figures)

List of Previous Saas-Fee Advanced Courses

1992 INTERACTING BINARIES
S.N. Shore, M. Livio, E.P.J. van den Heuvel
1991 THE GALACTIC INTERSTELLAR MEDIUM
W.B. Burton, B.G. Elmegreen, R. Genzel
1990 ACTIVE GALACTIC NUCLEI
R. Blandford, H. Netzer, L. Woltjer
1989 THE MILKY WAY AS A GALAXY
G. Gilmore, I. King, P. van der Kruit
1988 RADIATION IN MOVING GASEOUS MEDIA
H. Frisch, R.P. Kudritzki, H.W. Yorke
1987 LARGE SCALE STRUCTURES IN THE UNIVERSE
A.C. Fabian, M. Geller, A. Szalay
1986 NUCLEOSYNTHESIS AND CHEMICAL EVOLUTION
J. Audouze, C. Chiosi, S.E. Woosley
1985 HIGH RESOLUTION IN ASTRONOMY
R.S. Booth, J.W. Brault, A. Labeyrie
1984 PLANETS, THEIR ORIGIN, INTERIOR AND ATMOSPHERE
D. Gautier, W.B. Hubbard, H. Reeves
1983 ASTROPHYSICAL PROCESSES IN UPPER MAIN SEQUENCE STARS
A.N. Cox, S. Vauclair, J.P. Zahn
*1982 MORPHOLOGY AND DYNAMICS OF GALAXIES
J. Binney, J. Kormendy, S.D.M. White
1981 ACTIVITY AND OUTER ATMOSPHERES OF THE SUN AND STARS
F. Praderie, D.S. Spicer, G.L. Withbroe
*1980 STAR FORMATION
J. Appenzeller, J. Lequeux, J. Silk
*1979 EXTRAGALACTIC HIGH ENERGY ASTROPHYSICS
F. Pacini, C. Ryter, P.A. Strittmatter
*1978 OBSERVATIONAL COSMOLOGY
J.E. Gunn, M.S. Longair, M.J. Rees
*1977 ADVANCED STAGES IN STELLAR EVOLUTION
I. Iben Jr., A. Renzini, D.N. Schramm
*1976 GALAXIES
K. Freeman, R.C. Larson, B. Tinsley
*1975 ATOMIC AND MOLECULAR PROCESSES IN ASTROPHYSICS
A. Dalgarno, F. Masnou-Seeuws, R.V.P. McWhirter

---

* Out of print

*1974 MAGNETOHYDRODYNAMICS
L. Mestel, N.O. Weiss
*1973 DYNAMICAL STRUCTURE AND EVOLUTION OF STELLAR SYSTEMS
G. Contopoulos, M. Hénon, D. Lynden-Bell
*1972 INTERSTELLAR MATTER
N.C. Wrickramasinghe, F.D. Kahn, P.G. Mezger
*1971 THEORY OF THE STELLAR ATMOSPHERES
D. Mihalas, P. Pagel, P. Souffrin

Books up to 1989 may be ordered from: SAAS-FEE COURSES
GENEVA OBSERVATORY
ch. des Maillettes 51
CH-1290 Sauverny, Switzerland

Books from 1990 on may be ordered from Springer-Verlag.

Observations and Physical Processes in Binary Stars

Steven N. Shore

GHRS Science Team/Computer Sciences Corporation, Code 681, Goddard Space Flight Center, Greenbelt, MD 20771 USA and DEMIRM, Observatoire de Meudon

Abstract

Binary systems represent the best examples of real time astrophysics. These lectures deal with the interface between observation and theory in comparatively normal close binary stars. The first half discusses the observational techniques for determining periods, radial velocities, light curve data, and surface maps of stars and accretion disks. The second portion deals with a range of specific physical processes. Although the observational data is slanted toward the ultraviolet, these lectures emphasize the need for multiwavelength and multi-technique approaches.

1 Introduction

The observational study of binary stars traces back to the beginnings of astrophysics. It has been one of the constants, a source of insight into the structure and evolution of stars and also a pointer to many of the needed improvements in hydrodynamic and radiative transfer theory. The advantage granted by binaries is the occasional and happy accident of aspect. Our line of sight to the stars is sometimes in the privileged position of being close enough to the orbital plane to witness eclipses. With the ambiguity of inclination removed, we can derive information about the system, and its constituent stars, that cannot be gained any other way for single stars. We pay a price, of course. All of astronomy is filled with Hobson's choices. We may want to study long period systems, so that the stars are well separated, but we have few opportunities to observe change and it takes a very long time. If we concentrate on short period systems we can obtain our required data quickly. However, the insights gained about the stars may not represent the typical properties of normal stars.

The aim of this part of the course is to provide an overview of some of the tools and observational questions raised by binary star systems. With the exception of pulsating variables, there are few areas in modern astrophysics that present such diversity and require so many physical processes to be considered. And there are none that depend as heavily on the historical record for data. To know the interests and techniques of earlier workers in the field is also to have an access portal to the literature, and possibly to some interesting observational material and insights. In 1962, in the introduction to *Astronomy in the 20th Century*, Struve and Zebergs lamented that there was at that time too little idea of the historical context of astrophysical investigation. Let's first look back at the development of the field before moving into the proper subject of the course.[1]

2 A Brief Historical Prolegomena to the Study of Binaries

"I have learnt at the R(oyal) Society that the periodical occultation of the light of Algol happened last night at about 12 o'clock; the period is said to be 2 days, 21 hours, and the discovery is now said to have been made by a deaf and dumb man, the grandson of Sir John Goodricke, who for some years amused himself with astronomy. This is all I have made out; you may depend on any intelligence which I think likely either to amuse or instruct you."[2]. With this letter, Joseph Banks, the president of the Royal Society, communicated to William Herschel the discovery of the periodicity of Algol's variations[3] The discovery of a periodicity was singularly important. Although at the end of the eighteenth century there were many reports of what we now know to be novae and supernovae, and there were also variations discovered in stars like *o* Cet and P Cyg, they were not clocks. Even more important, however, was Goodricke's interpretation – that the variations are produced by a dark companion of comparable size that occults the visible star – formulated within the context of Newtonian mechanics. There was a mania at the time for looking for evidence of a plurality of worlds, systems like the Solar System, and this fit right into line. Herschel immediately verified the variations, but also began the search for periods in other variable stars. He soon found them, and in keeping with the temperament of the times attributed all of them to the same cause.

Progress during the nineteenth century was rapid[4], spurred by a growing ability to apply celestial mechanics to extra-solar system bodies. Statistical arguments from star counts were mustered to demonstrate the existence of bound visual binaries (Forbes 1769, Herschel 1802, Boole 1847). The orbital motions of several visual binaries were discovered by Herschel in 1803, who also remarked on the existence of binaries in which the components have different colors. Once the distance scale was available from parallax measurements, it became possible to get some idea of stellar masses from these observations. For instance, in 1844

[1]I apologize in advance to people whose work was not cited, but as Adams (1979) states, "There are a hundred billion stars in the Galaxy and only a limited amount of space in the book!". Also, I have used the first person throughout these lectures in order to maintain the spirit of the course.

[2]Letter from Joseph Banks to Wm. Herschel, 3 May 1783

[3]An often unnoticed independent discovery of Algol's periodicity was made by Palitzch, a farmer-astronomer (who is also credited with the recovery of Comet Halley 1759) (Herschel 1847).

[4]The best contemporary overview is the *4th* edition of Clerke (1902, ***A History of Astronomy During the Nineteenth Century*** (London: A.&C. Black)), which is also an exceptional source for a discussion of the observational techniques of the time. It doesn't hurt to realize how hard-won so many of the most basic results were in our business.

Bessel used the wobble in the proper motion of Sirius to predict the existence of an unseen companion, which was discovered by Clark in 1861 during tests of the University of Mississippi $18\frac{1}{2}$ inch refractor. The *Bonner Dürchmusterung* survey by Argelander, Schönfeld, and Krueger at Bonn (1859-62) resulted in the discovery of many eclipsing systems and visual measurements of orbits were carried out at a number of observatories. By the 1880's, double star research occupied a major niche in astronomy. Pickering (in 1880) even obtained a geometrical solution for Algol's photometric variations to determine that the ratio of the components' radii.

With the advent of astronomical spectroscopy, and the application of wavelength standards to spectroscopic plates, a new direction opened in binary star research. The photographer-researcher Fox Talbot proposed the idea of using photographic spectroscopy and radial velocity variations as a method for discovering unresolved binaries as early as 1871, but the idea did not come into prominence for nearly 15 years. The discovery of the unresolved companion of ζ UMa (Mizar) (Ca II K line doubling was found by Maury in 1887 at Harvard, and later Vogel at Potsdam realized the that the orbit must be eccentric) and, as a by-product, the first observations of lines from the interstellar medium, served to illustrate the power of the new methods. But the most important event was the demonstration of orbital motion for Algol by Vogel in 1889, which not only verified the Goodricke hypothesis but was immediately applied to the whole class of variable stars (some inappropriately, like Cepheid variables). The peculiar binaries were also noticed. Of particular note is Maury's extensive study of β Lyr, one of the first "astrophysical" studies of an extra-solar system body, which was sparked by her discovery of spectral line variations and line doubling (as she remarked, of the sort seen in β Aur).

Space observations of binary stars, especially in the vacuum ultraviolet, started with low resolution spectrophotometry from OAO-2, TD-1, and ANS and with high resolution spectra from Copernicus (OAO-3), but it really blossomed after the launch of the International Ultraviolet Explorer Satellite (IUE). Now, using the Faint Object Spectrograph (FOS) for low resolution spectra and the Goddard High Resolution Spectrograph (GHRS), both on the Hubble Space Telescope, even higher quality data can be obtained for binaries in both the optical and ultraviolet. X-ray data from UHURU, SAS-3, HEAO-A, and EINSTEIN were followed by EXOSAT, GINGA, and now ROSAT. Although more important for cataclysmic variables, hot coronae in the RS CVn stars were first observed by HEAO-A and extensively monitored by EXOSAT. Many of these satellites left rich archives that are still to be extensively mined. Radio stars have been studied extensively with the VLA and single dish radio telescopes, millimeter observations are now possible using IRAM, and even some VLBI experiments have been performed on radio stars (especially novae). Infrared observations have been important for studying both the stellar, and the circumstellar, environment and more will certainly be possible with the future missions like the Infrared Space Observatory (ISO). And in groundbased observations, high time resolution is available with microchannel multi-anode array (MAMA) detectors and high speed photometers, and very high signal to noise (S/N) spectral observations are almost trivial with CCDs and linear solid state arrays like reticons and digicons.

I will close this broad-brush historical overview here with a simple epistemological comment[5]. I'm emphasizing space missions partly out of personal prejudice, but also because they serve to highlight the fact that there is no longer any real excuse for wavelength parochialism. There is, for all practical purposes, no portion of the electromagnetic spectrum that is inaccessible and all should be used where appropriate to address the physical processes we will discuss.

3 Techniques

The first part of these lectures will deal with methods, although not in a totally unbiased way. I need to say at the outset that none of the basic techniques required in binary star research are especially novel. What is very different from one study to another, and in time, is the introduction of new wavelength domains and the accumulation of higher quality data for long periods of time. Let's start with the most basic information available for a binary system, the period, the radial velocity variation, and the light curve.

3.1 Periods and How to Find Them

Almost everything we know about a binary comes from having a good idea of the system's period. The problem is hardly unique to the study of binaries but it is especially important here because the aspect of the system's geometry depends on being able to phase together information often obtained from disparate sources.

Several techniques are available for period determination. Lafler and Kinman (1965) and later Stellingwerf (1978) have introduced a statistical procedure, phase dispersion minimization (PDM). This does not assume anything about the actual period or the shape of the variation of the measured quantity with time. Instead it assumes that there is some characteristic timescale in the data and that folding the data on a range of Δt will produce a minimum normalized dispersion per phase bin (with some sampling of the phase fine enough to provide reasonable coverage and coarse enough to provide sufficient data points). Consider that the observations are taken at times $\{t_i\}$ and have values $\{I_i\}$. There are N such points. Now if the measurements are drawn randomly from a periodic distribution, we won't necessarily densely sample the period. But the idea is that the scatter in the data will be smaller when binned according to some phase if this phase is computed with the correct period. That is, take the variance in the data, irrespective of phasing, to be $\sigma^2 = \Sigma(I_i - < I >)^2/(N-1)$, as usual. Now we follow Stellingwerf's procedure, since it is an especially clear and general method. Create M phase bins, chosen based on the guessed period the guessed period.

[5] A very complete compendium of the early literature on variable stars in general, and eclipsing binaries in particular, is available in Schneller (1952). This is also a wonderful source for archival observations that often are the key to understanding secular evolution of close binary systems.

These will have an individual variance s_j and will have a combined (now phased) variance between the bins of:

$$s^2 = \frac{\sum_j^M s_j^2}{\sum_j^M n_j - M} \tag{1}$$

An easy way, then, to compare whether we have the right period with the overall variance of the data is to see if the number-weighted dispersion in each phase bin is smaller than the overall variance in the full data set. This is determined using a statistic, Θ, that compares the variances. The aim is, by stepping through some set of periods, to minimize:

$$\Theta \equiv s^2/\sigma^2 \tag{2}$$

This statistic can also be weighted by the errors in the individual measurements if they are not identical. This is essentially the same statistic as that introduced by Lafler and Kinman. Regardless of which one is used, the PDM is an excellent first guess for the period, one that can be used to limit the search space for more sophisticated treatments.

If the data is evenly sampled (something that is actually quite rare in astronomy), a simple FFT can be used. But when the power spectrum of the observing window is complicated, in other words there are gaps in the data set that may be large compared with a single period, the interpolation that is normally used to fill in the missing data cannot be applied. Deeming (1975) and Scargle (1982) have discussed the effects of stochastic sampling on period determinations. They showed that the periodogram can be extended to random sampling. If the N measurements $\{X_i\}$ are made at times $\{t_i\}$, then the periodogram is given by:

$$P_X(\omega) = \frac{1}{2}\frac{\sum_{k=1}^N [X_k \cos\omega(t_k - \tau)]^2}{\sum_{k=1}^N \cos^2\omega(t_k - \tau)} + \frac{1}{2}\frac{\sum_{k=1}^N [X_k \sin\omega(t_k - \tau)]^2}{\sum_{k=1}^N \sin^2\omega(t_k - \tau)} \tag{3}$$

where the phase lag is defined by:

$$\tan 2\omega\tau = \frac{\sum_{i=1}^N \sin 2\omega t_i}{\sum_{i=1}^N \cos 2\omega t_i}. \tag{4}$$

The operation is quite simple. Choosing some frequency resolution, you step through the range of ω and calculate $P_X(\omega)$ for the measured quantity X. The peaks in this spectrum have the spectrum of the sampling window already included in the denominators. Scargle (1982) and Horne and Baliunas (1986) have developed several methods for evaluating the probability of a peak in $P_X(\omega)$ being real. The resolution is normally limited by the Nyquist frequency, but for irregular intervals no such frequency exists. Instead there is a range of such frequencies, depending on the portion of the dataset being studied (see Scargle 1982).

One important problem, so frequently encountered in binary star work as in studies of other types of variables, is that the sampling is on the same order as a harmonic of the period and aliasing is often a serious problem. Randomizing the sampling interval can sometimes break these aliases and assist in the period search. Another serious problem is that there are often large seasonal gaps, due to the annual visibility of the target. These often contribute uncertainty to short periods and can introduce considerable problems with any periods that are long compared with a day but short compared with the invisibility period for the star. There are no simple solutions to these problems, and in fact there don't appear to be any general solutions at all. You should also keep in mind that the period may not be stable on monthly or yearly timescales, albeit usually by only small amounts, and that large gaps along with intrinsic period variations can lead to erroneous conclusions. The problem of aliasing has been examined in some detail recently by Roberts *et al.* (1987). They introduce the CLEAN algorithm, which is a way of removing possible aliases that was first developed in radio astronomy. Rather than spend more time on this technique, I'll urge you to read the paper. The method is especially easy to program and can be used in conjunction with either PDM or FFT methods.

3.2 Light Travel Effects

Sometimes, it is not possible to directly observe the radial velocity of a star in a binary. But if it is a periodic variable, the phase delay caused by the finite light travel time across the system can substitute (see Irwin 1962). This technique was first exploited in the seventeenth century by Roemer to measure the speed of light, so it has a considerable lineage. For binary pulsars, where neither star can be seen optically but where nanosecond timing accuracy is often available, radial velocities are determined using pulse arrival since $\Omega = (1 + v_{rad}/c)\Omega_0$, where Ω is the rotation frequency of the neutron star. The most extensively studied system is PSR1913+16, a neutron star-helium dwarf system, which is free of the complicating effects of mass transfer and/or mass loss. Apsidal motion has been used to test general relativity and also to determine the mass of the neutron star (Damour and Taylor 1991, and references therein). Period changes in these systems can be attributed to accelerations, which in the case of PSR1913+16 result from galactic rotation. More recently, the motion of binaries in globular cluster potentials has been observed from timing data. In effect, the acceleration masks as an $\ddot{\Omega}$ term in the period evolution.

There are no examples of "double line" systems where both components are pulsars. About the closest one gets is in binary X-ray pulsars. In general, in fact, there are no optical counterparts. In a few systems, specifically PSR1957+21, the non-collapsed component is observed, but not yet with sufficient resolution to unambiguously determine the radial velocity curve of the star (although Hα variations have been detected for this system). The timing data is also more difficult for the pulsar because the accuracy of pulse timing for millisecond pulsars strains solutions for the Earth's orbit. An example of how this shows up is the recent detection of, and subsequent retraction of, the low mass companion to PSR1829-10. The timing residuals were the result of an improper removal of the Earth's barycentric motion. However, when the corrections are applied it is possible to obtain truly remarkable precision, as demonstrated by the detection of low mass companions for PSR1257+12 by Wolszczan and Frail (1992) and the confirmation by Backer *et al.* (1992). The binary pulsars are at the extreme limit of the precision that can be obtained in binary star work and this is certainly a field that will dramatically improve in the next few years. Optical measurements can also yield timing orbits, but in general pulsators are simply not sufficiently stable clocks to do this. In fact, among the pulsars only the millisecond variety are generally quiet enough that timing noise, seen in almost all regular single pulsars, doesn't dominate the residuals. As X-ray pulse timing improves, it will be increasingly possible to determine orbits for the companions separately from the optical counterpart in X-ray binaries (something that the X-ray Timing Explorer (XTE) is designed to do in the next few years).

As an illustration of how powerful simple timing and period determinations can be for relatively normal binaries, take a look at the study of β Cep by Pigulski and Boratyn (1992). By examining the O-C diagram (the comparison between the observed and calculated time of a particular phase in the variation) for this star, they have derived a possible orbit with a period of 91.6 years and with a high ($e = 0.65$) eccentricity. Their predicted radial velocity amplitude is only about 8 km s^{-1} A visual orbit is also available for this system from independent speckle and classical measurements, and the orbit agrees very well with the one derived from the period changes.

3.3 Radial Velocities

Two important pieces of information are immediately available once we have the radial velocity of the *two* stars as a function of time. The first is the effective temperature of each star, and possibly the luminosity class. This presumes that the spectra are individually extracted, that the blending isn't so severe that the line profiles are inextricably combined. The other is more physical. From the maxima of the radial velocities, the mass ratio is determined irrespective of the inclination of the system so that if K_j is the amplitude of the jth component, $K_1/K_2 = M_2/M_1$. For many applications, this is actually all that is needed along with the period. or instance, the ratio of the Roche lobes (as will be discussed shortly) determines much of the history of the system and this ratio depends only on the mass ratio (actually, it must not depend on any other properties since all of the physics defining the Roche surface has been compressed into the three body problem, a purely gravitational interaction).

Double line systems are the comparative exceptions rather than the rule. But there are ways of overcoming the obstacle presented by single line systems by using a range of wavelengths for the spectroscopy. As an example, consider a system consisting of an M giant and an A main sequence star. In the near UV, say around 3000Å, there is no light from the M star. If the absolute magnitudes of the two stars happen to be close, then longward of 7000Å all of the light comes from the M star. So if you observe the radial velocity variations in these two wavelength regions, you can separate the two stars. Well, taking into account the problem of obtaining observing time in the separate regimes this isn't always a trivial task.

This method has been applied with great success in particular to the study of Cepheid variables, especially SU Cyg (Evans 1988, Evans and Bolton 1990, Evans 1992 and references therein), where in the optical the radial velocity variations imposed by the pulsation of the luminous component make the determination of orbital motion extremely difficult. The alternative is that there may be lines that clearly come from the individual components, but this also presumes that there is a large difference in either the luminosity or the effective temperature of the stars. In the example I just used, the M star contains molecular lines while the A star shows Fe II and related species. The great advantage of the UV is that there are many interstellar lines that are largely uncontaminated by the stellar profiles that serve as fiducials for the velocity. In the optical, unfortunately, there are few that aren't also strongly absorbed in most stellar atmospheres (we'll come back to this point in a moment).

Radial velocity data have the added benefit of recording the line profiles for the stars throughout the orbit. However, it isn't always true that the spectrum is actually used for such measurements. The correlation method can be implemented as either a digital or analog process. In digital form, it requires that spectra and standard be obtained, but this has the disadvantage that the full spectrum has to be reduced before the method can be applied. The alternative, pioneered by Griffin (*see* Griffin and Gunn 1974) is to use a spectral template of approximately the right spectral type and photometrically determine the correlation function using the template as a filter and measuring the throughput as a function of spectral lag. This has the advantage that the detector can have an enormous dynamic range, far higher than usually achieved even with CCDs, and a single photometer can be used. The correlation radial velocity meter, implemented at ESO as CORAVEL, has even been duplicated by the extragalactic community as the *z-meter* (used for the CfA redshift survey). Again, it makes full multiplexed use of the spectrum but the price is the complete lack of any information about the actual spectrum.

The algorithm is mathematically quite simple. Assume that the spectra are given by $F_j(\lambda)$. Then the cross correlation function for the intensity is given by:

$$R_{ij}(\Delta\lambda) = \frac{1}{\mathcal{N}} \sum_{\lambda_k} F_i(\lambda_k) F_j(\lambda_k + \Delta\lambda) \tag{5}$$

where $\Delta\lambda$ is the wavelength lag between the spectra and $\mathcal{N}$ is the normalization factor (which can be omitted). The presumption is that the distribution of spectral lines is essentially random in wavelength and strength so that the spectral correlation function is essentially gaussian (fig. 1). Once the maximum is determined, it is possible to get the wavelength shift. Unfortunately, this works in the many line and single line limit, but it may fail in the "few" line regime, especially if there is a stellar wind and the lines show slightly different radial velocities. We will return to this point later. The method has the advantage of multiplexing the spectrum, using all of the lines in a given interval and it can achieve both very high signal-to-noise ratio for the final correlation statistic and also very high wavelength accuracy (provided the spectral resolution of the data is high to begin

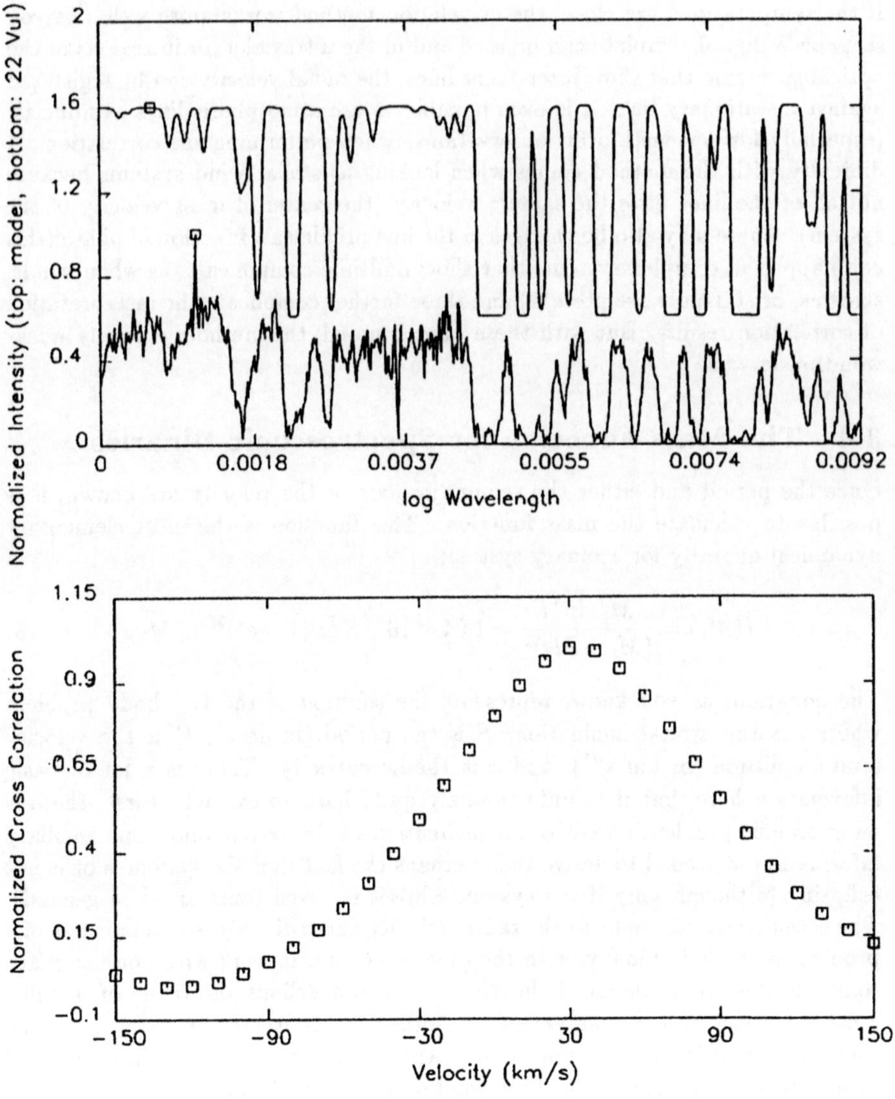

Figure 1: Top: A real spectrum, 22 Vul near minimum, and a synthetic template spectrum, (7000 K, iron peak ions only, turbulent broadening 20 km s^{-1}. Bottom: Cross correlation function of these spectra.

with). The correlation method also works especially well for single line systems with very low amplitude. In its original implementation, Griffin used a mask with a spectrum of a comparison radial velocity standard that was severely over-exposed and served as a differential filter for the dispersed light of the target star. He then imaged the entire transmitted spectrum onto a photometer and recorded the net signal as a function of shift between the stellar spectrum and the mask.

If the temperatures are close, the correlation method works quite well. As you see, now a digital template can be used and in the ultraviolet, or in regions of the optical spectrum that show interstellar lines, the radial velocity can be registered against a stationary line. It is even possible to use atmospheric lines as fiducials (especially for low earth orbit observations) when performing the correlation. A difficulty with the method comes when looking at stellar wind systems because not all of the lines have the same γ velocity (the center of mass velocity of the system). There may also be changes in the line profile as a function of phase (this can happen in complex systems that show multiple components, as when winds, streams, or starspots are present) and these further complicate the interpretation of correlation results. But with these *caveats* noted, the method generally works wondrously well.

3.4 The Mass Function for Spectroscopic Binaries

Once the period and either the semimajor axis or the velocity are known, it is possible to calculate the mass function. This function is the most elementary dynamical quantity for a binary system:

$$f(M_1) = \frac{M_1^3 \sin^3 i}{(M_1 + M_2)^2} = 1.04 \times 10^{-7} K^3 P (1 - e^2)^{3/2} \quad M_\odot. \tag{6}$$

The equation, as you know, represents the solution of the two body problem where i is the orbital inclination, P is the period (in days), K is the velocity semi-amplitude (in km s^{-1}), and e is the eccentricity. There is a lot of basic information here, but it is unfortunately quite hard to extract. First, there is an enormous problem posed by the inclination of the system and some ancillary information is needed to derive this, perhaps the fact that the system is or is not eclipsing (although only if the system eclipses is i well constrained in general). The eccentricity can be fit to the radial velocity curve directly, so there is less of a problem with that. Finally, with the (one) spectrum to work with you can make some guesses on the mass of the visible star and delimit the range of possible masses for the unseen component. In other words, there are many potential sources of error in $f(M)$ so approach it with some care. But with the exercise of proper care, it provides a lot of information about the binary masses.

Sometimes the simple determination of the mass and/or mass ratio can be of paramount importance. I'll just give a few examples here, but there are many many more available. Perhaps the best example of the application of $f(M)$ to a real system is Cyg X-1 (Gies and Bolton 1986). Only one star is seen, HDE 226868, an O star. This star is clearly massive, but its mass is only poorly known. The companion is an X-ray source. The system does not eclipse, shows time variations in both light and spectrum, and has few and only fairly broad lines in the optical with which to measure the radial velocity, many of which are contaminated by emission from the two stars. This mess has still to be sorted out properly. But from the fact that the period is 5.6 days, the velocity amplitude of the O star is about 75 km s^{-1}, the mass function $f(M) = 0.25 M_\odot$. To date, no apsidal motion has been reported (see below). Since the O star mass is *at least* 15 $M_\odot$, the companion must be at considerably greater than $4M_\odot$. This amazingly simple procedure thus leads to the profound result that the collapsed companion of the O star must be a black hole. This is fundamental astronomy – no deep theory is needed, but superb data from the latest technology and patience are indispensible.

Another example of how important radial velocity determinations can be comes form the study of Cepheid binaries. Here the problem is rendered more difficult by the pulsation of the visible star, as we've already discussed. But if one goes into the ultraviolet (and this is true for many classes of binaries) the companion can be studied without contamination from the giant. You see, this is a consequence of stellar evolution – after leaving the main sequence, most stars spend a considerable part of their lives cooler than their main sequence or less evolved companions. This means that for Cepheids, the main sequence stars are typically A or late B stars and can easily be separated out from the F, G, or K star that dominates the optical and near infrared.

For the Cepheid variables, there has been a long standing problem of the masses of the pulsating star. In particular, evolutionary masses have usually been larger than the corresponding pulsational masses determined from the light curves (see Cox 1985). A number of δ Cep stars have been discovered to be close binaries, in part by $O - C$ variations (see Szabdaos 1985), or based on the color variations of the stars around their pulsation cycle (*e.g.* Madore 1977), or by direct detection of the main sequence companions using ultraviolet spectrophotometry (*e.g.* Parsons 1981, Ferro and Madore (1986) provides an especially good example of this procedure). Using the UV, since the color excesses at minimum light are bluer than the corresponding position on the HR diagram indicates, the companion of these stars has been uncovered (Evans 1991, Evans and Bolton 1991, Evans 1992). With orbital determinations, since in general the companions prove to be normal main sequence stars, it is possible to place tighter constraints on the masses of the Cepheids. Several other Cepheids have been examined by more conventional methods, in particular W Sgr (Babel *et al.* 1989) and T Mon (Gieren 1989). For these systems, the orbital periods appear to be very long (about 1780 days for W Sgr, with $e \approx 0.5$ and about 150 years for T Mon). Neither useful for the Cepheid mass problem, but it indicates that the range of periods in these systems is similar to that found for other evolved binaries (Griffin 1992).

In fact, this last example illustrates one of the point I want to make in these lectures. Binary star research, like much of astrophysics, does not require extremely sophisticated theory or methodology to produce fundamental results. Many of the tools have been available for more than a century. Much of the observational data is prosaic and comparatively easy to obtain. Given current technology, almost all of the basic data, at least optically, can be obtained using small telescopes with good spectrographs and photometers. The real effort involves getting long runs of data, on obtaining multiwavelength coverage, on obtaining many *types* of observations for each interesting system, and carefully treating and reporting the observations, and this will be a recurrent theme throughout these lectures.

3.5 Resolvable Systems

The simplest way to discover resolved binaries near the diffraction limit (which for binaries closer than 100 pc would be about $300R_\odot$) is still occultation measurements. Timing has improved tremendously and since several clusters are in the lunar path, especially the Hyades, it's possible to determine orbits directly for a few otherwise unresolvable systems (see as an example Peterson *et al.* 1988, 1989). Here again one sees that very basic observations have important consequences.

Speckle interferometry works on the same principle as standard deconvolution, with the exception that it makes explicit use of the stochasticity of the atmospheric point spread function. The idea is that the central part of an image is coherent on short timescale. Thus a rapidly obtained stellar image is equivalent to a Fourier transform of the wavefront and the turbulence transfer function. The turbulence cells produce individual images in the image plane so that taking the power spectrum it's possible to remove these (Labeyrie 1978). The first speckle observations of a star were of the wide binary Capella (Gezari *et al.* 1972). Since then, observations have become more or less routine with this method. The biggest advance has been the use of photon-counting devices and fast arrays (Barry 1990, Bagnuolo *et al.* 1992) (fig. 2). Direct interferometric imaging is possible using phase-closure and multi-element aperture synthesis although this has not been applied to binaries just yet (*e.g.* Shao *et al.* 1988).

The Fine Guidance Sensors (FGS) on the *Hubble Space Telescope* can also be used for astrometry and binary star detection (Franz *et al.* 1992). The FGS consists of a Koester's prism interferometer that detects phase differences between the fringes in two orthogonal directions. By repeatedly scanning the two directions, an 'image" is constructed of the binary (the position angle, magnitude ratio, and separation are easily fitted by models) (fig. 3). The difficulty has been that the point spread function for the HST is not as advertised many years ago, as all of you know, and the complexity of the interferometric decomposition is substantially increased. Nonetheless, on the Hyades there is at least one simple result. The star H578 in the Hyades has been shown by independent speckle and FGS measurements to be a resolvable binary with separation of about 0.2 arcsec and a magnitude difference of $1.^m4$; the intercomparison is impressively close (0.02 arcsec) between the speckle and FGS results for the separation.

Radio interferometry has been the most successful to date in resolving the components of binaries. With the VLA and AT, routine resolutions at the 0.1 arcsec level are obtainable in reasonable times. A number of different systems can be studied in millimeter (for the few systems that have been examined) and at cm wavelengths (see Hjellming 1989).

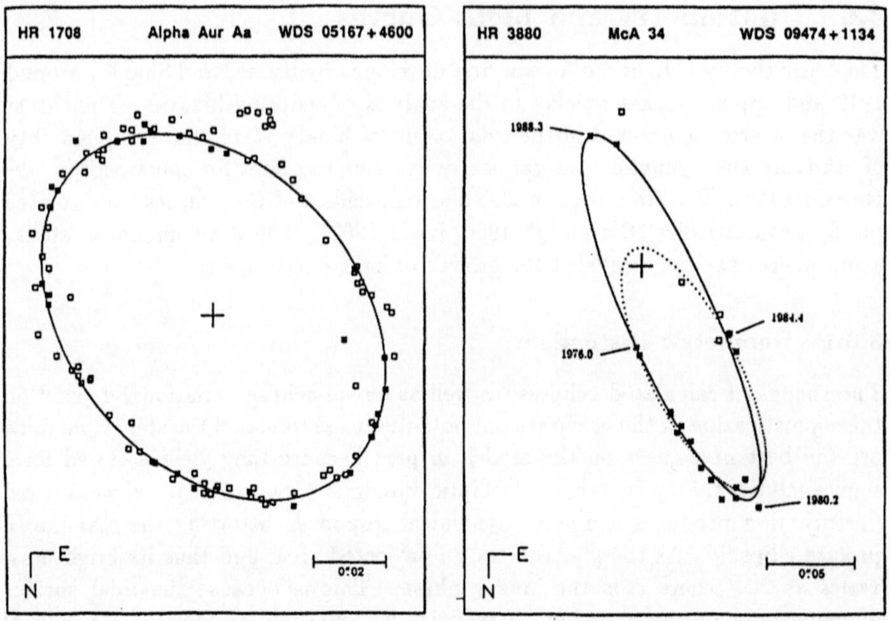

Figure 2: Orbits for Capella (left) and 19 Leo (right) determined using speckle interferometry (Barry 1990). Note that angular scale is 20 mas (left) and 50 mas (right). As a comparison, occultation measurements for 19 Leo give 0.031 arcsec for the separations. Also note that α Aur was the first binary whose orbit was obtained using the Mt. Wilson interferometer (in 1920).

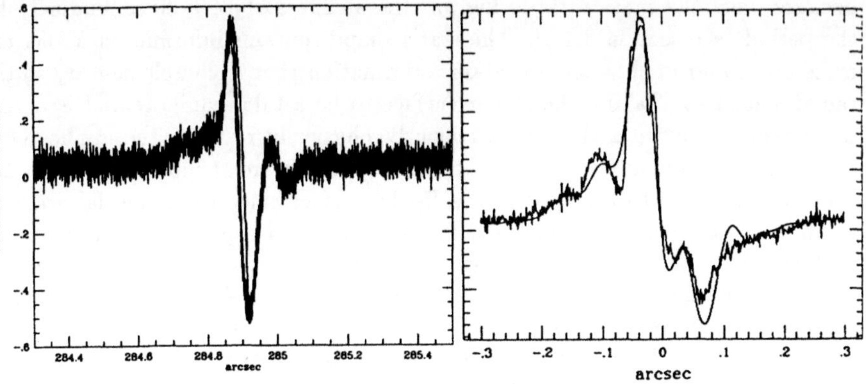

Figure 3: Sample FGS transfer function for a single star and for a binary.

3.6 Photometry and Light Curves

The basic theory of light curves was first developed by Russell and Shapley around 1912 and applied rather quickly to the analysis of eclipsing binaries. The thrust was the interest generated in the tidal origin of binary stars, and the possibility of studying the dynamical interaction of systems too faint for spectroscopy (see Darwin 1910). The stars were modeled as ellipsoids and the eclipses were treated purely geometrically (Binnindijk 1960, Irwin 1962). The development of stellar atmospheres theory paralleled the growth of light curve theory.

3.6.1 Geometric Distortions

The shapes of calculated eclipses, as well as out-of-eclipse variations, depend on the sophistication of the gravitational potential used to model the star. Spheroids are the best first guess on the model, in part because they yield a closed form semi-analytic theory in terms of elliptic integrals (Irwin 1962). A weak tidal perturbation produces a dipolar dynamical response, distorting the star into a prolate ellipsoid. As the star rotates its projected area, and thus its brightness, varies as 2ϕ, where ϕ is the binary phase. This is because the tidal surface depends only on $P_2(\phi)$, where here ϕ is the longitude relative to the line of centers. However, as the perturbation grows stronger – that is as the star grows larger with respect to its Roche surface (its tidal limit) – the distortion becomes progressively less symmetric about the rotational (and orbital) pole and goes more toward a variation that depends only linearly on ϕ.

Proper treatment of the full Roche distortion comes from including properly the equation for the surface normal of the Roche lobe-filling star. Actually, this is a better specified problem than for the ellipsoid, even though it seems harder, because once the mass ratio is known, the stellar surface is fixed uniquely by the period (see section 3.4.3). The star's shape contains information about the mass ratio, and in that sense contains information that is complementary to the radial velocities. Taking the stellar surface to be a tidal equipotential is a very important assumption that simultaneously physically restricts the applicability of the model and simplifies the light curve synthesis tremendously. The surface gravity is known at every point on a Roche surface since the potential gradient is known exactly everywhere once the mass radtio is specified. There are no ambiguities in the properties of any star in this potential because it is specified globally once and for all (the relative size of the star within its equipotential surface completely specifies the distortion of the star, for example). The most widely used code for this model is Wilson and Devinney's (1971) much-modified toolbox. An especially valuable feature is the code is that it permits the user to easily explore phenomenological combinations of different geometric objects, so that the epicycles can be easily adjusted before getting down to the physical picture.

3.6.2 Limb Darkening

Since, in general, stellar disks are unresolved, eclipses provide an invaluable check of stellar atmospheres calculations. Embodied in the Eddington-Barbier relation, the optical depth in a plane parallel atmosphere maps uniquely to a flux deficit as a function of distance from the center of the disk. In LTE, this means that the temperature gradient is mapped by the limb darkening. This happens through the Eddington-Barbier relation. The temperature as a function of optical depth, $T(\tau)$ yields, through the transfer equation, an intensity as a function of distance from the center of the visible disk, $I(\mu)$, where μ is the cosine of the angle from disk center. The phenomenological effect on an eclipse light curve is to produce a "softening" of the ingress and egress contact points. From the detailed modeling of the eclipses, one can derive a wavelength dependent limb darkening coefficient, u_λ, which is related to the gradient of the source function as a function of optical depth. Since the source function is the Planck function in LTE, the limb darkening is directly related to temperature, $T(\tau)$ (fig. 4). If the source function contains scattering terms the eclipse curve yields information about the density as well, as for the Wolf-Rayet stars (see below). Since the limb darkening effect is independent of phase for spherical stars, in detached spherical binaries this is equivalent to a direct check of the stellar atmosphere as a function of optical depth. The wavelength dependence of this coefficient provides a picture of the continuum opacity as a function of depth. Any geometric properties are modified

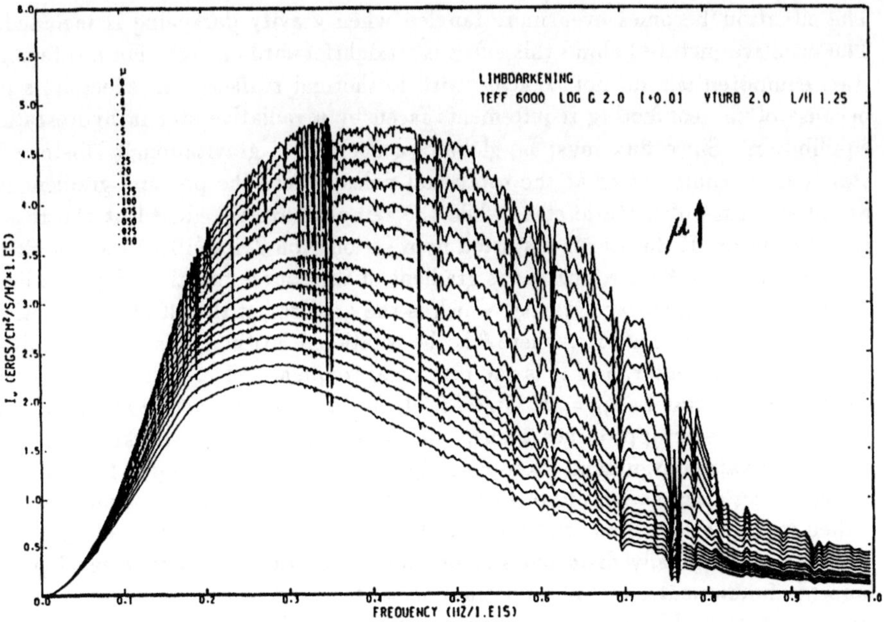

Figure 4: Sample spectra for use in limb darkening modeling (Kurucz 1992, *preprint*).

by the limb darkening, thus for ellipsoidal variables a uniform limb darkening alters the radius and axial ratios of the ellipse. The same is, of course, true for a spherical star. But when attempting to study the tidal interaction and mechanical adjustment of the components in close binaries, the process is rendered more difficult by the effects of the limb darkening. In addition, the detailed wavelength dependence of the opacity and the effects of sphericity conspire to affect the determination of precise geometric properties. Recent model atmosphere computations have been extensively used to provide a better handle on the opacity and on limb darkening coefficients (see figure 4).

Limb brightening may also occur at some wavelengths, and this further complicates life. Actually, the Sun is the best example of the effect. During a solar eclipse, the chromosphere contributes line emission that is not detected if the eclipse is measured only in continuum-sensitive filters (for instance, at 6500Å rather than at Hα). This shows up as an increased duration of the ingress and egress, and is one of the best ways to look for the presence of circumstellar emission from an extended atmosphere (*e.g.* Guinan 1989). Narrow bandpass filters, chosen for specific ions like Ca II, He I, or Hα, are the best ways to look for evidence of the extended emission regions in eclipsing systems. With the growth of multi-filter CCD imaging, it's easier now to perform such studies, and the fact that it can be done in several filters at once eliminates many of the uncertainties connected with earlier classical photoelectric photometry.

3.6.3 Gravity Darkening

The situation becomes even more tangled when gravity darkening is included. The intuitive picture behind this effect is straightforward enough. For a rotating star, equipotentials do not coincide with isothermal surfaces. In effect, this is because of the conflicting requirements faced by a radiative star in hydrostatic equilibrium. Since flux must be globally conserved, a gravitationally distorted star must become hotter at the rotational poles, where the pressure gradient is essentially that of a static star, while the temperature is reduced at the rotational equator. If the star is distorted by rotation, von Zeipl (1925) was the first to show that the flux, which is the gradient of the source function, depends linearly on the local surface gravity, which is the gradient of the effective potential $\psi_{eff} = \psi_{grav} + \frac{1}{2}r^2 \sin^2\theta\Omega^2$, where θ is the rotational latitude, r is the radial distance from the center, and Ω is the rotational frequency. The critical parameter for a rotating star is $\lambda_\Omega = R^3\Omega^2/GM$. If λ_Ω is of order unity, the equatorial region of the star approaches a cusp. In addition, it becomes darker from the flux conservation argument. Consequently, $T_{eff} \sim g^{1/4}$ if $F_{rad} \sim g$. This effect is called *gravity darkening*, although it has often been noted that it could also be called gravity *brightening* since the poles become hotter and brighter. The gravity darkening of a tidally distorted and/or rotationally distorted stellar envelope is another fundamental test of interiors models. Since the eclipse provides geometric information (like the relative stellar radii) and the out-of-eclipse variations indicate the shape of the star, the gravity darkening exponent can in principle be related to the structure of the equilibrium stellar shape.

The problems with this argument are manifold. First, the gravity darkening looks a lot like a limb darkening, except that it is equatorially symmetric, as opposed to true limb darkening which is radially symmetric about the center of the visible hemisphere. The gravity darkening is moderated by limb darkening, and the distortion of the star and the variation of T_{eff} over the surface feed back into calculations of u_λ, so multiwavelength photometric and spectral information must be combined when available to properly characterize both. Further, there is no unique way of calculating the gravity darkening law. You must have a precise idea of the geometry of the stellar surface before computing the emergent spectrum. The procedure is very much like the normal calculation of the emergent spectrum for a stellar atmosphere. Someone who does interiors computations (perhaps van den Heuvel or Livio) provides you with a two dimensional model having the run of physical parameters like mass as a function of r and θ. You then patch together a stellar surface and project the flux vector into the line of sight. To do this, the interior model must provide g as a function of latitude. From the von Zeipl law, you then get a local effective temperature and therefore you can solve the transfer equation along the local surface normal. However, be forewarned that the gravity darkening also depends on the details of the rotation law, because of the interlocking effects of geometry of the surface and the precise gradient of the effective gravitational potential (*e.g.* Smith and Worley 1974, Collins and Smith 1985).

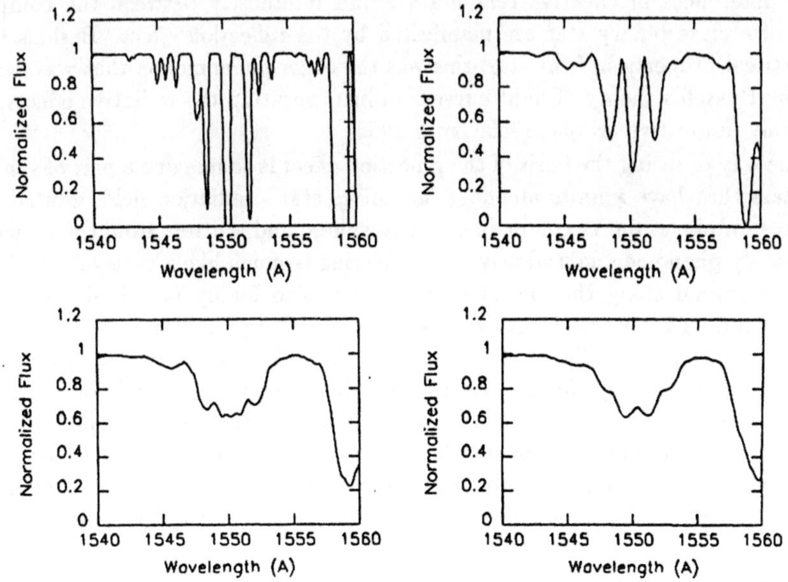

Figure 5: Line profiles for simple broadening function that assumes a linear limb darkening law. Rotational broadening of 0, 100, 200 km s^{-1}. The assumed behavior of the line profiles from a W UMa system may be similar (Lubow *et al.* 1977). The elliptical shape of the rotationally broadened profile reflects a uniform surface of a Riemann spheroidal shape (Chandrasekhar 1969).

The situation is probably quantitatively different, although qualitatively similar, for convective atmospheres. Because of the dependence of the convective velocity and the mixing length for convectively-transported flux on the local surface gravity, it can be shown that the gravity darkening law is of order $F_{rad} \sim g^{0.32}$ or $T_{eff} \sim g^{0.08}$, a much weaker dependence of the flux on the gravity (Lucy 1967). This is in part because, while the blob velocity increases due to buoyancy as the surface gravity goes up, the mixing length decreases because of the lower scale height with higher local gravity. Thus, a convective envelope should appear more uniform than a radiative atmosphere. Lubow, Shu, and Anderson (1977), for instance, argued that the gravity darkening for a fully convective atmosphere does not exist and that the surface should have a uniform effective temperature regardless of the rotation or distortion. Detailed studies of eclipses by fully convective stars are, however, very hard to do for detached systems because of the effects of limb darkening and other possible effects from the companion. For contact systems, the difficulty is increased by the physics of the common envelope. But, as we will shortly see, there are some interesting consequences for Doppler imaging of the surface (the resultant line profiles, fig. 5) that may be useful for spectral modeling and the use of line profiles to check on the proper form of the gravity darkening.

3.6.4 Reflection Effect

Large differences in effective temperature and luminosity between the components of a close binary star are manifested by the reflection effect. Besides the geometrical properties of the star, this was the other phenomenon that was dealt with by Russell's theory of light curves. In incorporating the radiative effects, it was truly innovative (Kopal 1959, Irwin 1962).

One way of seeing the basis of the reflection effect is to imagine a pair of spherical stars that have a finite albedo. The hotter star's radiation field, diluted by distance, impinges on its companion's photosphere and scatters from the surface. In a purely phenomenological way, the scattering is much like a passive reflection and is maximal along the line of centers. It is also highly polarized (and this may depend on both wavelength through the scattering function and also on the precise shape of the stellar surface) (fig. 6). Therefore, this scattered radiation introduces a phase-dependent term to the light curve that depends only on ϕ.[6] Since adiation is coming in from the outside, the optically thin parts of the the cooler star's atmosphere are heated by an extrinsically set source function. This irradiation alters the temperature gradient in these layers and changes the emergent spectrum. If the incoming radiation has a very different spectral distribution

---

[6]At this level, the theory was understood by Anaximander and his fellow Pre-Socratics when they described the origin of the phases of the Moon (I should add that Leonardo, as he often did, intuited the more complex effect when in the Codex Hammer he attributed the origin of the dusky illumination of the lunar disk as reflected sunlight from the Earth being reflected back by the Moon).

However, this is far too simplistic view of what's happening in a *stellar* system.

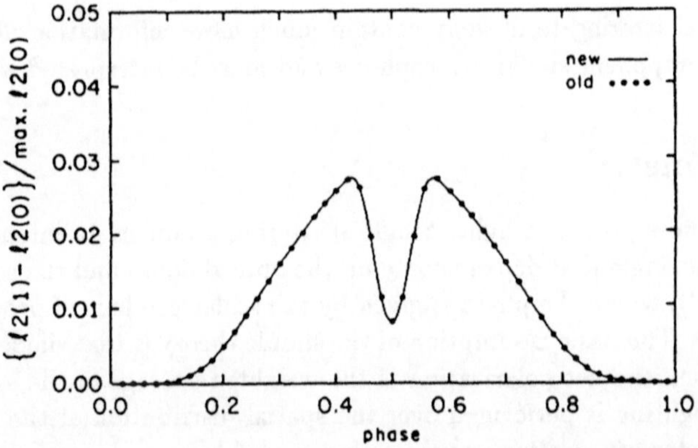

Figure 6: Reflection effect in calculated for an Algol system with gravity darkening and Roche geometry (Wilson 1990)

than the irradiated star, drastic changes can occur. If the atmosphere can radiatively cool, its line spectrum may be substantially changed, specifically because of an increase in the ionization of the gas. It is possible that the energy coming from outside has a greater flux than that coming up from below and strong temperature inversions can result. The best example I think we have is HZ Her, a close X-ray binary where the late-type component shows an A-type spectrum on the side of the side of the star facing the accreting neutron star and an F-type spectrum on the opposite hemisphere. Such strong temperature gradients drive currents, like those resulting from gravity darkening, but these cannot transport the energy efficiently over the whole surface (Kippenhahn and Thomas (1982) elegantly described this as the "match paradox"). It is an analogous problem to the chevron on Venus that is produced by the motion of the Sun, as seen in the slowly rotating frame of the Cytherian atmosphere, across the upper cloud decks. Phase resolved observations of binary systems with strong reflection effects are extremely important for studying the details of the thermal redistribution. In fact, the detailed examination of how the heat is transported is probably the only chance we'll get of analyzing the detailed radiation hydrodynamics of circulation currents.

Another example of the effect is the Hα emission region in the RS CVn star V471 Tauri. This region remains fixed at phase $0.^{p}5$ in the light curve, at the substellar point for the companion. The binary consists of a K subgiant and a hot white dwarf, and the WD illuminates the synchronously rotating hemisphere of the companion. The resulting temperature inversion does not seem to be sufficient to produce a detectable change in the optical light curve, that is the heating doesn't extend down into the photosphere. An even more extreme example of this type of effect is the pulsar PSR1957+21. The pulsar is ablating the red dwarf's atmosphere, reducing its mass by inducing a wind. Here, however, the chromosphere and photosphere are both affected. These and other examples should serve to indicate that the reflection effect, though phenomenologically equivalent to a scattering term, may contain much more information about the physics of the component stellar atmospheres and must be interpreted with care.

3.7 Polarimetry

Scattering produces polarized light. Single scattering produces an intensity as a function of phase angle that depends only on the optical depth and the scattering function $P(\hat{k} \cdot \hat{k}')$, where the phase is given by the scalar product of propagation vectors $\mathbf{k}$ and $\mathbf{k}'$. The basic assumption of the simple theory is that single scattering prevails. Then the net polarization is the weighted average of the scattering angles. The weighting is performed over the spatial distribution of the scatters. Thus, the geometry of a scattering volume is revealed by its polarized signal. For instance, a cloud of scatterers placed spherically symmetrically about a star will produce no net polarization, but let the distribution be spheroidal and the polarization will show the orientation of the axis of symmetry on the plane of the sky. The basic theory was developed by Brown and McLean (1977, see also Brown *et al.* 1978). Good reviews are provided by Kruszewski (1974) and Schulte-Ladbeck (1992). In a non-axisymmetric system, one with scatterers having a non-isotropic phase function, the Stokes components of linear polarization, Q and U, vary. If the system is a simple inclined disk, then the polarization executes a Lissijous figure in the (Q, U) plane, and the ratio of the two loops is essentially a measure of the inclination of the disk (fig. 7). Accretion disks also should produce a characteristic signature in linear polarization, but this will depend on aspect and on opacity (Drissen, *et al.* 1986).

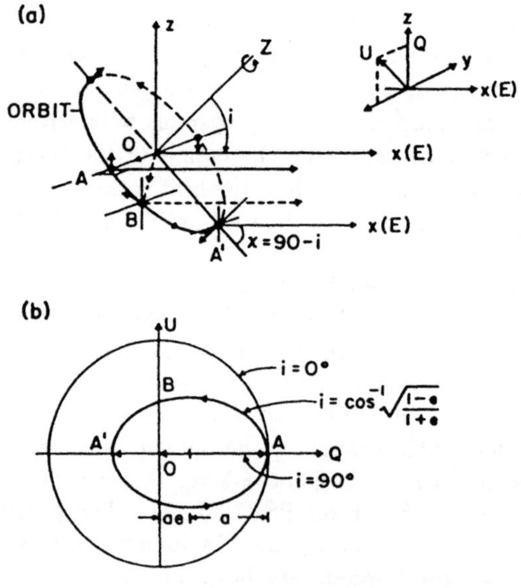

Figure 7: Polarization by single scattering in a close binary from a disk.

Much of the work on polarization in close binaries has been summarized by Koch *et al.* (1989) and by Corcoran (1988). Some systems show rather high linear polarization, notably AO Cas, SX Cas, AX Mon, ι Ori, and V453 Sco (all about 2%). All of these are with strong wind systems or show strong evidence from other diagnostics for thick accretion disks. Several other systems, Y Cyg, V Pup, W Ser, and υ Sgr have smaller intrinsic polarization, between 0.5% and 1%. These individual stars are very different kinds of systems. Interestingly, β Lyr shows about an order of magnitude lower net polarization, comparable to β Per and RY Sct. Still others, like U Cep, that are known to show large amplitude light curve variations, are not significantly polarized. The diversity of the observational material is still too large, and number of systems studied in detail is still too small, to make much sense out of what the polarimetry is telling us. But this is a very valuable technique, one that can be extended significantly now that CCD and reticon detectors can achieve such high S/N in relatively short times.

Polarization studies of Mira has shown that it has a variable position angle and linear polarization. Speckle observations of the envelope seem to show that the distortion of the giant's envelope follows the phase of the companion. There may be a tidal interaction between the two that produces a phase dependence of the polarization, hence its relevance for binary star research. But there may be other systems that show the same behavior (like α Ori, which also has a reputed companion that so far hasn't been detected photometrically).

The Wolf-Rayet stars, both single and binaries, have been among the best studied systems using spectropolarimetry and simultaneous spectroscopy (see Moffat and Robert 1992 and Schulte-Ladbeck 1992). Again, I emphasize that polarization results from broken symmetries, and therefore spectropolarimetry is an especially sensitive technique for studying the development of non-spherical structures in hot stellar winds. Since the lines are formed mainly by scattering, the development of discrete blobs in the winds produces time variability in different parts of the line profile (much like Doppler mapping but this time in polarization fraction and position angle). Phase mapping of binary orbits is not well developed to date. This is one area in which polarization can contribute wonderful data. As we will see, colliding winds should be quite unstable and display time variations in the structure of the shock front that may be accessible to polarimetric analysis.

There are several problems that need to be addressed by future studies using polarimetry. The first is the extension of the method to wavelengths outside the rather narrow range that has exploited thus far. Ultraviolet polarimetry is still available using the Faint Object Spectrograph (FOS) (although this will be compromised once the corrective optics are added to HST). Some work is planned for cataclysmics, but rapid variability studies should be possible for some mass-transferring systems that are not too bright (especially Wolf-Rayet binaries). There will be another flight of the Wisconsin Ultraviolet Photopolarimeter Experiment (WUPPE) instrument on the shuttle ASTRO payload and the utility of this polarimeter for the study of P Cyg systems has been well demonstrated.

Infrared polarimetry, especially of deeply imbedded sources, has been applied to pre-main sequence stars to look for circumstellar disks, and the same could be used for ζ Aur binaries to study the structure of the accretion regions discussed in other lectures at this school. In short, polarimetry is still a new technique in the astrophysicist's toolbox, but it is one of enormous potential and I hope that it will become increasingly frequently exploited in the next decade.

3.8 Doppler Imaging

Doppler imaging uses rotation to break the symmetry of a spatially unresolved surface as is designed to spectroscopically map a stellar surface. Accretion disks can also be modeled in this way because of the monotonic radial dependence of the velocity and other physical variables. The treatment of the problem has become quite sophisticated of late, but traces its roots to some of the first studies of stellar photospheres.

A paleolithic version of Doppler imaging is found in the observation by Rossiter and McLaughlin that the line profiles in detached eclipsing systems show systematic shifts as a function of eclipse phase. The reason for this is simply that for an unresolved surface, any extensive quantity (like the intensity of a spectral line) is a weighted sum over all parts of the stellar surface. In the limit, of course, this is a continuous integral with the weights being the local intensity given by some limb darkening law and surface shape. In effect, since the highest velocities in a rigidly rotating star, or for a star rotating with a differential rotation law like the Sun's, then the highest velocity occurs at the minimum projected surface area (along the line of sight). This is also the area with the greatest limb darkening so the intensity of a locally formed absorption line is even more diminished.

The fundamental assumption of Doppler imaging is thus best illustrated by eclipsing binaries, especially Algol systems. As alternate limbs are occulted, the flux-weighted radial velocity shifts in the opposite direction to the orbital motion. For some systems, this method still provides the best available way to determine intrinsic stellar rotational velocities, but before the advent of very high S/N spectra could not be exploited to its fullest extent. In principle, this method provides a less theoretically biased way of determining rotational velocities than profile modeling. Its utility is limited, though, by the fact that the eclipse conditions have to be "just right" (*e.g.* the stellar spectra can't be too badly blended, the surface brightness contrast between the components should be large, and the eclipse should be total). For this reason, eclipsing Algol type systems offer the best examples of the application of the Rossiter-McLaughlin for determining rotational velocities. A related application of eclipse dissection of the companion's geometry will be discussed in a moment.

Doppler imaging in its current incarnation is a descendent of the *oblique rotator* model, first developed by Deutsch (1970) for the magnetic chemically peculiar spectrum variables. First, let's outline the basis of the method in its analytical incarnation and then generalize it to the Doppler imaging problem. The principle behind the modern technique is that any quantity distributed over the surface of

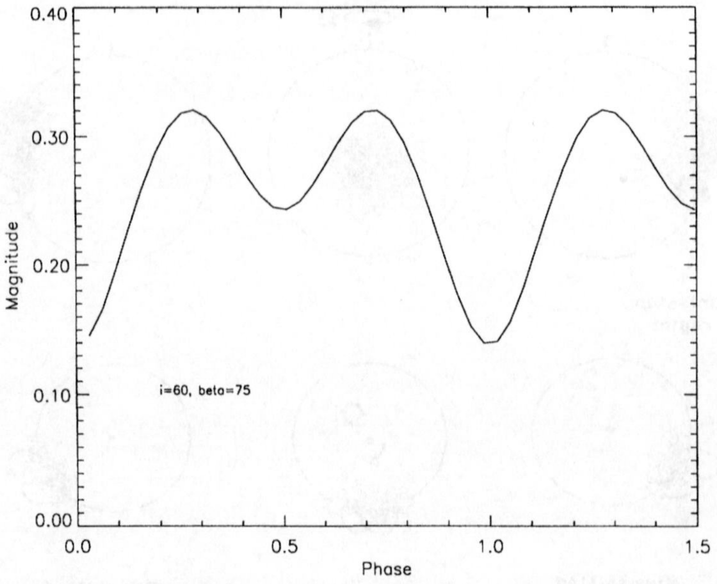

Figure 8: Oblique rotator model light curve as an example of Doppler imaging. Light curve for a spotted star with an expansion $\xi = a_0 + a_2 P_2(\theta)$ where a_0 and a_2 are constants.

a sphere can be represented in terms of spherical harmonics. That is, for some function $\xi(\theta, \phi)$ an expansion of the form:

$$\xi(\theta, \phi) = \sum_{lm} a_{lm} P_{lm}^{|m|}(\theta, \phi) e^{im\phi} \tag{7}$$

where θ is the co-latitude and ϕ is the co-longitude. It is assumed that there is some coordinate system in which the magnetic field is symmetric. If ξ is a flux distribution, it is a set of weights for any distributed quantity (see Megessier *et al.* 1979). For instance, if the surface has inhomogeneities such that its surface brightness varies in a non- axisymmetric way with position, then any physical property, like radial velocity or magnetic field, will also be weighted by ξ. For instance, assume that we have a line profile formed uniformly over the surface but weighted in its contribution by the local value of ξ. Then the mean radial velocity is given by:

$$< v(\Phi) >= \int d\Omega \xi v(\theta, \phi) \Lambda(\theta) = \sum_n A_n \cos n\Phi + B_n \sin n\Phi \tag{8}$$

where Λ is the limb darkening law and ξ depends on the phase in some frame (fig. 8). Now this may be applicable to the active star case, if there is some axis around which the system is symmetric that is oblique to the rotation axis. For instance, the best case for the model is the AM Her stars and intermediate polars, which have strong magnetic fields inclined at some angle β to the rotation axis. But the use of this method is more illustrative than immediately helpful.

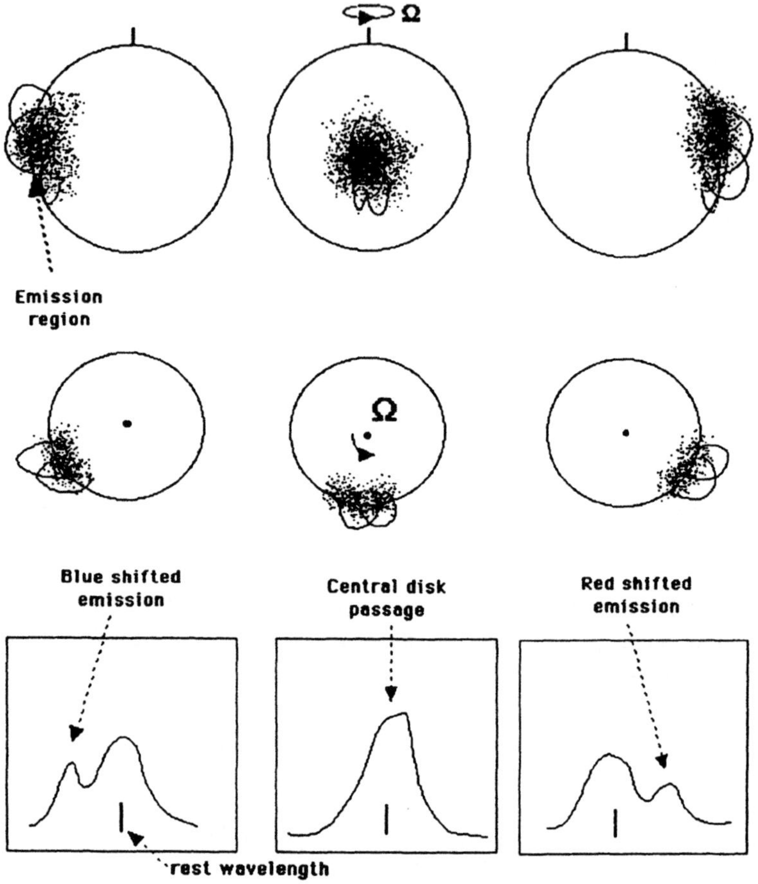

Figure 9: Example of Doppler imaging of an active star, showing effect of trapped plasma above a starspot on the emission line profile.

Where no preferred symmetry is available, more general use must be made of the Doppler imaging idea. Each point on the surface of a rotating star maps to a unique velocity, although the hemispheres may be redundant. Basic to the application of the method is the fact that any line profile maps onto the surface through some function like ξ. As the aspect changes, the profile becomes a function of orbital phase. A more immediate application of the method is found for the RS CVn and related binaries, those with active chromospheres and obvious isolated active regions. The individual emission regions each have some position (θ, ϕ) on the surface so that each maps into a particular radial velocity and phasing (much like the case of the spotted Ap stars) (fig. 9). There are uniqueness problems, of course, but the oblique rotator model and its later incarnation of doppler imaging has been quite successful in the modeling of active systems. Here the picture is one of magnetic loops that are localized in (θ, ϕ) and thus change velocity with phase without substantial change in amplitude until they become partially occulted (see, *e.g.* Neff *et al.* 1989) An example of this is provided by recent GHRS observations of the active system AR Lac (fig. 10). Notice that the emission at C IV changes substantially as a function of phase. In addition, the individual components (in this case a G and K star) contribute differently to the overall profile in each of the lines of the doublet. The orbital motion separates the stars unambiguously. Then rotation on the stellar surface easily separates the individual regions, either in velocity or just from the variations in the line strength (provided the regions aren't badly phased and occulted by the companion). If the system is eclipsing, then the knife-edge of the stellar limb dissects the surface even more effectively. Since the two stars are (more or less) spheres, the precise timing of the disappearances of spots/active regions reveals their latitude as well as longitude.

In a binary, there is one advantage over the single star case because the orbital plane provides an indication of the inclination of the system. For instance, in the RS CVn stars we know the masses of the components and we also know their rotational velocities. Therefore any point on the surface can be precisely associated with a latitude and longitude, especially in the eclipsing systems. And for accretion disks, we know the velocity field and the fact that the material is essentially planar in distribution.

The maximum entropy method (MEM) has been extensively applied by several groups (*cf.* Vogt 1981, Horne 1985, Marsh and Horne 1988 provide useful early discussions). The MEM is pretty easy to understand for stellar surfaces. For accretion disks, which may be a little harder to picture, the idea works as follows. Taking the circulation velocity to be $v_\phi \sim r^{-n}$, where $n = 1/2$ for Keplerian motion, we see that each point in the disk will map into some portion of the line profile. Subdivide the disk into annuli. The longest pathlength is through those annuli that lie closest to the gainer, but these are also the ones with the largest radial velocity. The resultant profile consequently has two peaks, more or less symmetrically placed about the rest wavelength, corresponding to the velocity of the innermost particle orbits. The fact that the emission measure, and also the continuum strength is a function of *both* the density and temperature, and that these two quantities depend differently on radius, sets up the conditions for doppler mapping of the disk.

Another consideration is that in a binary you know which star is involved. You can also produce a much more detailed picture if the system eclipses because the companion dissects the disk during the eclipse (the Rossiter-Mclaughlin effect again). Horne (1985) has used the maximum likelihood method most extensively to reconstruct images of the accretion disk, and here I follow his treatment. Since we know the lines of sight through a disk, we can assume that the integrated intensity is the sum over the individual contributions taken across the disk:

$$\mathcal{F}_\nu(\Phi) = \frac{\theta^2}{4N^2} \sum_{j=0}^{N^2-1} I_{\nu,j} V_j(\Phi) \tag{9}$$

where V_j is the volume (or surface) weighting factor for the N^2 elements into

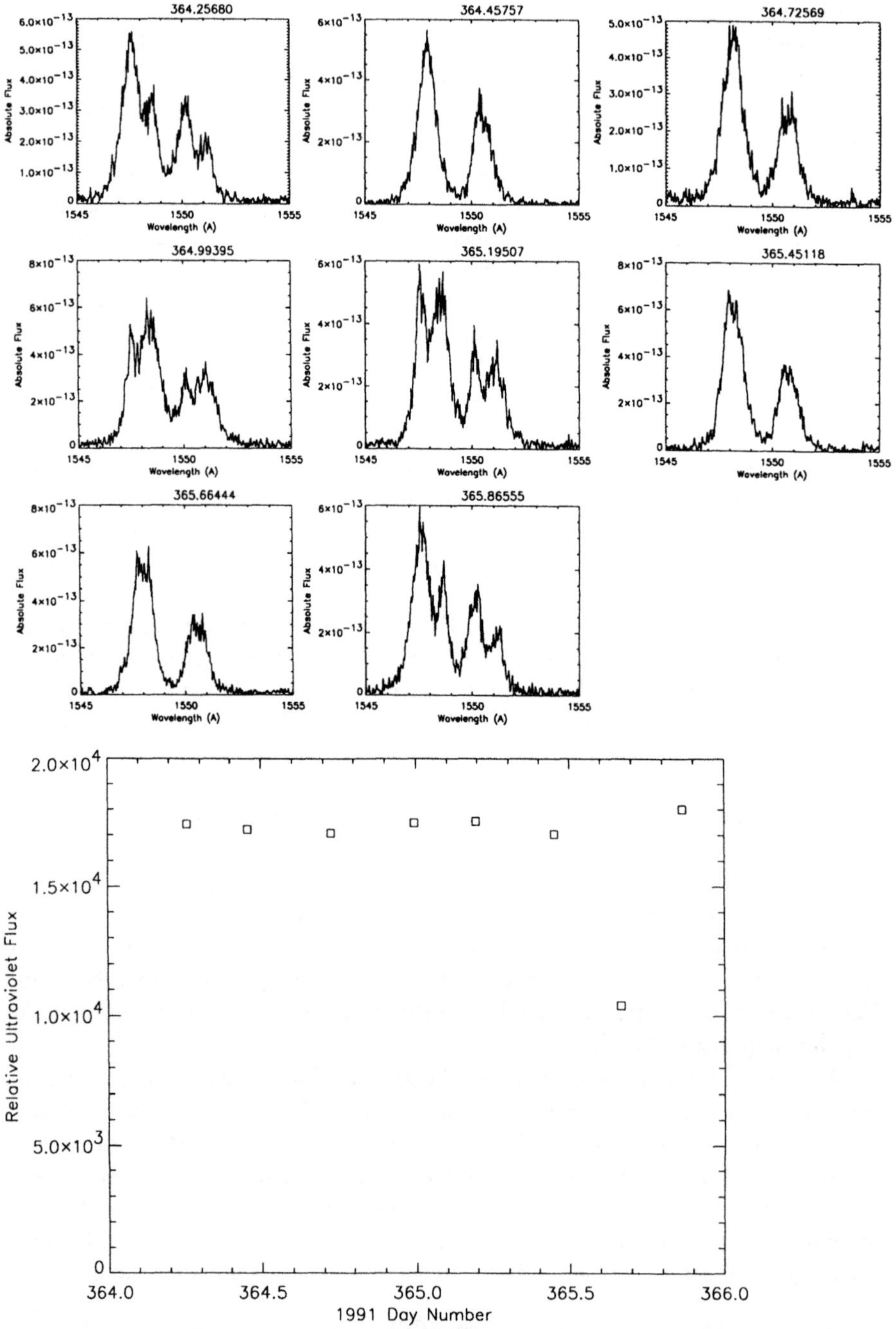

Figure 10: AR Lacertae, observed at C IV λ1550Å. Notice that the C IV from the two components are easily recognized and have different broadening. GHRS medium resolution (R=15000) spectrum, large aperture. The bottom figure is the integrated C IV count rate during the observational sequence.

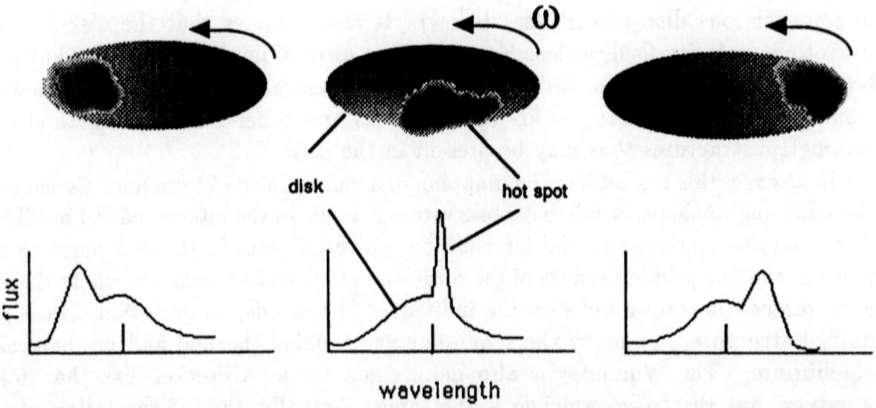

Figure 11: Cartoon example of disk doppler imaging including the effects of a hot spot.

which the disk has been divided. The angular diameter is θ and depends on the radius of the disk (which is assumed to be confined to within the Roche surface) and the separation of the stars. For some phase resolution $\Delta\Phi$, we have the equivalent of a point spread function that smears out the disk image. The reconstructed image is F and the input data is D so that we seek to minimize:

$$\chi^2 = \frac{1}{M}\sum_{k=0}^{M-1}[\frac{F_k - D_k}{\sigma_k}]^2 \tag{10}$$

This is done in the context of maximum entropy by defining:

$$S = -\sum_{j=0}^{N^2-1} I_j[\ln\frac{I_j}{D_j} - 1] \tag{11}$$

(Gull and Daniell 1978). To connect this with the oblique rotator model, you should keep in mind that MEM produces the best reconstruction of the surface distribution of the observed quantity consistent with the data. The Laplace series has the advantage that there is an analytical connection between the coefficients and geometric parameters of the surface. In fact, the oblique rotator method is like taking a spline smoothing of the MEM map.

The reconstructions generally require more than just photometry to break the ambiguity of azimuth, but it is easy to see how this works. Consider a hot spot, for instance (fig. 11). It is located at some position angle ϕ in the disk, which translates into a phase Φ. The broad emission that results from the disk, if the line is formed over the bulk of the disk underlies the narrower hot spot emission, which moves through the profile with phase in some velocity curve that is epicyclic about the orbital motion of the gainer. The success of this technique has been impressive for cataclysmics. It depends, however, on the model and is only as good as the parametric representation of the disk within the context of some viscous disk picture (see below). It also assumes that the disk has a monotonic velocity field, at least in its current form of implementation. That is, boundary layers muck up the process and are specifically avoided. Radial flows would also be a problem, as are spiral shocks and other extended bilaterally-symmetric structures that may be present in the disk.

In a sense, this is a lot like the mapping of a galaxy in the 21 cm line. Consider the following analogy. A full disk observation is made of the integrated H I profile. This contains no direct spatial information, but each point in the disk maps to a point in the line profile because of the rotation curve. Unlike a galaxy, where there is no direct connection between the individual H I clouds, an accretion disk is a unified structure, bound by the requirements of global thermal and mechanical equilibrium. The symmetry is also more exact for an accretion disk than for a galaxy, but the basic principle is the same. Actually, the interpretation of a galactic profile is harder because the rotation curves of many such systems are not necessarily monotonic; in fact, they're generally flat in the outer part of the galaxy and thus map degenerately onto the line profile. In a Keplerian disk, or even one with a boundary layer and turnover in the velocity curve, each point in the disk corresponds to a unique place in the integrated line profile or thermal profile. As we will see below, for an accretion disk, the temperature is a simple function of the radial distance from the gainer.

3.9 How to Discover Very Hot or Very Cold Companions of Low Luminosity: Multiwavelength Analyses

Sometimes the optical part of the spectrum suffices to answer all of the physical questions one might have about a binary system. Statistically, the mass function appears to be heavily weighted toward unity. However, this may in part be a selection effect. After all, most binaries are discovered by eclipses, multiple lines, or radial velocity variations. This tends to select for stars that are nearly equal in both mass and radius and also selects toward relatively short period. As the radial velocity studies have become progressively more sensitive, like those of Griffin (1991, and references therein) the mass functions for lower mass ratios have started to appear.

There is, however, one way, of finding companions that may have a very different luminosity, and presumably mass, from the visible star. Go to another wavelength region. The IR or UV are ideal places to look, just as the X-ray region is the best way to look for a collapsed companion in a close binary. A hypothetical state of stellar matter on the lower extension of the main sequence, the *brown dwarfs*, has not yet yielded unambiguous identification but is likely to exist. The best current candidate is HD 114762, which has a mass function $f(m) = 2.2 \pm 0.4 \times 10^{-6} \mathrm{M}_\odot$ with an 84 day period (Mazeh *et al.* 1990); the inferred mass of the companion is $M_2 \sin i = 0.011 \pm 0.001 \mathrm{M}_\odot$. It is also possible that HR 152, a giant, has a low mass companion with $M_2 \sin i = 0.025(M_1/M_\odot)\mathrm{M}_\odot$. These are stars that are more like massive planets, objects whose interiors are in a nearly liquid state with degenerate cores (not nuclear burning) and extremely low temperatures. They should show up in IR spectra longward of the peak of the main sequence counterpart. The problem here is one of sensitivity – these objects have extremely low luminosity, and very high S/N ratios are required in the data in order to detect them. They radiate near or longward of 10μ, depending on the choice of atmospheric opacity and equation of state. The method of looking for these has so far concentrated on systems that may have extremely large temperature differences without large luminosity differences, like brown dwarf – white dwarf binaries, but the evidence is still lacking for such systems.

In the case of white dwarfs, the problem of detection is a bit more complicated. A white dwarf is easily visible only during comparatively early stages after its formation, when it is extremely UV bright. Then, although optically faint because of the bolometric correction, it is easily seen in the vacuum ultraviolet. However, after it has cooled below about 10000 K, it will be far more difficult to detect directly with current instruments. We will return to this observational problem repeatedly in the second half of these lectures, but I should mention the excellent review by Johnson (1992) for the S and related stars.

Finally, as I've already mentioned in the discussion of radial velocities, sometimes you need to work in two wavelength regimes in order to determine the velocity curve of a binary. Remember, what is a single lined system in one wavelength regime may still be one in another, but the star being observed may be different (recall the discussion of the Cepheid variable problem, for instance). This is especially well demonstrated in a recent paper by Stockton and Fekel (1992). They examine several high mass ratio (low mass function) systems that are already known binaries. They show that modern detectors allow for differences exceeding an order of magnitude in brightness and that systems with mass functions greater than about 0.05 $M_\odot$ should generally be detectable if the system inclination isn't too close to face on.

4 Phenomena and Processes

4.1 Timescales: A Summary of Binary Star Observational Problems

The range of timescales found in the study of binary stars is so large that one often loses sight of the various processes involved and what their characteristic periods may be. In order to place these processes in perspective and to help fix them, let's juxtapose them with more mundane types of, events.

- Blink of an Eye (0.1 to 10 sec)
  - Hydrodynamic timescale for compact objects, especially for the inner portion of the accretion disk in the neighborhood of a neutron star or black hole.

- MHD timescale for active region loops and (scaling from the solar case) for flare initiation in solar-type active binaries.
- Thermal timescale for typical "chromospheric" densities (so the typical timescale for the formation of thermal inhomogeneities for optically thin, moderately dense plasmas).

- Lunchtime (of order minutes to hours, see Adams (1979) for more details)
  - Orbital timescale for cataclysmics. For the longest end of the timescale, the typical orbital timescale for low mass Roche lobe- filling main sequence stars (like contact binaries).
  - Dynamical timescale for stellar winds.
  - Dynamical and MHD timescale for accretion disks.
  - Pulsation timescale for main sequence stars, typical dynamical adjustment timescale.
- Good Cheese (months to a few years)
  - Orbital timescale for almost all binaries.
  - Pulsation timescale for most stellar types.
  - Dynamical timescale for driven circulation.
  - Too slow for typical MHD phenomena, unless on very large scale.
  - Typical interaction timescale in the interstellar medium, and typical radiative timescale for very low density plasmas.
- Graduate Student Thesis (years to decades)
  - Dynamical timescale for pre-main sequence stars, adjustment timescale during pre-main sequence accretion.
  - Typical interval between outbursts of recurrent novae (combination of accretion plus thermonuclear processes).
  - Thermal pulse timescale for AGB stars.
  - Symbiotic outbursts.
  - Large-scale dynamo activity in active stars.
  - Typical timescale for large scale accretion disk instabilities in black hole binaries and X-ray novae.
  - Period changes in rapid mass transfer stage of close binaries.
- Political Change (centuries to forever)
  - Common envelope phenomena.
  - General orbital evolution in mass transfer systems.

- Stellar evolution in general, especially for those stages involving either shell sources (RGB, AGB) or pre-main sequence contraction.
- Changing astrophysical models.

With this list, we see that some physical processes are directly accessible while others, like evolution, can only be ascertained statistically from large samples. It should also serve as a warning that making inferences on poorly sampled data can be a more dangerous pastime for some processes than for others.

4.2 Classical Orbits

In the spirit that it's always useful to start with the obvious, just a brief review of orbits will help. This is meant to highlight where many of the phenomena we will have to face differ from the classical view of a close binary star, that there are two, essentially non-interacting, stars orbiting the center of mass. The radial component of the momentum equation is:

$$\frac{d^2r}{dt^2} - \frac{v_\phi^2}{r} = a_r \tag{12}$$

and for the angular motion:

$$\frac{d}{dt}(r^2\frac{d\phi}{dt}) = a_\phi \tag{13}$$

and we will ignore the z component for the time being. In the absence of any angular torques, that is for centrally symmetric problems, the angular motion can be collapsed into a conserved quantity, the specific angular momentum, which is $j = rv_\phi = r^2\dot{\phi}$ and therefore the equation of motion is strictly radial for the central force problem:

$$\ddot{r} = -\frac{\partial\Phi}{\partial r} + \frac{j^2}{r^3} \tag{14}$$

Here the solution to the central field problem makes specific use of the gravitational force. The substitution of $a_r = -\partial\Phi/\partial r \rightarrow -GMr^{-2}$ completes the specification of the central field problem. Since there is a conserved quantity for the angular coordinate, we can substitute $d/dt = (j/r^2)d/d\phi$, so that defining $u = 1/r$ the equation for radial motion reduces to:

$$\frac{d^2u}{d\phi^2} + u = -\frac{GM}{j^2}. \tag{15}$$

This is the equation for a *harmonic oscillator* in the radial direction. Physically this is a very important result, indicating that the motion remains bounded within a finite range of radius only for a limited range of j and binding energy, and repeats periodically for the cyclically varying angle ϕ. In other words, for small enough values of j, the orbit closes on itself precisely; if j is large enough, the orbit is unbounded. The radial distance from a point mass is therefore $r = a(1-e^2)/[1+e\cos(\phi-\phi_0)]$ where we can now identify $e = (1-j^2/GMa)^{1/2}$ as the relation between the angular momentum and the eccentricity.

A capture orbit for a point mass is determined from the condition that an orbit has $e \leq 1$, while an escape orbit is the complementary set of parameters. The total energy of an orbit is given by $\frac{1}{2}(\dot{r}^2 + r^2\dot{\phi}^2) - GM/r = E$; for a bound particle in a circular orbit ($e = 0$) with radius a the binding energy is $E = -GM/2a$. This is the minimum energy orbit for a particle, since the angular momentum is minimal as well. For a given angular momentum, the increase in orbital energy increases the eccentricity and *vice versa.* For an orbit in which the angular momentum is changed in time, the loss induces a contraction of the semimajor axis to smaller values and a circularization of the orbit. This is the effect of atmospheric drag, or in complex gravitational systems of dynamical friction. For tides, the same is true – the loss of angular momentum produces a contraction and circularization of the orbit, with the attendant increase in the orbital frequency (Batten (1992) has used a similar argument to address the whole question of why there are binaries in the first place).

From these equations, we can immediately adduce the fact that the loss of mass from the system has a profound affect on the orbit. Period changes are intrinsically linked, as we will discuss shortly, to the loss of mass and angular momentum in winds and streams. Anything that changes the binding energy changes the semi-major axis, while anything that changes the angular momentum changes the eccentricity of the orbit. If the mass of the binary changes on a short enough timescale, the system can disrupt (and this is the presumed origin of the runaway stars first noted by Blaauw for the Orion association; see Gies and Bolton (1986)).

4.3 Perturbations, Roche Geometry, and Full System of Equations for the Evolution of the Orbital Elements

For nearly spherically symmetric bodies the potential reduces to a series in Legendre polynomials, P_{nm}, whose coefficients depend on the density (mass) distribution. When the central gravitating body is not spherical, additional terms are required for this power series, and the equations of motion reflect the presence of a torque. For the external gravitational field, this potential is generally expressed by the series in the radius, R, latitude, θ, and longitude ϕ:

$$\Phi(r,\theta,\phi) = -\frac{GM}{r} + \frac{GM}{r}[\sum_{n=2}^{\infty}[(\frac{R}{r})^n J_n P_{n0}(\cos\theta) +$$

$$\sum_{m=1}^{n}(\frac{R}{r})^n(C_{nm}\cos m\phi + S_{nm}\sin m\phi)P_{nm}(\cos\theta)]] \qquad (16)$$

The first term is the central potential, the rest can be thought of as containing terms for $\mathcal{R}$, the perturbation. Here R and M are the radius and mass of the body, respectively. The coefficients C_{nm} and S_{nm} are called *tesseral* ($m \neq n$) or *sectoral* harmonic coefficients ($m = n$), and the J_n are the *zonal* harmonic coefficients. For instance, the coefficient J_2 measures the gravitational quadrupole moment of the body and is responsible for the largest contribution to the precession of the orbit about a nonspherical central body. The specification of the potential provides important information about the internal mass distribution, and in this sense the study of the interior structure of a star through its external gravitational potential is a classic *inverse problem*, attempting to solve uniquely for the distribution in the density that gives rise to the observed gravitational field.

4.3.1 The Roche Potential

One extreme example of a non-spherical gravitational field is that of the *three body problem*. Here one assumes that a point mass orbits in the co-rotating frame of two (possibly unequal) masses. The motion of the large masses has a period P and is assumed to be circular about the center of mass. The third body, whose mass is negligible, orbits with any arbitrary inclination i to the (M_1, M_2) plane and with some initial energy, also called the Jacobi constant. The particle is constrained by minimum energy to move along zero velocity or equipotential surfaces if the body is initially at rest in the co-rotating frame. The precise form of the potential was first discussed by Roche, who also derived the limiting radius of a self-gravitating body being acted upon tidally by a perturber that now bears his name. The centrifugal term is at the rate of rotation of the massive components and is taken relative to the center of mass. The individual gravitational terms are referred to the locations of the massive bodies:

$$\Phi_{Roche}(x,y,z) = -\frac{1}{2}(x^2+y^2)\Omega^2$$
$$-\frac{GM_1}{((x-x_1)^2+y^2+z^2)^{1/2}} - \frac{GM_2}{((x-x_2)^2+y^2+z^2)^{1/2}} \qquad (17)$$

where x_j is the position of the jth body relative to the center of mass. Although this potential involves point masses, it can be generalized to finite bodies. We have already discussed one approximation to the Roche lobe. This is defined as the equipotential surface passing through the point where $\Phi'_{Roche} = 0$ along the line of centers (L_1). Scaled in terms of the separation, it is clearly a nonlinear function of mass ratio and a linear function of the separation.

There are several approximations available for the Roche radius as a function of mass and mass ratio. First, keep in mind that these are only approximations. It is also important to remember that the Roche surface is really defined only in the case of strictly synchronous rotation. Also, the radius that is given here, or in tables, is defined as the radius of a sphere whose volume is equivalent to that of a star filling its inner critical equipotential surface (yes, that is actually how it should be stated). It's thus useful to use R_{RL} when calculating scenarios, but it isn't useful in detailed dynamical modeling. But with these warnings in mind, the Roche radius is given by:

$$R_{RL} = aF(q) \qquad (18)$$

where F is some function of the mass ratio, q, and as always a is the separation of the components. For instance, from fitting the equipotential, Eggleton (1983) gives:

$$F(q) = 0.4q^{2/3}/(0.6q^{2/3} + \ln[1 + q^{1/3}]). \tag{19}$$

An alternative, simpler, approximation is given by:

$$F(q) = (\frac{q}{1+q})^{1/3} \tag{20}$$

(Paczynski 1971). The effect of radiation pressure will be discussed below, but it has the primary effect of reducing the masses of the components, depending on their luminosities.

4.3.2 Equations for Orbital Elements

Perturbations introduced by external bodies fall into several categories. For example, the effect of finite size of the central object in a two body problem introduces precession in an orbit that can be treated as a perturbation above the point central field. These orbital perturbations represent simple time-independent and periodic departures from the closed ellipse. They cause the orbiting body to evolve toward a stable trajectory if the central body is not rotating. Rotation of the central body introduces an additional characteristic timescale and can produce secular instabilities in the orbit. The basic starting point of a perturbation calculation is that one already knows what the orbit is for a particle. One is interested in finding out whether it is stable against small perturbations, due to other bodies, and what the evolution will be for the orbit. The tidal term results from the finite size of the star relative to its orbital radius. The differential gravitational acceleration across the body produces a torque that accelerates the perturber outward and slows the primary's rotation.

An orbit in the Kepler problem is described by a finite number of *elements*: e, the eccentricity, a, the semimajor axis, $a(1-e^2)$, the *semi-latus rectum*, and i, the inclination. The mean anomaly, $M = n\Delta t$ depends on the mean angular rate $n = 2\pi/P$ where P is the period, of the orbit about the center of the conic. The *eccentric anomaly*, E, measures the difference between the mean motion, referred to the osculating circle, and the motion about the center of the conic. The mean motion is related to the eccentric anomaly by the *Kepler equation*, $M = n\Delta t = E - e\sin E$ where Δt is the elapsed time from some initial epoch t_0. The *longitude of periastron*, ω, is measured in the plane of the orbit relative to the ascending node. Finally, the *true anomaly*, f, is given by:

$$\cos f = \frac{\cos E - e}{1 - e\cos E}; \quad \tan\frac{f}{2} = (\frac{1+e}{1-e})^{1/2}\tan\frac{E}{2} \tag{21}$$

This angular motion is referred to the longitude of closest approach and is given by $\phi - \omega$ where ϕ is the angular motion about the focus and the coordinate appropriate to the description of the angular momentum of the orbiting body.

Now, under the influence of an external force $\mathcal{R}$, called the *disturbing function*, all of the orbital elements can vary. In particular, torques change a and e, but also cause the orbit to precess so that Ω and ω vary in time (as we discussed for apsidal motion). The presence of additional mass in the system changes the orbital frequency, n, through changes in M. If the angular momentum changes it is possible for the total energy of the particle to change as well. The disturbing function has components:

$$R = \frac{\partial \mathcal{R}}{\partial r}; \quad S = \frac{1}{r}\frac{\partial \mathcal{R}}{\partial \phi}; \quad W = \frac{\partial \mathcal{R}}{\partial z}$$

for the radial, azimuthal, and vertical components of the perturbation. In effect, S is a torque, as is W (in the meridional plane). Since this is really ordinary celestial mechanics, it is worth noting that geodecists normally use the tidal perturbation for calculating the orbital evolution of satellites, so there is a considerable literature (outside of binary star research) that can be tapped for more insight into the problem. The full equations for the evolution of the orbital parameters under the action of these forces is given by:

$$\frac{da}{dt} = \frac{2}{a(1-e^2)^{1/2}}(Re\sin f + S\frac{a(1-e^2)}{r}) \tag{22}$$

$$\frac{de}{dt} = \frac{(1-e^2)^{1/2}}{na}[eR\sin f + [\frac{a(1-e^2)}{r} - \frac{r}{a}]S)] \tag{23}$$

$$\frac{di}{dt} = [na(1-e^2)^{1/2}]^{-1}\frac{r}{a}\cos(\omega + f)W \tag{24}$$

$$\sin i\frac{d\Omega}{dt} = \dot{i}\tan(\omega + f) \tag{25}$$

$$\frac{d\omega}{dt} = \frac{(1-e^2)^{1/2}}{nae}[-R\cos f + S[\frac{r}{a(1-e^2)} + 1]\sin f] - \frac{er\sin(f+\omega)\cot i}{a(1-e^2)}W] \tag{26}$$

$$\frac{dM}{dt} = n + \frac{1}{na}[(\frac{(1-e^2)\cos f}{e} - \frac{2r}{a})R - \frac{(1-e^2)\sin f}{nae}(1 + \frac{r}{a(1-e^2)})S] \tag{27}$$

Notice that the variation of the semimajor axis depends on both the radial and the torque, but because the orbit can be taken in the two body problem as planar, there is no dependence on W.

Changes in ω and Ω are equivalent to orbital precession. All of these may be periodic or secular, depending on the details of $\mathcal{R}$. General relativity produces a change in the gravitational field in the vicinity of a massive body. The distortion is equivalent to introducing an additional term in the harmonic oscillator equation:

$$\frac{d^2u}{d\phi^2} + u = \frac{GM}{j^2} + 3GMu^2 \tag{28}$$

so that the magnitude of the advance of the periastron is predicted to be $3GM/(a(1-e^2)) = 12\pi^2a^2/(c^2P^2(1-e^2))$, where c is the speed of light. Two types of GRT corrections are required for orbits. One is due to the precession of the orbit, the other is the radiation of gravitational waves. While not important in the vast majority of dynamical systems, it plays a role in binary star systems in which the components are massive, compact, and revolving with very short periods.

The rate of gravitational wave radiation varies as Ω^6 where Ω is the orbital frequency, so that for binaries like PSR1913+16 (see van den Heuvel's discussion in this volume) it produces a secular change in the semi-major axis that can be measured from several years of observation. These effects can be incorporated into the standard perturbation evolution equations through modifications to the disturbing function.

4.4 The Lagrangian Points and What They Mean

There are several (critical) points in the corotating frame of a binary system where the gravitational acceleration vanishes. These are called the *Lagrangian points.* They are all coplanar (fig. 12). Two lie perpendicular to the line of centers, L_4 and L_5, and are due to the balance between centrifugal and gravitational forces modified by the Coriolis acceleration. These points are actually potential *maxima,* and therefore local orbits are only quasi-stable. These points lose their stability for large mass ratio or in the presence of an eccentricity in the massive binary. Two Lagrangian points, L_2 and L_3, are potential minima and lie along the line of centers but beyond the two massive bodies. Finally, the most important point is L_1, the point of gravitational balance *between* the two bodies along the line of centers. This point corresponds to the size of the Roche radius, the separation at which the gravitational perturbation of the companion dominates over the self-gravitation of a deformable, compressible, self-gravitating body. Orbits are

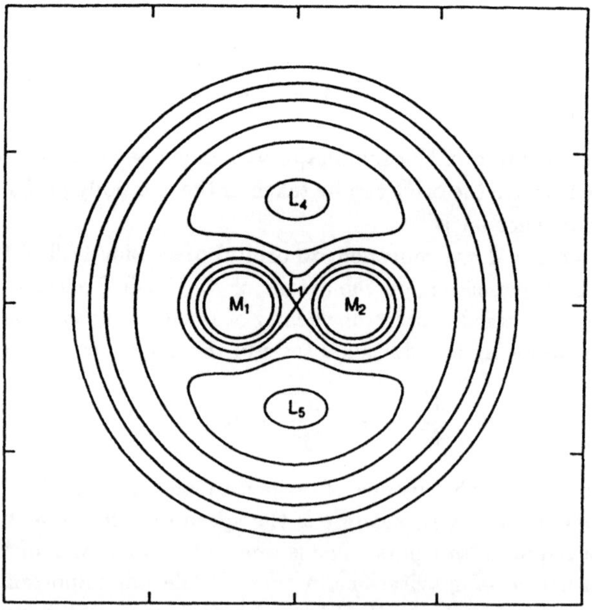

Figure 12: Roche lobes for a close binary with $q = 1$.

unconditionally unstable at L_1, L_2, and L_3. They are asymptotic at L_4 and L_5. If the central orbit is eccentric, the tidal component of the gravitational field at the Lagrangian points oscillates in the course of a single period. Depending on the local orbital frequency of a body about the L_4 or L_5 point, this oscillation may render the local orbit unstable on short timescale, transferring angular momentum to a trapped particle and sending it out of the system.

The classification system for close binaries is based on the Roche potential. If both stars are well separated, and neither photosphere is in contact with the Roche lobe, the binary is called *detached.* If one of the stars has the Roche radius, it is called *semi-detached,* and if both are at the Roche surface it is called a *contact* system. Most binaries are detached, even those which have strong winds and some evidence for accretion (like the ζ Aur systems). Only detached systems can be used, for instance, for mass determinations that lead to the calibration of the mass-luminosity relation (Popper 1980). Algol systems, and most mass transfer binaries, are of the semi-detached type. W UMa and common envelope systems (like those discussed by Livio as precursors of planetary nebulae) are contact binaries.

This taxonomic distinction, first introduced by Kuiper in the 1940's, is quite important. You see, the L_1 point represents more than just a stability point – it is the limiting surface that a star can maintain in hydrostatic equilibrium in the presence of a perturber. Therefore, any matter that manages to be pushed through this limiting radius will inevitably be transferred to the companion or eventually lost from the system. To see this, consider the stellar wind problem (see, for instance, Shore 1992). The effect of the L_1 point is to cause a switch in the sign of the gravitational acceleration on a parcel of gas. The one dimensional equation of motion for an isothermal gas near L_1 is:

$$(v^2 - c_s^2)\frac{dv}{dz} = -g_{eff} \tag{29}$$

where g_{eff} is the gradient of the effective potential. By definition, this vanishes at the inner Lagrangian point, and reverses sign on transferring mass through that point. In order for the flow to retain a monotonic velocity field, one with $dv/dz > 0$, say, the velocity must become supersonic immediately after the matter crosses the Roche surface at L_1. The material then funnels into a narrow stream and, hypersonically, heads toward the companion carrying the angular momentum of the L_1 point with it (see below, *accretion disks*).

Observationally, this is a difficult part of the flow to study in any detail. The main observable is the angle of the stream, which gives some information about the angular momentum. For some eclipsing systems it is possible to see the occultation of the emission line region near L_1 but in general the emission measures are low. Absorption that might be observed from the stream will introduce an artificial eccentricity to the binary orbit, but the effect is small. Usually, the existence of the stream is inferred from the characteristics of the accretion disk and the radius of the Roche lobe relative to the derived radius of the loser in a binary system.

4.5 Resonances

The best example of this resonance phenomenon is a child in a swing. The period of the system is given by the length of the supports for the swing. But the child controls the amplitude of the swing by kicking in phase, or in antiphase, with respect to the maximum velocity along the arc. Thus with the proper phasing, a periodic kick delivered at the minimum in the potential energy (maximum velocity) will produce an increasing amplitude. The same is true for bodies in gravitational orbits. These resonances grow without limit in the linear theory because the forces are always in phase thereby producing a secular acceleration. In near resonance, the tendency will be for the bodies to attract to the resonance and lock. If the two bodies are not in circular orbits, then the phasing of the kick is important and should the body be going at its maximum speed at the time the acceleration is delivered, its angular momentum will be increased and the eccentricity of the orbit should increase secularly. It is interesting that several binaries are known to display enhanced mass transfer during periastron passage, especially Vela X-1. It might also be possible to see resonances in systems with eccentric orbits and Mira variables as primaries, with enhanced phases of mass transfer when the mira is at maximal distension during a periastron passage of the companion. I'm tempted to speculate that this could act in some symbiotics, like R Aqr and MWC 560, but at the moment that is only speculation.

Resonances have also been invoked in cataclysmics, where they are used to explain superhump phenomena. A recent model (Whitehurst and King 1991) argues that the disks will develop strong spiral shocks and non-axisymmetric perturbations at this frequency. For a rotating coordinate system, we can introduce the *epicyclic frequency*, which is the rate at which bodies appear to revolve around the point of observation. Consider a moving observer in a coordinate system (r, ϕ) where ϕ is the azimuth and assume that this observer revolves about a central body in a gravitational field $\Phi(r)$ at a distance r_0 and with an angular frequency Ω (fig. 13). The radial velocity is therefore assumed to vanish for this observer and the angular speed of the frame is also constant. Now assume that we can create a co-moving locally cartesian coordinate system (ξ, η) around this orbit. A body with higher angular momentum than required for the circular orbit will therefore have a radial speed $\dot{\xi}$ and a variable angular speed $r_0\dot{\phi} = \dot{\eta}$. For a corotating frame of reference, the equations of motion are:

$$\ddot{\xi} = 2\Omega\dot{\eta} - \frac{\partial^2\Phi}{\partial r^2}\xi, \qquad \ddot{\eta} + 2\Omega\dot{\xi} = 0 \tag{30}$$

Contained within these equations is a characteristic frequency:

$$\kappa^2 = 3\Omega^2 + \frac{\partial^2\Phi}{\partial r^2} \tag{31}$$

which is the *epicyclic frequency*, representative of the dynamical shear in the corotating frame. Notice that it depends on the steepness of the gravitational field gradient. For instance, for a Keplerian orbit, $\Phi \sim r^{-1}$ and $\Phi'' > 0$ so that

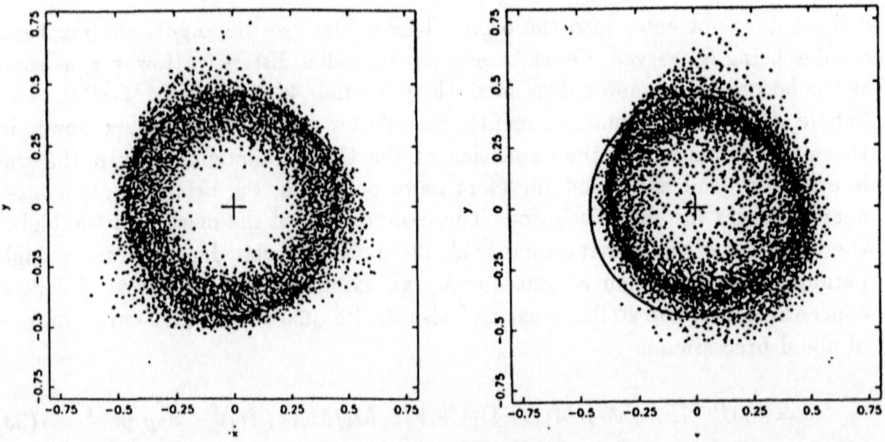

Figure 13: Example of orbital resonances leading to shocks in cataclysmic accretion disks (Lubow 1991). Circle indicates the 3:1 resonance radius. The mass ratio was $q = 0.2$. Two times are shown, showing the development of the instability.

the epicyclic frequency is greater than the orbital frequency. In terms of the orbital frequency of the reference circular orbit:

$$\kappa^2 = 4\Omega^2(1 + \frac{r}{2\Omega}\frac{d\Omega}{dr}) \tag{32}$$

since the reference (circular) orbit has $(d\Phi/dr)_0 = -r\Omega_0^2(r)$, Ω_0 being the local Keplerian frequency. The importance of this frequency for the resonance problem is that any force that is periodic on this timescale will produce a resonant interaction. For the more general problem of galactic structure, the points at which the perturbation frequency is the same as the epicyclic frequency are called the *Lindblad* resonances because of their connection with the structure of spiral galaxies.

4.6 Apsidal Motion

Apsidal motion – which, as I mentioned, was one of the first types of variability detected in binary orbits – can be studied both photometrically and dynamically. The most direct way to determine the internal structure of a self- gravitating body is to orbit a probe around it and observe its apsidal motion. The basic theory has not changed since Kopal (1959) worked it out. Rather than the usual recitation of the equations (for this, see Kopal (1965)), let me mention physically what's involved.

Take a completely spherical star. Then the orbit is given by the solution to the two body problem $r = a(1-e^2)/(1-e\cos E)$ where E is the eccentric anomaly given as the solution to the Kepler equation $E - e\sin E = nt$ where n is the mean motion, $2\pi/P$. Now suppose that the body is not spherical and point-like. For a point mass, the line of nodes is fixed (or arbitrary) and once defined does not enter into the orbit. This is because the angular momentum, besides being conserved, depends only on the radial distance. However, as soon as the body becomes more distended, the potential, $\Phi(\mathbf{r}) = -G\int dM(r')/|\mathbf{r}-\mathbf{r}'|$ (where $dM(r)$ is the mass element), contributes progressively higher power in the higher moments of the expansion of the Green function. Now if the star is centrally condensed, and therefore more point-like, the orbital motion more nearly mimics the closed solution. The more distended the mass, and the higher the moments of the gravitational field, the more the orbit departs from a single period. Apsidal motion is consequently an expression of the degree of central concentration of the stellar mass and also J_2, its quadrupole moment. The rate of nodal precession is:

$$\dot{\omega} = 360P^{-1}[F_1(e, M_2/M_1, r_1, \Omega_1) + F(e, M_1/M_2, r_2, \Omega_2)] \quad deg\ yr^{-1} \tag{33}$$

where an auxiliary function has been introduced to keep the notation more compact:

$$F_j(e,q,r,\Omega) = k_{2,j}r^5[15f_2(e)q + (\frac{\Omega}{\Omega_K})^2\frac{1+q}{(1-e^2)^2}]. \tag{34}$$

Here P is the orbital period in years, Ω is the rotational period of the star, Ω_K is the Keplerian frequency, and $f_2(e) \approx (1+\frac{3}{2}e^2+\frac{1}{8}e^4)/(1-e^2)$ (Kopal 1959, 1965). The $k_{2,j}$ parameter measures the mass internal concentration of the star :

$$k_2 \equiv \frac{16\pi}{5MR^5}\int_0^R \rho r^7 dr \tag{35}$$

for a spherical model (and for studying interiors models, this is the number we're trying to derive). In the relativistic case, this means that the gravitational field departs from Newtonian by some amount, in general given by $r_\star/r$, where $r_\star = GM/c^2$.

Examples of how such an analysis proceeds are found is Semeniuk and Paczynski (1968), Mathis and Odell (1973), and Guinan and Maloney (1985). This last study focuses on effects of relativistic corrections for an anomalous system, DI Her. One of the problems is that sometimes what seems to be apsidal motion is really a change in the orbital period, and spots and mass transfer complicate the situation. The effect of mass transfer on orbital solutions is well known – that the velocity curve can produce a very eccentric orbit for systems undergoing high rates of mass transfer while the light curve indicates that the system is circular, like RX Cas – and the complications that result from variable light curves can mask real apsidal motion while creating fake rates.

An interesting possibility has been raised recently, that the changes seen in apsidal motion and eccentricity are actually chaotic. Since this subject is quite *au courant*, I think it is too early to go into it in much detail. The idea is that tidally-generated oscillations in a close binary can manifest as variations in the rate of precession of the orbit, and also in variations in the eccentricity. In effect, this is due to a variable torque that, although of small amplitude, changes the angular momentum slightly. The observational data are simply too imprecise to test this picture, but with some systems, like binary pulsars that have companions that are not also neutron stars, this may be possible.

4.7 Period Changes

Let's start with the observational quantity, the O-C diagram. As I mentioned in the discussion of light travel time effects, this is the plot of the difference between the observed and computed time for a specific phase of, say, the light or radial velocity variations. For instance, in an eclipsing system, it is the delay or advance of the time of primary minimum, usually an easy quantity to observe. Now the phase is defined with respect to a period P by $\phi = (t - t_0)/P$, where t_0 is the epoch of reference. Now if the period is actually changing with time like $P(t) = P(t_0) + \dot{P}(t - t_0)$, the phase will also change via:

$$\phi = \frac{\Delta t}{P_0 + \dot{P}\Delta t} \approx \frac{\Delta t}{P_0}(1 - \frac{\dot{P}}{P_0}\Delta t) \tag{36}$$

so that the the phase delay, $\Delta\phi \equiv (O - C)$, is given by $\Delta\phi \approx \dot{P}(\Delta t/P_0)^2$, a *quadratic* in time. The effect of a wrong period is much simpler and easily distinguished from a real change. There are of course many reasons for the period to change, and there are also many ways that it can change, from steady to spasmodic. Here will will discuss some of the causes.

4.7.1 Mass Transfer and Mass Loss

Period changes can be used to deduce the rate of mass transfer or mass loss from a binary system by the following argument. Assume that the total angular momentum is given by $J = \mu a^2\omega$ where a is the semi-major axis and $\mu = M_1M_2/(M_1 + M_2)$ is the reduced mass of the system (assume that the orbital eccentricity is zero and remains such throughout the changes in the mass. Now take the tidal mass to remain fixed and assume mass transfer is conservative between the two components. If the angular momentum of the material and the system doesn't change, then if $\dot{M}_1 = -\dot{M}_2$, we obtain for the period change:

$$\frac{\delta P}{P} = 3\frac{M_1 - M_2}{M_1M_2}\delta M_1 \tag{37}$$

If there is no mass transfer but all of the material escapes the system, then

$$\frac{\delta P}{P} = 3\frac{2x - 1}{x(1 - x)}\frac{\delta M}{M}. \tag{38}$$

where $x = q/(1 + q)$, and M is the total binary mass. This equation actually masks an enormous variety of physical assumptions. For one thing, there is no guarantee that the mass transfer occurs conservatively. Here is the tie-in with the discussion by van den Heuvel, and one of the few places where it is actually possible to determine something about the empirical basis for the larger-scale scenarios of stellar evolution in close binaries. Consider what happens if we allow for some angular momentum *and* mass loss from the system as a whole. First, from Kepler's law:

$$2\frac{\dot{a}}{a} = -\frac{4}{3}\frac{\dot{\omega}}{\omega} + \frac{1}{3}\frac{\dot{M}}{M} \tag{39}$$

The angular momentum evolution is then given by:

$$-\frac{1}{3}\frac{\dot{\omega}}{\omega} = -\frac{1}{3}\frac{\dot{M}}{M} - \frac{\dot{\mu}}{\mu} + \frac{\dot{J}}{J} \tag{40}$$

Here notice that there are two contributors to the period change, the change in the center of mass *and* the change in the mass of the system as a whole (see especially the classic discussion is provided by Hadjidemetriou (1969) which is an especially complete treatment of the Roche geometry including the full treatment of the evolution of the orbital elements). The real theoretical burden falls on determining $\dot{J}$. This is very difficult to obtain observationally. If anything the problem is that determining the rate of angular momentum loss in the present context is circular – you observe the rate of period change. So you need to have a model to start with.

Not all period changes are secular and smooth. Such a change is detected in β Lyr, which changes its minimum time by about 10^{-5} cycle^{-1} β Lyr. Abrupt changes have been reported in W UMa systems, especially W UMa itself (see Batten 1973), and both steady period changes and small but detectable departures from a simple quadratic law are found in U Cep (*e.g.* Olson 1982, McClusky *et al.* 1988). An important point, noted by Binnindijk (1970), is that W UMa

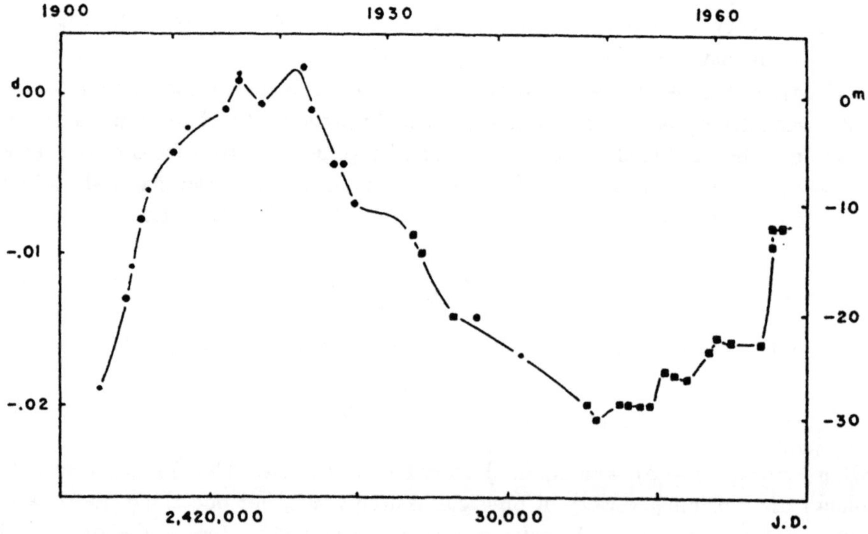

Figure 14: Period changes in W UMa (from Binnindijk 1970).

shows alternating exchange between the stars and mass loss from the system as a whole (fig. 14). Finally, the combination of mass loss, variable magnetic dynamo activity, and mass transfer between the components of a binary system all combine to produce a complex pattern of period changes. The period instabilities are thus very important indications of many different underlying physical processes in binaries, and unfortunately they are not unambiguous tools for analyzing them.

4.7.2 Gravitational Radiation

Posed as a problem to the reader in Landau and Lifshitz (1971) (section 106), the emission of gravitational waves is an interesting interplay between the angular momentum of the system nd its binding energy. The total energy of the binary is its binding energy. The emission of positive energy waves therefore drives the system toward coalescence because $E < 0$. Gravitational radiation depends (to lowest order) on the third time derivative of the mass quadrupole moment (Thorne 1980) and thus has a very steep dependence on the period, $\dot{E} \sim \omega^6 a^4$. Since $E \sim a^{-1}$ then the characteristic timescale varies like $t_{grav} \sim a^{-4}$. So far, the dwarf novae are the ones most likely, besides the short period binary pulsars, to evolve under the influence of gravitational waves; normal stars can never achieve the short periods at which gravitational wave emission plays any role.

4.7.3 Disks

Biermann and Hall (1973) proposed a simple picture that applies to the Algol systems. You store the excess angular momentum in a disk and spin the accreting star up, thereby transferring orbital to rotational angular momentum in a sort of double-entry bookkeeping. The energy dissipated by the process powers the disk and the angular momentum is eventually lost in the disk in a boundary layer that transfers the excess to the periphery of the disk (Pringle 1981; Frank, King, and Raine 1991). In this case, $\dot{J}$ is the rate of angular momentum increase of the accreter. Yet period changes are observed in stars that don't appear to be exchanging mass at all. Specifically, for many years Herczeg (1980) has been calling attention to a class of detached main sequence systems that undergo sporadic period changes. Gimenez (1992) has also called attention to these systems. So there must be other ways of changing the period without the specific exchange of material. The angular momentum carried by the stream at the the inner Lagrangian point is $\dot{J} = \dot{M}(GMxa)^{1/2}$ where x is the distance of L_1 from the center of mass. Thus, the highest specific angular momentum is carried by material coming from a low mass companion that is filling its Roche lobe, the kind of system that is most likely to form a disk around the gainer. Much of this angular momentum may reside in the disk and only slowly spin the underlying star up, depending on the details of the viscosity.

4.7.4 Winds and Spin

Two properties of close binary systems can conspire to produce an effect that mimics mass transfer. Take the stars to be close enough that they are tidally coupled, in fact assume that they're synchronous. Then the rotational and orbital angular momentum are strongly coupled. Now make one of the stars strongly magnetic and allow it to lose mass. A stellar wind is restrained to the Alfven radius which is given by $r_A = v_A/\omega_{rot}$, where

$$v_A = \frac{B(r_A)}{(4\pi\rho(r_A))^{1/2}} \tag{41}$$

is the Alfven speed. Then material lost from this point will carry an angular momentum. Call the stellar mass M_1 and the stellar radius R_1, although there is no restriction to the primary or secondary. Then:

$$\frac{\dot{J}}{J} = \frac{\dot{M}_1}{M_1}(\frac{r_A}{R_1})^2 \tag{42}$$

where we assume strong tidal coupling between the two stars so that $\omega_{rot} = \omega_{orb}$. In general the Alfven radius is large compared with the stellar radius so that the moment arm greatly amplifies the effects of the mass loss. But the consequences are profound, because the assumption that leads to this equation is that the two periods are equal. Then the change in the orbital angular momentum is driven by the tidal coupling between the stars. As the one slows down because of angular momentum loss, and also mass loss (although this is a minor effect in many cases), the period is forced to increase (for further discussion of magnetic braking see Mestel and Spruit 1989). Consequently, the system widens.

One system, V471 Tau, has been well studied among the RS CVn-type of active binaries to see whether there is a wind from the evolved star. In this case, we have a K subgiant and a white dwarf in a short (0.5 day) period binary. IUE spectra suggest that a wind has been observed directly from the K star with velocities of order 500 km s^{-1}, but this remains to be further studied (Mullan *et al.* 1989). Other studies of the environment of such systems using the companion as a "ζ Aur system in miniature" have not proven successful in detecting the outflows that seem to drive the period changes. AL Vel, a ζ Aur-type star, shows evidence of a wind that co-rotates out to about $\approx$10 stellar radii (Eaton *et al.* 1990). But unfortunately AL Vel is a relatively long period system that does not display large period changes.

Bolton (1989) has recently argued that the irregular period changes observed in Algol aren't caused by mass transfer at all, but are due to magnetic braking of one of the components and the subsequent tidal coupling between the stars. The idea, which combines many of the processes we are discussing here, is very pretty because it unifies the magnetic activity phenomenology between the Algols and the RS CVn stars. The idea is an extension of the one by Hall, Kreiner, and Shore (1980) (see Hall 1990). A strong couple exists between the stars in a semi-detached system because of the tide and therefore if one star spins down, the angular momentum will be drawn from the binary as a whole. As we saw above, the critical input to the period evolution equation is the torque, by whatever mechanism, that a member of the binary experiences. Regardless of the colatitude at which an active region happens to appear, constraining the superficial mass loss out to the Alfven surface inevitably produces spindown of the mass losing star. The spinup timescale for the system depends on the synchronization timescale (see next section).

4.8 Rotational Synchronization and Orbital Circularization

The problem of synchronization has been a vexing one for all who study close binaries. First there is the basic observational problem – it is quite difficult to determine the rotational velocity of a star uniquely (*e.g.* Collins *et al.* 1991). Specifically, depending on the differential rotation law, the shape of the star influences the relative weighting over the stellar surface. But there is an additional problem with binaries. One can usually assume, in complete ignorance of the facts, that the law for a solar type star is, say, like that of the Sun. However, there is some evidence that this is not the case in at least a few systems, the RS CVn stars, and this may be generally true.

Wilson (1989) has examined synchronism in Algol systems. Out of 33 systems, about 2/3 show synchronism for periods less than two weeks. In general, there is a linear relation between the orbital and rotational angular momenta of these systems. Without knowledge of the ages of these systems, it nevertheless seems that tidal synchronism is acting in the Algols. However, there is a hitch here. We don't know when the synchronism was established. Keep in mind that the Algols are systems that have undergone considerable interaction throughout their lives, both via mass transfer and mass loss from the system. The orbital angular momentum has generally been reduced, while the accretion of high specific angular momentum by the companion has tended to increase the rotation rate of the gainer. It is possible that a reverse transfer has occurred in these systems, and that the angular momenta have been equalized through processes that are unrelated to the details of the classical tidal coupling.

Claret and Gimenez (1992) looked at a sample of detached main sequence binaries with periods between 1/3 and 150 days and find that the circularization process takes place on a longer timescale than synchronization. Systematically, as other studies have found, stars in close binary systems rotate more slowly than their field counterparts, and, in general, synchronism appears to be pretty easily established within the main sequence lifetime of even fairly massive stars, of order 10 $M_\odot$. Be aware, however, that there are problems with their definition of synchronism for eccentric orbits, since they take the corotational velocity at periastron as their theoretical value.

Latham and collaborators have taken a different approach to the synchronization problem, focusing on a single cluster (Torres *et al.* 1992, Latham *et al.* 1992). They chose M67, an old galactic cluster that we will meet again in the context of the blue stragglers. They look at the distribution of periods and eccentricities for binaries discovered using correlation methods for periods less than about 10,000

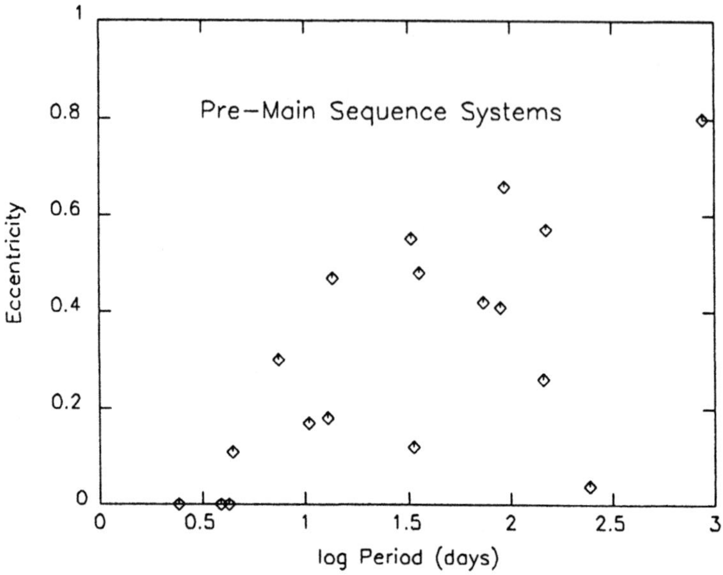

Figure 15: Orbital eccentricity for pre-main sequence binaries as a function of orbital period (Latham *et al.* 1992).

days (see Matheiu, Latham, and Griffin 1990) (fig. 15). In general, the systems with periods shorter than about 12 days show $e \approx 0$, while the largest eccentricities in their sample, of order 0.8, are present at *all* longer periods. In fact, the only short period system that shows an appreciable eccentricity is F190, a blue straggler! So at least in this one cluster, the evidence is that circularization can be achieved for close systems with periods of order a few weeks on a timescale of about 5 Gyr.

In summary, the statistics are as follows. For field Pop I stars (young disk stars) the circularization transition occurs at 8.5 days (Zahn and Bouchet 1989). For M67, an old galactic cluster typical of the thick disk population, the transition to circular orbits is at about 12.5 days, with similar results for blue stragglers in other clusters. For field halo stars the transition period is about 18.5 days. The evidence is that the longer the timescale over which interaction occurs, the longer the period of the system will be that displays synchronism. This seems to be true irrespective of the mass ratio (see Mathieu *et al.* 1992).

The β Cep stars provide an interesting means for testing the idea that main sequence, and near main sequence, stars with radiative envelopes are tidally synchronized. Modal analysis of these stars indicates that they are pulsating in very high order modes that can be correlated with the rotation frequency. Since the radii of these stars are approximately known, it is possible to determine whether they are synchronized or not. The simple answer is that they're not. At this stage it isn't clear precisely how the stars got into this condition, but it is probable that they are more rapidly rotating than their binary periods indicate they should. The same is true for the Be stars. In other words, there exists a class of

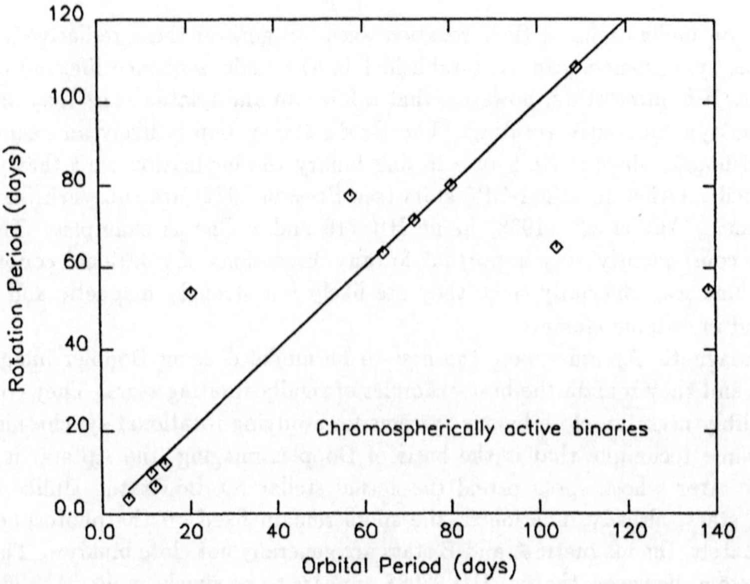

Figure 16: Comparison of orbital and rotation period from Hall (1986) as an example of using active binaries to test synchronization.

binaries for which it is not at all clear that tides have played any role in slowing the stars down and coupling the rotation.

Although still inconclusive, a wonderful test is available that makes use of active stars. Hall (1986) first introduced the idea that spotted binaries, like the RS CVn stars, can be used to test any synchronization mechanisms (fig. 16). These are stars with deep convective envelopes, and in general they are detached. The idea is then to look at the ratio of the rotation period to orbital period as a function of orbital period. In these systems, the correlation between the rotation time for the spots and the expected rotation due to synchronism is very good over the whole range of periods. The spot regions, however, do precess through the light curve, an indication that the active star(s) is not actually completely synchronized. A more complete discussion is provided by Hall and Henry (1991). The problem they and others face with this method is that the synchronism has to be defined in terms of a *corotating latitude*, a region in both hemispheres that actually is synchronous, in order for the comparison to be made exact. It is interesting then to note that the synchronism never occurs at the equator, although the most naive application of tidal theory suggests that this is where it should be found.

The Am stars are systematically slow rotators among main sequence A stars (Wolff 1983, Preston 1984, Dworetsky 1992). These are close binaries which show very sharp lines. They do not show spectral or light variations in most cases, but they have radii that are normal for their masses. Their orbital solutions can be combined with their masses and rotational velocities to provide

statistics on inclinations of their rotation axes. In general these radiative stars show that synchronism can be established in the main sequence lifetime of a $2M_\odot$ star. It is interesting, however, that a few Am and related stars may actually be *sub*synchronously rotating. The HgMn star χ Lup is likely an example of this although, since it isn't an eclipsing binary the inclination isn't that well determined[7]. Other so-called CP3 stars (see Preston 1974) are subsynchronous. For instance, Abt *et al.* (1968) name HR 710 and κ Cnc as examples. These stars are consequently very important for any discussions of rotational coupling in close binaries, especially since they are likely not strongly magnetic and are also found in datable clusters.

The magnetic Ap stars were the first to be modeled using Doppler imaging methods and they remain the best examples of rigidly rotating stars. They could be incredibly useful in close binary systems for studying rotational synchronism. By the same technique that is the basis of Doppler imaging, the Ap star is an oblique rotator whose spots reveal the actual stellar rotation rate. Unlike the RS CVn stars, already mentioned, the spots remain fixed on the photosphere. Unfortunately, the magnetic A and B stars are generally not close binaries. There is one system, however, that is. HD 98088 appears to be synchronous (Abt *et al.* 1968). This system is quite unusual, however, in presenting a synchronous but eccentric orbit. The mass ratio is near unity (the probable masses are around 1.7 and 2.2 $M_\odot$ and the eccentricity is about 0.2. As far as I know, there has not been a study of this 5.9 day period system with modern detectors. It is especially important to confirm the orbit because we know precisely the rotation period of the magnetic star.

Now for a little bit of theory. Consider the tidal radius, give approximately by $r_{tidal} \sim (\rho_2/\rho_1)R$ or:

$$\delta r \approx \frac{1}{4}\frac{M}{m}\left(\frac{R}{a}\right)^3 R. \tag{43}$$

Here a is the separation of the stars, R is the radius of the primary, M is its mass, and m is the mass of the companion. The mass of the tidally generated bulge is given by:

$$\delta M \approx \frac{M}{R}\delta r \tag{44}$$

The tidal torque is then approximated by:

$$\tau \approx -\frac{3GM\delta M}{R}\left(\frac{R}{a}\right)^3 \delta \tag{45}$$

where δ is the lag angle. Finally, we see that there is a *very* strong dependence of the torque on separation

$$\tau \sim \left(\frac{R}{a}\right)^6 \left(\frac{M}{m}\right)^2. \tag{46}$$

Now since, by Kepler's law, the separation is related to the period P of the system, we see that $\tau \sim P^{-4}$. It is quite surprising that this crude estimate is actually

[7]Dworetsky (1992, private communication) called this to my attention after the course, but it is a very important observation.

quite close to the actual dependence from a detailed calculation (Savonije and Papaloizou 1984). The full problem involves the redistribution of the tidal angular momentum through the envelope of the perturbed star, but I won't consider the details here. Instead, just note that the timescale for circularization of the orbit and for synchronism of the rotation and orbital periods depends strongly on the orbital period.

This classical estimate assumes that the star responds dissipatively only in the radiative tide. There are, however, other possibilities that may dominate the rate of internal angular momentum transport. The tidal force creates a potential gradient along the stellar surface, $(1/r)\partial\Phi/\partial\theta$. Since Φ is proportional, in lowest order, to $P_2(\cos\theta)$, this perturbation has a strong latitudinal dependence. Therefore, a current can be driven that redistributes angular momentum with latitude as well as depth. We shouldn't be surprised at this since it is the same effect observed in the Earth's oceans. But classical analyses of stars have often neglected the fact that they are hydrodynamically responsive entities, and such a tide can drive deep mixing. Tassoul (1987, 1990*a,b*) has outlined such a mechanism that considerably enhances the rate of synchronization in close systems. His argument is that a thin Ekman layer develops at the top of the star's envelope. Take the viscous stress to be $\tau = \mu\nabla\times\mathbf{v}$ where μ is the coefficient of viscosity. Assume that $\mu = 10^N\mu_{rad}$, where N is simply a free multiplier and μ_{rad} is the radiative viscosity (Tassoul 1978). Then:

$$\delta r = 10^{N/4}\left(\frac{\mu}{\rho_b\Omega_0}\right)^{1/4} \tag{47}$$

is the thickness of the Ekman layer, where ρ_b is the density in the surface layer and Ω_0 is the orbital frequency of the binary. He then estimates the spindown time to be:

$$t_{sd} = \frac{1}{2\Omega_0}\frac{R}{\delta r}\frac{M}{M'}\left(\frac{a}{R}\right)^3$$

or, in astronomically useful units:

$$t_{sd} \approx 5.4\times10^{(12-N)/4}\frac{1+q}{q}(L/L_\odot)^{-1/4}(M/M_\odot^{5/4}(R/R_\odot)^{-3}(P/day)^{11/4}\ yr. \tag{48}$$

The circularization timescale is given by:

$$t_{cir} = 9.4\times10^{(16-N)/4}\frac{(1+q)^{2/3}}{r_g^2}\left(\frac{L}{L_\odot}\right)^{1/4}\left(\frac{M}{M_\odot}\right)^{23/12}\left(\frac{R}{R_\odot}\right)^5\left(\frac{P}{day}\right)^{49/12}yr. \tag{49}$$

Here r_g is the fractional stellar gyration radius. The critical factor in this theory is the degree to which the circulation currents are turbulent in the boundary layer at the top of the envelope (Tassoul and Tassoul 1992). Especially clear discussions of some of the circulation problems for convective envelopes was developed by Scharlemann (1982) and for contact and close binaries by Smith and Smith (1981).

One interesting question that is very hard to address observationally is how synchronism is achieved in massive systems. In particular, binaries consisting of two O stars, like Y Cyg, are ideal for studying the tidal interaction of detached radiative envelope stars. The difficulty comes from the uncertainties in determining the rotational velocities of the stars. The lines are typically very broad, and in the most massive systems quite likely affected by the stellar wind, so it isn't even clear whether a zero-velocity surface is actually being observed. But the problem is certainly worth addressing.

4.9 Light Curve Instabilities: Messy Eaters and Massive Accretion Disks Among the β Lyrae Type Systems

As we've already seen, stable light curves provide valuable information about the shape and atmospheric structure of the components in a binary system. Light curve variations may signal more interesting complications. Secular changes often indicate the presence of circumstellar or circumsystem material. They may result from starspot activity on one or both of the components. If present, such changes must be dealt with in order to secure some of the more prosaic details of the binary system. But more important is the wide range of physical processes that can be revealed by something as simple as a change in the light curve.

The interacting binary β Lyrae has been known for over a century, and is one of a class of extreme mass-transferring systems that includes RX Cas (K1 III + A5e III, 32.3 days), SX Cas (B7e III + K3 III, 36.6 days), RZ Sct (B3 Ib + ?, 15.2 days), V367 Cyg (B8pe Ia + F4 III, 18.6 days)and W Ser (F5e(shell) Ib, 14.2 days) (Plavec 1982, 1983). These systems collectively have periods that are usually rather long for Roche-filling systems, about two weeks. They normally show a late-type component, although this star isn't always observable, and an earlier-type spectrum. However, this is where any resemblance to a normal star ends for the hotter companion. The spectrum is usually typical of a shell star, with a temperature of about 7000 to 10,000 K, in other words an A or F star. Often there are strong emission lines in the optical, along with the shell spectrum, and sometimes there are even P Cyg profiles on the hydrogen lines. Sometimes, although not always, He I lines are also present. The light variation is nearly continuous, indicative of an ellipsoidal variation of the optically dominant star. weak radio emission has been detected from a few (Dobias and Plavec 1985, Elias 1990).

These systems show strikingly ephemeral light variations. No two photometric curves ever precisely repeat, and in some cases, especially W Ser, the variations are so large from one period to the next as to render the light curve unrecognizable (fig. 17). Spectrum variations are also seen, with the absorption spectrum also showing components that don't necessarily follow the hotter star. In short, it appears as if these systems are dominated by something that varies on a hydrodynamic timescale. In the infrared, these systems often show some excess, indicating that bremsstrahlung arises from diffuse matter distributed both within and around the system. In the ultraviolet these stars show only emission or P Cyg profiles.

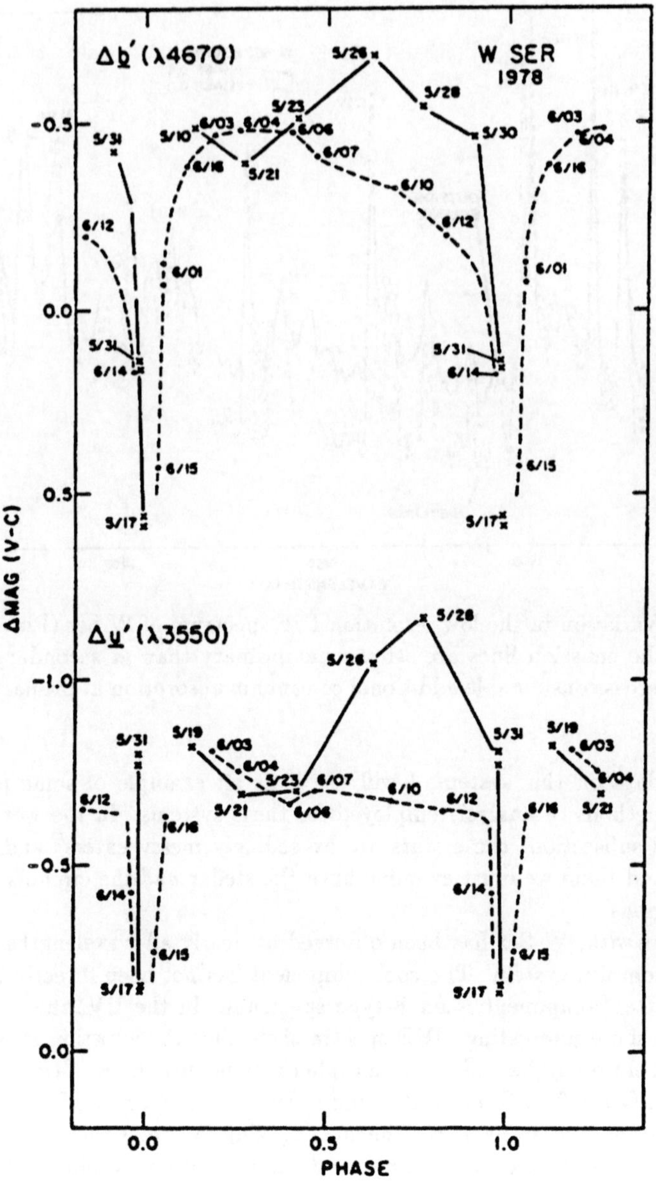

Figure 17: Examples of light curves for W Ser (Guinan 1989).

4.9.1 W Serpentis

The semi-detached system W Serpentis has been taken for some years as a prototype of the interacting systems in a rapid phase of mass transfer (Plavec 1982, Weiland 1984). In fact, some have gone so far as to create a new sub-class of interacting binaries, calling such systems "Serpentids". Since there is new high

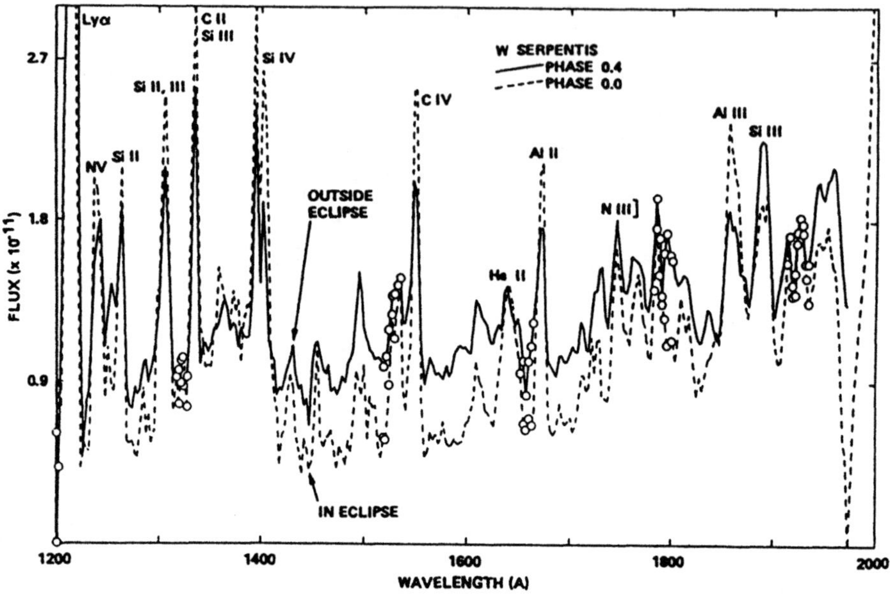

Figure 18: Variation in the low resolution IUE spectrum of W Ser (Plavec 1980). Note that the emission lines are stronger at primary than at secondary eclipse, and that there is considerable additional continuum absorption at primary eclipse.

resolution data on this system, I will use it as an example of some pitfalls in the usual methods of analysis employed for these systems. In the words of the title of this subsection, these stars are exceedingly messy eaters[8] and in order to understand them we must examine both the stellar *and* the circumstellar line forming regions.

To begin with, W Ser has been observed at nearly all wavelengths. It is an especially complex system. The cool component has not been directly observed, and the hotter component is an F-type spectrum. In the UV, the picture becomes even more interesting. IUE spectra show that the emission is present in what would normally be called chromospheric or nebular lines. Strong emission is seen from O I λ1300, Si IV λ1400, and C IV λ1550, all arising from plasma at temperatures of at least 20000 K. The most striking feature is that the emission is seen on top of a complex continuum, one dominated at low resolution by a shell-type absorption line spectrum. The emission is strongest at *primary* minimum, when the hot star is being occulted, and as such is anomalous. The original expectation was that this phase would be the one when the chromospheric emission would be suppressed by the eclipse. Instead, the line weakens about half an orbit later (fig. 18). The most likely reason for this is related to the mass transfer process. As I'll discuss below (accretion disks), a stream coming from the loser to the mass gainer in a semi-detached system both forms an extended accretion disk

[8]This happy description is due to Richard Wade.

and also impacts the outer parts of that disk at supersonic speeds. Consequently, a hot spot is formed in the outer disk. This region is then transported around the system with respect to the observer's line of sight and differentially absorbed by the disk. So when on the near side, this region is more clearly observed than when transported around to the far side of the disk (see Lyons *et al.* 1992).

4.9.2 β Lyrae

It is an interesting historical accident that the two anomalous prototypical binary stars, Algol and β Lyr, should have been discovered by Goodricke, the latter in 1794. Beta Lyrae is a relatively long period system for an Algol-type variable, 12.94 days, with a steady period increase. Its mass function is $f(M) = 8.4M_\odot$, with the hotter star a hydrogen poor B6 II star (Balachandran *et al.* 1986). Its radial velocity amplitude is quite large, 185 km s^{-1}, and this is rather nice in making the lines of any reputed secondary rather easy to identify. The light curve is dominated by a strong ellipsoidal variation, coupled with reflection effect, and it also displays large amplitude irregular variations that indicate the intrinsic instability of the circumstellar material within the system (fig. 19). Unlike Algol, this is a massive system, with the gainer (the unseen star) certainly being about 10 $M_\odot$. The mass of the evolved star appears to be 2 to 3 $M_\odot$. The early history of research on this star has been summarized in Sahade *et al.* (1959).

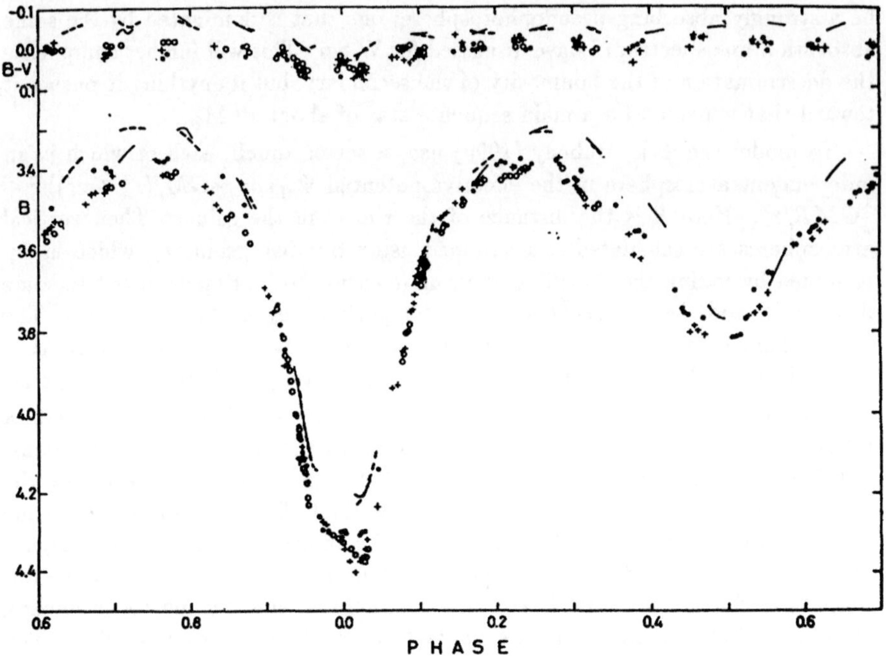

Figure 19: Variations of the primary eclipse light curve for β Lyr (Larsson-Leander 1970).

The spectroscopic and photometric antics of this star are well known and have been described many times. The main feature of light curve variation is at minimum light, which changes in depth by as much as 10% within a single orbit. Mass loss and transfer have been suspected in this system since the first optical spectra were obtained. The optical continuum shows the B giant absorption spectrum with Balmer and He I emission lines superimposed. The emission lines are variable in both flux and profile on timescales shorter than a single orbit. In the ultraviolet, P Cyg profiles are observed on many of the resonance lines. Plavec (1989) has pointed out that the emission line luminosity for β Lyr rivals the stellar luminosity in the UV. However, there is a substantial deficit in flux compared with a normal star, one that should be about B0.

The most complete recent attempt to model this vexing system has been by Hubeny and Plavec (1991). They assume that the gainer is shrouded by the material transferred during a rapid stage of Roche lobe overflow and that the radiation we are seeing comes from a rim on the accretion disk. This region, more or less like the torus once hypothesized for the system by Wilson (1974). They assume that the gainer is a B main sequence star, and that the star's continuum is reprocessed by the disk. In addition, there may be a small amount of heating from turbulence, and this is responsible for some of the continuum. They don't model the lines, though. The disk is thick, in keeping with the torus picture, and most of the energy we observe comes from the peripheral rim of the disk. Their model is quite successful in reproducing the basic features of the UV flux. I should remark, however, that in addition to the continuum, there must be a strongly absorbing pseudophotosphere, one that is dominated by the same absorption line spectrum I have discussed for W Ser. This will further complicate the determination of the luminosity of the secondary, but if anything it pushes it toward that expected for a main sequence star of about 10 $M_\odot$.

To model the disk, Hubeny (1990) uses a set of annuli, each of which is an independent atmosphere in the effective potential $\Psi_{eff} = -GM/(r^2 + z^2)^{1/2} + \frac{1}{2}GMR/r^2$. Here R is the distance of the rim from the gainer. Then vertical atmospheres are calculated and summed using the disk geometry, which is determined by taking the specific angular momentum to be Keplerian out to some distance R and then constant for the outer portion beyond the rim radius. The disk is somewhat like the non-Newtonian models that have been used in AGN disks. The rim therefore looks a lot like a heavily gravity-darkened flattened star.

One of the most interesting observations of β Lyr comes from the Voyager data (Polidan 1989). The spectrum above 1300Å shows the normal eclipse variations, but below 1200Å there is no evidence at all for an eclipse. Above 1300Å the variation is about 30 - 40% and look like normal ellipsoidal variations, while below 1100Å the variation is less than 10%. The observations are shown in Figure 20. Notice that the EUV continuum is well correlated in the sense that the low amplitude changes are strictly correlated. There is, however, no corresponding correlation between the Si IV region and the continuum below λ1200Å. The cause of this behavior remains an enigma. If the light curve in UV and optical wavelengths is due to a thick accretion disk, as argued by Hubeny and Plavec

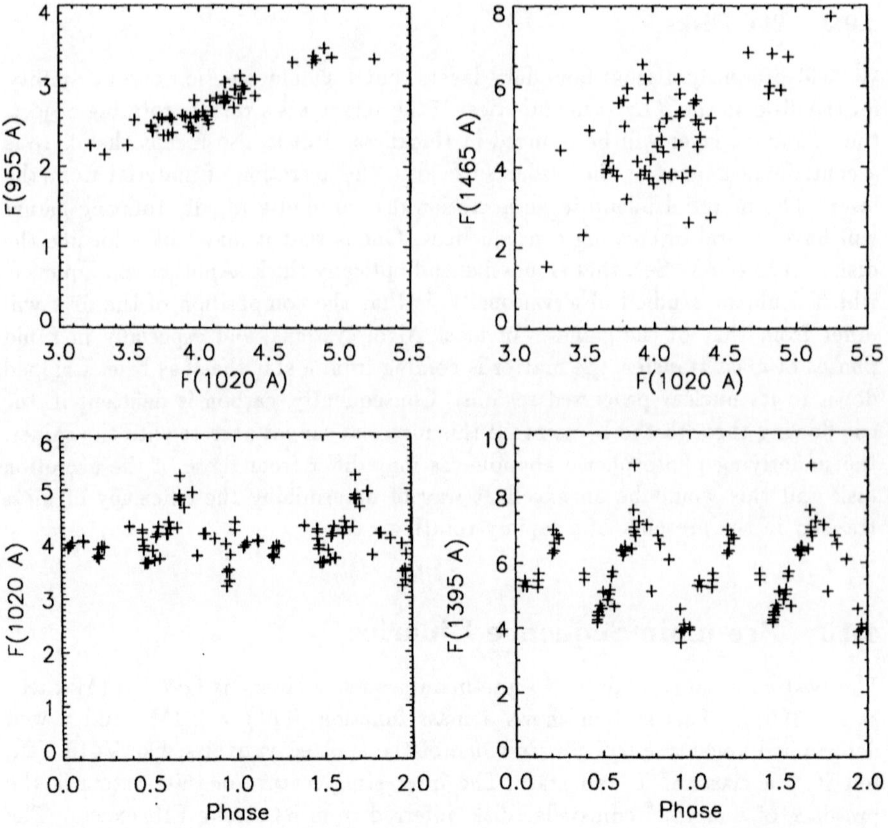

Figure 20: Voyager (900-1400Å) light curves for β Lyr (data supplied by R. Polidan).

(1991), then there should be modulation of the EUV at some level, at least by geometric eclipses. The possible explanation is that the primary modulation at longer wavelengths is due to iron peak and other atomic line absorbers that are absent below 1200Å. We will see more of this in the symbiotics, where the discrete nature of the line absorption from environmental ions is very important in modulating the continuum of the hotter star.

Another system in which something like this is likely observed is V356 Sgr (Polidan 1989), which is also extremely carbon deficient and likely N-rich. This star appears to be in a stage of rapid mass transfer as well, although not as much is known about its light curve instabilities or UV spectrum. The period is about 8.8 days, and the system consists of a B3 V and A2 II star. The B3 star is an extremely rapid rotator, about 350 km s^{-1}, much like U Cep. The large period change indicates a high rate of mass transfer, about $10^{-7} M_{\odot} yr^{-1}$, although not at all as large as for β Lyr. One interesting thing is that the resonance lines, especially N V λ1240 and Si IV λ1400, show redshifted emission components with large velocity widths and no evidence for absorption troughs.

4.9.3 The Disks

We will presently discuss boundary layers, but I should mention one possibility for the disks in the β Lyr-type binaries. If the accreter is a rapidly rotating object, the boundary layer will be reduced in thickness. But it also means that here is a centrifugal barrier at the stellar surface to the accretion of material from the loser. The material pileup in an accretion disk around a rapidly rotating gainer will have several interesting consequences. One is that it may look a lot like the disk in β Lyr or W Ser, that is massive and optically thick. Another consequence, which could be studied observationally, is that the composition of the disk will differ from that of the gainer. In most Algol systems, and especially in rapid phases of mass transfer, the matter is coming from a star that has been stripped down to its nuclear processed regions. Consequently, carbon is deficient in the gas flowing through the L_1 point. If this material cannot accrete onto the gainer, the underlying photospheric abundances may differ from those of the accretion disk and this would be an excellent way of determining the efficiency of mass transfer in the presence of a rapidly rotating gainer.

4.10 Pre-main Sequence Binaries

The best case study to date of a pre-main sequence binary is GW Ori (Matheiu *et al.* 1991). This system shows a mass function $f(M) = 0.1\mathrm{M}_\odot$ and a well determined velocity curve. It isn't one of the spotted variables, like V410 Tau, but it is a classical T Tau star. The interesting feature of this system is the presence of a large circumstellar disk, inferred from its strong FIR excess. The model employed by Matheiu *et al.* is similar in many ways to that used for β Pic, a circumstellar disk with a gap that has been cleared out of material. The idea is that the infrared excess observed in these systems comes from an enveloping disk that may surround the whole binary. Given the dimensions inferred from the disk temperatures and the periods of the encapsulated binaries, this picture is likely correct. Then the spectrum reflects the local distribution of dust around the binary. For instance, the near IR, about 10μ, comes from the innermost regions of the disk while the coldest material is located farthest away. This same thermal imaging has been applied successfully to circumstellar shells in OH/IR stars and other optically thick dust shells, so its application here may be justified. However, if the disk does not completely cover the central star – and the fact that we can see through the environment to the binary indicates that it probably doesn't – then the interpretation of dust temperature becomes more of a problem.

BM Orionis is a pre-main sequence system in the θ^1Ori subcluster in Ori OB1 (the Trapezium). It is an eclipsing binary with a period of 6.5 days. The optical primary is a ZAMS B star while the secondary appears to be above the main sequence. Fitting the star to pre-main sequence evolutionary tracks indicates that the star is just below the "'birthline" for pre-main sequence stars (Stahler 1983). Two interesting problems are posed by this binary. The first is that although the eclipse is total and the eclipse solution demonstrates that the radii of the stars are

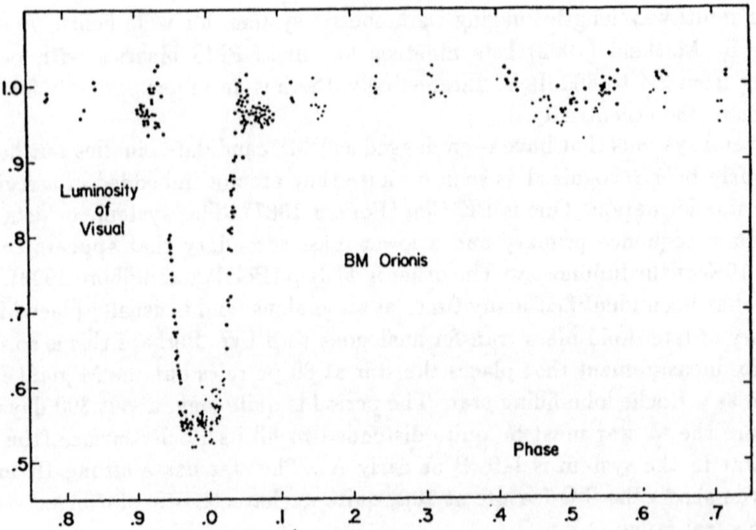

Figure 21: Light curve of BM Ori from Arnold and Hall (1976).

such that the optical primary should be completely obscured (Hall and Garrison 1969), it never is (Doremus 1970) (fig. 21). This first led Hall (1971) to propose that the secondary is flattened, perhaps a star surrounded by a disk. The optical depth of the surrounding material, were there any, would have to be quite large, comparable with what is detected in the β Lyr stars. Further observations by Popper and Plavec (1976) show that the secondary is a very peculiar spectrum. It has the spectrum of an A star. This will shortly begin to sound very familiar, since it is characteristic of stars surrounded by thick disks. In particular, the members of the β Lyr class show this characteristic. The orbital solution shows that the B star has 5.75 $M_\odot$ and the "A" star is 1.8 $M_\odot$. Since the eclipse is total, I have ignored i in the masses. Popper and Plavec obtain an axial ratio of about 3 or 4 for the secondary. I am dwelling on this system for two reasons: first, because it is an intrinsically interesting system for which there has been little modern work (*i.e.* CCD and infrared arrays) and which would certainly repay such investigations, and second, because it is the one PMS system for which we have essentially all of the requisite information to determine the properties of a near, but still clearly not, main sequence star.

Before the advent of CCDs, the search for eclipsing binaries in young clusters produced very few candidates and poor statistical coverage of them. Some photographic searches were conducted of young clusters (see, for instance, Koch and Perry (1975)), but the nature of any observed variability among the T Tau stars remained unconstrained. However, with the current large-format linear devices it's possible to do wide fields with high photometric accuracy, as work on old clusters has demonstrated. There have been few recent attempts to extend the original surveys, but such surveys are needed and extremely valuable. It would be especially useful to have multi-color photometry, perhaps with a 4-shooter or similar multi-wavelength imaging photometric system, for wide fields. A recent review by Matheiu (1992) lists nineteen low mass PMS binaries with periods ranging from 2.4 to 880 days. Interestingly, there is an upper envelope for these systems in the eccentricity.

Several systems that have been flagged as PMS candidate binaries but haven't previously been recognized as such because they are not imbedded in a region of active star formation. One is EK Cep (Popper 1987). This system consists of an A1.5 main sequence primary and a lower mass secondary that appears to have about 10% of the luminosity. The other is 17 Lep (Polidan and Shore 1992). This binary has been identified many times as anomalous, and is usually placed in the category of late rapid-mass transfer analogous to β Lyr. Perhaps this is so, but a parallax measurement that places the star at 80 pc rules out the M giant in the system as a Roche lobe-filling star. The period is quite long, about 300 days, and therefore the M star must be quite distended to fill its Roche surface (the early type star in the system is late B or early A). The star has a strong IR excess, and even shows the SiO feature at 10μ, quite uncharacteristic of evolved stars in this spectral range.

In general, the recognition of PMS status for a field binary is a real problem. The original definition of the Ae/Be stars by Herbig (see Herbig 1985) was that they are T Tau-like stars, that they are early-type stars, and that they are associated with nebular environments and dust. He explicitly said that the definition was arbitrary, and that it limited the search to this particular class of objects. He left open the possibility that such stars might be found elsewhere, and perhaps this is something worth examining again. Recently Perez and Danks (1992, *private communication*) have made a stab in this direction and more samples are clearly needed.

It should be borne in mind that any star on the main sequence with a period of less than about one week should, at some stage in its past, have been in contact with its companion. How the contact is broken, and what consequences such a stage has for the subsequent evolution of the binary, is presently an open question. It is, however, very important to note that the initial mass function (IMF) depends critically on the frequency of short period binaries. These stars are the only good handle we have on stellar absolute parameters and even in their present detached state, many close binaries must have been in contact systems before reaching the main sequence. Does this stage have observable counterparts?

4.11 Contact Binaries and Common Envelope Systems

The W UMa systems are the prototypical common envelope binaries, although they are very different from the systems that have been discussed in this course by Livio. These are otherwise normal main sequence stars, each of which is generating nuclear energy by either *p-p* or *CNO* core burning. They are short period systems, normally with orbital periods between 0.2 and 1 day. Essentially none of the late-type systems have mass ratios near unity while many of the early-type systems do. They come in two varieties. The A-type W UMa have deepest eclipse

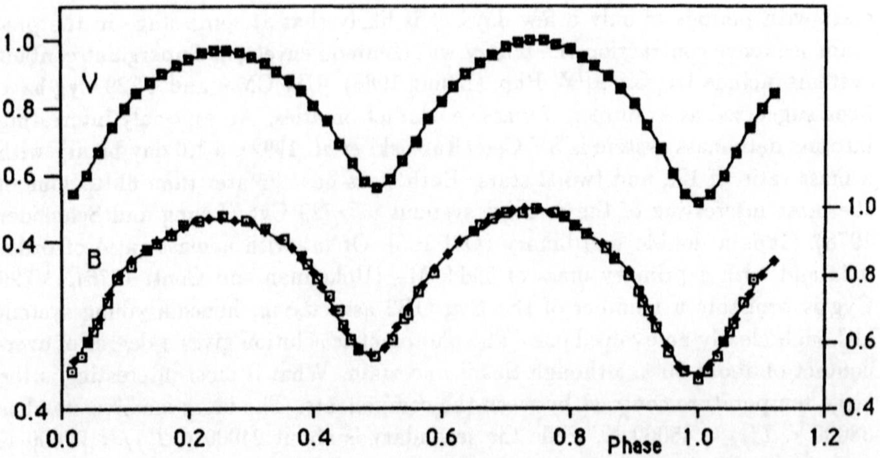

Figure 22: Light curves for a W-type W UMa system YY Eri (see Müyesseroglu *et al.* 1990). Note the virtual absence of color variations around the cycle.

at transit, while the W-type have them at occultation (see Rucinski (1985) for further discussion). Their light curves, as we've already discussed, are continuous. In general, they show no color variation through their orbital cycle (fig. 22). It is this feature that first led to the picture that they are stars imbedded in a common optically thick environment.

But having said these things, there is considerably more physics that remains unknown for the W UMa stars. The contact appears to be stable on timescales of order decades at least, but there are indications that short term period changes occur and that these are consistent with mass transfer between the stars and with system-wide mass loss. The stars show enhanced levels of activity, as we've already discussed, but they also show a tendency for this activity to concentrate on the more massive star. Since they show virtually no color variation, the surface temperature of the envelope is quite uniform, yet they typically have mass ratios much less than unity and therefore should not be barotropic. The contact point is likely unstable, and it isn't clear whether a discontinuity forms in the envelope (Shu and Lubow 1981) or a thermal oscillation is set up that occasionally breaks the contact (see discussion and references in Lucy and Wilson 1979). An important and long-standing problem with the W UMa systems is how the energy is transferred between the stars. In order to maintain the near isothermality of the envelope, a baroclinic flow must be occurring and you might suspect that such motion is strongly dissipative. But the unambiguous signature of the heating that is expected to result from mass and energy transport through the throat has not detected (see discussions by Guinan (1990) and Vilhu (1992)).

The contact systems of the W UMa type are only low mass systems. There is nothing, however, to *a priori* exclude the existence of contact systems among high mass stars, or even pre-main sequence systems. I should remind you here that if a system is near contact on the main sequence, as is the case for massive stars with periods of only a few days, it is likely that at some stage in the pre-main sequence contraction the binary was common envelope. Supergiant contact systems include UU Cnc, PW Pup, (Leung 1988). UW CMa, and V729 Cyg have been suggested as examples of massive contact binaries. An especialy interesting intermediate mass system is SV Cen (Rucinski *et al.* 1992), a 1.6 day binary with a mass ratio of 1.4, and two B stars. Both stars have greater than $6M_\odot$. One of the most interesting of the massive systems is V729 Cyg (Leung and Schneider 1978). It is a double line binary (O7f Ia + Of Ia) with a mass ratio of order 0.23 and with a primary mass of 59 ± 9 $M_\odot$ (Bohannan and Conti 1976). V729 Cyg is probably a member of the Cyg OB2 association, hence a young system although clearly an evolved one. The photometric solution gives a degree of over-contact of about 30%, although this is uncertain. What is most interesting is the large temperature contrast between the components. The more massive star has $38000 < T_{eff} < 45000$ K, while the secondary is about $24000 < T_{eff} < 29000$ K and the luminosity ratio is also large (Leung and Schneider 1978). As imprecise as these estimates are, it's clear that the temperatures for the system cannot be equal, in striking contrast to low mass contact binaries. As far as I know, this system and its relatives have not been re-investigated using more modern techniques. It is important to do this, because there was some disagreement between the photometric and spectroscopic orbits (although the disagreement involves quantitative, rather than qualitative, differences).

Shaw (1990) has discussed the interesting problem of near contact systems (for instance V1010 Oph, FO Vir, and V525 Sgr). It is not at all clear just how close does a star have to come to its tidal surface to initiate mass transfer? I mention this here because of the contact systems. At present, although light curve solutions provide some handle on the degree of overcontact in these binaries, we really don't know just how much is enough to establish thermal equilibrium of the common envelope.

4.12 Binary Central Stars of Planetary Nebulae

Perhaps the most extreme examples we have of compact objects in binary systems, aside from the binary pulsars that van den Heuvel has dealt with in his lectures, are the binary central stars found in some planetary nebulae (CPN stars for short). This is a rapidly developing area of research, aided by the recent improvements in detectors and reduction methods. The most complete review is provided by Bond and Livio (1990) and by Bond, Ciardullo, and Meakes (1992), which I summarize here. About 10% of CPN stars surveyed (and there are now 13 known binaries) show some kind of variability that can be directly attributed to binarity. There are several types of observed variations. Reflection effect and tidal distortion can be seen over a very wide range of inclination, and these will certainly be the most common. While the amplitude is a function of inclination, the currently available detectors have sufficient dynamic range that virtually all such stars should be observable.

Two are known to eclipse, Abell 46 = V477 Lyr and Abell 63 = UU Sge (see Bond *et al.* (1978) for a more thorough discussion of this star). In both cases, they look like V471 Tau consisting of a hot white dwarf and a G or K main sequence star. Another system, Abell 41, is similar but shows only a reflection-induced light curve. Ten have periods between 0.1 and 3 days. Another group, the Abell 35 stars (yes, even in this small a sample there is a clean distinction between subtypes) have G and K subgiants or giants as companions to the hot white dwarfs. These include Abell 35 = LW Hya (3.2 day period), LoTr 1 (6.6 days), and LoTr 5 = IN Com (5.9 days) (although Bond *et al.* (1992) suggest that the period of Abell 35 may be an alias of an 18 hour period). The amplitude of LoTr 1 may be intrinsically variable and hints of possible spot (or mass transfer?) activity on the cool companion.

These systems look a lot like cataclysmics, except that their periods are rather long. Only GK Per is known to have a period of order one day among the classical novae and no dwarf nova shows so long a period. However, these could be progenitors of these systems, since their periods are clearly too short to be symbiotics. There does not seem to be any direct connection with the planetary nebula morphology, but this may not be too surprizing since essentially all (that is, about 80%) of planetaries have some cylindrical or bipolar symmetry.

4.13 ζ Aurigae and Related Systems: Tomography of Stellar Atmospheres by Atmospheric Eclipses

I have repeatedly emphasized that the most important feature about binary stars is that they provide a clock and therefore can be used to provide time histories of a variety of hydrodynamic processes. These are, however, generally extrinsic to the stars themselves. In one class of binaries this feature provides a unique view of the intrinsic properties of at least one of the stars. These are the ζ Aurigae systems.

The ζ Aur stars are long period binaries consisting of a late-type supergiant and a main sequence star. The prototype system consists of a B and K star. If the evolved star has an M-type spectrum, it's usually classified as a VV Cep

Table 1. ζ Aur and Related Systems

Name	Spectrum	$\dot{M}$ ($M_\odot yr^{-1}$)	P(days)
AL Vel	K0 III + B8(V)	-	96
22 Vul	G3 Ib-II + B9 V	4×10^{-9}	249
ζ Aur	K4 Ib + B7 V	8×10^{-9}	972
32 Cyg	K5 Iab + B7 V	3×10^{-8}	1148
τ Per	(G5 III - G8 IIIa) + A2 V	-	1515
δ Sge	M3 II +B9 V	2×10^{-8}	3720
31 Cyg	K4 Ib + B4 V	4×10^{-8}	3784
VV Cep	M2 Ia + B9 V-IIIe	-	7450
KQ Pup	M2 Iab + B? V	-	9750
ϵ Aur	F0 Iab + ?	-	9890

binary. In at least one system, ϵ Aur, the evolved star is sufficiently distended as to be effectively invisible and it is even uncertain whether there is a normal star in the system. We shall shortly return to this especially exciting object. The ζ Aur binaries present a unique opportunity to probe the atmospheres of late-type evolved stars. Their primary importance is not so much their interaction, although there is some evidence that there is direct hydrodynamic interaction between the components. Rather, these systems consist of very disparate types of stars. Usually the evolved star is a G or K giant or subgiant, and the companion is an early-type main sequence star.

Individual studies have been done on almost every member of the class. Rather than listing them individually, I refer you to some excellent summaries by Wright (1970), Reimers (1989) and Carpenter (1992). In the visual and UV, the hot star serves as a probe, illuminating the outer atmosphere of its evolved companion. The general theory of atmospheric eclipses has a long history, beginning with the work of Kopal and Shapley (1950), although there are some initial oblique references to the problem in Schuster's original paper on radiative transfer. The physical picture is shown in Figure 23. Here an evolved supergiant eclipses a hot component. The early-type star therefore suffers a wavelength dependent extinction through:

$$\tau_\lambda = \int_{-\infty}^{\infty} \kappa_\lambda(r[s])\rho(r[s])ds \tag{50}$$

and therefore the emergent spectrum, in the absence of large scale flows, is $I = I_\star \exp(-\tau_\lambda$. Here τ_λ is the monochromatic optical depth. This is also a function of phase because as the line of sight passes through the atmosphere of the giant, individual convective cells are sampled and the temperature gradient is sampled. The atmosphere is structured by the requirements of radiative and hydrostatic equilibrium. Consequently, T, the temperature, is a function of τ, the mean optical depth of the atmosphere. The limb darkening required to produce the integrated spectrum of a star is the expression of this temperature gradient.

Now if we consider that the pencil beam from the companion probes to some maximum optical depth τ_{max} depending on the geometry of the eclipse, it is possible to study the details of the temperature gradient because the optical depth τ_λ along the line of sight translates into a map of τ as a function of depth. If the line of sight through the atmosphere is s, then there exists at each line of sight a maximum sampled atmospheric optical depth. Therefore there is a maximum temperature, and also a maximum turbulent velocity, assuming that the turbulence is never transsonic. The problem of deconvolving the atmospheric structure from the measurements along the line of sight is much like determining the limb darkening for a single star. Each line of sight sees what is in effect a different atmosphere. When near the limb, the path samples the temperature minimum, while a more nearly photospheric spectrum will be obtained near maximum eclipse. The point is that the star serving as the probe has a known spectrum and there is no interaction between the star and the late-type stellar atmosphere (fig. 23).

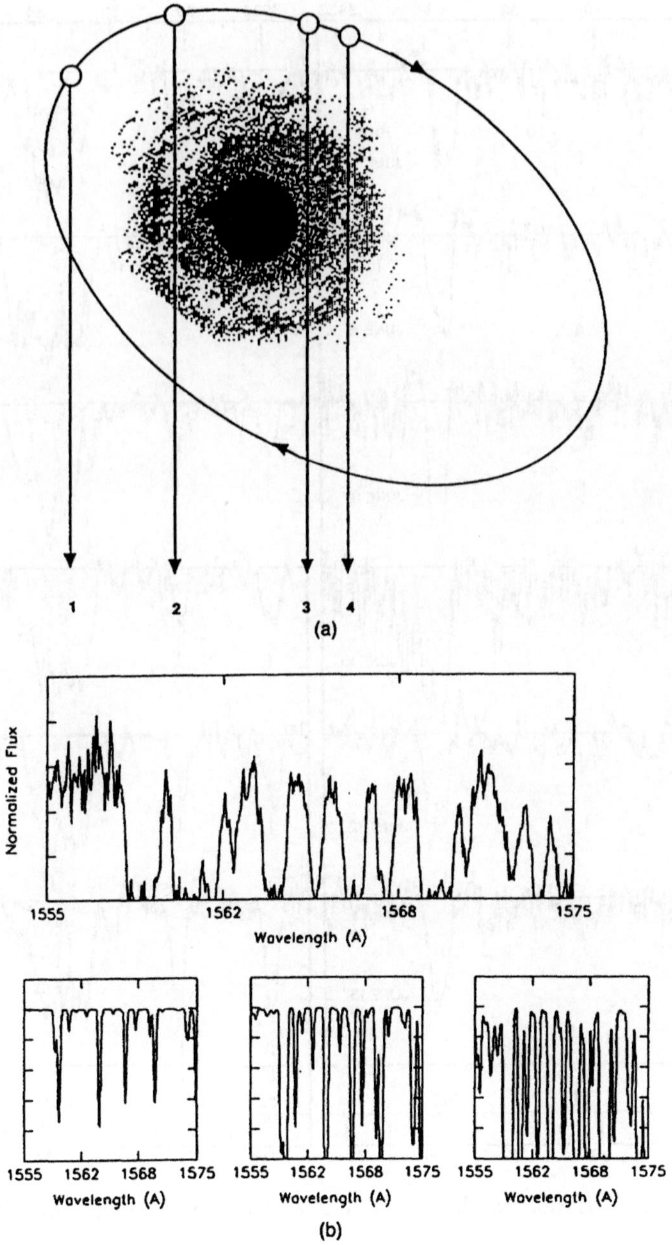

Figure 23: A schematic explanation for the atmospheric eclipse of a ζ Aur binary. Middle: 22 Vul near mid-eclipse. Bottom: Model spectra for increasing column density from 10^{21} to $10^{23}\mathrm{cm}^{-3}$ with $T_{ex} = 5000\mathrm{K}$ and $n_e = 10^{10}\mathrm{cm}^{-3}$.

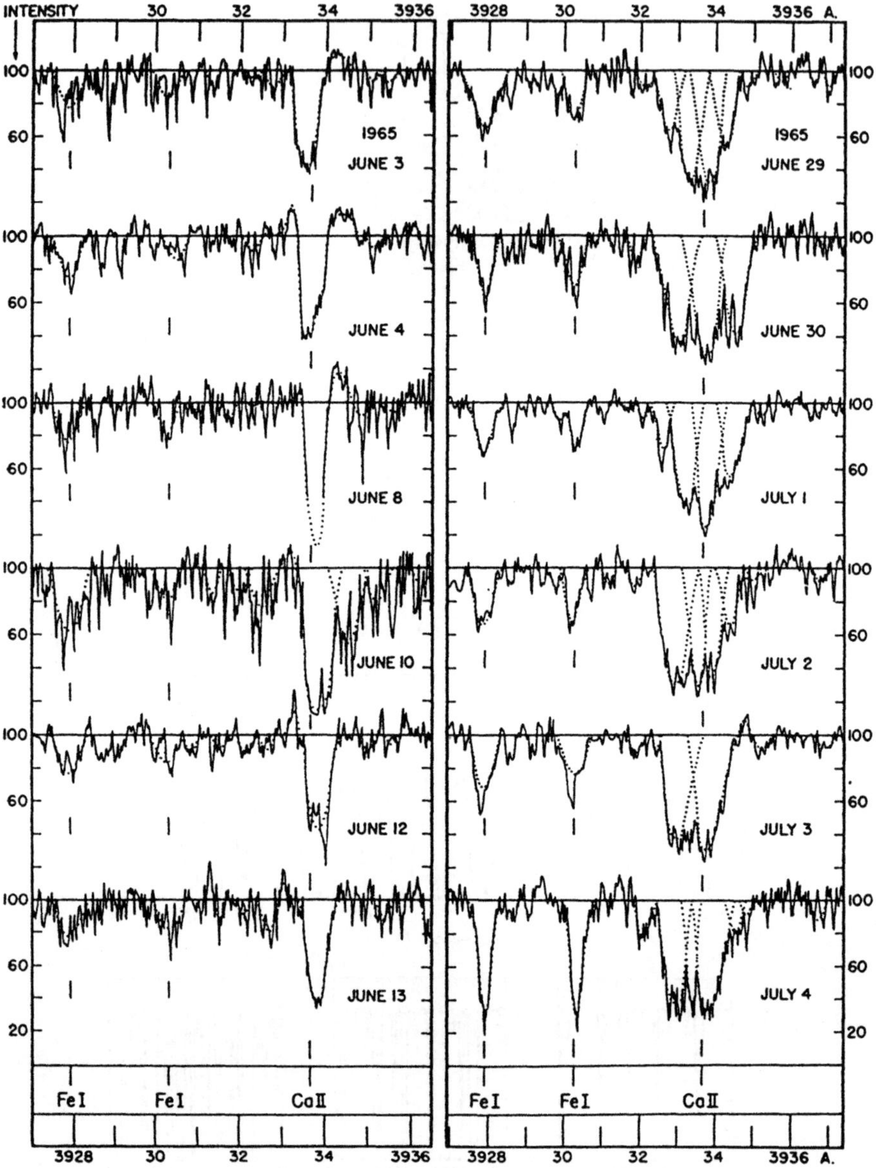

Figure 24: Variation in the Ca II K line strength through eclipse for 32 Cyg (Wright 1970).

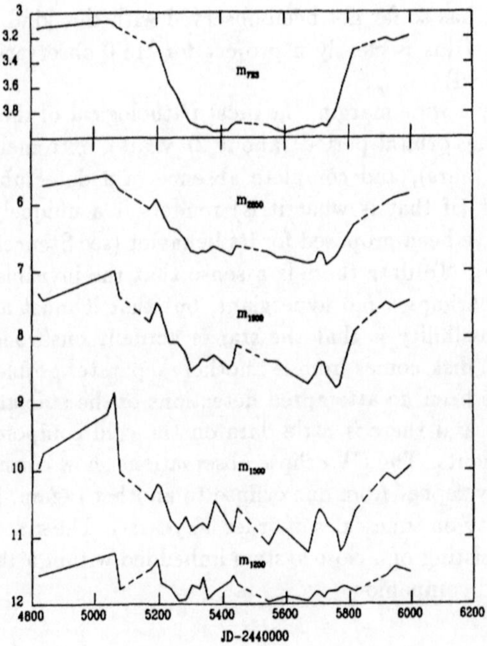

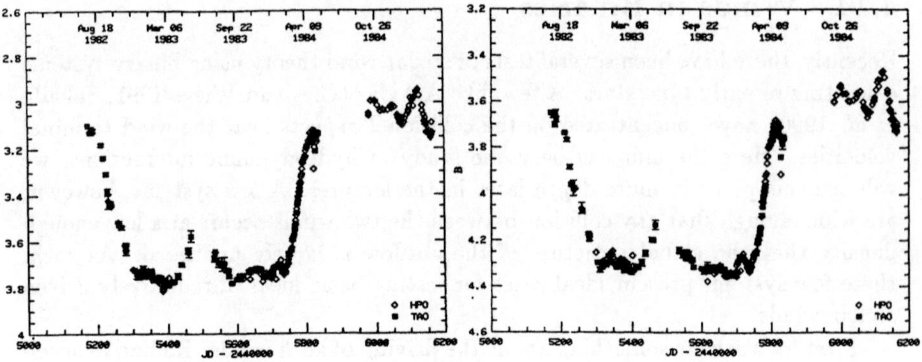

Figure 25: Compilation of optical and UV light curves for ϵ Aur: P. Schmidtke, optical; T. B. Ake, UV (see Stencel (1985).

Several systems have been especially well studied. For once, the prototype, ζ Aur, is truly representative of the phenomena. The other systems that have been well studied to date are 31 Cyg, 32 Cyg, τ Per, 22 Vul, AL Vel, and the extreme system ϵ Aur (fig. 24). Optical studies of such systems have been quite successful. A recent study of τ Per by Griffin *et al.* (1992) is a model of such an investigation. Here the absorption from the singly ionized iron peak species and the Ca II lines are used to study the details of the atmosphere. The eclipses, which are of comparatively short duration, show Ca II K line strength and profile variations on timescales of order hours. This system has also been observed in the ultraviolet but has so far not been observed with this kind of time resolution at high dispersion (this is clearly a project for HST observations, even though they are time critical).

ϵ Aurigae is, by a wide margin, the most pathological of the ζ Aur-type stars. It's enormously long orbital period (about 27 years), extremely long eclipse duration (about $1\frac{1}{2}$ years), and complete absence of a detectable spectrum from the red supergiant (if that is what it is) renders it a uniquely strange system. Many solutions have been proposed for its behavior (see Stencel (1985) for a brief sampling) (fig. 25). To date there is a sense that the invisible star must be an extended object, perhaps a red hypergiant, but that it must also have a bizarre geometry. One possibility is that the star is actually enshrouded in a disk (although where that disk comes from is another, separate, problem) (Carroll *et al.* 1991). There have been no attempted detections of the star at mm wavelengths (so far as I know) and there is little data on the cold component in the system from IR measurements. The UV eclipse observations show considerable structure, some of which may repeat from one eclipse to another (again, keep in mind that this implies stability on timescales of order 30 years!). The system is now thought to be a triple, consisting of a close system imbedded within a thick accretion disk and a very evolved companion.

4.14 Winds in Binaries

Recently, there have been several tests of stellar wind theory using binary systems consisting of early-type stars. A few, like AO Cas (Gies and Wiggs 1991, Sahade *et al.* 1990) have concentrated on the collisional aspects near the wind terminal velocities. Here the aim has been the study of hydrodynamic interactions, we will see this point in more depth later in the lectures. A few systems, however, are wide enough that any collision between the two winds occurs at a low enough density that the global structure of the outflow is largely unaltered. As such, these few systems present ideal cases for testing basic ideas of radiatively driven stellar winds.

First let's review something about the driving of such winds. Radiative acceleration takes place when photons are scattered or absorbed through lines. The oscillator strengths of such resonant transitions greatly exceed the continuum opacity so that the coupling between the lines and radiation is greater. The prehistory of the idea goes back to the first papers on radiative equilibrium in stellar atmospheres, a study by Milne in 1926, where it was realized that the rate of momentum transfer to an ion through a line depends on the oscillator strength for the transition. Consequently, the radiative acceleration is given by the sum over all transitions with the cross section weighted by the atomic excitation. Since the acceleration is continuous rather than ballistic, the ion continues to gain momentum from the radiation field as it moves outward until the opacity in the line is sufficiently reduced by abundance of the absorber so that it can no longer effectively absorb or until the velocity gradient saturates the line and the momentum transfer efficiency drops.

The former effect is the easiest to understand, and was exploited quickly in several contexts (none, unfortunately, related to binary stars, but see Vauclair (1983))[9]. The latter was first pointed out by Sobolev (1960). The optical depth in a line can be reduced along a path if the medium is moving. This is simply because the centroid of the line shifts in frequency as one moves through succeeding layers. If you think like a photon for a moment, this means that the probability of seeing an atomic absorber changes more rapidly in the presence of a velocity gradient than would be the case if the atmosphere were static. A photon emitted in the line center sees a reduced opacity in the layer above it and consequently has a greater probability of escaping. A photon emitted in the blue wing sees an increased opacity in the layer above it. The result is that the line is quite broad and presents a larger cross section as a function of height in the atmosphere than it would were it static. Taking the velocity gradient to be dv/ds, the change in the optical depth along a path ds is:

$$d\tau_\nu = \kappa_\nu \rho ds = \kappa_\nu \rho \frac{c d\nu}{\nu_0 dv/ds}. \tag{51}$$

Thus, as dv/ds increases the optical depth at frequency ν decreases. For each ion, and for each transition, the velocity gradient has a similar effect. However, lines can shadow each other if they are distributed densely enough. The result is that the integrated opacity along a line of sight depends on the velocity gradient. This clearly affects the terminal velocity of the wind as well. The radiative acceleration is modeled using a simple scheme, first introduced by Castor, Abbott, and Klein (1975) (see Kudritzki and Hummer 1990). They assumed that:

$$g_{rad} \sim \left(\frac{dv}{dr}\right)^\alpha \tag{52}$$

where α is called the *force multiplier*. If the wind is optically thin, physical picture is much simpler. Then the velocity gradient is irrelevant and each particle has an equal chance of seeing a radially-streaming photon and scattering it, thereby receiving a kick. The luminosity of the central source is all that matters here, so that $g = g_{rad}$ for an optically thin gas yields:

$$L_{Edd} = \frac{4\pi G M c^2}{\kappa} \tag{53}$$

for the *Eddington luminosity*. If $L \geq L_{Edd}$, radiation pressure dominates over gravitation in the dynamics of a single particle. The opacity, κ is approximately the Thomson scattering cross section, σ, for a fully ionized hot wind.

Aside from the role L_{Edd} plays in initiating mass outflow, it also serves to limit the rate of steady state mass accretion. In effect, the role of radiation is to reduce the mass of the gainer or loser in a binary. Since the Roche surface

[9]Actually the recent study of τ Per by Griffin *et al.* (1992) is of interest in this context. They show that the A star in this system is likely a λ Boo star. This class of weak metal line stars seems to require some form of elemental diffusion in order to explain the observed abundance pattern.

no longer matters when the mass ratio $q \leq 0.07$, there is a critical value of $\Gamma = L/L_{Edd}$. With radiation included, the mass of each component is replaced by a modified value, $M_j \rightarrow M_{0,j}(1 - \Gamma_j)$. Claims that the Roche surface does not exist if radiation is important exaggerate the physical situation – it's still there, but it is smaller. The problem is that the luminosity of an accreting body may depend on the mass accretion rate, so the flow is actually quite unstable to feedback whenever Γ approaches unity. In addition, if the luminosity is already high enough to affect the rate of accretion, it's entirely possible that the star will have a radiation-driven wind of its own, resulting in a direct hydrodynamic interaction between outflow and infall.

We know the masses of stars in a binary, and if the system eclipses we also have a handle on their radii. This means that we can predict what the escape velocity should be for the particular star, independent of any calibrations of the mass-radius-spectral type relation. Then from the direct observation of the P Cygni profiles that massive stars have, we can determine the ratio v_∞/v_{esc}. The terminal velocity is related to the reduced escape velocity by:

$$v_\infty = (\frac{\alpha}{1-\alpha})^{1/2} v_{esc}. \tag{54}$$

Here, if radiation is important, we must take $v_{esc} = GM(1-\Gamma)/R$. It's thus possible, at least in principle, to determine α directly from the observations of resonance line profiles around the binary cycle. Such an observational test has been attempted for a few systems by Corcoran and Heap (1990). The primary limitation is uncertainty in the model line profiles, partly because the systems are quite close and also because they are O stars.

I say *in principle* for all of the reasons that were stated at the outset. We are, after all, dealing with a binary system, and there is no guarantee that the value determined for the outflow from such systems will reflect the properties of single stars for which the theory has been developed. But we are stuck with having only binaries to work with for this test. If you look at an O star in a close binary, you can determine its mass independent of its spectrum. Now you can look in the UV at the resonance lines of the stars and determine, from the terminal velocity, what the force multiplier and mass are for the stars, and also what the mass loss rates are. The possibility is even there for looking at the effects of metallicity, since it's now possible to study the orbits of OB stars in the Magellanic Clouds (Bell *et al.* 1991, Niemala *et al.* 1992).

Another possibility is that we can employ the occultations provided by the companion to dissect the wind structure of a relatively normal star. This is especially easy to do if the companion is a much cooler star, a sort of inverse of the ζ Aur systems. It's well known that the winds of massive stars are unstable and prone to the formation of so-called *discrete absorption components* or DACs. These are presumed to represent shocks in the wind. Models differ as to their cause. They may be generated by photospheric pulsations generating advected waves that are amplified in the peripheral parts of the radiatively driven flow. Alternatively, they may be produced by an intrinsic instability that results from the flow being unconditionally unstable. The problem is that binary systems containing main sequence O stars do not also contain late-type stars of sufficient radius to act as knife-edges to the wind. Instead, they usually have other O stars as companions, when the star is not so evolved as to be a degenerate object. Thus in order to carry out this test, two conditions must be met. First, the O star in question must be especially active so that it is continually producing DACs. Second, the system must be observed around its orbital cycle and the occurrence of the DACs must be recorded for both stars. The velocity curve must be well enough known that any features in the line profile can be ascribed to one or the other star as a function of orbital phase. Obviously this is a very hard task to accomplish, and tests of this type have been attempted in only a few systems, specifically EM Car and AO Cas. In neither case is there unambiguous evidence that the sought-for effect has been detected.

The massive X-ray binaries present another interesting arena in which to study stellar winds. Here the XR source, a neutron star, illuminate the wind of the companion. Since this is ionization within a density gradient, an anisotropic ionization zone is formed around the XR source, in effect a little Strömgren sphere imbedded within the wind of the massive star. The first calculations of this effect were performed for HMXRB by Hatchett and McCray (1977), but actually the effect is precisely the same as the one responsible for the high ionization species in the symbiotics. Since the wind is being ionized, the terminal velocity measured for resonance lines of different ions should vary with phase. This is because the absorber is being destroyed by the impinging XR photons from the compact object. Recently, this mechanism has been used to explain some of the emission line variability in the symbiotics. In HMXRBs, it has been studied in several systems, for instance 4U1700-37 (HD 153919), 4U0900-40 (Vel X-1 = HD 77581), and SMC X-1 (Sk 160) (*c.f.* Cordova and Howarth 1989). Although the observations are difficult and still ambiguous, the mechanism must (at some level) be acting to produce profile variations. For some nearby late-type supergiant systems, like α Sco, it has even been directly imaged using the VLA and continuum measurements (Hjellming and Newell 1983). So as high S/N and high spectral resolution observations become more routine, it should be possible to get a better handle on this effect.

4.15 Colliding Winds in Binaries

If the binary system consists of two early type stars, as is seen for instance in Y Cyg or AO Cas, the components may both possess strong stellar winds. These outflows are likely to know about each other, and in fact may collide within the confines of the system. Such is the case of the WR+O binaries like V444 Cyg. It has also been suggested that HD 152270 shows this effect (Neutsch, Schmidt, and Seggewiss 1981). Alternatively, if the wind of one star overwhelms the other, direct interaction between the companion and the wind may take place with the wind actually impacting the surface. CQ Cep is probably an example of this. The wind is merely ionized by the companion, a situation often found within the symbiotic class. The wind of a very distended red giant may be dynamically shaped by a fast wind from the hot component. This is likely seen in a number of symbiotic novae, especially AG Peg. In order to treat these, we need first to discuss the general theory of the problem, and then to show some specific examples.

The best example of such systems is provided by close WR stars. The first analysis of such systems by Prilutskii and Usov (1976) suggested that wind collisions are capable of producing substantial X-ray emission. The theory is actually rather simple for their model and assumes blunt body flow. If a sphere is immersed in a supersonic flow, a bow shock forms at a distance $\Delta r = \frac{2}{\gamma-1}R$, where R is the stellar radius. Here the flow is brought to rest and dissipates energy collisionally, resulting in X-rays. Their original prediction assumed that emission would take place at the virial temperature $T_V = GM/kR \approx 10^7 K\ (M/M_\odot)(R/R_\odot)^{-1}$ and that it should be easily seen with 1-100 keV detectors. Subsequently, such emission was detected by Moffat *et al.* 1(982) from two WR systems, WR136 and V444 Cyg, the temperatures being of order a few million degrees. The observations initially indicated a possible phase-dependence to the emission measure, suggesting that internal conversion and absorption within the WR wind could produce modulation. More recent observations with ROSAT confirm this result (Corcoran *et al.* 1992).

Several WR stars are known to be binaries composed of a Wolf-Rayet star and an O star. What makes these systems especially interesting is that both of the components may have strong mass outflows and these may actually collide within the confines of the binary. The collision site may be deeply imbedded in the wind, and it may then be possible to see evidence for mechanical heating through the overionization of the material. There are several early suggestions that this might occur in close massive binaries. The first observational test was Heap's (1981) look at HD 47129 (Plaskett's star); new observations of this system were obtained recently by Sahade and Brandi (1991). A more detailed theoretical and observational treatment of wind-wind collisions was provided by Shore and Brown (1988) for V444 Cyg, and Shore and Corcoran (1992) have extended this to include CQ Cep, the only other O+WR star observable in the UV. Sahade and Brandi (1991) and Gies and Wiggs (1991) have added AO Cas, a system in which both stars are O giants. Two other systems, CQ Cep (WN7+O, period = 1.64 days) (Stickland *et al.* 1984) and HD 152270 (WC6+O5, period = 8.89 days) (Neutsch *et al.* 1981) do not appear to be interacting wind systems, perhaps because the O star is still on the main sequence. In both cases, the wind is deflected by the O star, and there may also be an ionization cavity, but none of the dynamical signatures seen in V444 Cyg are detected in these binaries. Numerical simulations have been produced for several WR systems by Luo, McRay, and Mac Low (1990) and Stevens *et al.* (1992). For symbiotics, there is reason to believe that the superwind of the compact object may interact with the red giant's outflow to produce helical structures in the surrounding nebula (Huang and Weigert 1982, Girard and Willson 1987).

First, let's look at a simple analysis of the interaction region. The model involves the equilibrium surface for two supersonic winds, with mass loss rates $\dot{M}_1$

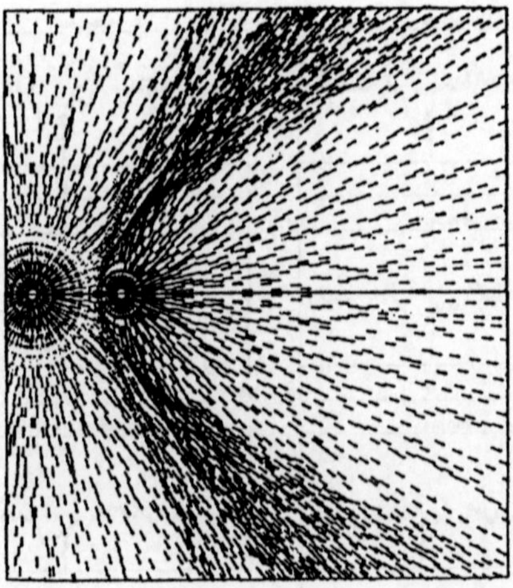

Figure 26: Colliding adiabatic winds for a model WR+O system with $\lambda = 0.1$ using smoothed particle hydrodynamics (Libersky, Petschek and Shore 1989, *unpublished*). This model is in the early stages of the collision so the structure is not yet in equilibrium.

and $\dot{M}_2$ and terminal velocities $v_{\infty,1}$ and $v_{\infty,2}$. The stagnation point is defined by momentum balance so is governed by the ratio $\lambda = d_1/d_2 = (\dot{M}_1 v_{\infty,1}/\dot{M}_2 v_{\infty,2})^{1/2}$, which is also the ratio of the distances of the shock from each of the component stars. Then taking momentum conservation across the front gives the basic shape of the stagnation surface:

$$\frac{d\eta}{d\xi} = \frac{1}{2}\frac{\dot{M}_1 v_1 r_2^2 \sin 2\alpha_1 + \dot{M}_2 v_2 r_1^2 \sin 2\alpha_2}{\dot{M}_1 v_1 r_2^2 \sin^2 \alpha_1 - \dot{M}_2 v_2 r_1^2 \sin^2 \alpha_2} \tag{55}$$

where r_j is the distance along the front from the jth star, $\sin \alpha_1 = \xi/r_1$, $\sin \alpha_2 = (a - \xi)/r_2$, and a is the separation of the system (fig. 26). In realistic models for mass loss, the flow is accelerated outward through some velocity law of the form $v_j(r_j) = v_{\infty,j}(1 - (R/r)_j)^\beta$ (taking R to be the radius of the star). Additional analytic analysis of the problem has recently been presented by Usov (1992), who calculated the width of the shock front.

Although momentum conservation specifies the shape of the stagnation surface, it is not enough to detail the shape of the shock front itself. The reason is that the flow is decelerated by a standing shock at some distance from the $\eta(\xi)$ surface. In this front, the wind is first heated and then accelerated. Depending on the flow timescale and the optical depth of the wind, this region may radiate X-rays and also strong emission lines. An approximate scaling law for the emission from the collision region takes bremsstrahlung as the emission mechanism and assumes that the region is optically thin, $\epsilon_{ff} \sim n_e^2 R^3 T^{1/2}$. The temperature scales as $T \sim v^2$, the emission measure, EM, scales as $EM \times \Lambda(T) \sim n_e^2 V \Lambda(T) \sim \dot{M}^2 v^{n-1} r^{-1}$ so that $L_{XR} \sim \dot{M}^{1/2} v^{-3/2} f(\lambda)$. Here the last factor, $f(\lambda)$, represents

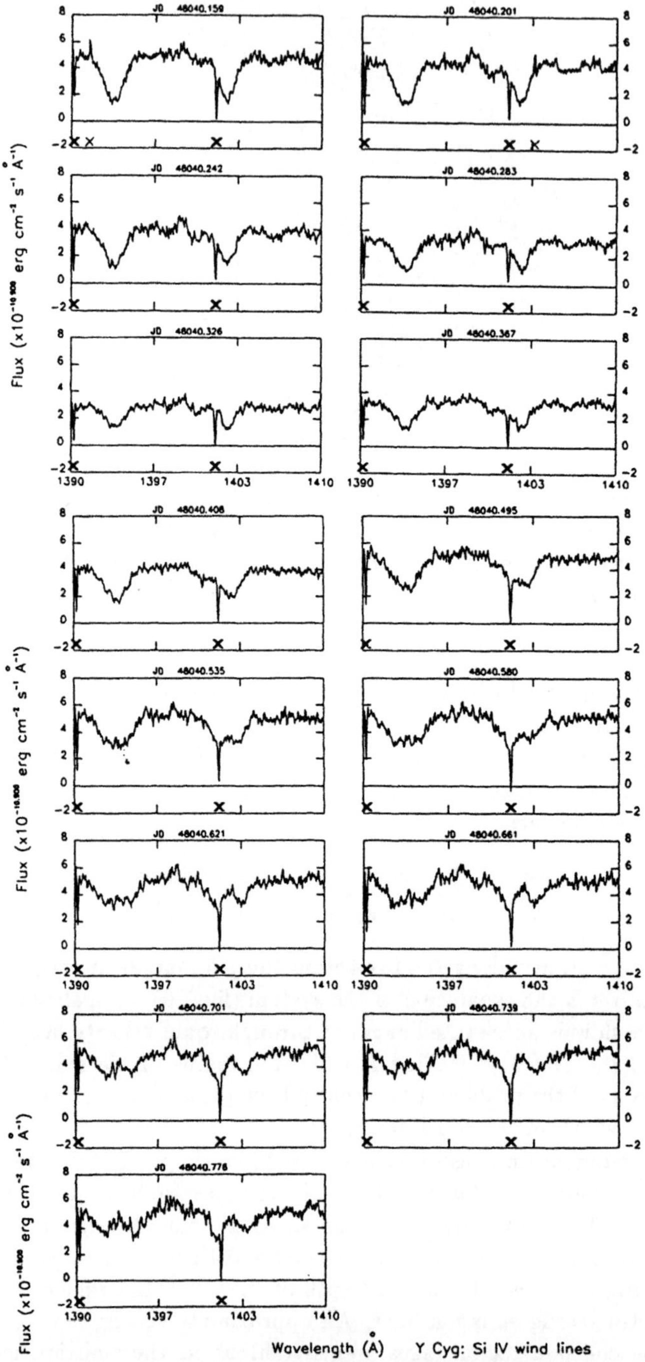

Figure 27: Variations of the Si IV ultraviolet resonance lines during a part of the orbit of Y Cyg (from primary eclipse through first quadrature). High resolution IUE spectra.

the functional dependence on the ratio of the momenta of the two winds. Depending on the precise dependence of the volume emissivity on the temperature, different dependences can be derived for the velocity scaling.

Recent two-dimensional hydrodynamic computations by Stevens *et al.* (1992) show that the interaction regions are prone to shear and thermal instabilities and that these result in convected blobs along the shock front even for steady state winds. It is interesting that many O star binaries should show this effect, but to date few have been studied. One such case in point is the old workhorse Y Cyg (fig. 27). The system consists of two presumably identical O9 or B0 stars, yet the UV variations show something else is happening here. Spectra taken at opposite phases around secondary minimum so not look at all alike. The Si IV profile, an unsaturated wind line, shows strong variations. The radial velocity of the system is large enough that the two components clearly separate at velocity extrema, yet they show different profiles. The C IV line is useless for this purpose because the profile is saturated.

Now what would we expect from the X-rays in such a system? Since the interaction produces high temperatures, of order 10^8K, we would anticipate that copious X-ray production should result. But the interaction region will be deeply imbedded within the wind and likely will be rather small. The emissivity of optically thin bremsstrahlung varies as $\epsilon \sim \dot{M}^2 L^{-1}$, where L is the characteristic size of the collision zone. This region will be seen modulated by the winds of the two stars, so two things should vary, the intensity of the X-rays and the spectral index (also called the hardness ratio). When the region is eclipsed, if only electron scattering is important the intensity of the X-rays decreases without a change in the spectral index. This is not quite like an ordinary ζ Aur eclipse, since in the giants, line opacity dominates the change in intensity and the effective spectral index changes. Here, the WR wind is usually assumed to be completely ionized and therefore, unless there are continuous absorbers among the high ionization species of the wind, the Thomson cross section is all that matters. The two winds do not interpenetrate, the shock sees to that, so the density along a line of sight changes as the system revolves. Also, and this may be important for WR+O binaries, the *chemical composition* also depends on the line of sight – the abundances in the O star wind are different from those in the WR. The change in the abundance of the CNO absorbers, and perhaps of heavier species, should make a difference in the 1-100 keV regime, but detailed models have not been computed to study this effect (although Stevens *et al.* (1992) have made a stab at this).

The optical light curve of V444 Cyg can be understood in terms of electron scattering. Cherepatschuck (1967, see also Eaton *et al.* (1985)) has developed a method for calculating the mass loss rate for the WR on the basis of the modulation of the O star by a simple inversion method, much as we've just discussed for the X-rays. The subsequent work by Koenigsberger and Auer (1985) has identified Fe IV-VI as important absorbing species in the optical, which complicates the analysis of the density and temperature stratification of the wind. But the basic idea is the same. The intensity is a map of the density along a line of sight, and therefore the density can be obtained by direct inversion via an Abel-type

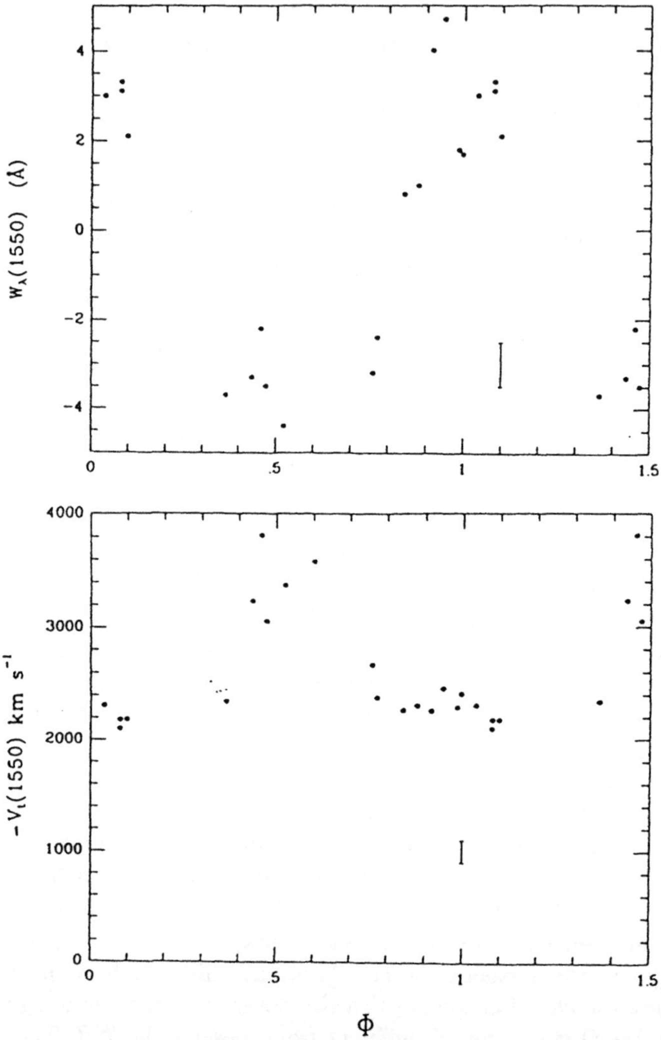

Figure 28: Variation in the terminal velocity and equivalent width for the C IV lines in V444 Cyg (Shore and Brown 1988).

equation (since the column density is all that comes into the optical depth, along with the diffuse scattering of photons within the wind).

One spectroscopic signature of the wind interaction, discussed by Shore and Brown (1988), Gies and Wiggs (1990), and Shore and Corcoran (1992), is connected with the fact that material is accelerated outward along the interaction front by the wind collision. This produces a cone of accelerated material that is anisotropic. Therefore, the resonance lines form extended blue absorption wings at the terminal velocity extending to very high velocities. These wings only appear when the system is viewed at the right phase and outline the opening angle of the conical shock that lies in front of the companion to the WR star (fig. 28).

If the two winds are both dynamical the acceleration is more pronounced than if one of the stars presents a blunt body in the flow. There is also the possibility that small scale instabilities will be observed in the front, perhaps appearing as the blobs reported in polarization measurements. Specifically, Stevens *et al.* found that the front is unstable to thin shell effects, and that these advect out on short timescale and actually grow in the transported gas (real instabilties!).

4.16 Accretion Wakes and Bow Shocks in Wind Accreters

Ultraviolet observations of the ζ Aur stars often show Mg II in emission and sometimes indicate P Cyg profiles (fig. 29). These have been interpreted as evidence for bow shocks formed by wind accretion (Chapman 1981, Ahmad 1986). The main difference between these wide binaries and the WR systems is that the gainer is a blunt body within the wind of the companion. The heated region should produce the same sort of shock cone that is observed in V444 Cyg while at the same time mimicking the line profiles of CQ Cep. An important difference is that the accreter is negligibly small compared with the radius of curvature of the wind at the accretion site and also compared with the binary separation. The wind accretion rate is simply the capture cross section for streaming near a compact star $\sigma \sim (GM/v_W^2)^2$ where v_W is the aberrated wind speed. Then the rate of mass accretion from the wind scales as v_W^{-3}. The angle of the cone is given by $\beta \sim a_s/v_W$, which is the inverse Mach number. The cone is aberrated by the motion of the companion, and theoretically there should be some intrinsic instability of the accretion wake due to the flip-flop instability (Livio, *this course*). But the main thing is that the wind accretion should show some standing shock, and may even show a disk. An interesting contrast in Mg II profiles is shown in Ake and Parsons (1988) comparing the ζ Aur systems HR 2554 and 22 Vul.

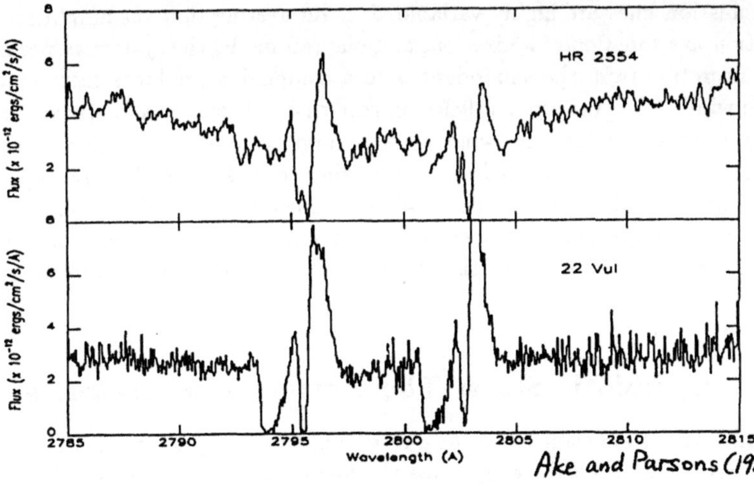

Figure 29: Mg II profiles in HR 2554 (Ahmad 1986).

The evidence for disks in wind accretion comes mainly from recurrent novae and symbiotics, as we've already discussed. However, one important observation is that T CrB and Mira, very different systems in their behavior but somewhat related in the UV, both show spectroscopic disk-like spectra. By this I mean that there are high ionization lines, C IV and N V, and that these are variable on short timescales. It is also possible that there is some dynamical broadening of the emission lines in T CrB. Finally, the white dwarfs in these systems show power-law-like continua, and this is characteristic of disk, rather than wind, accretion. Recently, Cannizzo and Kenyon (1992) have modeled the flickering and outburst behavior for T CrB using a disk, but the details of the formation of the structure are left as an exercise to the reader. Nonetheless, the evidence is mounting that disks do form in some wind-accreting binaries.

Mira, *o* Ceti, may be an example of how the presence of a companion can affect the envelope structure and mass loss of a giant. Speckle interferometric measurements by Karovska *et al.* (1991) show that the giant in the Mira system has a deformed outer envelope. Recently, Quirrenbach *et al.* (1992) have confirmed this with the Mk III optical interferometer. Karovska *et al.* argue that the deformation tracks the orbit of the companion, which has a period of about 400 years. While not at all certain, the envelope distortion may be the product of tidal interaction between the components. Perhaps this distortion can also produce variable mass loss in the system? We don't have enough data on the variability of this system in the UV and EUV to be certain. However, it is well known that the gainer in the Mira system is quite variable in the UV and perhaps this is connected with the changes in the mass loss rate from the long period variable.

One unusual system, S 18/SMC, may be an example of a wind accreter where the loser is a massive hot star (Shore *et al.* 1987). This system has an O supergiant as its primary. It is very luminous (about $10^5 L_\odot$), and it shows high excitation and high ionization lines in the UV (no photospheric lines are detected). The emission lines are highly variable. It is interesting that the amplitude of the variation is a function of ionization, and one can model the system assuming that wind accretion from the supergiant onto a companion produces extra ionization in the wind. Zickgraf *et al.* (1989) suggest that this star is a B[e] star, one that shows evidence for bimodal winds and/or large scale disks. There may be other systems like this, and it would be interesting to look for early-type supergiants that display wind accretion effects where the companion isn't a compact object (there are examples of wind accretion, modulated by orbital phase, in some high mass X-ray binaries but none that I know of where the gainer is a main sequence star).

4.17 Symbiotic Stars: The Universal Laboratories

It is only about a decade since the question of the binary nature of the symbiotic stars was finally settled, really since the launch of *IUE*. The original definition, reflected in their name, is that these stars present contradictory pictures to the

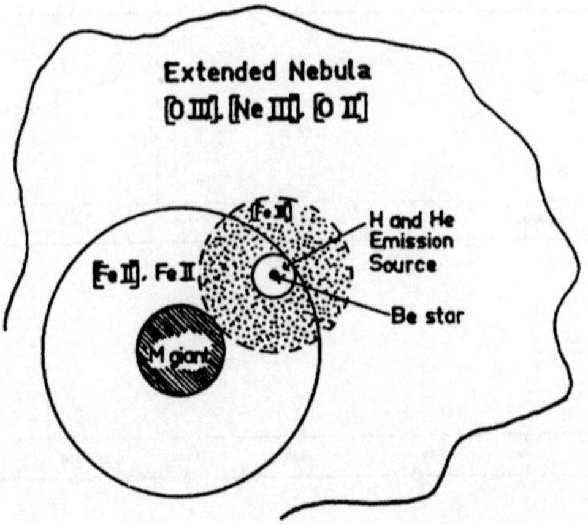

Figure 30: Schematic of a typical symbiotic star from Payne-Gaposchkin (1957).

observer: they show the continua of M giants with superposed emission lines from highly ionized species. It is interesting to note that the earliest discussions of the class anticipated the currently held picture. In Payne-Gaposchkin's summary (1957), we find the diagram of these systems consisting of a wide binary with a shared extended atmosphere (fig. 30). A hot ionizing source acts on the extended envelope of a red giant or supergiant to produce a cavity of ionization. The current observational picture has expanded considerably since Payne-Gaposchkin's summary, and a masterful historical and empirical overview is provided by Kenyon (1986). An excellent catalog, along with a collection of optical spectra and finder charts, is found in Allen (1984). The most recent compilation of observations and conjectures is Mikolajewska *et al.* (1988). Most of these systems resemble stars like Mira or α Sco in having essentially no hydrodynamic interaction, at least in the sense that we would have in an Algol system. Yet they also resemble the recurrent nova class that we have already discussed, those with supergiant companions. And a few may be undergoing mass accretion from a stream lost by the giant companion, which fills its Roche surface. I should remark that the most fascinating part of the original work on these systems is that it all took place before the idea that red giants were mass-losing objects had been firmly established!

First, let's outline the basic picture. The symbiotics, which are unequivocally binary stars, are empirically divided into two broad taxa. The *S-type* (for "star") show infrared spectra consistent with the photosphere of the cool star (like EG And, V443 Her, Z And, and CI Cyg, all binaries) while the *D- type* show silicate features of other signatures of dust (like HM Sge, RR Tel, V1016 Cyg, none of which are known to be periodic radial velocity variables) (see Allen 1984, Kenyon 1986, Vogel 1990, Whitelock and Munari 1992); about 20% of the symbiotics are D-type. The optical spectrum usually looks like a normal M star, usually a giant, with strong emission lines superimposed on the photospheric spectrum (Blair *et al.* 1983). The binaries are all long orbital period systems, with periods

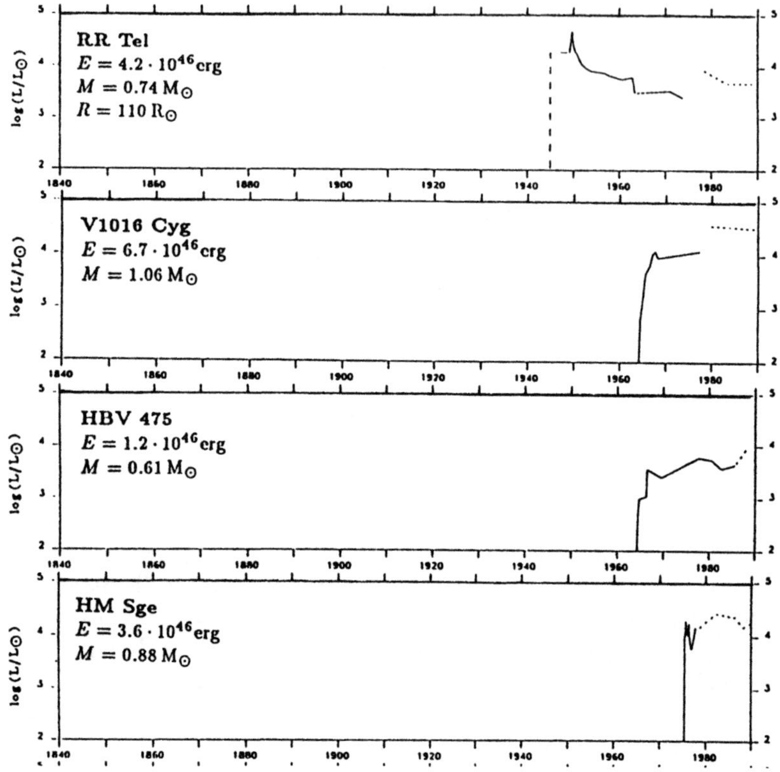

Figure 31: Light curves of several symbiotic novae from Mürset and Nussbaumer (1992). These are reconstructed bolometric light curves, using the optical and UV spectra to constrain the EUV portion of the flux.

ranging from about one year for R Aqr to several years for CH Cyg and CI Cyg. Some systems contain a Mira variable (notably R Aqr, RR Tel, V1016 Cyg, and BI Cru) (Kenyon 1992). Often the systems detected at centimeter wavelengths show some kind of radio structure that changes with time on scales of arcseconds, usually anisotropic or, to use the current buzz word, "jetlike". These binaries are known to show variations in both light and spectrum over timescales of decades, sometimes on timescales of centuries. Several show rapid brightening, like V1016 Cyg and RR Tel, that takes decades to mature (Nussbaumer 1992). Others show more rapid outbursts, like HBV 475 and AG Peg (fig. 31). None can be said to be steady. R Aquarii, for example, contains a Mira and is surrounded by extensive nebulosity. The optical H II region was observed in the last century while radio mapping of the environment has been accomplished in the past few years (Kafatos *et al.* 1986, Michalitsianos and Kafatos 1988). This system also has a very well-known jet that has recently been imaged by HST (Burgarella *et al.* 1992). AG Peg is also surrounded by a compact radio nebula, as is RX Pup. Most recently, Bang *et al.* (1992) have reported detection of a large (about 0.6

pc) ring around V1016 Cyg which accounts for much of the cold dust detected with IRAS. AG Dra has been detected as an X-ray source with both *Einstein* and ROSAT (Stencel *et al.* 1992). Its emission measure may have increased from $\log EM = 55.4$ to 55.8 while the temperature seems to have decreased from about $\log T_{XR} = 6.0$ to 5.8 over about a decade. These authors also point out that the emission from the X-ray source is severely altered by circumstellar absorption, so once again the environment *must* be properly included in order to obtain correct parameters for the stars.

Table 2. Some Symbiotic and Related Systems

System	Spectrum	Type	P (days)	e	f(M) ($M_\odot$)
T CrB	M4	RN	228	0.01	0.30
RS Oph	M1	RN	≈400	–	–
EG And	M2	S	481	0.04	0.02
AG Dra	K5	S	554	≈0	0.01
V443 Her	M5	S	600	≈0	<0.01
AX Per	M:	S	681	≈0	0.03
Z And	M3	S	757	0.2	0.02
AG Peg	M2	S	819	0.12	0.02
CI Cyg	M4	S	855	0.15	0.03
HBV 475	M:	S	964	0.3:	21:

The ultraviolet is where most of the interesting action happens. First of all, we see in all of these systems strong continua, rising to the shorter wavelengths. The UV thus provides the essential demonstration that the peculiar spectral morphology first noticed in the optical is really the result of the binary nature of the stars – there is indeed another radiation source in the vicinity besides the cool giant (fig. 32). In general, the continuum is consistent with a power law or a blackbody-type distribution, although more likely than not this is an expression of our ignorance of the true nature of the spectral distribution. Color temperatures are often very high, over 5×10^4K. The luminosities of these systems are not all that large in the UV, or for that matter in the optical, typically of order 1000 to less than $10^4 L_\odot$. The range again indicates our basic inability to determine the far-UV component of the spectrum where much of the luminosity is emerging. The UV is, however, dominated by a strong emission line spectrum of nearly all of the usual ions found in hot, low density plasmas: the resonance lines of C IV λ1550, O IV, S IV, and Si IV λ1400, and N V λ1240, as well as the excited state and intercombination lines of He II λ1640, N III] λ1750, N IV] λ1490, O III] λ1665, Si III] λ1895, C III] λ1910, O III λ3130. The UV and optical share several species, notably O III and He II. However, optical spectra of a few of these systems also show strong coronal species like Fe V-VII, Mg V, Ne IV-V, and other lines usually observed in planetary nebulae. The UV also shows lines of these species.

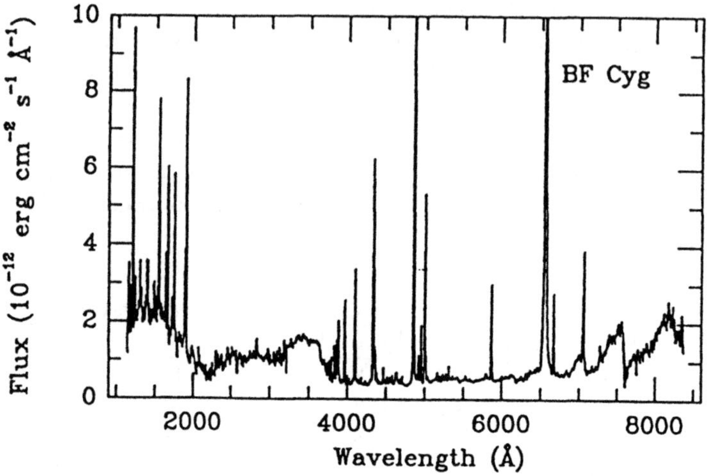

Figure 32: Composite continuum of BF Cyg. Notice the strong UV emission line spectrum (especially resonance emission transitions), the Balmer discontinuity in emission at 3647Å, the strong Balmer emission line spectrum, and the TiO absorption features in the red continuum (Kenyon 1992).

An interesting feature of symbiotic spectra is the presence of strong emission lines from very low excitation and low ionization states. The ubiquitous lines of permitted and forbidden transitions of the Fe-peak show up. In addition, ionized iron peak fluorescent lines often appear in the spectra both in the optical and ultraviolet (Carpenter and Johansson 1988). The detection of O I λ1641, a line pumped by the ground state transition, shows that considerable neutral gas is present even in those stars that show S-type spectra (see below, discussion of the RS Oph outburst). This line was first pointed out in CH Cyg (Hack 1982) and has since been detected in several symbiotics, including RR Tel (Shore and Aufdenberg 1991, Aufdenberg 1992). N I lines have also been detected in the UV in RR Tel and several other symbiotics, all pumped by EUV radiation (we will return to this in a moment).

Several symbiotics are eclipsing systems, notably EG And, PU Vul, and Z And. Much of the eclipse is caused by selective extinction of the companion by the line spectrum of the giant, much like a ζ Aur system. The hot component in the eclipsing symbiotic binaries probes the cool giant's atmosphere. A superb example of the analysis of one of these eclipses is provided by Vogel *et al.* (1992), who model EG And in eclipse. It has recently been realized that most of the short wavelength extinction (shortward of 1500Å) is due to Rayleigh scattering in the M star's atmosphere (Vogel 1990). This opacity is especially important near the wavelength of Lyα. It produces a net reddening of the hot components continuum, most strongly near λ1216Å. Because the density of the neutral species (hydrogen) depends only on the mass loss rate the optical depth is given by:

$$\tau_\lambda(s) = \sigma_\lambda \int_{-\infty}^{\infty} N_H dz = \frac{2\sigma_\lambda \dot{M}}{4\pi\mu m_p v_\infty} \int_0^\infty dr/[r\frac{v(r)}{v_\infty}(r^2 - s^2)^{1/2}] \tag{56}$$

where σ_λ is the Rayleigh scattering coefficient and N_H is the neutral hydrogen column density (Vogel *et al.* 1992). For EG And, the radius derived for the M2.4 II star is 75±10 $R_\odot$, the first time the radius of an M giant has been directly determined. The main point is that the phenomenology of the atmospheric eclipses of the symbiotic systems is essentially the same as that observed for other highly evolved systems. The main advantage is that they have far hotter UV bright components and so should permit the study of the atmospheric structure in EUV transitions (now that the Extreme Ultraviolet Explorer (EUVE) has been successfully launched). For any such stars seen out of the plane, the EUV variations should be even more dramatic than for the continuum around 1200Å because of H I continuum absorption.

The selective absorption helps to explain many of the peculiar properties of the continua of the hot component in these systems (figs. 33, 34). For instance, in Z And, at intervals of about two years, the strongest emission lines like He II λ1640Å disappear and the spectrum takes on the character of a chromosphere, showing only relatively low ionization species like Fe II. While other systems show this effect as well, like PU Vul, SY Mus, and EG And, Z And is one of the few for which a large number of high resolution spectra have been obtained over IUE's lifetime. Now on looking at the higher resolution data, we see more clearly what is happening. There is another opacity in the system, one that is always present but on occasion becomes the dominant one – a strong absorption forest of ionized

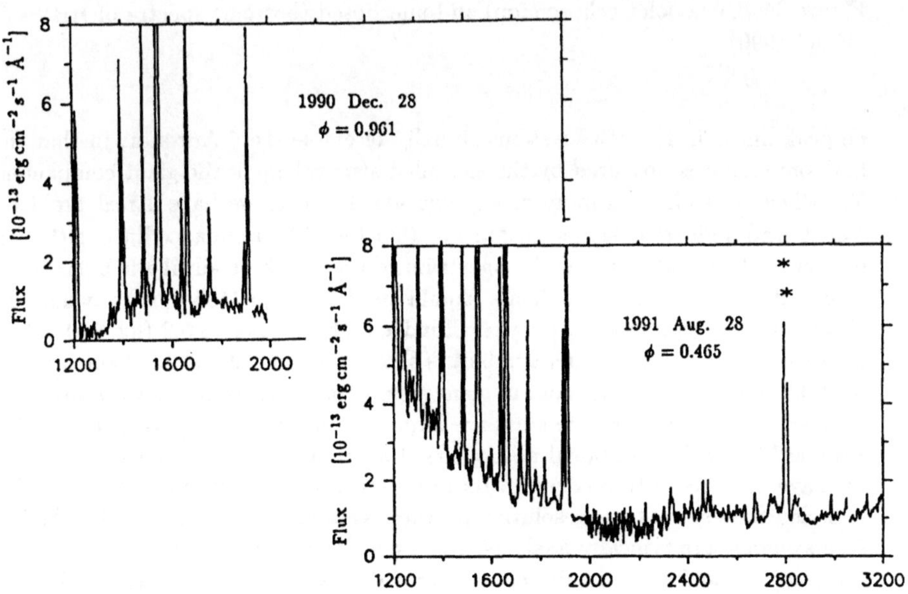

Figure 33: Ultraviolet eclipse spectra of EG And (Vogel *et al.* (1992).

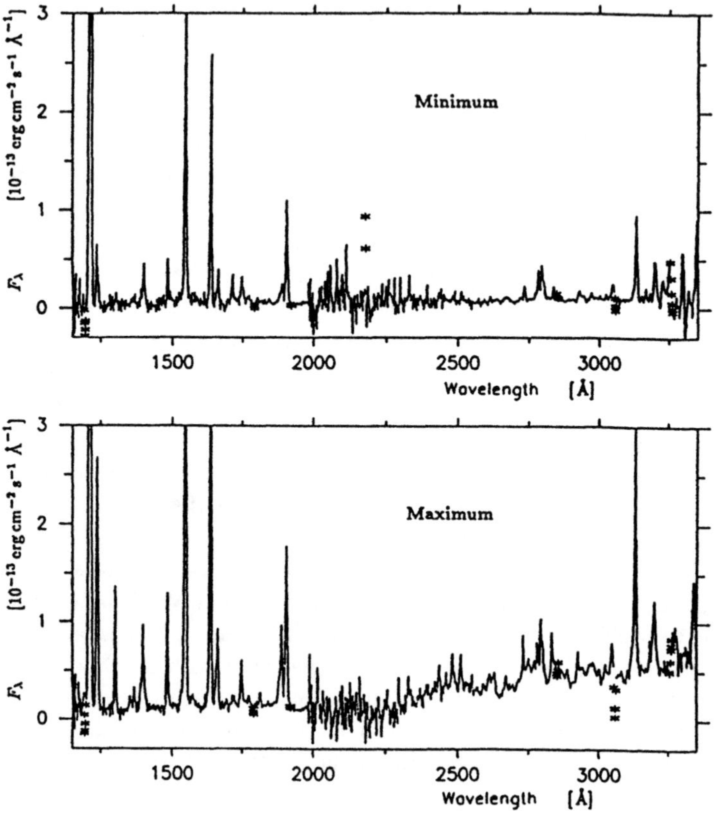

Figure 34: Ultraviolet eclipse (top) and uneclipsed (bottom) spectra of HBV 475 (Vogel 1990)

Fe-peak lines. In fact, the systems actually resemble the ζ Aur stars in that the hot component is obscured by the extended atmosphere of the giant companion. Yet when we look even more closely, we see that here we have a real problem. The Z And high resolution spectra show that the C IV resonance lines, although narrow and typical of a *nebular* spectrum, do not look at all like nebular lines. The ratio of the two components should be $F(1548)/F(1550) \approx 2$, while the observed radio can become quite small indeed, often closer to 0.2 (fig. 35). More detailed examination of other symbiotics shows the same effect. Michalitsianos *et al.* (1988) have pointed this out in many star, notably RX Pup. Now the situation is clearer – the fact that the stars are immersed in a mutually obscuring wind is responsible for the additional absorption. I mention this effect especially because we have only recently recognized its importance and omnipresent nature. The emission lines seen at low resolution for these systems, and I suspect, also the VV Cep systems, are typically analyzed using nebular codes. These assume that the hot source is ionizing a small nebula within the red giant wind, roughly elliptical in shape although the precise structure depends on the density and velocity law

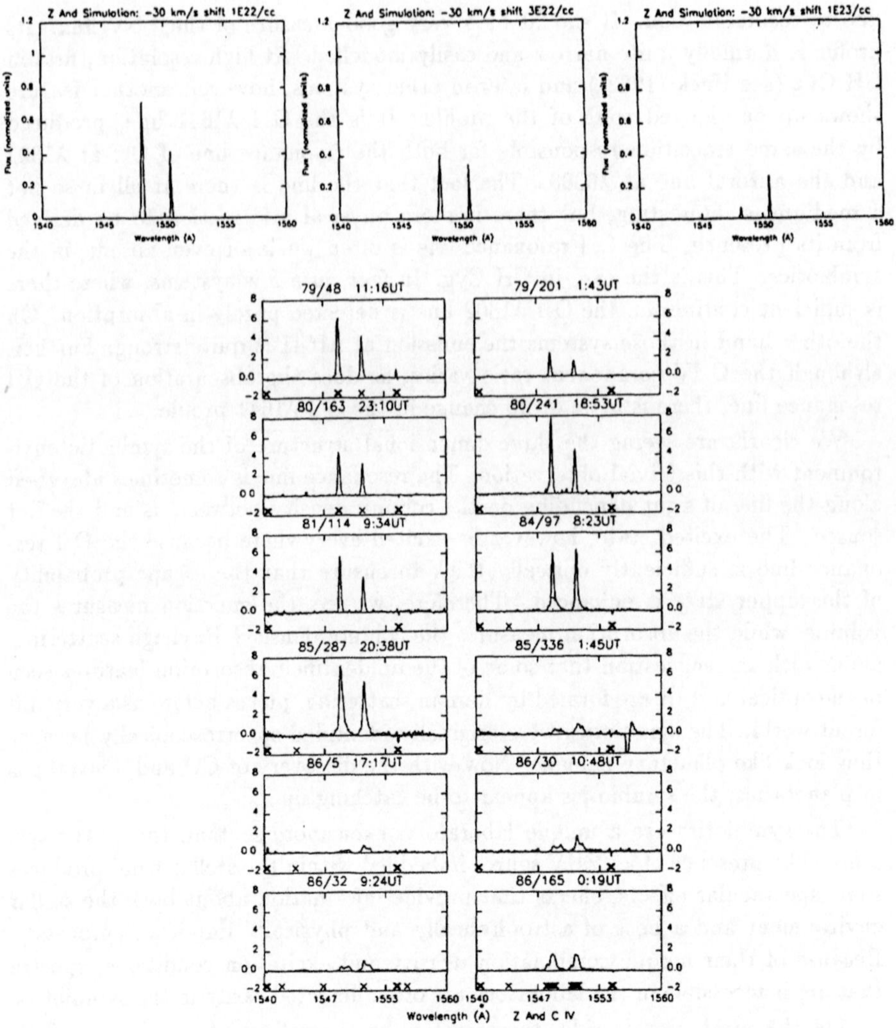

Figure 35: Example of two extreme model C IV profiles simulating an eclipse for Z And compared with the eclipse and out-of-eclipse spectra taken with IUE. The model assumes that the C IV lines are 30 km s^{-1} wide and in the nebular ratio, absorbed by a shell with variable density shifted at the terminal velocity (60 km s^{-1}) of the red giant wind.

for the red giant. However, the line ratios obtained from low resolution spectra cannot be correct. The Fe curtain obscures the emission line fluxes and alters the nebular ratios. In addition, the central source of the FUV is itself covered by the intervening matter so that the incident flux on the nebula is $F_\lambda \exp(-\tau_\lambda)$.

Another indication that the surrounding material must be included in analyses of symbiotics is shown by the case of the He II λ1640 line. Normally this line is due to recombination. It should be a very good measure of the FUV flux. Its profile is normally quite narrow and easily modelled. At high resolution, first in CH Cyg (see Hack (1982)) and later in other systems, however, another feature shows up on the red wing of the profile. It is the O I λ1641 line, produced by the same transition responsible for both the resonance line of O I at λ1302 and the auroral line at λ6300. The fact that the line is there at all in so hot a medium is interesting, but there is more physical information to be derived from its presence. The O I resonance line is often weak, or even absent, in the symbiotics. This is the case in CH Cyg. In fact, in a few systems, where there is sufficient continuum, the O I λ1302 line is detected purely in absorption. On the other hand in those systems the emission at λ1641 is quite strong. Further, although the C IV component ratio varies, as does the obscuration of the O I resonance line, there is little or no change in the O I λ1641 profile.

We clearly are seeing the three-dimensional structure of the symbiotic environment with this trivial observation. The resonance line is sometimes absorbed along the line of sight depending on the column density between us and the hot source. The excited state, however, is excited everywhere because the O I resonance line is sufficiently optically thick to ensure that the escape probability of this upper state is enhanced. Therefore, we see, the emission measures the volume, while the absorption measures the column density! Rayleigh scattering, along with the suggestion that some of the unidentified absorption features seen in the optical and IR are formed by Raman scattering, places before us a very different world. The symbiotics were originally identified spectroscopically because they look like planetary nebulae. Now with the discovery of CO and neutral gas in planetaries, the symbiotics appear to be catching up.

The symbiotics are a unique laboratory, even more so than the ζ Aur systems. The presence of an EUV source imbedded within the stellar wind produces some spectacular effects, effects that provide information about both the stellar environment and a host of astrophysically and physically important processes. Because of their unique combination density and excitation conditions, spectra that are inaccessible in the laboratory are often detected easily in the symbiotics.

On the most prosaic side, these systems are excellent atomic lamps. This feature of the symbiotics has been beautifully exploited by the Zurich group in their efforts to study atomic spectra. The Fe II spectrum of RR Tel has, for example, been used to supplement laboratory studies in the UV (see Fawcett 1988). Forbidden and intercombination lines of ions that are terribly hard to produce in the lab are easily accessible in the rarified environments of the outer red giant wind in symbiotics. In fact, except for the solar corona, no astrophysical objects other than planetary nebulae provide so wide a range of ionization and excitation with which to study atomic structure (fig. 36). Nearly all atomic species from neutral to eight times ionized of most elements with $Z < 40$ are observable, and all in optically thin emission lines.

The observations of the RR Tel eruption provides a dramatic illustration of this point and also highlight the development of a symbiotic nova (Thackeray 1977, Penston *et al.* 1983). During the outburst, progressively higher ioniza-

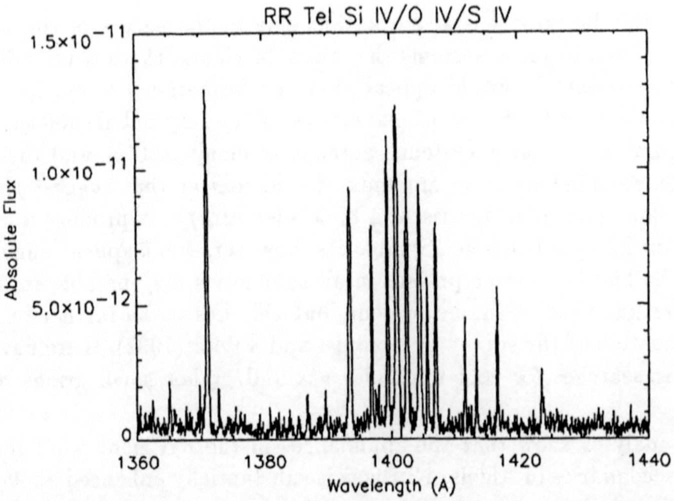

Figure 36: IUE high resolution spectrum of RR Tel (SWP 20242) of the region around λ1400ÅThis is an excellent example of how the emission lines from the red giant envelope provide many useful nebular diagnostics. Note the wealth of emission lines from O IV, Si IV, and S IV, ideally suited for comparison with laboratory data.

tion states of Fe were observed, beginning with Fe II and [Fe II] and progressing through the forbidden coronal-like forbidden lines up to Fe VIII. In the UV, [Mg V], [Mg VII], [Ne V] and related coronal species have been observed as the envelope H II region has expanded. V1016 Cyg (Nussbaumer and Schild 1982) has also provided an excellent example of a symbiotic nova that was well covered in the UV and optical. Both of these symbiotic stars showed relatively slow, radiatively dominated, outbursts. They also both serve to illustrate the complex nature of the circumstellar environment surrounding the hot component. In both cases, there is evidence from the O I λ1641 line that substantial neutral material remains in the envelope. There is also clear evidence for N I emission in RR Tel, and probably in other symbiotics and even recurrent novae (Shore and Aufdenberg 1992, *in preparation*), the lines being excited by EUV radiation around 900 - 912Å. To date, I should add, there have been no searches for molecular emission, although recently Seaquist and Taylor (1992) have studied the emission from several symbiotics in the millimeter and submillimeter regimes. Their study shows that the spectrum is that of optically thin thermal bremsstrahlung with temperatures of order 10^4K to several times that. They did not find any indication of FIR excesses from cold material (IRAS was also silent on this point), nor have there been any reports of CO emission. This last may be the result of insufficient density to excite the CO (which is produced by H_2 collisional excitation in densities of order 100 cm^{-3} or greater). To date, IUE spectra have not had the sensitivity to observe continuum absorption in the UV A-X system (around λλ1400-1650Å). But Vogel's (1990) model for Rayleigh scattering points strongly to the need to look for molecular absorption components from the extended environment.

I mention this because it is an important clue to the nature of the wind of the red giant. Even in those systems that show M giants, there is no indication of molecular material. It would appear that the formation process for symbiotics really is different from that of planetaries. Planetary nebula nuclei are, as we've seen, sometimes binary systems, perhaps as many as 10%, and these H II regions are surrounded by large amounts of cold matter that suggest that the common envelope phase (as discussed in Livio's lectures) may produce a massive cold supergiant-like photosphere. Symbiotics, however, don't appear engulfed in cold material. The hot source present in all symbiotics may be able to destroy any molecular gas that might be present, but this has to be further explored. With the exception of the survey by Seaquist and Taylor (1992), there have been no systematic searches for cold molecular gas and/or hot small grains for any symbitoics.

Nebular analyses show that the abundances in the red giant wind resemble those of the secondaries in Algols. Nitrogen is substantially enhanced above solar, and the O/N and C/N ratios are subsolar (Nussbaumer *et al.* (1988). Perhaps this is the reason why we don't see CO molecular emission in these binaries. We seem to be seeing material that has been mixed from a region that has undergone CNO processing. There have been no abundance analyses of the hot components in these systems, but such studies will be important in detailing the accretion process. Not all of the symbiotics appear to be wind accreters. Specifically, CI Cyg and AX Per have been modelled as Roche lobe-filling binaries (case C mass transfer). Here the abundances of the accreted material could serve as a tracer of the mass transfer process.

What is the physical structure of symbiotics? This is a hard question to get a handle on. Their spectrophotometric variations seem to point to accretion disk instabilities. The timescales range from minutes, to years. When short term changes are observed, they usually involve line strengths but not as clearly line profiles (although few have been studied sufficiently in the UV at high resolution and the optical spectra are equivocal on this point). There are several systems, notably AG Dra, HBV 475, CH Cyg, and AG Peg, that display broad emission with typical velocity widths of order a few hundred km s^{-1}, but these may be due to a wind rather than a disk. I know of no clear instance where the UV shows P Cyg profiles, or where any optical emission lines show double peaked profiles, as seen in cataclysmics. The systems undergo outbursts that seem to involve actual mass ejection, as opposed to the sorts of disk instabilites seen in the FU Ori or T Tau stars. About 20 percent of the known symbiotics contain Mira variables and this should have observable consequences for the mass transfer, but the details are not known. Two systems, MWC 560 as in MWC 560 (M. Karovska 1992, *private communication*) and CH Cyg (Mikolajewski *et al.* 1990) show similar timescales, of order 1000 sec. This is interpreted as a signature of a polar acreter in CH Cyg, and it may be in MWC 560 as well, but there is considerable work on both line and continuum variability yet to do to check this. Also, there could be radio or X-ray pulsations that reuslt from the accretion.

4.18 Recurrent Novae and Symbiotics

Recurrent novae are so named because they explode at least once in the typical astronomer's lifetime. Several are very frequent repeaters, like U Sco (about once per decade), while others have very long latency periods like V745 Sco (about 60 years). It is this wide range that makes them especially difficult to recognize before the fact. Many are faint, because of their relative rarity among the nova classes. The ones most likely to explode have evolved companions, often supergiants, and the white dwarf in these systems is often quite massive. This means that they come from a comparatively small population in the galaxy. Others, like T Pyx and U Sco, and LMC 1990 Nr. 2, appear to have main sequence dwarfs or subgiants and very short periods and could be more numerous.

There are not many known recurrent novae, but among this select group is a well defined subset that consists of white dwarfs and M giants: T CrB, RS Oph, V3890 Sgr (Nova Sgr 1962), and V725 Sco (Nova Sco 1936). These systems show nearly identical outbursts in the optical; they are all very fast novae. T CrB showed some unusual photometric excursions after optical maximum, but there were, unfortunately, no UV observatories in the 1940's to track what was happening. RS Oph (the 1985 eruption), V3890 Sgr (the 1990 outburst), and V725 Sco (the 1989 outburst) have all been followed to some extent by IUE. I need to emphasize here that the optical only provides a piece of the picture, and one that is rapidly dominated by the M star as the nova fades. The ultraviolet is required to really obtain a complete understanding of the outburst. What emerges from the observations is that the M giant recurrents are remarkably similar to the symbiotic novae, except that the timescale for the event is ten to a few hundred times faster for the recurrents than for the symbiotics. The recurrents do not seem to eject much mass, typically less than $10^{-5}M_{\odot}$. Generally, their spectra are optically thin from the onset of the ejection. The UV line profiles rapidly develop the narrow emission line signature of the ionized envelope of the companion. They show He II and other very high ionization species within a few days to weeks of outburst, and usually the optical coronal lines are observed in a matter of weeks (Payne-Gaposchkin 1957, Duerbeck 1987).

The other subclass possesses a low mass main sequence star as a companion. These include T Pyx, U Sco, and the recent extragalactic recurrent LMC 1990 #2. In general, they show the same low mass ejecta as the M giant + WD systems, but the ejection velocities are typically higher. They look a lot like classical novae, but there is some evidence that the white dwarf in the recurrents may be more massive than for normal novae (Starrfield *et al.* 1989) and at least LMC 1990 #2 was super-Eddington at maximum (Shore *et al.* 1991). Since Livio has discussed these systems in some detail, I'll concentrate on the best studied of the giant recurrents as a contrast to a symbiotic outburst.

The recurrents that have red giants as the mass loser are quite similar to VV Cep systems, except that the gainer happens to be a white dwarf instead of a main sequence star. Only a few are well studied, and several have only been recognized in the past two years. These are T CrB, RS Oph, V745 Sco, and V3890 Sgr. In all cases, the evolved star is an early M giant, with an absolute magnitude of about

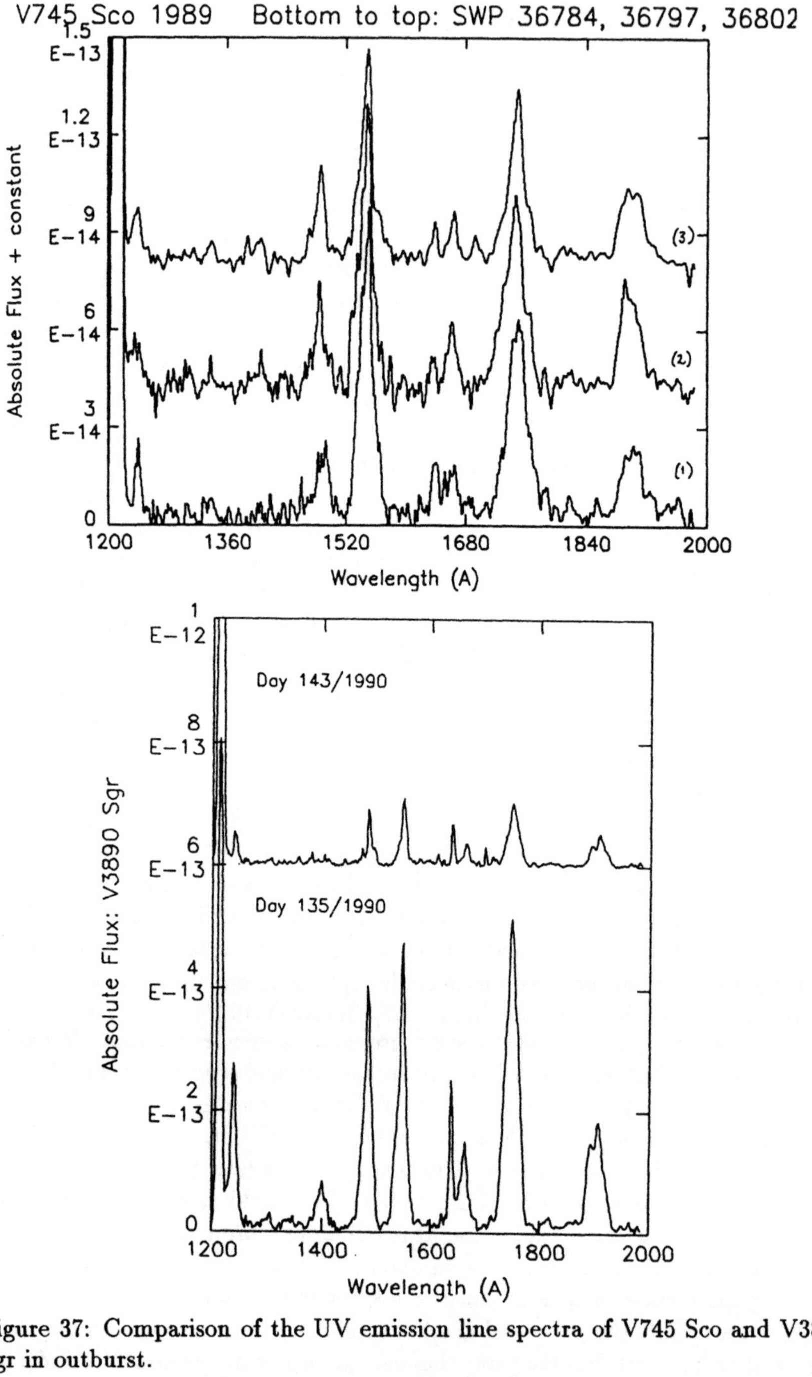

Figure 37: Comparison of the UV emission line spectra of V745 Sco and V3890 Sgr in outburst.

- 1 or -2, and the accreter is a white dwarf in all cases. There is little evidence for an accretion disk of any substance formed during quiescence. Rather, the UV spectrum of the WD is a continuum, sometimes with relatively high excitation lines superimposed. The optical spectrum looks a lot like a symbiotic, showing He II and hydrogen Balmer lines as well as low density tracers like [O III], all sitting on top of an M star photospheric spectrum. This spectrum is established very early in the outburst, usually within a few weeks of optical maximum as soon as the WD and ejecta have faded sufficiently.

During an outburst, *the UV spectrum never becomes optically thick* (fig. 37). This is a phenomenological result only, though, because it is entirely possible that enough mass might be ejected to form a dense shell. However, from the detailed analysis of several systems, specifically the most recent ones like V745 Sco and V3890 Sgr, it appears that the ejecta masses must be substantially smaller than found for a classical nova in outburst. The mass ejection in RS Oph 1985 showed collimation. VLBI imaging showed that the ejecta took the form of oppositely directed plumes (although the baseline coverage was not sufficient to tell whether there might have been jets as well it seems unlikely). However, from the fact that the two most recent members of the class, V745 Sco and V3890 Sgr, also showed similar spectral development, it is possible that the same properties of the ejecta can be inferred for them.

There is a more serious question that remains to be addressed: why are there systems that look so nearly identical – the symbiotics and the M giant recurrent novae – and yet behave so differently? It is true that the symbiotic novae share much of the same phenomenology as the few recurrents with giant companions, but the resemblance is more one of process than timescale. The symbiotic novae are long-lived objects, taking years or even centuries to complete their ejection events. They are dominated by winds that show similar velocities to those observed in the ejecta of the recurrents. But the symbiotic novae are clearly *not* explosive and, most important, they show no evidence for the nucleosynthesis products that are observed in the recurrents In the recurrents, similar abundance patterns are seen to those in classical novae. I think this is one of the most interesting theoretical questions posed by the symbiotics and one that requires more input from observations. Specifically, it is important to detail the *quantitative* differences between the two classes of objects in order to understand why one is able to reach steady state mass loss while the other explosively ejects material and turns off.

4.19 RS Ophiuchi 1985: The Anatomy of an Outburst within a Red Giant Wind

I'm going to dwell at some length on the recurrent nova RS Oph, which is an especially illuminating system for many reasons. It can be thought of as a symbiotic in fast forward, or as a very slow classical nova for reasons that we'll see shortly. Since most of the information is available for the 1985 outburst only, I'll concentrate on that. You should also consult the excellent interim report on research on the outburst in Bode and Evans (1986).

Since the published analyses of the UV spectrum in outburst have been rather sketchy, let me concentrate on this part of the data for a moment. The initial outburst spectrum was dominated by the ejecta, but this quickly gave way to a symbiotic-like morphology derived from the ionization of the environment. The initial rise took about 60 days. All of the resonance lines normally associated with the hot plasma were observed, all with velocity widths of order 1800 km s^{-1}. Actually, this is quite slow for a nova, more in line with the widths seen in symbiotics like AG Peg. Lines of N V, Si III λ1300, C IV, He II, C III]λ1900, C II 2320, and Mg II were seen, all broad and well resolved even in low resolution IUE spectra. I'll return to these in moment. On about day 90 (the outburst was first seen on day 25, 1985) a dramatic change appeared. Narrow emission lines appeared superimposed on the ejecta lines. As the ejecta faded, these lines continued to increase in strength for a short time and then themselves declined. The ionization of the spectrum increased during this phase: as the Mg II disappeared, [Mg V] appeared. In the ensuing weeks, quite a number of coronal lines showed up, first in the UV and then in the optical. Iron coronal lines, [Fe VII] and [Fe X], appeared. These were also seen in the UV, which increased to [Mg VII]. The narrow components of N III], and Si III] and C III] indicated that the electron density in the line forming region was continually decreasing. This is consistent with the low resolution behavior, which showed that the emission lines became unresolved just after UV maximum. Notably, the O III Bowen fluorescent lines increased in strength along with the narrow He II component.

The most dramatic changes were observed at C IV (fig. 38). Here the ejecta appeared as broad emission lines which were soon cut by numerous narrow absorption features. The absorption lines gradually became weaker as the outburst proceeded, and about 60 - 90 days after optical maximum the ejecta were surmounted by narrow nebular emission lines. It seems that the H II region had finally expanded sufficiently that the emission measure of the envelope in C IV dominated over the absorption. At this stage, the envelope was so ionized that the iron line absorbing column seen against the ejecta was substantially reduced. In addition, the ejecta were expanding into progressively lower density regions of the envelope and the result is that there was a reduced column of neutral and low ionization matter in the way. Finally, the ejecta faded leaving only the emission from a region with $n_e \approx 10^5 \text{cm}^{-3}$.

Radio observations at the time of the outburst show a standard nova light curve. That is, the rise time was a few weeks and the decay lasted for several months. The light curve has been interpreted as a shock expanding into the red giant wind (Spoelstra *et al.* 1987). But an alternative explanation is that we're seeing the ionization wave from the explosion as it propagated into the red giant wind, like the radio light curve seen for SN 1987A. If so, then the centimeter wavelength behavior of RS Oph is actually a very good model for the symbiotics, which also show radio emission during outburst. The radio emission was sufficiently strong that it was possible to obtain a 21 cm H I absorption measurement for RS Oph. This shows that its distance is about 1.6 kpc (Hjellming *et al.*, in Bode and Evans 1987).

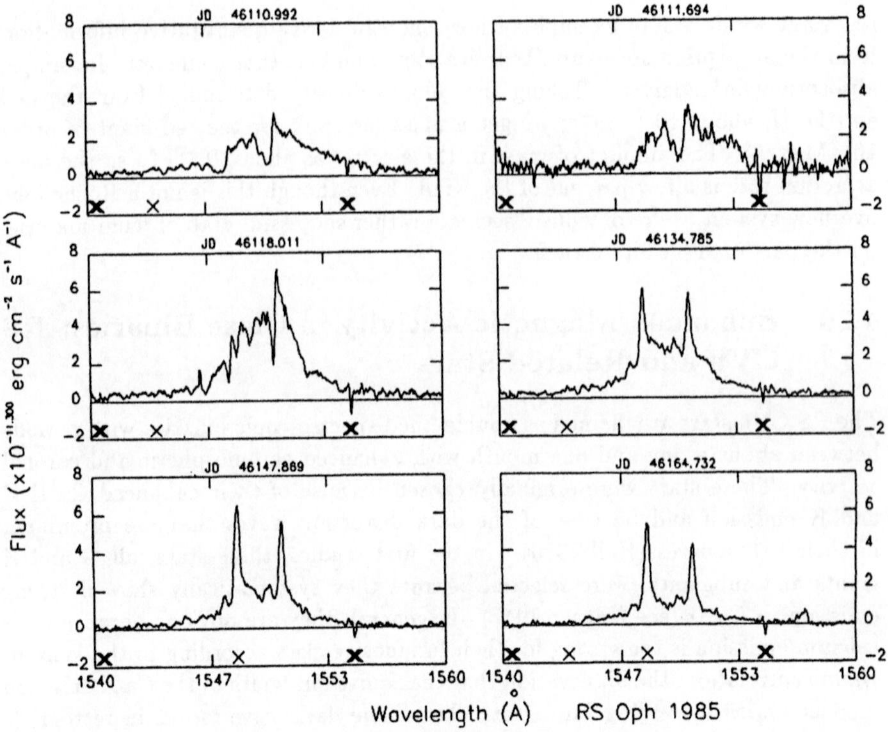

Figure 38: Development of C IV λλ1548,1550 in RS Oph 1985. Julian dates are given atop each profile as JD2440000+date.

Even more interesting, however, is that the nova was imaged using the European VLBI network (Porcas *et al.*, in Bode and Evans 1987). The ejecta were clearly elongated – one is tempted to say jet-like but there is no evidence to support continuous ejection – with an aspect ratio greater than 3. The expansion velocity of the bipolar lobes was about 3000 km s^{-1}, so that combined with the radial velocity, we have for the first time an image of the ejecta in three dimensions. It is important to keep in mind that this nova did not produce spherical ejecta and there was also never any indication of P Cyg profiles in the UV or optical spectrum. Recall that symbiotics (R Aqr, CH Cyg, and HM Sge) also show this kind of ejection of plasma. Unlike T CrB (Cannizzo and Kenyon 1992), however, there was no compelling evidence for an accretion disk in the system either before or after outburst, so perhaps this isn't the only way that the environment can shape the ejecta (this point has also been discussed by Livio for common envelope ejection of planetary nebulae). The ejecta, even in the relatively early stages of the outburst, show normal abundances with the exception of enhanced CNO. Helium is likely normal or only very slightly overabundant, although some wind material might have been swept up into the expanding material.

As the H II region expanded, it showed that the nebular lines never became narrower than about 30 km s^{-1}, which is probably the expansion velocity of the red giant wind. As an example of how one can derive quantitative information from the absorption spectrum, here are some numbers that come out of a simple absorption line analysis. Taking the column density determined from the O I and Fe II, about 10^{17} cm^{-2}, we get a mass loss rate for the red giant of order $10^{-7} M_\odot yr^{-1}$. The amount of mass in the ejecta was about $10^{-6} M_\odot$ so the mass accretion rate is a few percent of the wind. Even though this is not a Roche lobe overflow system, we're obviously seeing a rather successful grab of wind material on the part of the white dwarf.

4.20 Enhanced Magnetic Activity in Close Binaries: RS CVn and Related Stars

The RS CVn stars are defined as double-lined spectroscopic binaries with periods between about a day and one month with enhanced chromospheric and coronal activity. These stars were originally chosen because of their enhanced Ca II H and K emission and because of the dark distortion waves that are prominent in their light curves (Hall 1976). In the first studies, these stars, all G and K giants and subgiants, were selected because they systematically showed strong emission in Ca II (see Hiltner 1947). In general, they are obvious because their calcium emission is too strong for their luminosity class according to the Wilson-Bappu correlation, the observation that the equivalent width of the Ca II emission increases with increasing luminosity class. The dark wave moves in retrograde through the light curve, indicating that it has a shorter period than the orbital period. The amplitude of the wave is a function of time.

Several systems have now been studied for a few decades (Stassmeier *et al.* 1988). The periods for the amplitude variation are of order decades, and the migration of the wave seems to accelerate during the declining stages of the cycle. Although the light curve distortions were originally thought to be the result of circumstellar matter in the binary (see Batten 1973), it is now clear that these regions are more closely akin to starspots. The nature of the activity was made clear by the 1977 flare in HR 1099 = V711 Tau, a radio and X-ray event that lasted for nearly two weeks and reached a peak flux of order 1 Jy at 20 cm (fig. 39). This flare was the first extensively observed event, a complex series of outbursts that showed strong Hα and Ca II K enhancements

The first detailed study of the RS CVn systems by Hall (1976) made very effective use of the observations of Ca II emission to point out the existence of a class of stars that have the following properties. They are close systems, with periods of order a few weeks or less. Their rotation rates are higher than field stars of the same spectral type, presumably because of the tidal interaction between the components. The strong Ca II emission is sometimes contributed by both stars. The optical primaries have spectral types between late G and late K, usually subgiants and these binaries frequently have near unity mass ratio. They show variable light curves on long timescales, although the light curves appear to be stable on single orbital periods. The most important point is that the stars systematically show dark waves that move in retrograde through the light curves.

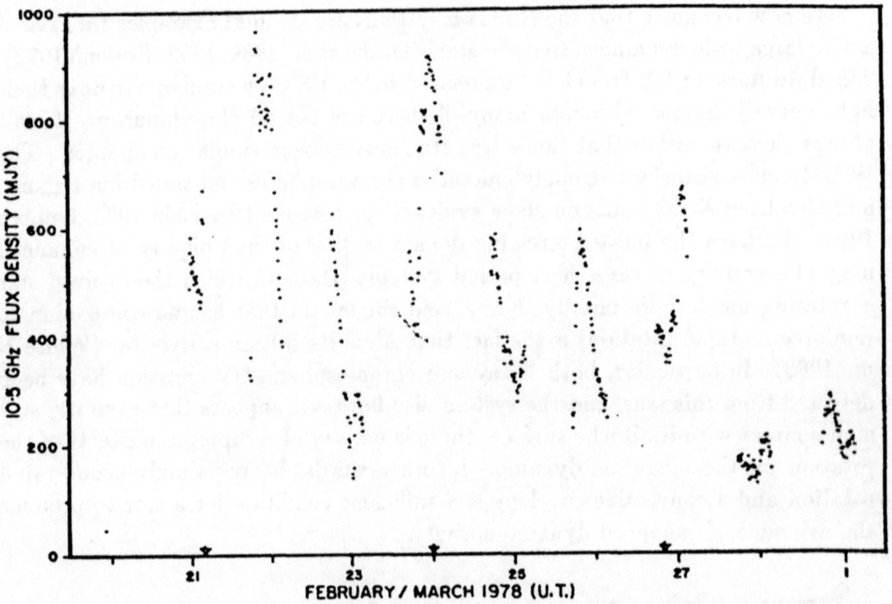

Figure 39: The 1977 radio flare in HR 1099 (Feldman *et al.* 1978). Note the complex structure of the time sequence.

The $B-V$ does not substantially change during a cycle, except for the change due to eclipses, and this places severe upper limits to the temperature of the active region. The Ca II and other chromospheric emission lines like Mg II and C II vary in antiphase with the optical light, even for non-eclipsing systems (this has been especially well studied in AR Lac, II Peg, and RS CVn). The light curve distortion covers a large phase of the orbit, indicating that the region must be a large fraction of the stellar surface. The relative stability of the distortion over several orbits argued initially against a hydrodynamically ordered source for the obscuration.

Many RS CVn binaries are strong X-ray and radio sources. Their X-ray luminosities lie in the range $10^{30} \leq L_X \leq 10^{32}$ erg s^{-1}. The normal stars lie below the lower range by at least an order of magnitude. The luminosities in X-ray and at 6 cm scale with the stellar luminosity, $L_{XR} \sim L_{bol}$; a similar result comes from the radio data (Drake *et al.* 1992). There is a positive correlation between both emissions and the stellar rotational velocity, but this is less well established and does not appear to be linear. The RS CVn stars possess strong, hotter than normal coronae with temperatures extending up to about 10^8K. These stars show no organized strong magnetic fields, but spots on their photospheres reach very large covering fractions, of order 20 to 50 percent, indicating that complex fields must be present, and they show indirect evidence of strong magnetic fields from Zeeman line broadening (see *e.g.* Donati *et al.* 1990). Some systems, specifically V711 Tau (= HR 1099), λ And, UX Ari, and AR Lac, show circularly polarized radio emission, indicative of coronal loops.

We now recognize that the RS CVn systems are the best examples we have of active large scale dynamo activity in stars (Drake *et al.* 1989, 1992, Rodonò 1992). The dMe stars, or BY Dra class, are related to the RS CVn stars by virtue of their light curve behavior, although many of these are not all close binaries. It isn't always the case either that the active star must have a similar companion. The W UMa class also show strongly enhanced chromospheres and transition regions, and also from X-ray emission show evidence for coronae (Rucinski 1985, Guinan 1990). Perhaps the most interesting demonstration of the ubiquity of enhanced magnetic activity in very short period systems (those in which the evolved star is rotating much more rapidly than a field single star that has no companion to reinforce its rapid rotation) is the fact that Algol itself is an active star (White *et al.* 1986). In particular, both X-ray and chromospheric UV emission have been detected from this star, and the system also flares. It appears that even if a star is in contact with its Roche surface, there is no way of escaping the effects of the rotation on the magnetic dynamo. In other words, having simultaneous rapid rotation and a convective envelope is a sufficient condition for a star to produce the evidence of enhanced dynamo activity.

Table 3. Some Representative RS CVn and Related Systems

Eclipsing				
Name	HD/DM	P_{orb} (d)	Spectral Types	M_{hot}/M_{cool}
V471 Tau	+16°516	0.52	K0 V + DA	0.8
DH Leo	86590	1.07	K0V + dM	1.4
AR Lac	210334	1.98	G2 IV + K0 IV	$\approx$1
HR 5110	118216	2.61	F2 IV + K2 IV	1.9
V711 Tau	22468	2.84	G5 IV + K1 IV	0.8
SZ Psc	219113	3.97	K1 IV + F8 V	$\geq$0.8
Z Her	163930	3.99	F4 IV-V + K0 IV	1.1
RT Lac	209813	5.07	G9 IV + K1 IV	0.5
RS CVn	114519	4.80	F5 IV + K0 IV	0.96
FF Aqr	-3°5357	9.21	sd0 + G8 III-IV	0.25
Non-eclipsing				$f(M)$ ($M_\odot$)
II Peg	224085	6.72	K2-3 IV-V	0.004
σ Gem	62044	19.60	K1 III	0.08
λ And	222107	20.52	G8 IV-III	0.0006

I will mention at the outset V471 Tau = BD+16°516, a K dwarf - WD binary first discovered as an eclipsing variable by Young and Nelson (1972) (see also Bois, Lanning, and Mochnocki 1988). The eclipses last for about 30 min, with ingress and egress taking about one minute. This is a very difficult system to observe, and it's a small miracle that we know about it. Before proceeding further I should add, however, that its cousin is HZ 9, an M dwarf + WD system in which the white dwarf is considerably cooler than V471 Tau, about 17000 K as opposed

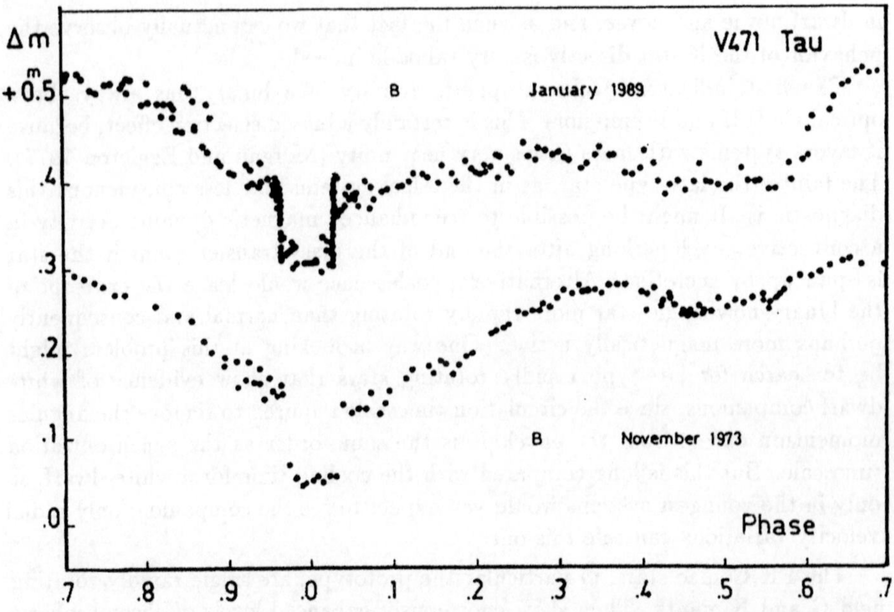

Figure 40: Light curve (B) for V471 Tau between 1973 and 1989 (Ibanoglu 1990).

to about 30000 K for V471 Tau (Lanning and Pesch 1981). The M star is not known to be either an especially rapid rotator or especially active, although it shows strong Hα emission for its spectral type. It's extremely interesting as a test because HZ 9's orbital period is the same as V417 Tau's to within a few percent.

V471 Tau, precisely because of its weirdness, is a key system for understanding the dynamo process (fig. 40). It is one of the few RS CVn systems with a white dwarf or with a low luminosity companion that can be detected in eclipse and the high temperature of the dwarf indicates that it has formed relatively recently. It has an exceedingly short period, 0.52 days, and therefore has the highest rotation frequency of the class. The fact that it looks so much like both cataclysmics and RS CVns emphasizes the broad application of the dynamo picture to close binaries with convective envelope star.

The white dwarf in the V471 Tau system is especially interesting because there is some evidence that the binary is a member of the Hyades although this isn't a sure thing (Bois *et al.* 1988). At an age of about 1 Gyr, the temperature of the white dwarf is incompatible with the expected cooling time. While there is no evidence that the K star fills its Roche surface, the system resembles cataclysmics and dwarf novae and may be a precursor to those systems. The optical light varies continuously outside eclipse eclipse, dominated by a dark wave that moves through the lightcurve in phase retrograde. As in the RS CVn systems, this was first interpreted as circumstellar material. In fact, it is certainly an active region – the K star is an XR source and has recently been detected by the Wide Field Camera EUV survey by ROSAT (Pounds *et al.* 1992, preprint). The behavior observed in V471 Tau is therefore much like that expected for the mass losers in dwarf novae and novae, and as such the fact that we can actually observe the behavior of the K star directly is very valuable indeed.

The first indicator of chromospheric activity in a binary has always been optical Ca II H and K emission. This is certainly a biased selection effect, because it favors systems with mass ratios very near unity (Morgan and Eggleton 1977). The fainter the late-type star, as in the Algol systems, the less conspicuous this diagnostic is. It might be possible to see enhanced magnetic dynamo activity in a convective envelope long after the end of the mass transfer event if the star is spun up by accretion. Alternatively, coalescence would leave the envelope of the binary-now-single star more rapidly rotating than normal and consequently perhaps more magnetically active. One way of looking at this problem might be to search for late-type rapidly rotating stars that show evidence of white dwarf companions, since the circulation timescale required to remove the angular momentum excess from the envelope is the same order as the synchronization timescale. But this is long compared with the cooling time for a white dwarf, so only in the youngest systems would you expect to see the companion; only radial velocity variations can rule this out.

The FK Comae stars, in particular the prototype, are single rapidly rotating field G and K giants. They show enormously enhanced levels of chromospheric and coronal activity, large X-ray fluxes, variable line profiles, and all of the usual signatures of a hyperactive dynamo. The problem is that these stars are all apparently single, unlike the RS CVn stars that they so closely resemble. FK Com has been studied by several groups and no periodic radial velocity or photometric variations have yet been discovered; the current upper limit for v_{rad} variations is about 5 km s^{-1} (McCarthy and Ramsey 1984, Ramsey *et al.* 1992, *preprint*). Even quite low mass companions are unlikely in those systems that have been well observed, like FK Com and AB Dor, given their high rotation velocities.

These stars may be post-main sequence coalesced systems, like merged W UMa descendents (Walter and Basri 1982). The common envelope phase can lead to such an end-state. However, the orbital angular momentum of such a system would either have to be dissipated or redistributed in the newly re-formed star. The convective envelope will therefore be rapidly rotating when formed and probably remain so for a considerable time. Perhaps one can use the spindown calculations already discussed to obtain an estimate of the turbulent viscosity, based on the idea that the rotation is relic from the coalescence. FK Com, whose spectroscopic and photometric behaviors mimic those seen in the active binaries, shows highly variable radio emission and flaring. So if single FK Com supports the view that a convective envelope if spun up sufficiently will produce magnetically active chromospheres and coronae. But in binary systems, the process is much easier to study.

In a broader theoretical point of view, the basic information about enhanced magnetic activity comes from the observation of strong chromospheres and other high temperature plasmas in close binaries with late-type components. X-ray observations, especially with the Einstein Observatory, have been used as an independent way of getting a handle on the dynamo activity. The argument that the more rapidly rotating stars should have more vigorous dynamos and consequently stronger coronae is well worn (especially if the coronal plasma is mainly in large loop structures). In general, for both binaries and also for rapidly rotating (apparently) single late-type stars, the strongest Ca II H and K line emission and the highest X-ray fluxes are associated with shortest period systems (Noyes *et al.* 1984). But let's examine dynamo theory for a moment in order to see what sorts of scaling relations might exist for the activity.

The simplest way to for a star to generate a magnetic field is if it is rotating. Cowling's theorem – also known as the *anti-dynamo* theorem – states that no stationary (time-independent) axisymmetric dynamo is possible. But shear is essential to break the spherical symmetry. The symmetry is further broken, and the effect of the rotation translated into a poloidal field, through the combined action of circulation and turbulence (Parker 1979). An initially axisymmetric field is sheared by differential rotation, and if it is initially cylindrical (B_z) or poloidal (B_r, B_θ), then an azimuthal field (B_ϕ) results. Here r and θ are the radius and latitude, respectively. A poloidal field results from a toroidal potential field, $\mathbf{B}_p = \nabla \times \mathbf{A}_\phi$, so that the toroidal magnetic field results from a distortion of the poloidal field. Finally, in order to convert the toroidal field back into a toroidal potential, some additional symmetry breaking is required. Turbulence in a rotating medium has vorticity which is parallel to the local angular velocity vector and is neither radial nor even hemispherically symmetric.

In an electrically conducting fluid, buoyant turbulent cells produce a helical twist to the toroidal field, and induce a poloidal conversion. This is the basis of the $\alpha - \Omega$ dynamo model. The electromotive force is schematically given by $\mathcal{E} \sim \alpha \mathbf{B}$, where α is related to the velocity correlation function and essentially measures the amplitude of velocity fluctuations in the fluid. The assumption is that the turbulent fluctuations in the fluid velocity field drive buoyant motions (through the density fluctuations or temperature) and that these translate locally into some kind of electromotive force. In a turbulent medium, the velocity breaks into a mean component, $\mathbf{V}$, plus a fluctuating part, $\mathbf{u}$ (which has a vanishing mean value, but for which $< u^2 >$ does not vanish). Here the brackets, $< \cdots >$, represent ensemble averages over the turbulent spectrum of the eddies in the fluid. The magnetic field evolution depends on both the mean field $\mathbf{B}_0$ and the fluctuating part $\mathbf{b}$ and the dynamo equation becomes $\partial \mathbf{B}_0 / \partial t = \nabla \times (\mathbf{V} \times \mathbf{B}_0) + \nabla \times < \mathbf{u} \times \mathbf{b} > + \eta \nabla^2 \mathbf{B}_0$, where $\mathcal{E} = < \mathbf{u} \times \mathbf{b} > = \alpha \mathbf{B}$. Thus α represents the fluctuations in the fluid, and describes schematically how this feeds back into the magnetic field strength. The evolution of the fluctuating part of the field is given by $\partial \mathbf{b} / \partial t = \nabla \times (\mathbf{u} \times \mathbf{B} + \mathbf{V} \times \mathbf{b}) + \nabla \times (\mathbf{u} \times \mathbf{b} - < \mathbf{u} \times \mathbf{b} >) + \eta \nabla^2 \mathbf{b}$. Then $\alpha \sim l_0 < u^2 > / \eta$ and therefore depends on the velocity fluctuation spectrum. The turbulence therefore controls the small scale structure, and the differential rotation (shear) controls the large scale, ordered field and provides the symmetry breaking necessary to generating the dipole.

As usual, lacking nearly all vital information we resort to scaling. An estimate of the strength of the field, and an approximate scaling law, results from the quantitative side of this picture. Differential field stretching causes poloidal to toroidal conversion, which takes place at a rate $\delta v_\phi/L$. Vorticial motion of rising convective eddies transforms toroidal to locally poloidal field at a rate Γ, which is a pseudoscalar quantity whose sign depends on the hemisphere. The dynamo equations simplify by dimensional analysis. For the poloidal field, which is given by a vector potential field:

$$\mathbf{B}_p = \nabla \times \mathbf{A}_\phi \rightarrow \frac{\Delta A_\phi}{L}. \tag{57}$$

The rate at which poloidal field dissipates by turbulence and diffusion is, in the steady state, balanced by the conversion of toroidal to poloidal flux:

$$-\eta \nabla^2 \mathbf{A}_\phi \approx \Gamma \mathbf{B}_\phi \rightarrow -\eta \frac{\Delta A_\phi}{L^2} \sim -\Gamma B_\phi \tag{58}$$

Finally, the poloidal field is wrapped to form the azimuthal field at a rate that depends on the shear $\Delta\Omega$

$$-\eta \nabla^2 \mathbf{B}_\phi \approx \nabla \times (v_\phi \times \mathbf{B_p}) \rightarrow -\eta \frac{B_p}{L^2} \sim \frac{\Delta v_\phi}{L} B_\phi \sim \Delta\Omega B_p. \tag{59}$$

The dimensionless number that results, called the *Dynamo Number*, is essentially the ratio of the rotational to the buoyant timescales:

$$N_D = \frac{\Gamma \Delta\Omega L^2}{\eta^2}. \tag{60}$$

Notice that N_D is independent of the magnetic field, and that in the steady state case it is of order unity. For large values, the field will not be steady (what is usually meant by an active dynamo). Now the rate Γ has the dimension T^{-1}, the magnetic diffusivity varies as L^2T^{-1}. Assuming that turbulence and buoyancy (convection, for example) provide a characteristic timescale τ_c, then $N_D \sim \Delta\Omega\tau_c$. This is the astronomer's version of the *Rossby* number. To remind you of its role, the Rossby number measures the importance of rotation vs. any turbulent or ordered motions in the system (in essence, it is the ratio of the Coriolis to inertial force). This is why there is so much interest currently in the correlation between dynamo activity and *Ro*.

The chromospheric and coronal indicators, like Ca II H and K emission, Mg II λ2800 emission, and L_{XR}/L_{bol}, scale with P/τ_c. The precise relation varies between different studies, but the surface fluxes appear to vary roughly quadratically with τ_c/P. This number has the same form as the classical Rossby number. However, notice that the empirical scaling depends on the rotation period but that the dynamo number depends on the differential rotation rate, $\Delta\Omega$, and not on Ω. Spot migration has been used to obtain estimates of the differential rotation law for these stars. In general, the active region produces a dark wave that moves in retrograde through the optical light curve, indicating that the active region rotates faster than the binary period and that there is some region on the stellar surface that is rotating faster than synchronously. Shore and Hall (1980) pointed out that this behavior yields a differential rotation law that is more nearly rigid than for the Sun, but there are few systems that have been well studied for long periods of time (Strassmeier *et al.* 1988). The problem is that the spots are relatively shortlived and secular changes in the wave amplitude can obscure the migration over the period of the dynamo cycle. So for the moment, although the Rossby number usually used for studies of these systems uses the rotation frequency of the binary assuming synchronism, this isn't really physically consistent and more work needs to be done to derive the differential rotation laws for these stars.

As we discussed a moment ago, the general result for close binaries is that the shorter period systems show more activity (Gibson 1985, White *et al.* 1986). Nearly from the outset of UV and X-ray work on binaries, Algol has been recognized as a weak RS CVn system. It's interesting to note that the sporadic episodes of Ca II emission in Algol, so unusual from the viewpoint of an accretion disk, can be easily understood in terms of chromospheric activity of the giant (Bolton 1989, Richards 1990). It si also important that the radio spectrum differs from the β Lry like systems (Elias 1990) and is markedly nonthermal. While RT Lac is the only RS CVn-type system with an Algol-like inverted mass ratio, many systems are very near filling their Roche surfaces and are likely the precursors of an Algol-type activity stage (Huenemoerder 1988, Huenemoerder and Barden 1986). There is considerable evidence for activity on nearly all W UMa stars that have been observed with IUE (*e.g.* Guinan 1990, Vilhu 1992). Flaring is observed on essentially all types of stars that show enhanced activity, and both radio and X-ray flaring are not uncommon among the Algol and related short period systems.

Once more, binaries allow for a critical test of the correlation of Rossby number with dynamo activity. The synchronism of the stellar rotation and the precise determination of the radius removes the uncertainty in Ω and R. The synchronism of close active binaries in clusters would provide the age term so that some measure of τ_{conv}, the eddy turnover time, can be derived (Goldman and Mazeh 1991).

The RS CVn stars, at least the first examples, were all of a piece. They presented a remarkably unified picture of evolution, and there were some initial speculations that the base of the giant branch possessed some special properties for binary-induced dynamo activity. In a pioneering study, Morgan and Eggelton (1977) showed that this is actually an accident of the discovery criteria. Specifically, the evolutionary lifetimes made it especially likely that stars would be discovered in relatively short period systems at this stage of evolution if q is of order unity. They argued, and later work has certainly borne out their conclusions, that the RS CVn phenomenon should be far more widespread and that if the taxonomic criteria were relaxed more would turn up. Eggleton (1992) provides a marvelous description of the population simulation method for determining the role of observational bias and selection effects in a wide variety of binary systems. So here again is a warning – as in many areas of astronomical work selection effects can be misleading and enter into the samples in some subtle ways.

4.21 Accretion Disks

The basis of all current treatments of accretion disk structure is that viscous dissipation acts to remove angular momentum and to redistribute heat throughout the flow. The precise nature of the viscosity is where most treatments differ. All of the basic paradigm was originated in the first papers by Shakura and Sunyaev (1973) and Novikov and Thorne (1972) (see *e.g.* Pringle 1981, Frank, King, and Raine 1991 for more details). Some of this material has also been covered by Livio, but is included here in part to keep these lectures self-contained and also because there are differences between the disks in cataclysmics and those in more "normal" systems.

For a dissipative flow with a viscous coefficient η and a stress tensor t_{ij} the dissipation rate is given by $\dot{E} = -\frac{1}{2}\eta\sigma_{ij}t_{ij}$ where σ_{ij} is the shear tensor. For Keplerian flows like those we expect in binaries, this shear is especially simple:

$$\sigma_{r\phi} = \frac{\partial v_\phi}{\partial r} - \frac{v_\phi}{r} = -\frac{3}{2}\frac{v_\phi}{r} = -\frac{3}{2}(\frac{GM}{r^3})^{1/2}. \tag{61}$$

The dissipation rate consequently depends on the the stress tensor, and for disks depends on the shear. The argument goes as follows. Any off-diagonal stresses, be they due to shocks, magnetic fields, or shear-generated turbulent instabilities, will be some fraction of the pressure. Any supersonic flows quickly dissipate energy, so it can be assumed that the turbulence is subsonic, even if magnetic fields are involved. Only the $t_{r\phi}$ component of the stress tensor is needed for the calculation of disk structure and energetics. Then taking the so-called α-disk approximation (Shakura and Sunyaev 1973) we have $t_{r\phi} = \alpha p$ where p is the gas pressure (although it can also be the total pressure if radiation plays a role). The rate of work for a thin disk, *i.e.* one which has an area $2\pi r^2$, is $\dot{E} = W = 2\pi r^2\omega(r)t_{r\phi}$, where $\omega = v_\phi/r$. If we assume Keplerian motion in the disk, and this is a very important physical assumption, then the shear is fixed and especially simple. But first, notice that the shear is given by:

$$\sigma_{r,\phi} = \frac{dv_\phi}{dr} - \frac{v_\phi}{r} = r\Omega' \tag{62}$$

for an axisymmetric disk. Then the stress tensor is given by $t_{r\phi} = \eta\sigma_{r\phi}$ so that the torque is:

$$\tau = 2\pi r^3\eta\Sigma\Omega' \tag{63}$$

For a Keplerian disk, τ is proportional to $\Sigma r^{1/2}\eta$. This extremely elegant representation is the source for all of the recent work on accretion disk instabilities. You see that the torque depends on both the viscosity and the surface density. An increase in Σ, which is also related to the rate of inflow of material, leads directly to an increase in the couple so that the rate of mass transfer and of angular momentum loss by the accreting material both go up. Now we are in a position to solve the continuity equation:

$$\frac{\partial \Sigma}{\partial t} + \frac{1}{r}\frac{\partial}{\partial r}(r\Sigma v_r) = 0 \tag{64}$$

coupled with the momentum equation:

$$\Sigma \frac{v_r}{r}\frac{\partial v_\phi}{\partial r} = \frac{\partial \tau}{\partial r}. \tag{65}$$

This means that we have a solution for the mass conservation equation, since we assume that the Keplerian disk never breaks its circulation law (and is therefore time independent). This is quite a strange assumption, when viewed in light of the possible timescales for the processes. The disk is assumed to transfer angular momentum on the viscous timescale, but the vertical structure is assumed to change on a dynamical timescale (which is always much shorter).

The simplest representation of the viscosity is $\eta = \alpha a_s H\Sigma$ where H is the scale height. Thus clearly η depends on T as well as Σ:

$$\frac{\partial \Sigma}{\partial t} + \frac{3}{r}\frac{\partial}{\partial r}(r^{1/2}\frac{\partial}{\partial r}r^{1/2}\eta\Sigma). \tag{66}$$

Here is where the first diagnostic procedure enters. We can directly determine the velocity profile of a disk in an eclipsing system as the companion occults different portions of the disk. This is extremely important, because the rate of shear determines the dissipation and changes in the circulation velocity can be directly noted. The rate of energy dissipation can also be mapped by several techniques, and therefore it should be possible to check the rate of energy generation by turbulent processes. In addition, the turbulent velocity can be estimated by using the absorption lines formed within the disk against the dissipation region. Well, at least in principle. In practice, the problem is more convoluted than this. In general, the disks are not optically thin. Thus we generally cannot directly observe disk absorption against the photosphere of the accreter[10]. That's a pity because any less direct means is subject to the problem of height-dependent shear and the confusion of vertical structure with the shear term.

We can determine Σ directly from some types of measurements, specifically the behavior of the light curve during eclipse. We can also determine the departures from Keplerian motion. Notice also that if we are wrong and the disk really behaves like a boundary layer, we can always substitute this into the continuity equation because we haven't made any specification of the velocity law until the last step (the one that gave us a factor of 3 in the divergence term of the continuity equation). Notice that the shear enters into the rate of mass transfer through:

$$\dot{M} = 2\pi r \frac{\partial}{\partial r}\eta\Sigma \tag{67}$$

[10]The exceptional cases may be Z Cha and OY Car, for which there is evidence of an absorption spectrum from the ionized iron peak species in low resolution ultraviolet spectra. These dwarf novae show edge-on accretion disks.

so that a change in the viscosity results also in a change in the mass accretion rate. Herein lies the source for the bimodal disk instability model, first discussed by Meyer-Hoffmeister (1983) and Cannizzo and Wheeler (1984) and since elaborated by many authors. The reason is that the structure is determined by the vertical energy transfer, which depends on the superadiabatic temperature gradient. This in turn depends on the opacity and the fact that there is a double-valuedness to the opacity because hydrogen ionization causes the disk to have two stable states, one at low and one at high temperature.

Disks are unstable for a variety of reasons. The first is that the peripheral parts of the disk may be very distended, and the disk may know about the gravitational perturbation of the companion, not just the central gravitational field of the gainer. In this case, specific resonances can occur in the disk, analogs of the Lindblad resonances in density wave theory for spirals. Actually, there were first pointed out not in accretion disks but in planetary ring systems, but the ring systems may be self-gravitating, and they are certainly not structured by radiation and convection. In an accretion disk, the perturbation grows at specific resonances (*e.g.* 1:1, 2:1, 3:1) and at radii where κ, the epicyclic frequency, equals multiples of the orbital period (depending ion the details of viscosity and departures from Keplerian motion this may occur at very different points). The resonances take place in a supersonic medium, so individual particle orbits are being systematically deviated in a nonaxisymmetric way, causing the flow to slam into itself at supersonic speeds. This sets up spiral shocks in the disk, and these shocks tend to send material spiralling inward. In fact, they act to increase the viscosity in two ways. The first is that they dissipate energy locally, since a strong shock increases the specific entropy of the gas. The second is that they turn the flow because of shock refraction and send it inward. The net result is to enhance the mass accretion rate, and at the same time heat the disk. Models by Spuit (1987) and Matsuda *et al.* (1988) have been elaborated on by a number of authors (Spruit (1990) provides a good summary). However, there is no direct evidence for these spiral shocks in the disk light curves or in their spectra. In fact, three dimensional accretion calculations show that the spirals are less organized than the two dimensional models seemed to indicate.

The second reason for disk instability is far more interesting because it is explicitly time dependent. We have already seen that the time dependent behavior of the disk is governed by a parabolic equation that has the form of a diffusion equation for the viscous torque. The viscosity, ν, depends on the surface density, Σ, and temperature, T. The rate of inflow and, as a consequence of mass accretion, depends on the viscosity and the angular momentum gradient in the disk. The disk's vertical structure, hence its surface density, depends on being able to maintain radiative, or at least thermal, equilibrium and thus depends on the opacity, κ. In the critical range of temperature between about 7000 and 15000 K, the opacity changes rapidly because of the ionization of neutral hydrogen, the primary chemical constituent of the disk material. Some metals are also important in this range, especially ionized iron peak species, but hydrogen is overwhelmingly the most important continuous opacity source. Now suppose that The temperature is increased for a disk at around 10000 K. Hydrogen ionizes and the opacity drops. Therefore the disk vertically cools and collapses, and locally the surface density increases. If the temperature is lowered, the medium turns neutral the opacity again drops. If the disk is cooler than this initially, then the opacity increases on being heated and the disk puffs up, lowering the surface density and therefore lowering the opacity. Thus there is the possibility of having the same condition met within the disk that is seen in the envelope of a pulsating star, first described by Eddington on the analog of a steam engine: when the medium expands it cools down and when the medium is compressed it heats up. This drives a mechanical response to the thermal conditions that is sufficient to maintain the instability.

So far, I have concentrated on the type of disk that is observed in cataclysmics. Besides these, Algols, β Lyr-type binaries, and Be stars show disks. Since they haven't been discussed so far, let me focus for a moment on the Be stars. A few Be stars *are* binaries, 88 Her, HR 2142, AX Mon, 17 Lep, HD 218393, HD 173219, and ϕ Per (Peters 1976, Harmanec 1987) and some of these are among the most variable of the class. These are the ones that modeling studies should focus on, in my opinion. One problem is that often when a shell or Be star is found to be a binary, it's removed from the category and studied separately. The most complete survey for binaries among the Be stars has been carried out recently by Baade (1992) who used the very clever technique of looking for the lines of the secondary in the red and concluded that either the mass ratios are very much smaller than unity or the Be stars simply don't have companions as a rule. The latter seems most likely. However, as Baade has noted, there seem to be two somewhat distinct groups of Be stars. The early B emission line stars look like the classical type with high rotation and appropriate inclination dependent emission line profiles (and disk around the brightest of these, γ Cas, has actually been resolved by optical interferometers). The other group, the later B emission line stars, may actually be binaries since many of their physical properties seem to differ from the hotter counterparts. In particular, their UV spectra look very much like the Algol systems that have rapidly rotating photospheres (like U Cep). This is likely to be a very fruitful area for future research.

There is one phenomenon that has been pointed to as a sure sign of an accretion disk, the V/R variations. These are periodic changes in the emission line profile, especially at Hα which show systematic variations in the ratio of the blue to red components of the emission. Various models have been proposed over the years, including the presence of precessing elliptical disks (suggested by Huang) and possibly even spiral shocks. Since photospheric profiles of massive stars show such behavior as a result of nonradial pulsation, Baade (1992) has suggested that the same may be true for the Be stars. One is tempted to think that there could be an interaction between the underlying star and any surrounding accretion disk in a binary system, or that tides set up by a companion could be responsible, but much more work remains to be done here.

Applications of viscous accretion disks to comparatively normal systems has been slow in coming. Lee *et al.* (1991) are among the first to use them for Be stars, and Shore and King (1986) have discussed the boundary layer around main sequence stars. Prendergast and Taam (1974) performed some simulations of the disk in U Cep, showing that the boundary layer vanishes in the limit of critical rotation for the gainer. But otherwise, there hasn't been much done. Perhaps this is because of the feeling that since the disks are optically thin and there doesn't seem to be turbulent heating within them they are not as interesting as the ones around cataclysmics. This is not to say that they aren't likely to exhibit interesting phenomena. In spite of their low opacities, the disks will still respond dynamically the presence of a companion. All of the tidal effects derived for compact accretion disks will likely also happen for Be star disks. Spiral waves, disk flapping, and turbulence (although not a significant heat source) should all be observable and may be involved in explaining some of the behavior of the Be and shell stars, especially those in close binaries.

4.22 Boundary Layers in Accretion Flows

The hard part of mass transfer isn't getting the material from one star to another. In a close binary, gravity does the work for you. The difficulty comes when you have to get the material to accrete. Then, although it would seem that gravity still dominates, there are other effects that mitigate its influence. The first is the fact that any matter passing through the L_1 point carries excess angular momentum that must be dissipated. Once it is in orbit around the companion, being in a Keplerian orbit it will remain stable and circulating. Unless some internal viscosity acts to slow its motion, it will not flow inward. We have already discussed the structure of the accretion disk and the fact that the radial motion is a slow drift. We need, though, to consider what is actually happening at the inner zone, the so-called *boundary layer* (Schlichting 1965). This layer is at the surface of the star that is receiving the material, and therefore it may be susceptible of observational analysis.

4.22.1 Boundary Layers: Theory

The first point is that the inner boundary condition used for the original α-disk calculations was a black hole. This choice was not just because it is a site of X-ray emission and thus especially interesting for disk calculations. It also means that there is an infinitely greedy sink of essentially zero radius at the innermost part of the disk. Any material that manages to make it within $6m$, where $m = GM/c^2$, will fall in with no further interaction with the disk material. A neutron star is similar in being extremely small but is now a solid (really!) surface, so the accreting matter sees a wall at its inner edge. For a normal star, the problem is exacerbated by the large size of the accreter with respect to the outer edge of the disk. And with a normal star, we have the chance to study its atmosphere, at least down to the photosphere.

A boundary layer of an accretion disk is defined as that region wherein the radial velocity gradient of the circulation velocity is positive. Matter at the base must co-rotate with the stellar surface. At the top of the boundary layer, it must co-rotate with the Keplerian motion of the disk as a whole. Therefore, somewhere in the disk, if the material is to actually settle onto the surface of the gainer, the sign of the radial component of the shear must reverse sign. This is extremely important because the gradient of v_ϕ may become quite large, leading to a strongly dissipative zone that may actually be quite geometrically thin. The turbulent viscous dissipation is proportional to the square of the shear so that the sign change does not alter the dissipative character of the layer, just the rate. It is also possible that by Rayleigh's theorem, this zone may become turbulent and thus the energy dissipation can be still further enhanced (see *e.g.* Greenspan 1968). The basic equations that detail the physics are as follows. Energy transport in the layer is described by:

$$\frac{1}{r}\frac{\partial}{\partial r}r\Sigma v_r(\frac{1}{2}v_\phi^2 - \frac{GM}{r}) + \frac{\partial \mathbf{F}\cdot\hat{\mathbf{z}}}{\partial z} = 0. \tag{68}$$

We assume that $v_\phi = v_K$, the convected mass flux is $\frac{1}{2}\Sigma v_r v_K^2$. The momentum equation for the orbital velocity is:

$$\frac{1}{r}\frac{\partial}{\partial r}r^2\Sigma\Omega v_r = \frac{1}{r}\frac{\partial}{\partial r}(r^2\nu\Sigma\frac{\partial\Omega}{\partial r}). \tag{69}$$

Finally, we can make the assumption that $\dot{M}$ is constant in the layer. This may not be justified for any specific system if the boundary layer is unstable or if the underlying star is rapidly rotating, but here we adopt Kley (1991) to obtain:

$$-\nu\Sigma r^3\Omega\frac{\partial\Omega}{\partial r} = \frac{\dot{M}}{2\pi}[r^2\Omega^2 - \beta\Omega\Omega_K(r_0)r_0^2]. \tag{70}$$

In Kley's treatment, Ω, the orbital frequency, is assumed to be some fraction, β, of the Keplerian frequency. So here, once the surface density is specified, we have a complete solution for the circulation. A boundary layer is described by the condition that the radial motion is dominated by the shear and the rate of circulation. The pressure gradient is given by $\partial p/\partial r = \rho v_\phi^2/r$. Therefore, the approximate thickness of the layer is $\Delta r \approx a_s R/v_\phi(R)$ the same estimate as the disk scale height. Actually, this isn't too surprising since the same condition applied for the pressure gradient in the boundary layer as in the vertical structure of the disk. The boundary layer for an astrophysical flow is very different than most terrestrial flows because the infall . The If the layer is optically thick, most of the binding energy of the material is released in it so that:

$$4\pi R\Delta R\sigma T_{BL}^4 \approx \frac{1}{2}\frac{GM}{R}\dot{M} \tag{71}$$

and $T_{BL} = T_\star(R/\Delta r)^{1/4} \approx (v_\phi^2\dot{M})^{2/9}$, and the shock temperature is given by $T_S \sim GM/kR$, the virial temperature. All of this radiation is reprocessed in the layer if it is optically thick, so to an external observer it looks as if we are seeing a stellar surface, albeit one with extreme gravity darkening and a strange spectral distribution.

4.22.2 Boundary Layers: Observations

It is also important to note that the accreted material is very hot. It passes through an accretion shock on the way onto the stellar surface and therefore has a much higher specific entropy than the envelope. Kippenhahn (1974) long ago pointed out that this is a problem for understanding the comparatively normal abundances of the accreters in Algol systems. It has become clearer that the gainers in Algols are CNO enhanced (see Yoon and Honeycutt 1992, and refs. therein).

But there is another problem that is equally interesting. A boundary layer formed around a rotating star comes to rest with a moving surface and thus needs to lose less angular momentum. This means that the layer can be thinner, or if the surface is rotating at breakup, it disappears. Thus for systems like U Cep, where it appears that the stream impacts the surface, it can penetrate to some depth and then spin up the layer. Some of the mixing at the inner boundary may also be driven by the Kelvin-Helmholtz Instability for a shear flow (Wu and Wang 1991). The layer will be dominated by Ekman-type flow, so that $\Delta r = \sqrt{(\eta \omega_K(R)^{-1} E^{-1})}$ where E is the Ekman number for the layer. This is the same result reported by Kley (1991) although that treatment does not use the Ekman number, opting instead for a constant of proportionality, C_{BL}. The Ekman number in the boundary layer should be of order unity; Kley finds that $C_{BL} = 1.90$. Since the viscous coefficient is of order $a_s H_p$ we obtain $\Delta r \sim (a_s/v_K)^{1/2} h^{1/2} R$, where $h \equiv H_p/R$. For a typical compact object, h is very small, but for main sequence stars it can be fairly large since $H_p \sim g^{-1}$. Further analytic treatment of the boundary layer around a rapidly rotating accreter can be found in Colpi *et al.* (1991).

What are the observational signatures of a boundary layer? The obvious first answer is a departure from Keplerian motion in the circulating circumstellar material. This has been observed in some systems, see for instance Kaitchuck and Park (1988) on TZ Eri. But this is only part of the story. An optically thin layer should be quite hot, and the attempts to find these regions in systems with compact (WD or neutron star) gainers has been an obsession with those who work on X-ray binaries. There are serious problems with the theory in this case, since temperatures of the accretion zone should be of order the virial temperature, at least 10^7K, for compact objects, and this simply isn't seen. The alternative, that the layer is optically thick, is equally problematic.. But when the gainer is a main sequence star, it may be possible to place further constraints on the nature of the layer.

Let me repeat one point for emphasis: if accretion takes place from a disk, a boundary layer *must* exist. It is possible that the underlying star is rapidly rotating, but in the case of some systems like U Cep, the apparent rapid rotation of the gainer may simply be a manifestation of an optically thick, relatively low temperature, boundary layer. In fact, it is hard to tell the difference between a thick disk and such a layer if the mass accretion rate is high enough. The outer envelope of the gainer, if spun up to Keplerian velocities or nearly so, will look precisely like such a layer. The the layer may not have a large spatial extent.

The envelope mass increases rapidly with depth because of the increasing density and the accreted material is viscously decelerated. The resulting torque creates the equivalent of a detached photosphere, an optically thick surface that comes to corotation within a density scale height.

The angular momentum carried by material at the L_1 point determines the distance of closest approach to the gainer. The easiest way to estimate the likelihood of disk formation is to ask what the radius of the equivalent circular orbit at the companion will be, assuming that the angular momentum is conserved in the transfer in the stream. Call the orbital frequency Ω_b, and the distance of the inner Lagrangian point from the center of mass is $f_{RL}a$ in terms of the semimajor axis, then the specific angular momentum in the stream is $j_{RL} = f_{RL}^2 a^2 \Omega$ at L_1. This is equated to the specific angular momentum, $j_{circ} = r_c^2 \omega(r_c)$ at the gainer, where r_c is the radius of a circular orbit about the gainer. Since the orbit is circular, this is $j_{circ} = (GM_2 x_c)^{1/2} a^{1/2}$, where $x = r_c/a$. The orbital frequency there depends only on M_{gainer} and r, while the L_1 angular momentum is some fraction, f, of the orbital value $\mu a^2 \Omega$ so that $r = (xa)^4 \Omega^2/(GM)$. Since x depends on the mass ratio and the orbital period, we obtain:

$$x_c = \frac{f_{RL}^2 \mu^2}{GM_2} a^3 \Omega_b^2 \sim \frac{f_{RL}^2 \mu^2 M}{M_2} \tag{72}$$

for the radius of the equivalent circular orbit (see also Shu and Lubow (1981)). Notice that this is only a weak function of Ω_b, which enters through the fractional radius of the Roche lobe. Now the observation consequences here are profound.

If the gainer is a white dwarf, there is always enough room in the system to form a disk. But a main sequence gainer may be so large that the stream impacts the stellar surface before it has a chance to form a true, separate, disk structure. Should this happen, the angular momentum is deposited in the outer layers of the gainer, spinning up the outer envelope and producing the spectrum of a turbulent, optically thick, disk rotating at the *breakup* frequency for the gainer. In other words, when we look at the spectrum, we would expect to see lines with rotational broadening at essentially the Keplerian velocity for the gainer. If the mass transfer rate is $\dot{M}$ and the accretion is some fraction, f, then $\Delta J_2 = \tau_{RL} f \dot{M} \mu a^2 \Omega_b$ is the amount of angular momentum gained by the companion, taking τ_{RL} to be the timescale for the loser to maintain contact with the Roche lobe. The depth to which the gainer can be spun up follows from calculating the mass fraction that can reach ω_K, the surface Keplerian velocity.

This is precisely what many Algol systems appear to show. The prototype is U Cephei, which is also well known for its light curve instability and occasional Hα flaring. This system also displays sporadic mass loss episodes. You can see from the UV variations that the spectrum is very complex. The stellar photosphere displays very broad lines, mainly the spectrum of the iron peak absorbers, with complicated red and blueshifted absorption lines from intrasystem gas. There is clearly a stream in this system, one that impacts the surface, so the distribution of matter in these "impact systems" is more complicated than even the β Lyr-type binaries. An accretion disk, especially one like the β Lyr systems possess, won't look like the U Cep or Algol-type spectra because they have equivalent rotational velocities at the disk periphery that are considerably slower – they look more like shell stars. However, the temperature gradient in these disks may cause them to look like they have very rapidly rotating photospheres underneath the shell, and this underlying spectrum may be just as weird as those seen in U Cep and related objects. In the boundary layer picture, then, these systems are not really such rapidly rotating objects at all – just their envelopes have been spun up. Within the layer the effective acceleration approaches the stellar surface gravity, but at the top of the layer it is like that of a supergiant. This is because of the combined effects of pressure gradient and centrifugal acceleration. Thus the luminosity within the layer is given by the accretion shock while the spectrum is produced by the local conditions within the pseudophotosphere that results in the boundary layer. Some work has been done on this problem over the years (for instance, Shore and King 1986). The optically thick boundary layer may be unstable for several reasons, will show turbulent velocities that are possibly supersonic (because of the shear flow at the top of the layer) and may even be time variable.

There may be a few examples of stars that display boundary layer-type spectra. I think that the active binaries R Arae (Barba and Sahade 1992) and HD 207739 (Parsons, Holm, and Kondo 1983) are examples of this phenomenon. R Arae is a short period extreme mass transfer system with a period of 4.4 days (see McClusky and Sahade 1989). It shows extremely variable UV and optical lines, and what looks like a variable disk and/or stream. The system V1914 Cyg = HD 207739 is a ζ Aur-type system with a period of 140.8 days, a mass function $f(M)=5M_\odot$, and B1 I+cF or F8 II+? (optical). The mass ratio $q \approx 0.6$. There may be a 50 day periodicity that Parsons *et al.* attribute to mass loss by the star and phase modulated mass transfer. HD 207739 shows episodic mass accretion (Parsons *et al.* 1992). The problem with analysing such systems is that the model envelopes and atmospheres are currently inadequate.

A severe problem arises with the interpretation of the environment around Be stars and also accreting systems, coming from some naive assumptions about the nature of the absorption spectrum they should form. Several studies have pointed out that the C IV and Si IV resonance lines are found even in late-type (B9.5 Vne or A0 Vne) systems. Hubeny, Stefl, and Harmanec (1985) pointed out the dangers in this interpretation by simulating the spectrum of a rapidly rotating photosphere with a realistic absorption line list. There are an enormous number of lines in both of these regions from the singly ionized iron peak species. With large rotational broadening, these can simulate the presence of the resonance lines. For instance, fig. 41 displays the variations of the λ1550Å region and fig. 42 shows sample slab models to point out where the iron absorption from circumstellar gas may absorb. The U Cep spectra show similar features and, in fact, the variations of these lines around the cycle may show the variations in column density rather than high temperature regions. Algol shows a similar behavior (few other binaries have been observed systematically over a single orbit in the UV). So the *caveat* to the observer is clear: just because something looks like a high temperature region doesn't mean that it is if the rotational broadening is great enough.

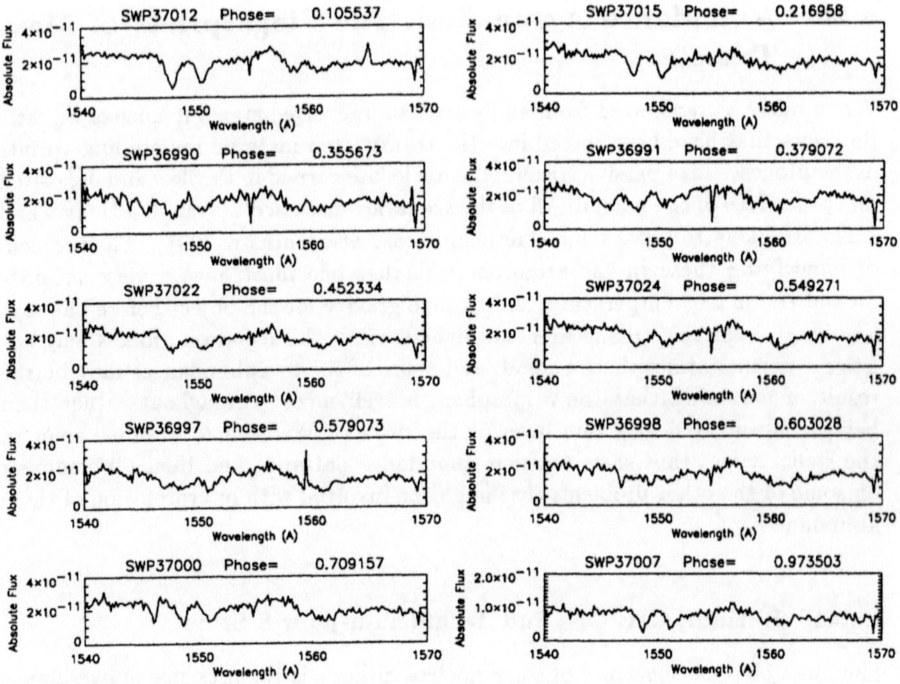

Figure 41: Phase variations of U Cep around λ1550Å. This sequence of IUE observations was obtained by Koch *et al.* (1989, IUE archival data).

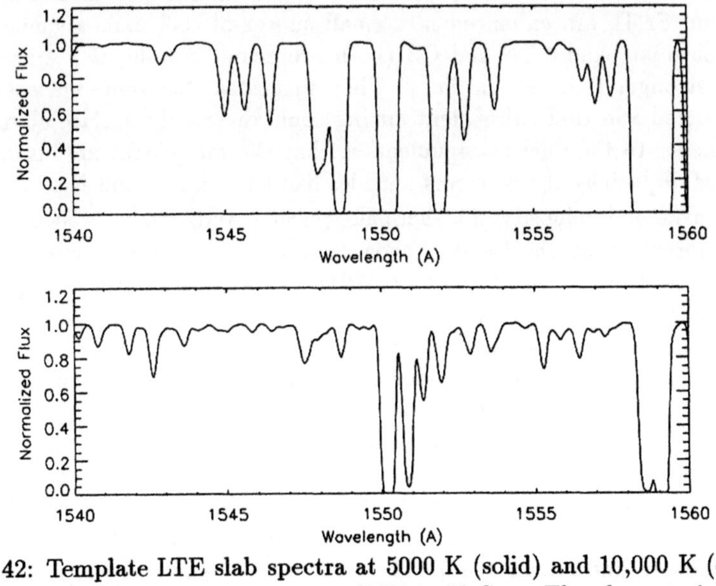

Figure 42: Template LTE slab spectra at 5000 K (solid) and 10,000 K (dashed) for comparison with the variations at C IV in U Cep. The characteristic lines at λλ1546, 1548, and 1550Å are due to singly and doubly ionized iron peak transitions.

4.23 Abundance Anomalies: The Fingerprints of Mass Transfer

When mass is transferred from an evolved to unevolved star, the nucleosynthetic products that have been mixed into the transported material remain fingerprints of the process. Like passive tracers, the nuclei are carried in the flow and deposited on the surface of the gainer. All of the scenarios discussed by van den Heuvel and Livio are likely to leave some clue behind (see also Tutukov 1991). The problem of identifying these in the atmosphere is that one must have a very accurate idea of the mass, temperature, and surface gravity for the object before one says anything about the enhancements or deficiencies. The accretion process may not leave a normal atmosphere behind, and some of the overabundances may be the result of assuming that the atmosphere is well mixed throughout rather than being concentrated in a thin layer at the surface. We need to examine some of the stellar types that show peculiar abundance patterns, and then I'll comment on some of the other problems that might be involved with interpretation of these abundances.

4.23.1 Barium, CH, CN, and Technicium-poor S Stars

The most obvious signature of some nucleosynthesis is the presence of overabundances of s-processed elements. I say most obvious because these were the first such mass transfer systems discovered, the *barium* stars. Bidelman and Keenan (1951) first noticed that the spectral lines of several heavy elements, especially Ba II and Sr II, are enhanced in a small subset of cool main sequence stars. Two molecular bands, CN and CH (both prominent classification criteria) also showed stronger than normal lines. They suggested that some nucleosynthesis had occurred and that subsequent mixing could distort the surface abundances. The clincher to the nuclear arguments is that the rare earths are also strongly enhanced, especially the s-process nuclides like La, Ce, Sm, and Nd.

The average luminosity of a thermally pulsing AGB star is $\geq 3000\ \mathrm{L}_\odot$. However, observationally the Ba stars show far lower luminosities, more like 100 to 1000 $L_\odot$ (McClure *et al.* 1974, Iben 1991). As further studies have pushed to lower luminosity for cluster and field stars, the Ba-like anomalies, specifically the CH subclass, have appeared (Bond and Luck 1991). Finally, Tomkin *et al.* (1989) has found that the s-process anomalies are detectable even in main sequence F stars.

As late as the 1960s there were debates about the process responsible for this enhancement, usually at least an order of magnitude (Smith 1992). The final clue to the origin came with the discovery of radial velocity variations that demonstrate all members of the class are binaries (McClure, Fletcher, and Nemic 1980; Jorissen and Mayor 1988). The companion's spectrum was not observed, however, until the launch of IUE when it became clear that the companion is, in all cases, a white dwarf. Thus, the presence of the Ba anomaly is the result of mass accretion from an evolved companion. The inconsistency of the various nuclear signatures is the result of the mixing process and the fact that the mass transfer takes place rapidly and the process is thus not in equilibrium.

The orbital eccentricities and mass functions show an interesting distribution. First, the eccentricities are more heavily concentrated toward circular orbits than G and K giants (McClure and Woodworth 1990). The mean eccentricity, $< e >$, is 0.29 for the normal G and K giants (41 stars), while $< e >=0.18$ for the Ba stars (16). The Ba stars are more strongly peaked toward zero eccentricity than the normal giants, and Richardson-Lucy rectification supports this. In mass function, the Ba stars are confined to $f(M) < 0.6\mathrm{M}_\odot$ with a mean around $0.15\mathrm{M}_\odot$. The normal giants show a much broader distribution, as do the CH stars.

The S stars are defined as carbon-rich stars with strongly enhanced s- process abundances and unusually large $^{13}\mathrm{C}/^{12}\mathrm{C}$ ratios (in a Population I mixture, this ratio is about 1%). In the S stars it is as high as 25%. Merrill (1952) discovered the evidence for ^{99}Tc I in the spectrum of many of these stars, thereby opening the first window to direct neutron processing in a stellar interior, since this is a short-lived species, $t_{1/2} = 2 \times 10^5$ years, and must be rapidly mixed from the processing site into the envelope in order to be observable (Lattanzio 1989, Iben 1991). The neutron source, ^{22}Ne, must be synthesized after the completion of the CNO process, mainly during shell flashes and the so-called deep dredge-up stages of red giant and AGB evolution. Since the timescale for shell flashes and the subsequent convectively driven mixing is of order 10^5 years or less, Tc can continually be remixed into the atmosphere. However, a substantial subset, upwards of 25% of all S stars, show no Tc at all. If some form of mixing is responsible for the observed abundances, then this is quite unexpected. The answer is that many, if not all, of the Tc-deficient S stars are binaries with WD companions, much like the Ba and CH stars. Johnson, Ake, and Ameen (1991) argue that all of the Tc-deficient S stars in their sample (5 out of 5) and only one of the Tc-excess (*i.e.* normal) S stars (out of 8) are binaries with white dwarf companions. The typical period is of order several years. Johnson (1992) assembles the additional evidence, including the fact that the Tc-deficient S stars tend to be weak IR sources (as opposed to the normal AGB star) and that these stars also show strong He I λ10830Åemission, indicative of a hot ionizing source (no normal S star shows this line). Ultraviolet observations also indicate the presence of enhanced emission activity. While some of this can probably be attributed to wind accretion by the companion, since these systems look a great deal like symbiotics, some may be due to stellar continuum from a recently produced white dwarf.

The discovery of *one* eclipsing system in a class of purported binaries is always important, but for the Ba stars it has been especially so. HD 46407 has been announced recently as a 458 day eclipsing system (Jorissen *et al.* 1992). The primary eclipse lasts for about 10 days, and the secondary has a phase duration of order $0.^p4$ or so, when the giant is in front of the hot source. The light curve isn't stable, and the period isn't that well known. Perhaps some of the variation seen here is the result of the flip-flop instability producing a variable circumstellar environment? More observations are needed, along with the usual patience required when working on this class of binaries.

An interesting test of mass transfer might be the observation of Li in the atmospheres of the main sequence gainer. It is well known that large overabundances of light elements, especially ^{6}Li, can be produced in the AGB stage of evolution, when the products of incomplete CNO are mixed up into the atmospheres of giants (for instance, the C and S stars WX Cyg, HR 8714, and WZ Cas Keenan and McNeil 1976, Querci 1986). The transfer of this material to the main sequence star would be an indication of the nucleosynthetic signature from the loser, which would then become a white dwarf. Since the mixing and destruction timescales are approximately the main sequence lifetime of an intermediate (*i.e.* 1.5 or 2 $M_\odot$) star, any transferred lithium would serve as a clock of the timescale for the mixing event. Unfortunately, this would not be an easy test, but it would be a good one since there is no other way for the lithium to disappear other than by deep mixing (unlike Tc which, in Lewis Carroll's description from *Hunting of the Snark*, "softly and suddenly vanishes away" through decay) and the timescales are well known from observation of main sequence cluster stars.

Luck and Bond (1991) have reexamined the abundances in the subgiant CH stars, the analog of the Ba stars, and find that the abundance pattern fits that of a Pop II star polluted by s-processing. Again, there seems to be a mounting body of evidence that the mass transfer process is surprisingly efficient, and that the stars do not mix very deeply. How the envelope manages to maintain its status is anyone's guess, and I'll leave this as a problem for the reader.

4.23.2 Am Stars

The metallic line A stars are all close binaries (Wolff 1983, Lane and Lester 1987). It isn't clear, however, what this implies. There is no compelling evidence that these stars systematically have evolved (WD) companions, and they are frequently double-lined binaries. They systematically show short periods and very slow rotation, something we have come to associate with young systems that have undergone synchronization and circularization following formation. It is interesting, however, that the abundance enhancements for one system, χ Lupi, show a pattern of increasing abundance with increasing Z (Leckrone and Wahlgren 1992). The dominant model for these stars has not changed over more than two decades. Radiatively driven differential element segregation can occur in a sufficiently quiescent atmosphere (see Vauclair 1983). The radiative acceleration is exactly the same as for a stellar wind, only in this case it can be thought of as a would-be stellar wind that fails to reach the sound speed. The force is transferred through absorption lines exactly as in the full wind case, except that this time collisions within the atmosphere act to slow the diffusing ion. Equilibrium occurs when the majority of the absorption lines saturate, leaving only the weakest transitions that cannot continue to accelerate the ion. The key is that the atmosphere must be extremely quiet. Diffusion requires that $v_D \approx g_{rad}\tau_c$ where g_{rad} is the radiative acceleration and τ_c is the mean collision time (Michaud 1980) . The typical velocities are of order 10^{-2}cm s^{-1} so that $v_{turb} \leq v_D$ is required. The equilibrium abundance is reached when $g = g_{rad}$ for the species. When the species reaches a noble gas configuration, it is assumed to stop diffusing (in other words, the radiative acceleration drops and therefore $g_{rad}\tau_c$ turns negative, leading to an abundance enhancement at small optical depth). Including thermal diffusion somewhat eases the requirements for stability, but the basic order of magnitude argument is unchanged.

In the Ap and Bp stars, there is little trouble in reaching this extreme level of stability because the strong magnetic fields guarantee that at low optical depth $v_A > a_s$ and therefore $\beta < 1$. On the other hand, no Am star is known to be strongly magnetic and severe upper limits are placed by the fact that in χ Lup, the rotational velocity is less than a few km s^{-1} (Leckrone and Wahlgren find that ξ_t, the microturbulent velocity, is 1 km s^{-1}. I mention these stars only to indicate that binarity seems to play a role, in that the star is rotating substantially slower than field single stars of the same mass and evolutionary state, but there is no reason to believe that the mass transfer between the components is responsible for the abundances, at least not in all systems. The kicker in all this is that Sirius is an Am star. And the overall abundance pattern in the Am, and especially the HgMn group, is uncomfortably like that observed in the Ba stars; see Takada-Hidai (1991) for a superb compilation.

The abundance pattern is amazingly like what one would assume for s-process or high neutron flux interior processing of the iron peak. Iron itself is always the least enhanced element, sometimes being at solar, sometimes possibly even lower than solar values. Then the abundance increases with increasing Z. If I can speculate for a moment, it is possible that there are *two* processes acting here. If the mass of the companion has a range, then the core temperatures will have a range, as will the densities. The nuclear processing that would result would not be the same from one star to another because of the mass of the companion. If this material is transferred to an otherwise stable atmosphere, subsequent diffusion could easily result (it is not possible, after all, to turn off the radiative process) but the point is that the initial conditions for the abundances would not be the same as in the standard scenario. In particular, any species that is the product uniquely of peculiar neutron irradiation, like the heaviest isotopes of Hg, would be transported in the accreted material and might then be sorted out in the subsequent diffusion. Of course, at the moment this is entirely speculation because we do not have any Ba stars that have been examined in the UV for the presence of elements like Hg (especially the rare isotope ^{204}Hg) or some of the heavier species. It would be very exciting to find such overabundances in the systems that show no Tc because this would be another indication that the mass transfer had taken place. I should add that there is no evidence, so far as I know, of Tc in the HgMn stars which, as we have seen, is not an indication that no nucleosynthesis is involved in the observed abundances. It just means that nothing has been irradiated *internally*, but it leaves open the possibility of an *extrinsic* source. An important test is to see if the Ba stars are also peculiar in their abundances of the HgMn signatures, Pt and Hg. Such a test has not been made.

Binarity may be the efficient or even proximate, but not the final, cause of the anomalies. Thus, weird abundances are not uniquely a signature of mass transfer.

4.23.3 Algols

Let us take the Algols to be representative of early Roche lobe overflow binaries. That is, the evolved star is neither on nor after the AGB stage. Yet even in these systems, the CNO abundances are systematically altered. The observed pattern is like the one that you would expect from incomplete CNO processing in the red giant. In CNO, hydrogen converts to helium through ^{12}C processing to ^{14}N. The longer the process proceeds, the closer you get to equilibrium abundances, with a He/H ratio higher than normal and an increasing N/O ratio (carbon is always more depleted than oxygen).

Detailed study of the cool components in the Algols shows that carbon is 0.25 to 0.5 dex lower than normal G and K giants (Pathasarthy *et al.* 1983; Balachandran *et al.* 1986, Yoon and Honeycutt 1992). In one really extreme system, V356 Sgr, there is virtually no carbon at all in the emission line spectrum (Polidan 1988). The UV spectra show that the absence of carbon can't merely be an excitation or ionization effect and here it is being observed independently of the photosphere of the gainer. The material may be nitrogen rich, so some of the products of nucleosynthesis may have been mixed into the mass that has been transferred between the contact component (the giant) and the main sequence gainer.

Now an interesting possible study is to use IR and UV simultaneously on these binaries to determine the complete picture of the mass transfer. In the near IR, say shortward of 1μ, the red component is observable without contamination from the hotter companion. The gainer dominates the blue and UV. Emission lines can be observed in the UV and in the mid-IR, especially some of the fine structure lines of CNO material. High resolution observations of the UV therefore provide profiles for both the photospheric matter (the absorption lines) and the surrounding transferred stuff (from the emission lines). One such study has been attempted by Tomkin (1989) for R CMa, a 1.13 day Algol system with an F main sequence gainer and a very small mass ratio, $q \approx 0.15$. The abundances on the main sequence star are essentially solar. The intrinsic abundances in the transferred material be quite depleted in carbon, by about a factor of almost 300 relative to solar, and nitrogen should be up by a factor of about 7. Oxygen should be nearly solar. The fact that the abundances look about normal, with C being perhaps slightly down, indicates that the mixing on accretion must be quite efficient and that the process is not now taking place at a very high rate. But if none of the present day mass is actually being deposited onto the gainer's photosphere, then the resulting abundances would be the same. In other words, there is still some slop in the interpretation of semi-detached system abundances.

The emission line profiles are also important because for an eclipsing system this provides a sort of map of how the material is being accreted onto the gainer (recalling the discussion of Doppler imaging for accretion disks). The IR then provides the detailed information about the loser. The degree of chemical alteration is a clue to how far the star has been stripped of its mass. As a check of evolutionary scenarios, we need to know this. So not just the way that the mass has been transferred but also limits on the amount that has been lost from the system (although not on the angular momentum carried by that matter) can be derived from the abundances in Algols.

This problem of the abundances of the stars and gas in semi-detached binaries is of *major* importance to the study of binary star evolution. Now that good detectors are available over a very wide range of wavelengths, the spectroscopy of these systems should be pursued vigorously. And with the growing number of both fully-line blanketed LTE and non-LTE model atmospheres, the abundance determinations for these relatively cool gainers becomes more tractable.

4.23.4 Blue Stragglers in Clusters

Quite a number of years ago, a cloud appeared on the horizon of stellar evolution theory in the form of the *blue stragglers.* Identified in clusters, these are defined as members that lie on the main sequence above the point of hydrogen core exhaustion for the cluster (Mermilloid 1982). Interest focused on them early in the history of close binary star work precisely because of the Algol problem. The primary in an Algol system should be moved up the main sequence because of mass accretion, perhaps above the turnoff point, a sort of "born again" status. Given the limited technology of the early 1970s, it is understandable that this suggestion fell into disfavor. There was simply no way to check on it. The alternative suggestion, first proposed by Wheeler (1979), was really quite clever. If a star can be maintained in a homogeneous state, it will remain on the main sequence longer and evolve more slowly than the rest of the cluster. Several mechanisms were thought to be responsible, including magnetic buoyant mixing and circulation induced by rapid rotation. However, the stars are frequently slow rotators and there were none of the other signs of magnetic activity connected with the stars.

With the advent of CCD spectrographs, the problem has been attacked again in earnest. Mathys (1991) has provided one of the few comprehensive studies of a single cluster, M67, an old (3.5 GYr) Population I galactic cluster. Of the 11 blue stragglers identified by Matheiu and Latham (1986), one is definitely a binary with a period of 4 days (Peterson *et al.* 1984) and three more show radial velocity variations that are consistent with binary motion but with no clearly identified periods. The rest of the stars are probably constant to within a few km s^{-1}. Two of the stars, F153 and F185, have abundances that are quite similar to both the Ba *and* the HgMn and the classical Am stars (compare Mathys (1991), fig. 43, with Takada-Hidai (1991), fig. 44), with increasing abundance for higher Z and Y, Zr, and Ba being about 10 times solar. The projected rotational velocity for F153 is very low, <10 km s^{-1} while F184 is just a bit slow for an A star (80 km s^{-1}). It should be noted that F190, the only clearly identified binary, has v_{rot} = 90km s^{-1}. None of the systems, I should add, are clearly double-lined spectroscopic binaries, a very telling point.

NGC 5466 is another cluster whose blue stragglers have been well studied. Several are binaries, three are eclipsing systems, and two seem to be contact systems (Nemec and Harris 1987, Milone *et al.* 1992). Milone *et al.* (1992) find

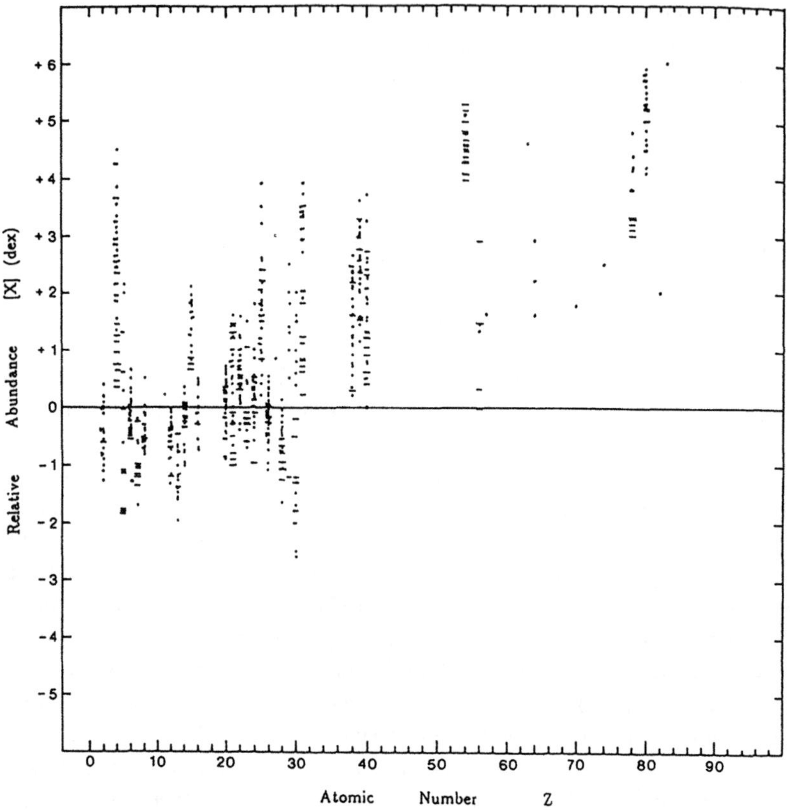

Figure 43: Abundance pattern in HgMn stars as a function of atomic number, differential abudances relative to solar (Takada-Hidai 1991).

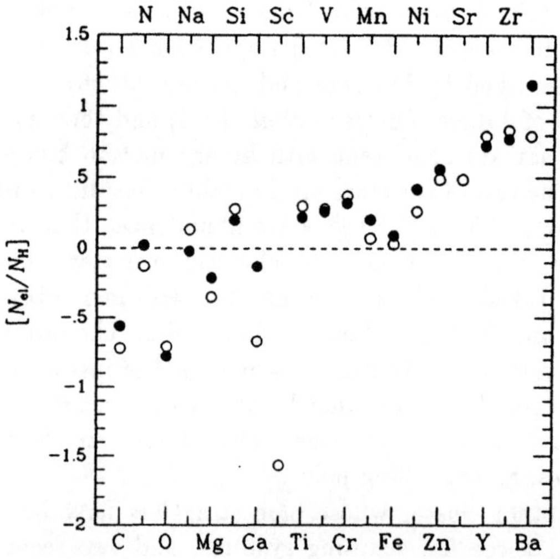

Figure 44: Abundance pattern for blue stragglers in M67 (Mathys 1991).

that as many as 7 of the nine blue stragglers in their sample show radial velocity variations in excess of the errors (about 1 km s^{-1}). It is interesting to note that the rotational velocities of the stars do not correlate with the amplitude of the variations, and that even some extremely rapid rotators (about 125 km s^{-1}) show more than 3σ radial velocity variations. They point to the possibility that some of the binaries may be the result of tidal capture since formation, rather than birth pairs.

Unfortunately, none of the cluster blue stragglers has been studied in the ultraviolet to see whether there is a white dwarf in the system (an especially hard measurement with IUE because of the scattered light below 1600Å). However, the absence of such a companion would not in itself constitute an *experimentalis crucis* because of the rapid cooling rate of a hot white dwarf. Since the timescale for evolution of the blue stragglers from the main sequence is over 1 Gyr, even with their increased mass, any compact companion would be cold and comparatively dark by that time.

4.23.5 Analysis of the Efficiency of Mass Transfer

The indications all point in the same direction: the Ba stars, Tc-deficient S stars, blue stragglers, and perhaps even some Am stars, are the products of mass transfer from an evolved red giant or AGB star. We are facing a daunting prospect here. These stars are obviously anomalous in their envelope abundances. What about the more subtle problems of inconsistent abundances that may still be lurking in the analyses of apparently single stars, based on single spectra or only a limited sample of objects and only a limited wavelength range? This is not to say that there are not truly peculiar abundances observed in *single* stars. The carbon stars appear fully capable of producing all of their abundances by purely internal means, and the isotopic tracers of mixing conform with this view. The Ap stars are usually not binaries and are clearly possessed of strong surface magnetic fields. The He strong and He weak stars fall in the same category. So we are clearly dealing with a mixed bag, for which we must include careful analysis of the effects of mass transfer.

It is now appropriate to mention one possible use of the abundance anomalies observed in post-mass transfer systems like Ba stars. The gainer can initially be taken as a Population I main sequence star, with essentially solar abundances, onto which we are dumping nuclear waste. Therefore, the abundances determined from the S stars can be used as the input for the accreted material. If a mass $\epsilon\Delta M$ is accreted, the observed abundance is weighted by the amount of mass transferred and the rate at which it is incorporated into the deep envelope of the gainer through mixing. Thus, the observed overabundance ϵ' is a mass weighted quantity $\Delta M/m$, where m is the mass fraction of the envelope into which ΔM is mixed. The fact that we see the anomaly at all, merely as a product of mass transfer, argues either that the material cannot be thoroughly stirred – most of the gained matter must reside in the upper envelope – or that most of the mass from the loser must be deposited on the gainer. The turbulent velocity necessary to thoroughly mix the envelope on a timescale comparable with the ^{99}Tc lifetime is much less than 1 km s^{-1}, of the same order as the turbulent velocities derived even for radiative envelope main sequence stars. The mass transfer would seem to be very efficient and the mass gained must be comparable with the mass of the original star. This assertion is supported by the position of the blue stragglers with respect to cluster turn-offs but has problems of its own, mainly those connected with the stability of the accretion process on such a short timescale for the gainer.

4.24 Magnetic Coupling Between Stars

As we have already discussed, magnetic coupling is a very widely invoked mechanism for sustaining angular momentum loss while maintaining synchronism. Its chief advantage is a sort of "adiabatic" action, slow enough to allow for equilibrium changes in the system yet rapid enough to be observable. It also has the advantage that it gets both stars involved in the action. That is, one of the stars can be losing mass while the other can be losing angular momentum.

In cataclysmics, this latter feature is especially important. The AM Her stars, or *polars*, are cataclysmic variables with a magnetized white dwarf (gainer) and a Roche lobe-filling red dwarf (see Frank *et al.* 1991). The magnetic field is oblique to the rotation axis and forms an accretion column. A magnetosphere is present around the gainer and here the accretion disk pressure (if a disk forms at all) is balanced by the magnetic field. To get some idea for the scaling, take the case of spherical accretion. Then the infall velocity provides the ram pressure that compresses the magnetosphere. The steady mass accretion provides the estimate of the density, $\rho_\star = \dot{M}/(4\pi R_\star^2 v_\star)$ and the ram pressure is $p = \rho_\star v_\star^2$. Since the accretion velocity is approximately the free fall speed, $p \approx G^{1/2}(4\pi)^{-1}\dot{M}M^{1/2}R_\star^{-5/2}$. Thus, the magnetospheric boundary condition given by $B_\star^2 \approx 4\pi\rho_\star v_\star^2$ yields a radius for the magnetosphere of:

$$R_\star \sim M^{10/7}\dot{M}^{-2/7}\mu^{4/7} \tag{73}$$

where μ is the magnetic moment of the accreting star (*e.g.* White and Stella 1988). The estimate is really only valid for spherical accretion. In a shear flow, the infall rate must be modified to include the effects of circulation. Then the slow inward drift produced by the accretion of matter produces a different dependence of the magnetospheric radius on the flow parameters:

$$v_A(R_\star) = v_K(R_\star) \rightarrow \frac{B_0 R_0^3}{4\pi\rho_\star^{1/2} R_\star^3} = (\frac{GM}{R_\star})^{1/2} \tag{74}$$

where B_0 is the surface field. Thus:

$$R_\star \sim \rho_\star^{-1/5} M_\star^{-1/5} B_0^{4/5}. \tag{75}$$

Now since the luminosity and mass accretion rate are related, and we can assume steady state accretion, we can eliminate the density by assuming that $L_X \sim \rho_\star R_\star$. Then $R_\star \sim L_X^{-1/4} B_0$.

The AM Her stars are the most extreme examples of magnetic control of the accretion process. In some systems, the field can completely disrupt the disk, while in others it funnels matter directly to the poles. The magnetic fields can be directly measured, through circular polarization of the emission lines – like Hα – and through the quadratic Zeeman effect. But a deeper reason for being interested in these systems is that they are certainly the most extreme examples of what must be a very general phenomenon, the interaction of flows with stellar magnetic fields in close binaries.

Now why dwell on this at such length? For one thing, magnetic disks seem to be very important in a wide variety of objects, especially evolved systems like cataclysmics (see Belvedere 1989). Another reason is that the net result of having a magnetic field on the star is that angular momentum can be very efficiently transferred at the magnetospheric boundary to the underlying star. The torquing provided by the disk, which has an enormous specific angular momentum (since the orbital velocity is assumed to be Keplerian it is the largest stable angular momentum per unit mass allowed in a gravitational field) and therefore when the stuff finally gets onto the star, it spins the companion up very efficiently. This means that if a strong tidal coupling exists between the components, the accretion will drive the change in separation of the stars through mass transfer but also through angular momentum losses. This is an observable effect that we've already discussed briefly in the section on period changes.

4.25 Twisted Disks, Stream Wagging, and Other Perversities

There are a few problems that I have not had time to cover here in any great depth, but which still need to be mentioned. Most of these involve bizarre hydrodynamical behavior in mass transferring systems. As a preamble, it's useful to recall that after a paper at the 1979 IAU Symposium No. 88 session on Algol systems, Jim Pringle remarked "May I congratulate you on being the first person today to present a mass transfer stream that actually obeys Newton's laws?" (p. 226). This is a far more common situation than you might think in close binaries. We don't really know too much about the details of the initial conditions at L_1 for many streams. Worse, there are examples of flow variations that take place on timescales of single orbits. The same is true of the disks, and even the three dimensional structures of accretion regions in relatively normal binaries seem to be variable on single orbits.

4.25.1 Twisted Disks

One of the most puzzling problems in accretion disk theory is that of the so-called *twisted disk.* A twisted disk can be thought of as an accretion disk with a non- axisymmetric vertical structure. The first evidence for such a picture was provided by Her X-1 = HZ Her, a relatively low mass X-ray binary close binary with a 1.4 day orbital period and 1.7 sec XR pulse period. The X-rays show a fairly steady 36 day modulation during which time the source seems to turn off, independent of the phase in the orbit. There is no indication that the mass accretion rate changes on this timescale, and the orbit is circular so that (unlike Cir X-1) orbit-modulated accretion is not taking place. The model has been for some time that the accretion disk is warped and that a slaved precession of the disk, tied to the accreting neutron star, produces period occultations of the gainer (Petterson 1975). The structure of the accretion disk is presumed to be frozen in, reflecting the effects of a strong surface magnetic field on the neutron star.

The X-rays impinging on the companion, which in this case is an F star, are intense enough to produce dramatic spectral variations between the illuminated and shielded hemisphere of the loser. It's likely that this is the main driver for the mass transfer since the heating is sufficient to drive the mass loss over the edge of stability and enhance Roche lobe overflow. Now if the system is really synchronous and the neutron star is precessing, what are the effects of the illumination? As the gainer precesses, it illuminates a cone that varies in colatitude on the companion. Prendergast (1991, *private communication*) has pointed out that this variation in the site of the heating may have a profound effect on the dynamics of the emergent stream. The effect of the neutron star on its companion has been studied by a number of people, notably Anderson (1976). Her X-1 also shows a long term variation, one that does not seem to be at all steady but which hasn't received as much attention. On timescales of about a decade, the optical brightness of the system drops by several magnitudes and then irregularly recovers. The feedback between the accretion, the XR emissivity of the neutron star, and the mass loss from the main sequence companion are all tied together in a nonlinear fashion. I'll speculate that there must be chaos at some level in this system, but just how it enters is still an interesting problem for the future.

An especially interesting application of twisted disks has been to the extreme ζ Aur-type binary ϵ Aur. The light curve and mass function for this system have been interpreted as the product of a very unusual system, an F0 supergiant orbited with a period of 27 years by a close binary system consisting of two A stars with a period of about 4 days. This idea comes from a persuasive analogy between ϵ Aur and the triple system HD 157978/9, the latter consisting of a wide (1170 day), eccentric ($e \approx 0.45$), orbit with a close A star binary (with a mass ratio near unity). Eggleton and Pringle then argue that this binary is the precursor to ϵ Aur. The idea is that the disk hypothesized in ϵ Aur is the product of Roche lobe overflow by the supergiant. The mass accretion onto the close binary smothers it in an enshrouding disk. Kumar (1987) has proposed that the close binary, if inclined to the plane of the long period orbit, produces an inclined disk that will perform radially dependent forced precession. The picture is quite different from the one proposed for HZ Her because magnetic fields play no role here. The structure of the disk, whether viscous or inviscid, is the result of the large effective quadrupole moment of the inclined binary. Thus, the proposed twisted disk for this system (and perhaps other very wide binaries with mass transfer) is actually a pretty straightforward consequence of the gravitational field of the accreting system. The tilted orbital planes drives the precession since the gravitational field is no longer planar and axisymmetric (well, at least it isn't planar anymore).

4.25.2 Transient Disks and Stream Wagging

Stream wagging, if it exists, is the change in the direction for the stream at the L_1 point. Bolton (1989) has discussed the effects of stream wagging in Algol, and Gillett *et al.* (1989) and Richards (1992) have reported out of eclipse variations of H_α that can be interpreted as either the presence of an intermittent disk or as stream flopping. The direction of the stream at the L_1 point seems to switch from leading to trailing. Since such motions are not possible without a non-mechanical cause, so far as we know, the variation of the streams in these systems points to some magnetic field activity. The loser in Algol is well known for its RS CVn characteristics, as I mentioned earlier, so perhaps what we are seeing is some phasing of the active regions with the position of the inner Lagrangian point. Then the transfer of matter could be affected by the fact that the sound speed is actually, in the presence of a magnetic field, the magnetoacoustic speed, which is always higher than a_s, and therefore the magnetic field steers the matter between the stars. It can be shown that the field can survive the transfer in the stream without reconnection and therefore will be carried into the accretion region of the gainer and possibly acted on by the disk material.

There are serious problems with this kind of observation. In general, it demands excellent S/N ratios . In the case of very weak profiles, the model that you choose for the photospheric profile is critically important in determining which features can be taken as real. The disk emission must be extracted from the absorption profile by a careful subtraction and there are many dangers with this technique, especially if the match to the photospheric profile is not precise. Kaitchuck (1992) has discussed evidence for transient disks in some Algol systems. Photometric methods may have more success (eclipse timing can be more sensitive than spectroscopy for some applications, like looking for very weak disks) but this is still an unresolved problem.

5 Exeunt

Truly the gods have not revealed to mortals all things from the beginning; but mortals by long seeking discover what is better.
(Xenephones of Colophon, ca. -530)

Here we end this overview of some of the problems at the boundary between observation and theory for binary stars. This is a biased survey, to be sure, but it is representative of the scope of the problems still to be examined more completely. I don't know of any other area of astrophysical research that offers so much to the observer and theorist alike. Here we have real-time physics at its best, and frequently at its most frustrating. To study binaries we require virtually the entire arsenal of astronomical observational techniques – photometry, spectroscopy, polarimetry – and access to the entire electromagnetic spectrum. And to understand them we need to use the theoretical machinery of radiative transfer, hydrodynamics, MHD, high energy processes, and even nucleosynthesis. These lectures have barely scratched the surface of the field. The depths remain to be mined, and I leave that as a final exercise for the reader. *Acknowledgements:* I wish to thank the organizers of this Saas-Fee school, Harry Nussbaumer and Astrid Orr, for their invitation to present these lectures. Their tireless work before, during, and after the school was awe inspiring. My sincere thanks to Mario Livio and Ed van den Heuvel for their comments, questions, and discussions during the school. The students were magnificent and asked many penetrating questions and were most encouraging in their interest. I want to take this opportunity to thank my office-mate, Tom Ake, for wonderful and insightful discussions over the past few years, and to acknowledge my enormous debt to Tom Bolton, who initiated me into this field. I also warmly thank Bruce Altner, Jason Aufdenberg, Doug Brown, Mike Corcoran, Doug Gies, Ed Guinan, Sally Heap, Ivan Hubeny, Yoji Kondo, Bob Matheiu, Andy Michalitsianos, Ron Polidan, Mirek Plavec, Nichol St. Louis, Werner Schmutz, George Sonneborn, Sumner Starrfield, and Manfred Vogel for valuable discussions about many aspects of this course and related research on binary stars. This work has been supported, in part, by NASA (for both IUE and GHRS research).

References

Abt, H. A., Conti, P. S., Deutsch, A. J., and Wallerstein, G. 1968, *ApJ*, **153**, 177.

Adam, J., Innes, D. E., Shaviv, G., Storzer, R., and Wehrse, R. 1991, in *Theory of Accretion Disks* (eds. Meyer, F., Duschel, W. J., Frank, J., and Meyer-Hoffmeister, E.) (Dordrecht: Kluwer) p. 403.

Adams, D. 1979, *The Hitch-hicker's Guide to the Galaxy* (NY: Harmony Books)

Ahmad, I. A. 1986, *ApJ*, **301**, 275.

Ake, T. B. III, Johnson, H. R., and Ameen, M. H. 1991, *ApJ*, **383**, 842.

Ake, T. B. III and Parsons, S. B. 1988, in *A Decade of UV Astronomy with the IUE Satellite* (ed. Rolfe, E. J.) (ESA SP-281, vol. 1), p. 213.

Alexander, M. E. 1987, *MNRAS*, **227**, 843; see also 1988, *MNRAS*, **235**, 1367.

Allen, D. A. 1984, *Proc. Astr. Soc. Australia*, **5**, 369.

Andersen, J. 1991, *A& A Rev.*, **3**, 91.

Anderson, L. 1975, *ApJ*, **202**, 232.

Arnold, C. N. and Hall, D. S. 1976, *Acta Astr.*, **26**, 91.

Baade, D. 1992, in *Evolutionary Processes in Close Binary Systems - IAU Symp. 151* (eds. Kondo, Y., Sistero, R., and Polidan, R.) (Dordrecht: Kluwer), p. 147.

Babel, J., Burki, G., Mayor, M., Waelkens, C., and Chmielewski, Y. 1989, *A&A*, **216**, 125.

Backer, D., Sallman, S., and Foster, R. 1992, *Nature*, **358**, 24.

Bagnuolo, W. G. and Gies, D. R. 1991, *ApJ*, **376**, 266.

Bagnuolo, W. G. *et al.* 1992, *AJ*, **103**, 1399.

Balachandran, S. *et al.* 1986, *MNRAS*, **219**, 479.

Baliunas, S. and Guinan, E. F. 1985, *ApJ*, **294**, 207.
Bang, M. K., Bode, M. F., Ivison, R. J., and Meaburn, J. 1992, *MNRAS*, **256**, 59P.
Barba, R. and Sahade, J. 1992, in *Evolutionary Processes in Close Binary Systems – IAU Symp. 151* (eds. Kondo, Y., Sistero, R., and Polidan, R.) (Dordrecht: Kluwer), p. 327.
Barry, D. J. 1990, in *Active Close Binaries* (ed. Ibanoglu, C.) (Dordrecht: Kluwer) p. 729
Batten, A. H. 1973, *Binary and Multiple Systems of Stars* (NY: Pergamon).
Batten, A. H. 1992, in *Evolutionary Processes in Close Binary Systems – IAU Symp. 151* (eds. Kondo, Y., Sistero, R., and Polidan, R.) (Dordrecht: Kluwer), p. 235.
Bell, S. A., Hill, G., Hilditch, R. W., Clausen, J. V., Reynolds, A. P., and Gimenez, A. 1991, *MNRAS*, **250**, 119.
Belvedere, G. (ed.) 1989, *Magnetic Fields and Accretion Disks in Astrophysics* (Dordrecht: Kluwer).
Bidelman, W. P. and Keenan, P. C. 1951, *ApJ*, **114**, 473.
Biermann, P. and Hall, D. S. 1973, *A&A*, **27**, 239.
Binnendijk, L. 1960, *Properties of Double Stars* (Phil.: Univ. Penna. Press).
Binnendijk, L. 1970, *Vistas in Astr.*, **12**, 217.
Blair, W. P., Stencel, R. E., Feibelman, W. A., and Michalitsianos, A. G. 1983, *ApJS*, **53**, 573.
Bode, M. and Evans, A. (eds.) 1987, *The 1985 Outburst of RS Ophiuchi* (Manchester: Manchester Univ. Press).
Bohannan, B. and Conti, P. 1976, *Ap. J.*, **204**, 797.
Bois, B., Lanning, H., and Mochnocki, S. W. 1988, *AJ*, **96**, 157.
Bolton, C. T. 1989, *Space Sci. Rev.*, **50**, 311.
Bond, H. E., Ciardullo, R., and Meakes, M. G. 1992, in *Evolutionary Processes in Close Binary Systems – IAU Symp. 151* (eds. Kondo, Y., Sistero, R., and Polidan, R.) (Dordrecht: Kluwer), p. 517.
Bond, H. E., Liller, W., and Mannery, E. J. 1978, *Ap. J.*, **223**, 252
Bond, H. E. and Livio, M. 1990, *Ap. J.*, **355**, 568.
Bradstreet, D. H. and Guinan, E. F. 1990, in *Active Close Binaries* (ed. Ibanoglu, C.) (Dordrecht: Kluwer) p. 467.
Brouwer, D. and Clemence, G. M. 1961, *Methods of Celestial Mechanics* (NY: Academic Press).
Brown, J. C. and McLean, I. S. 1977, *A&A*, **57**, 141.
Brown, J. C., McLean, I. S., and Emslie, A. G. 1978, *A&A*, **68**, 415.
Burgarella, D., Paresce, F., and Vogel, M. 1992, in *Nonisotropic and Variable Outflows from Stars: ASP Conference Series Vol. 22* (ed. Drissen, L., Leitherer, C., and Nota, A.) (San Francisco: ASP Press), p. 298.
Cannizzo, J. K. and Kenyon, S. J. 1992, *ApJL*, **386**, L17.
Cannizzo, J. K. and Wheeler, J. C. 1984, *ApJS*, **55**, 364.
Carpenter, K. G. 1992, in *Evolutionary Processes in Close Binary Systems – IAU Symp. 151* (eds. Kondo, Y., Sistero, R., and Polidan, R.) (Dordrecht: Kluwer), p. 51.

Carpenter, K. G. and Johansson, S. 1988, in *Advances in Astrophysics* (ed. Rolfe, E. J.) (ESA SP-281), p. 349.
Carpenter, K. G., Pesce, J. E., Stencel, R. F., Brown, A. Johansson, S., and Wing, R. F. 1988, *ApJS*, **68**, 345.
Carroll, S. M., Guinan, E. F., McCook, G. P., and Donahue, R.A. 1991, *ApJ*, **367**, 278.
Catalano, S. 1990, in *Active Close Binaries* (ed. Ibanoglu, C.) (Dordrecht: Reidel), p. 347.
Chandrasekhar, S. 1969, *Ellipsoidal Figures in Equilibrium* (NY: Dover).
Chapman, R. D. 1981, *ApJ*, **248**, 1043.
Collins, G. W. II, Truax, R. J., and Cranmer, S. R. 1991, *ApJS*, **77**, 541.
Collins, G. W. II and Smith, R. C. 1985, *MNRAS*, **213**, 519.
Colpi, M., Nannurelli, M., and Calvani, M. 1991, *MNRAS*, **253**, 55.
Corcoran, M. F. 1988, *PhD Thesis, Astronomy*, Univ. of Penn.
Corcoran, M. F. and Heap, S. R. 1990, in *Properties of Hot Luminous Stars* (ed. Garmany, C. D.) (San Francisco: ASP Press) p. 294.
Cordova, F. and Howard, I. 1989, in *Exploring the Universe with the IUE Satellite* (ed. Kondo, Y.) (Dordrecht: Kluwer), p. 395.
Cox, A. N. 1985, in *Cepheids: Theory and Observation - IAU Coll. 82* (ed. Madore, B. F.) (Cambridge: Cambridge Univ. Press).
Damour, T. and Taylor, J. H. 1991, *ApJ*, **366**, 501.
Darwin, G. H. 1910, *Collected Scientific Papers*, **3** (Cambridge: Cambridge Univ. Press), p. 436.
Deeming, T. J. 1975, *Ap. Space Sci*, **36**, 137.
Duerbeck, H. W. 1987, *Space Sci. Rev.*, **45**, 1.
Deutsch, A. J. 1970, *ApJ*, **159**, 985.
Dobias, J. J. and Plavec, M. 1985, *AJ*, **90**, 773.
Donati, J.-F., Semel, M., Rees, D. A., Taylor, K. and Robinson, R. D. 1990, *A&A*, **232**, L1.
Doremus, C. 1970, *PASP*, **82**, 745.
Drake, S. A., Simon, T., and Linsky, J. L. 1989, *ApJS*, **71**, 905.
Drake, S. A., Simon, T., and Linsky, J. L. 1992, *ApJ*, *in press.*
Drissen, L., Lamontagne, R., Moffat, A., Bastien, P., and Seguin, M. 1986, *ApJ*, **304**, 188.
Duschl, W. J. 1989, *A&A*, **225**, 105.
Dworetsky, M. M. in *Chemically Peculiar and Normal Main Sequence Stars: IAU Colloq. 138* (eds. Castelli, F. and Dworetsky, M. M.) (San Francisco: ASP Press), *in press.*
Eaton, J. A. 1988, *ApJ*, **333**, 288.
Eaton, J., Cherapatchuck, A., and Khaliullin, Kh. 1985, *ApJ*, **297**, 266.
Eaton, J. A., Kondo, Y., McClusky, G. F. and Shore, S. N. 1990, *AJ*, **100**, 799.
Eggelton, P. 1992, in *Evolutionary Processes in Close Binary Systems - IAU Symp. 151* (eds. Kondo, Y., Sistero, R., and Polidan, R.) (Dordrecht: Kluwer), p. 167.
Eggelton, P. and Pringle, J. 1985, *ApJ*, **288**, 275.

Elias, N. M. 1990, *ApJ*, **352**, 300.
Evans, N. R. 1988, *ApJS*, **66**, 343.
Evans, N. R. 1991, *ApJ*, **372**, 597.
Evans, N. R. 1992, *ApJ*, **389**, 657.
Evans, N. R. and Bolton, C. T. 1990, *ApJ*, **356**, 630.
Falk, A. E. and Wehlau, W. H. 1974, *ApJ*, **192**, 409.
Fawcett, B. C. 1988, *Atom. Data Nucl. Data Tables*, **40**, 1.
Feibelman, W. A., Bruhweiler, F. C., and Johansson, S. 1991, *ApJ*, **373**, 649.
Feldman, P. *et al.* 1978, *AJ*, **83**, 1471.
Frank, J., King, A. R., and Raine, D. J. 1991, *Accretion Power in Astrophysics* (Cambridge: Cambridge Univ. Press).
Franz, O. G., Wasserman, L. H., Nelan, E., Lattanzi, M. G., Bucciarelli, B., and Taff, L. G. 1992, *AJ*, **103**, 190.
Ferro, A. A. and Madore, B. F. 1986, *ApJ*, **302**, 767.
Garcia, M. R. 1986, *AJ*, **91**, 1400.
Gieren, W. P. 1989, *A&A*, **216**, 135.
Gezari, D. Y., Labeyrie, A., and Stachnick, R. V. 1972, *ApJ*, **173**, L1.
Gibson, D. M. 1985, in *Radio Stars* (eds. Hjellming, R. M. and Gibson, D. M.) (Dordrecht: Reidel), p. 225.
Gies, D. R. and Bolton, C. T. 1986, *ApJ*, **304**, 371.
Gies, D. R. and Bolton, C. T. 1986, *ApJS*, **61**, 419.
Gies, D. R. and Wiggs, M. S. 1991, *ApJ*, **375**, 321.
Gillett, D., Moucher, M., and North, P. 1989, *A&A*, **219**, 219.
Gimenez, A. 1990, in *Active Close Binaries* (ed. Ibanoglu, C.) (Doredrecht: Kluwer), p. 137.
Girard, T. and Willson, L. A. 1987, *A&A*, **183**, 247.
Goldman, I. and Mazeh, T. 1991, *ApJ*, **376**, 260.
Greenspan, H. P. 1968, *Theory of Rotating Fluids* (Cambridge: Cambridge Univ. Press).
Griffin, R. F. 1991, *Observatory*, **11**, 291.
Griffin, R. F. and Gunn, J. E. 1974, *ApJ*, **191**, 545; see also Griffin *et al.* 1988, *AJ*, **96**, 172.
Griffin, R. E. M., Schroeder, K.-P., and Griffin, R. F. 1992, *A& A*, **254**, 289.
Guinan, E. F. 1989, *Space Sci. Rev.*, **50**, 35.
Guinan, E. F. and Carroll, S. M. 1990, in *Active Close Binaries* (ed. Ibanoglu, C.) (Dordrecht: Kluwer), p. 7.
Guinan, E. F. and Maloney, D. 1985, *AJ*, **90**, 1519.
Gull, S. F. and Daniell, G. J. 1978, *Nature*, **272**, 686.
Hack, M. 1982, in *Advances in Ultraviolet Astronomy: Four Years of IUE Research* (eds. Kondo, Y., and Mead, J. M.) (NASA CP-2238), p. 82.
Hadjidemetriou, J. D. 1969, *Ap. Space Sci.*, **3**, 31.
Hall, D. S. 1971, *IAU Colloq. 15, Bamberg Variable Star Colloquium*, p. 217.
Hall, D. S. 1976, in *IAU Coll. 29*, p. 287.
Hall, D. S. 1986, in *Cool Stars, Stellar Systems, and the Sun* (eds. Zeilek, M. and Gibson, D.) (Berlin: Springer-Verlag) p. 40.

Hall, D. S. 1989, *Ap. Space Sci.*, **50**, 219.
Hall, D. S., and Garrison, L. M. 1969, *PASP*, **81**, 771.
Hall, D. S., and Henry, G. W. 1990, in *Active Close Binaries* (ed. Ibanoglu, C.) (Dordrecht: Reidel), p. 287.
Hall, D. S. and Kreiner, J. 1980, *Acta Astr.*, **30**, 387.
Hall, D. S., Kreiner, J., and Shore, S. N. 1980, in *Close Binary Stars: Observations and Interpretation – IAU Symp. 88* (eds. Plavec, M. J., Popper, D. M., and Ulrich, R. K.) (Dordrecht: Reidel) p. 383.
Harmanec, P. 1987, in *Physics of Be Stars* (ed. Slettebak, A., and Snow, T.) (Cambridge: Cambridge Univ. Press) p. 339.
Hatchett, S. and McCray, R. A. 1977, *ApJ*, **211**, 552.
Heap, S. R. 1981, in *The Universe at Ultraviolet Wavelengths: The First Two Years of IUE* (ed. Chapman, R. D.) (NASA CP-2171) p. 485.
Herbig, G. 1985, in *Birth and Infancy of Stars: Les Houches 41* (eds. Lucas, R., Omont, A., and Stora, R.) (Amsterdam: North-Holland).
Herschel, J. 1847, *Treatise on Astronomy* (London: Longman's).
Hiltner, W. A. 1947, *ApJ*, **166**, 481.
Hjellming, R. M. 1989, in *Galactic and Extragalactic Radio Astronomy* (eds. Verschuur, G. and Kellerman, K.) (Berlin: Springer-Verlag).
Hjellming, R. M. and Newell, R. T. 1983, *ApJ*, **275**, 704.
Hoare, M. G. and Drew, J. E. 1991, *MNRAS*, **249**, 452.
Honma, F., Matsumoto, R., Kato, S., and Abramiowicz, M. A. 1991, *PASJ*, **43**, 261.
Horne, J. H. and Baliunas, S. L. 1986, *ApJ*, **302**, 757.
Horne, K. 1985, *MNRAS*, **213**, 129.
Huang, R. and Weigert, A. 1982, *A&A*, **112**, 281.
Hubeny, I. 1990, *ApJ*, **351**, 632.
Hubeny, I. and Plavec, M. 1991, *AJ*, **102**, 1156.
Hubeny, I., Stefl, S., and Harmanec, P. 1985, *Bull Astr. Inst. Czech.*, **36**, 214.
Huenemoerder, D. P. 1988, *PASP*, **100**, 600
Huenemoerder, D. P. and Barden, C. C. 1986, *AJ*, **91**, 583:
Hutchings, J. B. and Cowley, A. 1976, *ApJ*, **206**, 490. HD 47129
Iben, I. Jr. 1991, *ApJS*, **76**, 55.
Irwin, J. B. 1962 in *Astronomical Techniques* (ed. Hiltner,W. A.) (Chicago: Univ. of Chicago Press) p. 584.
Jensen, K. S., Clausen, J. V., and Gimenez, A. 1988, *ApJS*, **74**, 331.
Johansson, S. and Carpenter, K. G. 1988, in *Advances in Ultraviolet Astronomy* (ed. Rolfe, E. J.) (ESA SP-281), p. 361.
Johnson, H. R. 1992 in *Evolutionary Processes in Close Binary Systems – IAU Symp. 151* (eds. Kondo, Y., Sistero, R., and Polidan, R.) (Dordrecht: Kluwer), p. 157.
Jorissen, A. and Mayor, M. 1988, *A& A*, **198**, 187.
Jorissen, A., Manfroid, J., and Sterken, C. 1992 in *Evolutionary Processes in Close Binary Systems – IAU Symp. 151* (eds. Kondo, Y., Sistero, R., and Polidan, R.) (Dordrecht: Kluwer), p. 411.

Kafatos, M., Michalitsianos, A. G., and Hollis, J. M. 1986, *ApJS*, **62**, 853.
Kaitchuck, R. H. 1989, *Space Sci. Rev.*, **50**, 51.
Kaitchuck, R. H. and Park, E. A. 1988, *ApJ*, **325**, 225.
Kallrath, J., Milone, E. F., and Stagg, C. R. 1992, *ApJ*, **389**, 590.
Kaluznk and Shara 1987, *ApJ*, **314**, 585.
Karovska, M., Nisenson, P., Papaliolis, C., and Boyle, R. P. 1991, *ApJL*, **374**, L51.
Keenan, P. and McNeil, R. 1976, *An Atlas of Spectra of Cooler Stars: Types G, K, M, and S* (Columbus: Ohio State University Press).
Kenyon, S. J. *et al.*, 1991, *AJ*, **101**, 637. - CI Cyg
Kenyon, S. J. 1986, *The Symbiotic Stars* (Cambridge: Cambridge Univ. Press).
Khaliullin, Kh. F. and Cherepaschuck, A. M. 1976, *Sov. Astron. - AJ*, **20**, 186.
King, A. R. and Lasota, J.-P. 1991, *ApJ*, **378**, 674.
King, A. R., Regev, O. and Wynn, G. A. 1991, *MNRAS*, **251**, 30P.
King, A. R. and Shaviv, G. 1984, *Nature*, **308**, 519.
Kippenhahn, R. 1974, in *Late Stages of Stellar Evolution* (Dordrecht: Reidel) p. 21.
Kippenhahn, R. and Thomas, H.-C. 1981, in *Fundamental Problems in the Theory of Stellar Evolution* (ed. Sugimoto, D., Lamb, D. Q., and Schramm, D. N.) (Dordrecht: Reidel) p. 237.
Kippenhahn, R. and Thomas, H.-C. 1982, *A&A*, **114**, 77.
Kley, W. 1991, *A&A*, **247**, 95.
Koch, R. H. and Perry, P. M. 1974, *AJ*, **79**, 379.
Koch, R. H., Elias, N. M., Corcoran, M. F., and Holenstein, B. D. 1989, *Space Sci. Rev.*, **50**, 63.
Koch, R. H. and Pfeiffer, R. J. 1990, in *Active Close Binaries* (ed. Ibanoglu, C.), (Dordrecht: Kluwer).
Koenigsberger, G. and Auer, L. 1985, *ApJ*, **297**, 255: see also Koenigsberger, G. and Auer, L. 1992, in Nonisotropic and Variable Outflows from Stars: ASP Conference Series Vol. 22 (ed. Drissen, L., Leitherer, C., and Nota, A.) (San Francisco: ASP Press), p. 239.
Kopal, Z. 1959, *Close Binary Stars* (NY: Wiley).
Kopal, Z. 1965, *Adv. Astr.*, **3**, 89.
Kruszewski, A. 1974, in *Planets, Stars, and Nebulae Studied with Photopolarimetry* (ed. Gehrels, T.) (Tucson: Univ. Arizona Press) p. 845.
Kudritzki, R. P. and Hummer, D. G. 1990, *ARAA*, **28**, 303.
Kumar, S. 1987, *MNRAS*, **225**, 823.
Kumar, S. 1990, *MNRAS*, **245**, 670.
Lafler, J. and Kinman, T. D. 1965, *ApJS*, **11**, 216.
Landau, L. D. and Lifshitz, E. M. 1971, *Classical Theory of Fields* (NY: Pergamon).
Lane, M. C. and Lester, J. B. 1987, *ApJS*, **65**, 137.
Lanning, H. and Pesch, P. 1981, *ApJ*, **244**, 280.
Larsson-Leander, G. 1970, *Vistas in Astr.*, **12**, 183.
Latham, D. W., *et al.* 1988, *AJ*, **96**, 1908.

Lattanzio, J. 1989, in *Evolution of Peculiar Red Giants* (ed. Johnson, H. R. and Zuckerman, B.) (Cambridge: Cambridge Univ. Press), p. 161.
Lecar, M., Wheeler, J. C., and McKee, C. F. 1976, *ApJ*, **205**, 556.: tidal circularization of Her X-1
Leckrone, D. L. and Wahlgren, G. M. 1992, *private communication.*
Lee, U., Saio, H., and Osaki, Y. 1991, *MNRAS*, **250**, 432.
Leung, K-C., and Schneider, D. P. 1978, *ApJ*, **224**, 565: V729 Cyg
Leung, K-C., and Schneider, D. P. 1978, *ApJ*, **222**, 924: UW CMa
Linsky, J. L. 1989, in *Multiwavelength Astronomy* (ed. Cordova, F.) (Cambridge: Cambridge Univ. Press), p. 49.
Lu, P. K. 1991, *AJ*, **101**, 2229.
Lubow, S. H. and Shu, F. H. 1975, *ApJ*, **198**, 383.
Lubow, S. H. and Shu, F. H. 1981, *ARAA*, **19**, 277.
Lubow, S. H. 1991, *ApJ*, **381**, 259, 268.
Luck, R. E. and Bond, H. E. 1991, *ApJS*, **77**, 515.
Lucy, L. B. 1967, *ZAp*, **65**, 89.
Lucy, L. B. and Wilson, R. E. 1979, *ApJ*, **231**, 502.
Lucy, L. B. and Sweeny, M. A. 1971, *AJ*, **76**, 547.
Luo, D., McCray, R., and Mac Low, M.-M. 1990, *ApJ*, **362**, 267.
Lyons, R. W., Beaver, E. A., Rosenblatt, E. I., Weiland, J. L., and Shore, S. N. 1991, *BAAS*, **23**, 1415.
McCarthy, J. K. and Ramsey, L. W 1984, *ApJ*, **283**, 200.
McClure, R. D. 1989, in *Evolution of Peculiar Red Giants* (ed. Johnson, H. and Zuckerman, B.) (Cambridge: Cambridge Univ. Press) p. 196.
McClure, R. D., Fletcher, J. M., and Nemec, J. M. 1980, *ApJL*, **238**, L35.
McClure, R. D. and Woodworth, A. W. 1990, *ApJ*, **352**, 709.
McClusky, G. E. and Sahade, J. 1989, in *Exploring the Universe with the IUE Satellite* (ed. Kondo, Y.) (Dordrecht: Kluwer), 427.
McClusky, G. E., Kondo, Y., and Olson, E. C. 1988, *ApJ*, **332**, 1019.
Madore, B. F. 1977, *MNRAS*, **178**, 505.
Marsh, T. R. and Horne, K. 1988, *MNRAS*, **235**, 269.
Matheiu, R. D., Adams, F. C., and Latham, D. W. 1991, *AJ*, **101**, 2185.
Matheiu, R. D., Duquennoy, A., Latham, D. W., Mayor, M., Mazeh, T., and Mermilliod, J.-C. 1992, in *Binary Stars as Tracers of Stellar Evolution* (eds. Duquennoy, A. and Mayor, M.) (Cambridge: Cambridge Univ. Press), *in press.*
Mathieu, R. D. and Latham, D. W. 1986, *AJ*, **92**, 1364.
Mathis, J. S. and Odell, A. P. 1973, *ApJ*, **180**, 517.
Mathys, G. 1991, *A&A*, **245**, 467.
Mazeh, T., Latham, D. W., Stefanik, R. P., Torres, G., and Wasserman, E. 1990, in *Active Close Binaries* (ed. Ibanoglu, C.) (Dordrecht: Kluwer), p. 267.
Megessier, C., Khokhlova, V. I., and Ryabchikov, T. A. 1979, *A&A*, **71**, 295.
Mermilloid, J.-C. 1982, *A&A*, **109**, 37.
Merrill, P. 1952, *ApJ*, **116**, 21.
Mestel, L. and Spruit H. C. 1987, *MNRAS*, **226**, 57.

Michalitsianos, A. G. and Kafatos, M. 1988, in *The Symbiotic Phenomenon – IAU Colloqu. 103* (ed. Micholajewska, J. *et al.*) (Dordrecht: Kluwer), p. 235.

Michalitsianos, A. G., Kafatos, M., Fahey, R. P., Viotti, R., Cassatella, A. and Altamore, A. 1988, *ApJ*, **381**, 477.

Michaud, G. 1980, *ApJ*, **85**, 589.

Mikolajewska, J., Friedjung, M., Kenyon, S., and Viotti, R. (eds.) 1988, *The Symbiotic Phenomenon – IAU Colloq. 103* (Dordrecht: Kluwer).

Milone, E. F., Stagg, C. R., and Kallrath, J. 1992, in *Evolutionary Processes in Close Binary Systems – IAU Symp. 151* (eds. Kondo, Y., Sistero, R., and Polidan, R.) (Dordrecht: Kluwer), p. 475.

Moffat, A., Firmani, F., McLean, I. S., and Segewiss, W. 1982, in *Wolf-Rayet Stars: Observations, Physics, and Evolution – IAU Symp. 99* (eds. DeLoore, C. and Willis, A.) (Dordrecht: Reidel), p. 577.

Moffat, A. F. J. and Robert, C. 1992, in *Nonisotropic and Variable Outflows from Stars: ASP Conference Series Vol. 22* (ed. Drissen, L., Leitherer, C., and Nota, A.) (San Francisco: ASP Press), p. 203.

Morgan, J. G. and Eggelton, P. P. 1979, *MNRAS*, **187**, 661: discovery statistics for RS CVn-type stars.

Mullan, D. J., Sion, E., Bruhweiler, F., and Carpenter, K. 1989, *ApJL*, **339**, L33.

Mürset, U. and Nussbaumer, H. 1992, in *Evolutionary Processes in Close Binary Systems – IAU Symp. 151* (eds. Kondo, Y., Sistero, R., and Polidan, R.) (Dordrecht: Kluwer), p. 435.

Müyesseroglu, Z., Demircan, O., Derman, E., and Selam, S. 1990, in *Active Close Binaries* (ed. Ibanoglu, C.) (Dordrecht: Kluwer), p. 277.

Nather, R. *et al.* 1981, *ApJ*, **244**, 269.

Neff, J. E., Walter, F. M., Rodonò, M., and Linsky, J. L. 1989, *A&A*, **215**, 79.

Nemec, J. M. and Harris, H. C. 1987, *ApJ*, **316**, 172.

Neutsch, W., Schmidt, H., and Seggewiss, W. 1981, *Acta Astr.*, **31**, 197.

Novikov, I. and Thorne, K. 1972, in *Black Holes* (ed. deWitt, C. and deWitt, B.) (NY: Gordon and Breach).

Noyes, R. W., Weiss, N. O., and Vaughan, A. H. 1984, *ApJ*, **287**, 769.

Nussbaumer, H. 1990, in *Evolution in Astrophysics: IUE Astronomy in the Era of New Space Missions* (ed. Rolfe, E. J.) (ESA SP-310), p. 87.

Nussbaumer, H. and Schild, H. 1981, *A&A*, **101**, 118.

Nussbaumer, H., Schild, H., Schmid, H. M., and Vogel, M. 1988, *A&A*, **198**, 179.

Nussbaumer, H. and Stencel, R. 1989, in *Exploring the Universe with the IUE Satellite* (ed. Kondo, Y.) (dordrecht: Kluwer) p. 203.

Nussbaumer, H. and Vogel, M. 1987, *A&A*, **182**, 51.

Nussbaumer, H. and Vogel, M. 1991, *A&A*, **248**, 81.

Olson, E. C. 1982, *ApJ*, **259**, 702.

Parker, E. 1979, *Cosmical Magnetic Fields* (New York: Oxford Univ. Press).

Parsons, S. B. 1981, *ApJ*, **247**, 560.

Parsons, S. B., Dempsey, R. G., and Bopp, B. W. 1992, in *Evolutionary Processes in Close Binary Systems – IAU Symp. 151* (eds. Kondo, Y., Sistero, R., and Polidan, R.) (Dordrecht: Kluwer), p. 341.

Parsons, Holm, A. V., and Kondo, Y. 1983, *ApJL*, **264**, L19.
Pathasarthy, M., Lambert, D. L., and Tomkin, J. 1983, *MNRAS*, **203**, 1063.
Payne-Gaposchkin, C. 1957, *The Galactic Novae* (Amsterdam: North-Holland).
Peale, S. J. 1976, *Ann. Rev. Astron. Astrophys.*, **14**, 215.
Penston, M. V. and Allen, D. A. 1985, *MNRAS*, **212**, 939.
Penston, M. V., Benvenutti, P., Cassatella, A., Heck, A., Selvelli, P., Macchetto, F., Ponz, D., Jordan, C., Cramer, N., Ruefner, F., and Manfroid, J., 1983, *MNRAS*, **202**, 833.
Peters, G. J. 1976, in *Be and Shell Stars* (ed. Slettebak, A.) (Dordrecht: Reidel) p. 417.
Peters, G. J. and Polidan, R. S. 1984, *ApJ*, **283**, 745.
Peterson, D. M. and Solensky, R. 1988, *ApJ*, **333**, 256.
Peterson, D. M., Baron, R., Dunham, E. W., Mink, D., Aldering, G., Klavetter, J., and organ, R. 1989, *AJ*, **98**, 2156.
Pettersen, J. 1975, *ApJL*, **201**, L61.
Pigulski, A. and Boratyn, D. A. 1992, *A&A*, **253**, 178.
Plavec, M. 1980, in *Close Binary Stars: Observations and Interpretation – IAU Symp. 88* (eds. Plavec, M. J., Popper, D. M., and Ulrich, R. K.) (Dordrecht: Reidel) p. 251.
Plavec, M. 1989, *Space Sci. Rev.*, **50**, 95.
Plavec, M. J. 1990, in *Active Close Binaries* (ed. Ibanoglu, C.) (Dordrecht: Kluwer), p. 49.
Plavec, M. 1992, in *Nonisotropic and Variable Mass Outflows from Stars: ASP Conf. Series 22* (eds. Drissen, L., Leitherer, C., and Nota, A.) (San Francisco: ASP Press) p. 47.
Polidan, R. S. 1988, in *A Decade of UV Astronomy with the IUE Satellite* (ed. Rolfe, E. J.) (ESA SP-281, vol. 1), p. 205.
Polidan, R. and Shore, S. N. 1992, *BAAS*, *in press.*
Popper, D. M. 1980, *ARAA*, **18**, 45.
Popper, D. M. 1987, *ApJ*, **313**, L81.
Popper, D. M. and Hill, G. 1991, *AJ*, **101**, 600.
Popper, D. M. and Plavec, M. 1976, *ApJ*, **205**, 462.
Preston, G. W. 1974, *AARA*, **12**, 257.
Pringle, J. E. 1981, *ARAA*, **19**, 137.
Pringle, J. E. and Savonije, G. J. 1979, *MNRAS*, **187**, 777.
Prilutskii, O. F., and Usov, V. V. 1976, *Sov. Astr. - AJ*, **20**, 2.
Querci, F. 1986, in *The M-type Stars* (eds. Johnson, H. R. and Querci, F.) (NASA SP-492), p. 9.
Quirrenbach, A., Mozurkewich, D., Armstrong, J. T., Johnston, K. J., Colavita, M. M., and Shao, M. 1992, *A&A*, **259**, L19.
Regev, O. 1983, *A&A*, **126**, 146.
Regev, O. 1991, in *Structure and Emission Properties of Accretion Disks* (eds. Bertout, C., Collin, S., Lasota, J.-P., and Van, J. Tran Thanh) (Paris: Editions Frontieres) p. 311.

Reimers, D. 1989, in *FGK Stars and T Tauri Stars* (eds. Cram, L. E. and Kuhi, L. V.) (NASA SP-502) p. 53.
Richards, M. T. 1990, *ApJ*, **350**, 372.
Richards, M. T. 1992, *ApJ*, **387**, 329.
Richards, M. T., Bolton, C. T., and Mochnochi, S. W. 1989, *Space Sci. Rev.*, **50**, 358.
Richards, M. T., Mochnocki, S. W., and Bolton, C. T. 1988, *AJ*, **96**, 326.
Roberts, D. H., Lehar, J., and Dreher, J. W. 1987, *AJ*, **93**, 968.
Rodonò, M. 1992, in *Evolutionary Processes in Close Binary Systems - IAU Symp. 151* (eds. Kondo, Y., Sistero, R., and Polidan, R.) (Dordrecht: Kluwer), p. 71.
Rucinski, S. 1985, in *Interacting Binaries* (eds. Eggleton, P. P. and Pringle, J. E.) (Dordrecht: Reidel), p. 13.
Rucinski, S. M., Baade, D., Lu, W. X., and Udalski, A. 1992, *AJ*, **103**, 573.
Sahade, J. and Brandi, E. 1991, *ApJ*, **379**, 706.
Sahade, J., Huang, S.-S., Struve, O., and Zbergs, V. 1959, *Trans. Amer. Phil. Soc. NS*, **49**, 1.
Savonije, G. J. and Papaliouzou, J. 1984, in *Interacting Binaries* (eds. Eggleton, P. E. and Pringle, J. E.) (Dordrecht: Reidel).
Scargle, J. D. 1982, *ApJ*, **263**, 835.
Scharlemann, E. T. 1978, *ApJ*, **219**, 617.
Scharlemann, E. T. 1982, *ApJ*, **253**, 298.
Schneider, D. P., and Leung, K-C., 1978, *ApJ*, **223**, 202: AO Cas
Schneller, H. 1952, *Geschichte und Liter. des Lichtwechsels der Veränderl. Sterne: 1916 - 1950* (2nd Ed., 2 vols.) (Berlin: Akademie Verlag).
Schulte-Ladbeck, R. E. 1989, *AJ*, **97**, 1471.
Schulte-Ladbeck, R., Meade, M. R., and Hillier, D. J. 1992, in *Variable and Non-Isotropic Outflows from Stars* (eds. Drissen, L., Leitherer, C., and Nota, A.) (San Francisco: ASP Press), p. 118.
Seaquist, E. R. and Taylor, A. R. 1992, *ApJ*, **387**, 624.
Semeniuk, I and Paczynski, B. 1968, *Acta Astr.*, **18**, 33.
Shlichting, T. 1965, *Boundary Layer Theory* (NY: McGraw-Hill).
Shakura, N. and Sunyaev, R. A. 1973, *A&A*, **23**, 1.
Shao, M. *et al.* 1988, *A& A*, **193**, 357.
Shaw, J. S. 1990, in *Active Close Binaries* (ed. Ibanoglu, C.) (Dordrecht: Kluwer), p. 241.
Shore, S. N. 1988, in *A Decade of UV Astronomy with the IUE Satellite* (ed. Rolfe, E. J.) (ESA SP-281, vol. 1), p. 67.
Shore, S. N. 1992, *An Introduction to Astrophysical Hydrodynamics* (San Diego: Academic Press), pp. 452.
Shore, S. N. and Aufdenberg, J. 1991, *BAAS*, **23**, 942
Shore, S. N. and Brown, D. N. 1988, *ApJ*, **334**, 1021.
Shore, S. N. and Corcoran, M. F. 1992, in *Evolutionary Processes in Close Binary Systems - IAU Symp. 151* (eds. Kondo, Y., Sistero, R., and Polidan, R.) (Dordrecht: Kluwer), p. 359.

Shore, S. N. and Hall, D. S. 1980, in *Close Binaries: Theory and Observation – IAU Symp. 88* (ed. Popper, Plavec, and Ulrich) (Dordrecht: Reidel), p. 389.
Shore, S. N. and King, A. R. 1987, *A&A*, **154**, 263.
Shore, S. N., Sanduleak, N., and Allen, D. A. 1987,*A&A*, **179**, 59.
Smith, R. C. and Smith, D. H. 1981, *MNRAS*, **194**, 583.
Smith, R. C. and Worley, R. 1974, *MNRAS*, **167**, 199.
Smith, V. V. 1992, in *Evolutionary Processes in Close Binary Systems – IAU Symp. 151* (eds. Kondo, Y., Sistero, R., and Polidan, R.) (Dordrecht: Kluwer), p. 103.
Spoelstra, T. A. T., Taylor, A. R., Pooley, G. G., Evans, A., and Albinson, J. S. 1987, *MNRAS*, **224**, 791.
Spruit, H. C. 1987, *A&A*, **184**, 173.
Spruit, H. C., Matsuda, T., Inoue, M. and Sawada, K. 1987, *MNRAS*, **229**, 517.
Stahler, S. W. 1983, *ApJ*, **274**, 822.
Starrfield, S. and Snijders, M. A. J. 1989, in *Exploring the Universe with the IUE Satellite* (ed. Kondo, Y.) (Dordrecht: Kluwer), p. 377.
Stassmeier, K. G., Hall, D. S., Zeilik, M., Nelson, E., Eker, Z., and Fekel, F. 1988, *A& A Suppl*, **72**, 291.
Stellingwerf, R. F. 1978, *ApJ*, **224**, 953.
Stencel, R. (ed.) 1985, *The 1982-1984 Eclipse of Epsilon Aurigae* (NASA CP-2384).
Stencel, R. E., Brugel, E. W., Kenyon, S. J., Bickert, K. F., Fleming, T. A., Schmitt, J. H. M. S., *et al.* 1992, in *Cool Stars, Stellar Systems, and the Sun* (San Francisco: ASP Press), *in press.*
Stevens, I. R., Blondin, J. M., and Pollack, A. M. T. 1992, *ApJ*, **386**, 265.
Stickland, D. J. *et al.*, 1984, *A&A*, **143**, 45.
Stockton, R. A. and Fekel, F. C. 1992, *MNRAS*, **256**, 575.
Szabados, L. 1985 in *Cepheids: Theory and Observation – IAU Coll. 82* (ed. Madore, B. F.) (Cambridge: Cambridge Univ. Press), p. 75.
Takada-Hidai, K. 1991, in *Evoltuion of Stars: The Photospheric Abundance Connection – IAU Symp. 145* (eds. Michaud, G. and Tutukov, A.) (Dordrecht: Kluwer) p. 137.
Tassoul, J.-L. 1978, *Theory of Rotating Stars* (Princeton: Princeton Univ. Press).
Tassoul, J.-L. 1987, *ApJ*, **322**, 856; 1988, *ApJL*, **324**, L71; 1990, *ApJ*, **358**, 196.
Tassoul, J.-L. and Tassoul, M. 1992, *ApJ*, **395**, 259.
Thackeray, A. D. 1977, *Mem. RAS*, **83**, 1.
Thorne, K. 1980, *Rev. Mod. Phys.*, **52**, 285.
Tomkin, J. 1989, *Space Sci. Rev*, **50**, 245.
Tomkin, J., Lambert, D. L., Edvardsson, B., Gustofsson, B., and Nissen, P. E. 1989, *A&A*, **219**, L15.
Torres, G., *et al.* 1992, in *Evolutionary Processes in Close Binary Systems – IAU Symp. 151* (eds. Kondo, Y., Sistero, R., and Polidan, R.) (Dordrecht: Kluwer), p. 491.

Tutukov, A. V. 1991, in *Evolution of Stars: The Photospheric Abundance Connection – IAU Symp. 145* (eds. Michaud, G. and Tutukov, A.) (Dordrecht: Kluwer) p. 351.
Umana, G., Catalano, S., and Rodono, M. 1991, *A&A*, **249**, 217.
Usov, V. V. 1992, *Ap.J.*, **389**, 635.
Vauclair, S. 1983, in *Astrophysical Processes in Upper main Sequence Stars: 13th Saas-Fee Course* (Geneva: SSAS) p. 167.
Vilhu, O. 1992, in *Evolutionary Processes in Close Binary Systems – IAU Symp. 151* (eds. Kondo, Y., Sistero, R., and Polidan, R.) (Dordrecht: Kluwer), p. 61.
Vogel, M. 1990, *PhD Thesis*, ETH/Zurich.
Vogel, M. 1990, in *Evolution in Astrophysics: IUE Astronomy in the Era of New Space Missions* (ed. Rolfe, E.) (ESA SP-310), p. 393.
Vogel, M., Nussbaumer, H., and Monier, R. 1992, *A&A, in press.*
Vogt, S. 1981, *ApJ*, **250**, 327.
Wallerstein, G., Willson, L. A., Salzer, J., and Brugel, E. 1984, *A&A*, **133**, 137.
Walter, F. M. and Basri, G. 1982, *ApJ*, **260**, 735.
Weiland, J. 1984, *PhD Thesis, Astronomy: UCLA.*
Wheeler, J. C. 1979, *Comm. Ap.*, **8**, 133.
Whitelock, P. A. and Munari, U. 1992, *A&A*, **255**, 171.
White, N. E. and Stella, L. 1988, *MNRAS*, **231**, 325.
White, N. E. *et al.* 1986, *ApJ*, **301**, 26.
Whitehurst, R. and King, A. 1991, *MNRAS*, **249**, 25.
Williams, R. E. 1980, *ApJ*, **235**, 939.
Wilson, R. E. 1989, *Space Sci. Rev.*, **50**, 235.
Wilson, R. E. 1990, *ApJ*, **356**, 630.
Wilson, R. E. and Devinney, E. 1971, *ApJ*, **166**, 605.
Wolff, S. C. 1983, *The A-type Stars: Problems and Perspectives* (NASA SP-463).
Wolszczan, A. and Frail, D. A. 1992, *Nature*, **355**, 145.
Wright, K. O. 1970, *Vistas in Astr.*, **12**, 147.
Wu, D. and Wang, D. 1991, *MNRAS*, **250**, 760.
Yoon, T. S. and Honeycutt, K. 1992, in *Evolutionary Processes in Close Binary Systems – IAU Symp. 151* (eds. Kondo, Y., Sistero, R., and Polidan, R.) (Dordrecht: Kluwer), p. 355.
Zahn, J.-P. 1977, *A&A*, **57**, 383.
Zahn, J.-P. and Bouchet, L. 1989, *A&A*, **223**, 112.

Topics in the Theory of Cataclysmic Variables and X-Ray Binaries

Mario Livio

Space Telescope Science Institute, 3700 San Martin Drive, Baltimore, MD 21218 and Dept. of Physics, Technion, Haifa 32000, Israel

1 Introduction

In the present series of lectures I discuss various aspects of the accretion process onto compact objects. I will concentrate mainly on theoretical topics associated with cataclysmic variables (CVs). This class of objects, which consist typically of a white dwarf and a low mass main sequence star in a short period binary system, offers probably the best opportunity for attempts to understand the accretion process in close binaries. In particular, CVs allow observers (in many occasions) to examine separately the accretion disk (or even specific parts of it) and the binary components. The understanding of physical processes associated with accretion disks that has resulted from the study of CVs and low mass x-ray binaries is also extremely important for processes related to star formation and Active Galactic Nuclei. The fact that the timescales involved in CVs range typically from seconds (for processes related to the white dwarf) to months (*e.g.,* the recurrence time of dwarf nova eruptions) has made them relatively easy targets for multi-wavelength observations. Furthermore, the fact that CVs experience eruptions and outbursts that are quite dramatic, helped to make them popular targets even among amateur astronomers. Consequently, our understanding of the physical processes occurring in CVs and their accretion disks significantly exceeds our understanding of other systems, which are also thought to be powered by accretion.

I will make no attempt to cover all the theoretical topics that are associated with CV physics. I refer the interested reader to a few excellent references that discuss some of the aspects not covered in this series of lectures [1]–[4]. I also apologize in advance to the more specialist readers for the style of this manuscript, which tends to be more pedagogical, rather than conforming to the usual style of review articles. One of the consequences of this style is the fact that I do not cite all of the relevant references (this is true in particular for references related to observational material, since I discuss mainly theoretical aspects). Rather, I give some of the more basic references and the reader is directed to references found therein.

While I tried to make these lectures as self-contained as possible, I have to admit that the level of knowledge that is required is not uniform throughout the entire manuscript. This is partly a consequence of the desire to remain as concise as possible.

In sections 2–7, I discuss the physics of mass transfer, the accretion process and accretion disks. Section 8 discusses dwarf nova eruptions. Section 9 deals with some aspects of common envelope evolution. Section 10 presents several topics in nova theory and section 11 discusses topics related to the evolution of CVs and explores the possible connections of CVs to accretion induced collapses and Type I supernovae.

Throughout the lectures, I have tried to give a part which is more basic and represents more or less established ideas and a part which discusses more controversial ideas. In the latter part, I have sometimes included my own opinions and speculations. These should always be taken with a grain of salt, since they are intended to be understood as thought provoking tools rather than to represent eternal truths.

2 Roche Equipotentials

Let us examine a binary system in which two stars of masses M_1 and M_2 rotate about their center of mass in a circular orbit. In the Roche approximation it is further assumed that the gravitational field generated by the two stars is like that of two point masses. This approximation is quite good in the case of CVs (especially for the white dwarf component, since its radius is very small compared to the binary separation). It is also assumed that the stars corotate with the binary (see below). Under these conditions, the potential (gravitational plus centrifugal) can be written (in a coordinate system which corotates with the binary, with the origin at the center of mass) as

$$\phi = -\frac{GM_1}{|\boldsymbol{r}-\boldsymbol{r}_1|} - \frac{GM_2}{|\boldsymbol{r}-\boldsymbol{r}_2|} - \frac{1}{2}(\boldsymbol{\Omega}_B \wedge \boldsymbol{r})^2. \tag{1}$$

Here $\boldsymbol{r}_1$ and $\boldsymbol{r}_2$ denote the position vectors of the two star centers and $\boldsymbol{\Omega}_B$ is the angular velocity vector of the binary system. If we now denote by $\boldsymbol{v}$ the velocity of a fluid element relative to the rotating frame, then the equation of motion (the Euler equation) for this element reads

$$\frac{\partial \boldsymbol{v}}{\partial t} + (\boldsymbol{v}\cdot\nabla)\boldsymbol{v} = -\nabla\phi - \frac{1}{\rho}\nabla p - 2\boldsymbol{\Omega}_B \wedge \boldsymbol{v}, \tag{2}$$

where ϕ is given by eq. (1). The last term on the RHS of eq. (2) is the Coriolis acceleration.

The angular velocity of the binary Ω_B is related to the separation between the binary components, a, by Kepler's law

$$\Omega_B = \left[\frac{G(M_1+M_2)}{a^3}\right]^{1/2} . \tag{3}$$

In CVs, the separation between the two stars is sufficiently small, so that the secondary star (typically a low mass main sequence star) is forced to corotate with the orbit by tidal forces. This is a consequence of the following process: in the absence of corotation, the tidal forces vary in strength, inducing pulsations which are damped by some form of viscosity. The energy thus dissipated causes the secondary to corotate.

Corotation implies that we can set $\boldsymbol{v} = 0$ in eq. (2), for a fluid element of the secondary star. The surface of the secondary is determined by the condition $p = 0$, and thus, from eq. (2), by a surface satisfying $\phi = \text{const}$. It therefore becomes important to determine the shapes of the Roche equipotentials, which are given by $\phi = \text{const}$.

It is more convenient to work in a Cartesian coordinate system, with the origin at the center of the primary (with the z axis in the direction of the axis of rotation and the x-axis along the line joining the two star centers). In these coordinates the equipotentials take the form

$$-\frac{GM_1}{(x^2+y^2+z^2)^{1/2}} - \frac{GM_2}{[(x-a)^2+y^2+z^2]^{1/2}} - \frac{1}{2}\Omega_B^2[(x-\mu a)^2+y^2] = \text{const.} \tag{4}$$

where $\mu = M_2/(M_1+M_2)$. From eq. (4) it is clear that the shape of the equipotentials is determined by the mass ratio $q = M_2/M_1$, while the scale is determined by the binary separation. An example of the equipotentials in the orbital plane ($z = 0$) is given in Fig. 1, while a three-dimensional plot of the equipotential surfaces is presented in Fig. 2. It is easy to understand the qualitative properties of the equipotentials. Close to each star, the potential is dominated by the gravitational potential of that star, thus the surfaces are almost spherical. As we move farther from a stellar center, two effects start to become important: (i) the tidal effect of the companion, which causes an elongation in the direction of the companion (the x-axis) and (ii) flattening by the centrifugal force. Consequently, the surfaces are distorted in a way that their largest dimension is along the line of centers.

The most important equipotential surface, from the point of view of binary stars evolution, is the critical surface with a figure-of-eight cross section which passes through the inner Lagrangian point L_1. The two cusped volumes which are enclosed by this critical surface are called the Roche lobes of the respective stars. A systems in which both components are within their Roche lobes is called *detached*. If one component fills its Roche lobe the system is called *semi-detached* and when both components overflow their Roche lobes the binary is called a *contact* binary.

All of the observational data on CVs suggests that they are all semi-detached systems, with the secondary filling its Roche lobe. The fact that the secondary stars in CVs are corotating with the orbital motion is partly a consequence of the fact that they are lobe filling, because for such stars the tidal effects (which bring about corotation) become very strong.

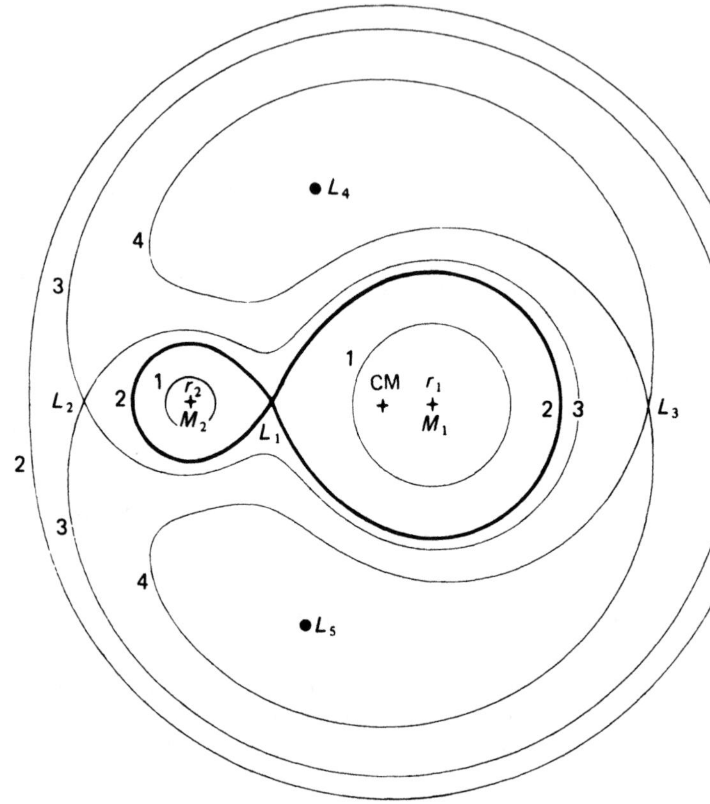

Fig. 1. Sections in the orbital plane of Roche equipotentials, for a binary system with mass ratio $q = M_1/M_2 = 0.2$. The Lagrangian points are indicated. The equipotentials are labeled in order of increasing ϕ (from [2]).

A suitable measure of the size of the Roche lobe is given by the radius of a sphere which has the same volume as the lobe. This can be approximated [5] by (for all values of q)

$$\frac{R_{RL}}{a} = \frac{0.49q^{2/3}}{0.6q^{2/3} + \ln(1 + q^{1/3})}. \tag{5}$$

Other approximations often used are [6]

$$\frac{R_{RL}}{a} = \begin{cases} 0.38 + 0.20 \log q & 0.5 \le q < 20 \\ 0.462 \left(\dfrac{q}{1+q}\right)^{1/3} & 0 < q < 0.5 \end{cases} . \tag{6}$$

It is interesting to note that for $q < 0.5$, eq. (6) implies (together with eq. (3)) that for such lobe-filling secondaries, $P\bar{\rho}^{1/2}$ is a constant (where P is the orbital period and $\bar{\rho}$ is the mean secondary density $\bar{\rho} = \frac{M_2}{\frac{4\pi}{3}R_2^3}$).

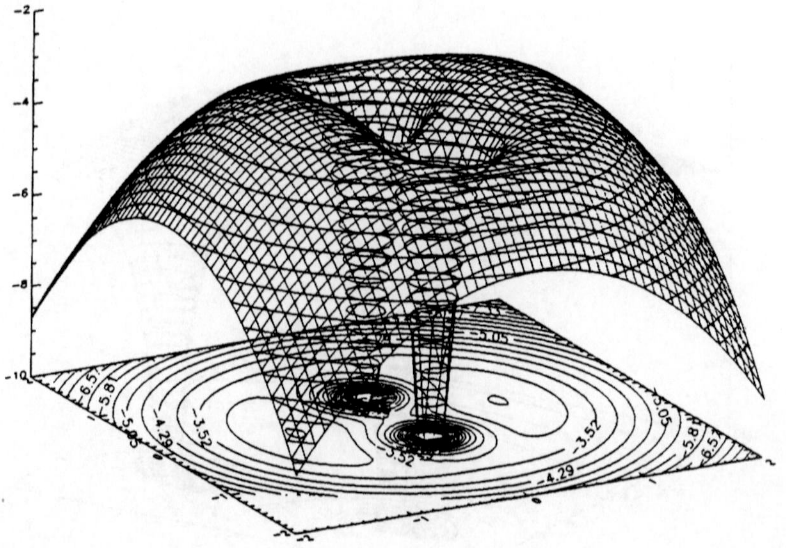

Fig. 2. A three-dimensional representation of the mapping of the equatorial plane into the Jacobi integral of motion for equal masses (from Kluźniak [5]).

There exists observational evidence that in CVs with orbital periods of less than about 9 hours, the secondary stars are essentially unevolved low mass main sequence stars [7]. Such stars obey the approximate mass-radius relation [8]

$$\frac{R}{R_\odot} = \frac{M}{M_\odot}. \tag{7}$$

Using eqs. (3), (7) and eq. (6) for $q < 0.5$, it is easy to show that an approximate relation between the secondary mass and the orbital period can be obtained

$$\frac{M_2}{M_\odot} \simeq 0.11 \left(\frac{P}{1\ \mathrm{hr}}\right). \tag{8}$$

This relation is approximately obeyed by CVs with short orbital periods (less than about 7 hours). This is a useful relation since it allows an approximate determination of the secondary's mass in systems in which the orbital period is known (see [7] for a detailed discussion).

3 Mass Transfer

The importance of the Roche lobe lies in the fact that stars that fill their Roche lobe start to transfer mass through L_1 to their companion (the situation is schematically presented in Figs. 3 and 4). Mass transfer can be triggered either

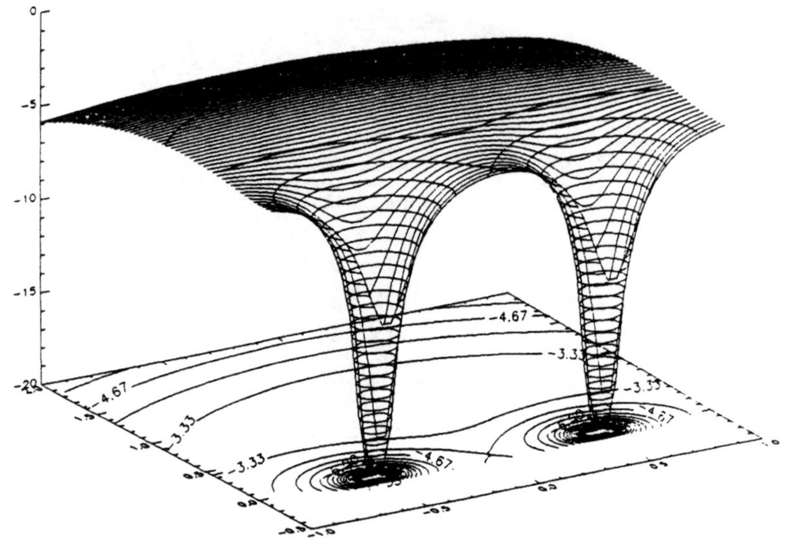

Fig. 3. A cut of the equipotential wells of Fig. 2 (from Kluźniak [5]).

by the expansion of a star (due to its own evolution) to the point that it fills its Roche lobe, or by orbital contraction due to systemic angular momentum losses (see sections 9 and 11).

The rate of mass transfer depends largely on the relations among the following parameters [9]

$$\zeta_L \equiv \frac{d\ln R_{RL}}{d\ln M}$$
$$\zeta_S \equiv \left(\frac{\partial \ln R}{\partial \ln M}\right)_{x,s} \tag{9}$$
$$\zeta_{TE} \equiv \left(\frac{\partial \ln R_{EQ}}{\partial \ln M}\right)_x .$$

Here ζ_L represents the change in the Roche lobe radius with respect to mass (of the lobe filling star), ζ_S represents the change in the radius of the star with respect to its mass (for a fixed composition profile and specific entropy) and ζ_{TE} represents the change in the radius of a star in thermal equilibrium with respect to its mass (for a fixed composition profile). The following regimes can be identified: (1) $\zeta_S < \zeta_L$, dynamical timescale mass transfer; (2) $\zeta_{TE} < \zeta_L < \zeta_S$, thermal timescale mass transfer; (3) $\zeta_L < \zeta_S$, ζ_{TE}, orbital evolution or nuclear timescale. In the first case, the mass losing component is unable to contract as fast as its Roche lobe as it is losing mass, therefore, mass transfer is enhanced. This results in a mass transfer on a dynamical timescale. In the second case, the star is unable to remain in thermal equilibrium as it loses mass (because that would lead to a too high mass-loss rate). Its departure from thermal equilibrium will allow the star to keep contact with the Roche lobe and transfer mass

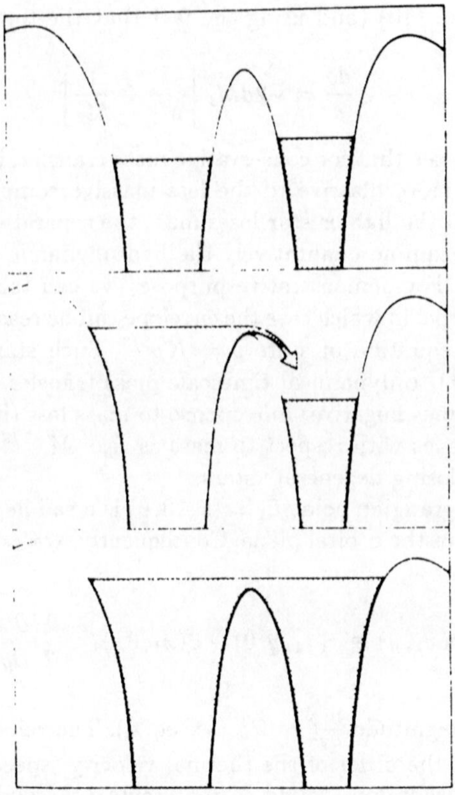

Fig. 4. A schematic representation of the equipotentials as a function of distance along the line joining the two stars. (a) a detached binary, (b) semi-detached and (c) a contact binary (from [10]).

on a thermal timescale. In the third case, the mass transfer process is not self-fed, but rather it occurs as a result of the expansion of the lobe-filling star (as a consequence of its evolution) or because of the fact that angular momentum losses from the system decrease the size of the Roche lobe. In this case mass transfer occurs on the nuclear timescale or the timescale determined by the angular momentum losses. The two mechanisms that are thought to be responsible for systemic angular momentum loss in CVs are gravitational radiation and magnetic braking (see section 11).

The following thing should be noted. If the mass transfer is conservative, namely, no mass is lost from the system and no angular momentum is lost from the orbital motion, then we have $M_1 + M_2 = \text{const.}$ or, $\dot{M}_1 = -\dot{M}_2$, and

$$J^2 = \frac{G M_1^2 M_2^2 a}{M_1 + M_2} = \text{const.} \qquad (10)$$

Differentiating eq. (10) (and using the fact that the total mass is conserved) we obtain

$$\frac{da}{a} = -2dM_1 \left[\frac{1}{M_1} - \frac{1}{M_2}\right]. \tag{11}$$

From eq. (11) we see that for conservative mass transfer, if mass is being transferred from the more massive to the less massive component the separation decreases while if the lighter star loses mass the separation increases.

Let us now examine qualitatively the hydrodynamic problem of mass loss through L_1 [10]. For demonstrative purposes we can take a star with a deep convective envelope, in which case the envelope can be reasonably approximated by a polytropic equation of state $p = K\rho^{5/3}$. Such stars are susceptible (in principle at least) to dynamical timescale mass transfer because they tend to expand (ζ_S becomes negative) in response to mass loss (it is easy to show that their radius behaves with respect to mass as $R \propto M^{-1/3}$). A similar situation occurs for mass-losing degenerate stars.

The inner Lagrangian point $L_1 = (x_1, 0, 0)$ is a saddle point of the potential $\phi(x, y, 0)$ (eq. 4) in the orbital plane. Consequently, $\nabla\phi = 0$ at L_1, which allows us to write

$$\Delta\phi(x_1, y) \equiv \phi(x_1, y, 0) - \phi(x_1, 0, 0) \simeq \frac{1}{2} \left.\frac{\partial^2\phi}{\partial y^2}\right|_{L_1} y^2. \tag{12}$$

To an order of magnitude $\frac{\partial^2\phi}{\partial y^2} \sim \Omega_B^2$ (see eq. 4). The velocity of the gas flowing through L_1 is of the order of the thermal velocity (speed of sound) C_S. This is because the flow region around L_1 represents a nozzle leading into a volume with very low density. The width of the nozzle, W, can be estimated by the requirement that a gas element should have a kinetic energy that is sufficient to overcome the potential difference (from point $(x_1,\ W)$ to $(x_1,\ 0)$), namely (see eq. 12)

$$C_S^2 \simeq \Delta\phi(x_1, W) \simeq \Omega_B^2 W^2 \tag{13}$$

from which we obtain

$$W \sim \frac{C_S}{\Omega_B}. \tag{14}$$

We would like to be able to express W also in terms of quantities of the lobe filling component. To this goal, we note that from hydrostatic equilibrium in the star's atmosphere (z is a vertical distance)

$$-\frac{GM}{R^2} = \frac{1}{\rho}\frac{dp}{dz} = \frac{p}{\rho}\frac{1}{p}\frac{dp}{dz} \simeq \frac{C_S^2}{H} \tag{15}$$

where H is the pressure scaleheight and R is the star's radius.

Since in a lobe-filling star we have (to within an order of magnitude) $R \sim a$, eq. (15) can be written as $\Omega_B^2 \sim C_S^2/(HR)$, which together with eq. (14) gives

$$W \sim \sqrt{HR}. \tag{16}$$

We can now estimate the mass transfer rate. This is given by the mass flux ($\sim \rho C_S$) times the area of the nozzle ($\sim W^2$),

$$\dot{M} \sim \rho C_S W^2|_{L_1} \sim \frac{(\rho C_S^3)}{\Omega_B^2}|_{L_1}. \quad (17)$$

Let us define the degree of Roche lobe overfill ΔR, (namely, by how much the star's radius exceeds R_{RL}, given *e.g.*, by eq. 5). For the assumed equation of state ($p \propto \rho^{5/3}$) $\rho \sim C_S^3$. From eq. (15) we obtain that for a depth ΔR below the stellar surface we have

$$\frac{C_S^2}{\Delta R} \sim \frac{GM}{R^2} \sim \Omega_B^2 R. \quad (18)$$

Therefore, substituting for ρ, C_S and W in eq. (15) we obtain

$$\dot{M} \sim \left(\frac{\Delta R}{R}\right)^3. \quad (19)$$

More precisely, it can be shown that [11]

$$\dot{M} \simeq -AM\Omega_B \left(\frac{\Delta R}{R}\right)^3 \quad (20)$$

where A is a dimensionless coefficient which depends on the density profile of the mass losing component and somewhat on the mass ratio q. For typical systems A is of the order of a few. Relation (20) has been verified numerically using a two-dimensional hydrodynamic calculation of the flow through L_1 [12]. From eq. (20) we see that in order to obtain a transfer rate of $\sim 10^{16}$ g s^{-1} in a 3 hr orbital period CV, the degree of overfill that in required is of order $\Delta R/R \sim 3 \times 10^{-5}$.

In the case of thermal timescale mass transfer, the mass-loss rate is regulated essentially by the departure from thermal equilibrium imposed on the envelope of the star by the mass loss process. The maximum mass transfer rate thus obtained is of order

$$\dot{M}_{max} \simeq -\frac{M}{\tau_{KH}} \quad (21)$$

where τ_{KH} is the (initial) Kelvin-Helmholtz timescale of the mass-losing component

$$\tau_{KH} \simeq \xi \frac{GM^2}{RL} \quad (22)$$

where ξ is a parameter of order unity.

4 The Formation of an Accretion Disk

After leaving the L_1 point (with the thermal velocity) the gas starts to accelerate towards the accreting star (a white dwarf in the case of CVs). Since the dimensions of a white dwarf are typically much smaller ($R_{WD} \sim 5 \times 10^8$ cm) than the orbital separation (which is of the order of a solar radius), the gas acquires velocities along its path (at a distance r from the white dwarf), satisfying

$$V_S \sim \left(\frac{GM_{WD}}{r}\right)^{1/2} > \left(\frac{GM_{WD}}{a}\right)^{1/2} \simeq 4.4 \times 10^7 \left(\frac{M_{WD}}{M_\odot}\right)^{1/2} \left(\frac{a}{R_\odot}\right)^{-1/2} \text{cm s}^{-1}. \tag{23}$$

At the same time, the speed of sound in the gas (at temperatures not exceeding 10^4 K, typical for the secondaries in CVs) is of order

$$C_S \sim \left(\frac{kT}{m_p}\right)^{1/2} \sim 10^6 \left(\frac{T}{10^4\ K}\right)^{1/2} \text{cm s}^{-1}. \tag{24}$$

Thus, the Mach number of the flow after leaving the L_1 region is very high $\mathcal{M} \gtrsim 40$. This situation has two consequences: (i) The gas flow through the nozzle from L_1 forms a relatively narrow stream (since the stream expands on a timescale which is much longer than the flow timescale). (ii) Along the stream, pressure forces can be neglected and the stream path can be calculated using ballistic trajectories of test particles.

The equation of motion is then given by eq. (2), with the pressure gradient term on the RHS neglected. This eq. can be integrated numerically to give a stream trajectory like the one presented in Fig. 5 (see [13], [14]). Since the white dwarf radius is very small, the stream's closest approach to the accreting star, R_{min}, is larger than the white dwarf radius. Therefore, the stream misses the white dwarf and strikes itself. This produces two strong shocks (in the two colliding streams) which radiate the thermal energy that was generated by dissipation of kinetic energy in the collision process. Since angular momentum with respect to the white dwarf is roughly conserved (the tidal interaction with the secondary is negligible), the steady state configuration that will result, is a circular orbit (minimum energy at fixed angular momentum). This orbit has the property that the tangential velocity of the incoming stream is parallel (and equal) to the speed of the simple (circular) periodic orbit at the point of impact (Fig. 6). Because of the fact that the radius of the ring thus obtained is small compared to the separation (for typical CV systems), we can obtain a reasonable estimate for the radius (the "circularization radius," R_c) by equating the specific angular momentum of a mass element at L_1 to that of a mass element in the ring. We thus obtain

$$(GM_{WD}R_c)^{1/2} \simeq X_1^2 \Omega_B, \tag{25}$$

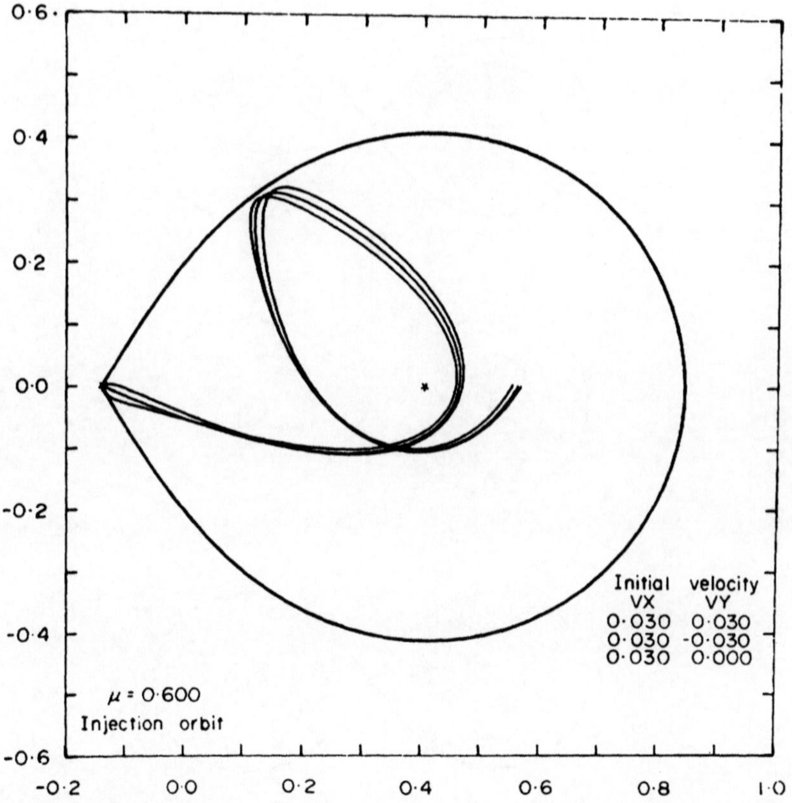

Fig. 5. Particle trajectories in the orbital plane, after passing the L_1 point with a low velocity, for a system with $q = 2/3$ (from [13]).

where X_1 is the distance of L_1 from the white dwarf. In the Roche geometry, X_1 is given approximately by [15]

$$X_1 \simeq (0.5 - 0.227 \log q)a. \tag{26}$$

Substituting into eq. (25) we obtain

$$R_c \simeq (1+q)(0.5 - 0.227 \log q)^4 a. \tag{27}$$

For typical CVs, we obtain from eqs. (27) and (5) that R_c is smaller than the radius of the white dwarf's Roche lobe by a factor 2–3.

The ring that forms at R_c experiences shearing at a rate $Rd\Omega/dR$ ($\Omega \simeq (GM_{WD}/R^3)^{1/2}$). Viscosity or other mechanisms that are capable of transporting angular momentum (see section 6) act to reduce the shear, thereby dissipating kinetic energy into heat, which is radiated away by the gas. The inner edge of the ring therefore spreads inward, as the orbiting gas elements spiral-in. Similarly, due to the fact that the angular momentum of the inward

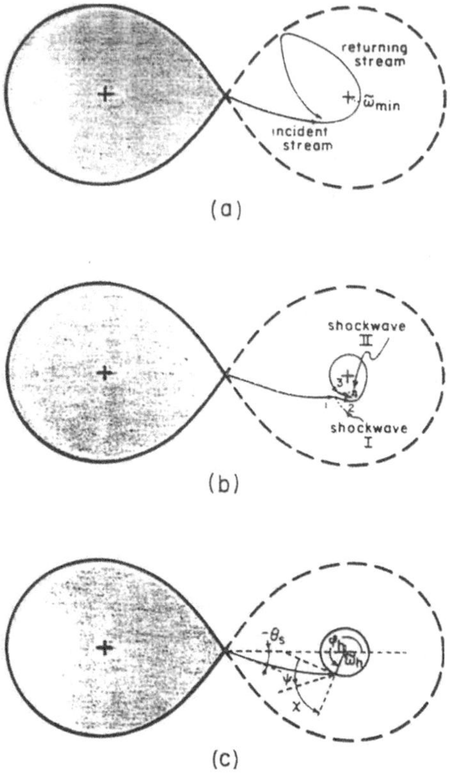

Fig. 6. Stream trajectory and the circularization of the formed ring in the orbital plane for equal masses (from [14]).

moving material is transported to the outer parts of the ring, the ring spreads also outwards. In this way, an accretion disk is formed around the white dwarf. The outer edge of the disk reaches roughly the tidal radius, where the tides that are induced by the secondary star truncate the disk (and the angular momentum is transferred back into the orbit). Alternatively, the disk could (in principle) lose angular momentum via a magnetically coupled wind (see section 6).

Two-dimensional hydrodynamic simulations have demonstrated the process of disk formation and the fact that relatively little mass is lost from the system (e.g. [16]) (Fig. 7). Observations of CVs and low mass x-ray binaries (which contain a neutron star instead of a white dwarf) indicate disk radii which are of the order of 80% of the accreting star's Roche lobe [17, 18], similar to radii obtained in the numerical calculations.

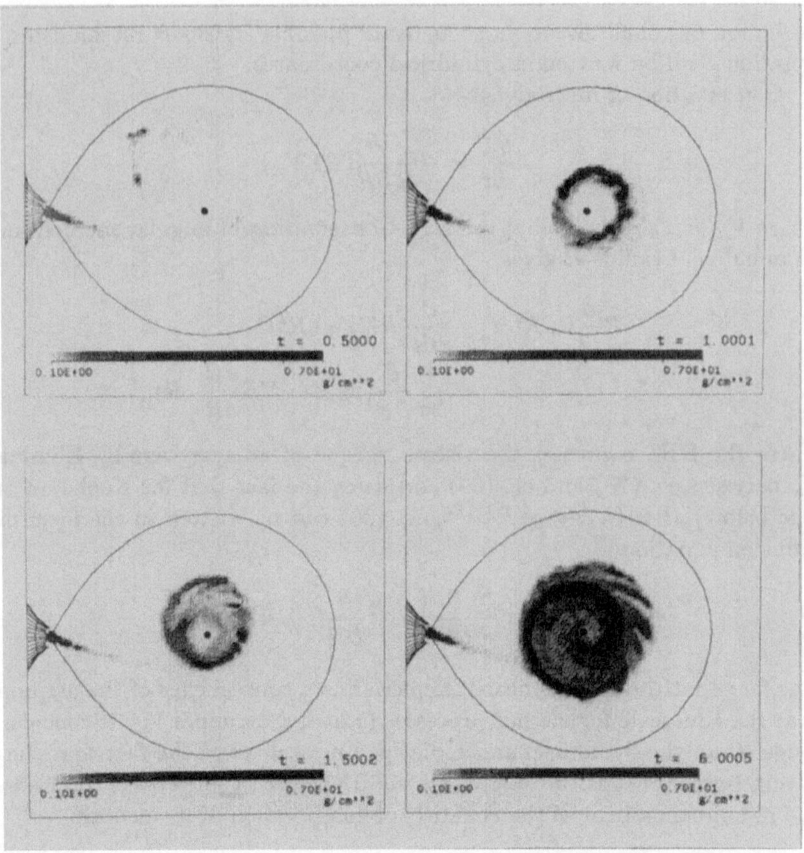

Fig. 7. The time evolution of the surface density in a 2D simulation of the formation of an accretion disk. In this calculation $M_1 = M_\odot$, $M_2 = 0.5M_\odot$ and the orbital period is about 0.2 day (from [16]).

5 Standard Steady Accretion Disk Models

We consider a disk which is geometrically thin, $H/R \ll 1$ (where H is the disk half thickness). Under such conditions, it is convenient to define a surface density by integrating the gas density over the disk thickness

$$\Sigma(R,t) = \int \rho dz. \tag{28}$$

As explained in the previous section, angular momentum has to be transported outwards in the disk in order to allow the material to be accreted. The mechanism by which this angular momentum transport takes place is one of the major unsolved problems in accretion disk theory (see section 6).

For the moment, I shall assume that irrespective of the actual transport mechanism, it manifests itself in the form of an effective kinematic viscosity ν.

In section 6, I shall discuss in some detail possible transport mechanisms. The equations will be written in cylindrical coordinates.

Conservation of mass gives

$$\frac{\partial \Sigma}{\partial t} = -\frac{1}{R}\frac{\partial}{\partial R}(R\Sigma V_R), \tag{29}$$

where $V_R(R, t)$ is the radial velocity. Conservation of angular momentum for an annulus of radius R gives

$$\begin{aligned} R\frac{\partial}{\partial t}(\Sigma R^2 \Omega) &+ \frac{\partial}{\partial R}(R\Sigma V_R \cdot R^2 \Omega) \\ &= \frac{1}{2\pi}\frac{\partial}{\partial R}\left(2\pi R \cdot \nu \Sigma R \frac{d\Omega}{dR} \cdot R\right), \end{aligned} \tag{30}$$

where the RHS expresses the viscous torque of adjacent annuli. Eliminating V_R between eq. (29) and eq. (30) and using the fact that for Keplerian orbits (see below) $\Omega = (GM_{WD}/R^3)^{1/2}$, eq. (30) can be written in the form of the diffusion equation

$$\frac{\partial \Sigma}{\partial t} = \frac{3}{R}\frac{\partial}{\partial R}\left[R^{1/2}\frac{\partial}{\partial R}(\nu \Sigma R^{1/2})\right]. \tag{31}$$

The fact that the orbit is nearly Keplerian is a consequence of the assumption that the timescale for viscous processes (which determines V_R) is considerably longer than the dynamical and radiative timescales and the fact that the disk is thin (which means that $C_S \ll V_\phi$, eq. 34 below). Under these conditions, in the radial component of the equation of conservation of momentum

$$V_R\frac{\partial V_R}{\partial R} - \frac{V_\phi^2}{R} = -\frac{1}{\rho}\frac{\partial p}{\partial R} - \frac{GM_{WD}}{R^2} \tag{32}$$

both the pressure gradient term and $V_R \partial V_R/\partial R$ can be neglected (see discussion of timescales in Section 8).

In the vertical direction (perpendicular to the disk plane), we assume hydrostatic equilibrium to hold

$$-\frac{1}{\rho}\frac{\partial p}{\partial Z} = \frac{GM_{WD}Z}{(R^2+Z^2)^{3/2}} \simeq \frac{GM_{WD}Z}{R^3}, \tag{33}$$

(where the last equality uses the fact that the disk is thin $Z/R \ll 1$). If we take the pressure to be approximately $p \simeq \rho C_S^2$ (where C_S is the speed of sound), then eq. (33) gives to an order of magnitude

$$\frac{H}{R} \sim \frac{C_S}{V_\phi}. \tag{34}$$

In a steady state ($(\partial/\partial t) \equiv 0$), eqs. (29) and (30) can be integrated to give

$$2\pi R\Sigma(-V_R) = \dot{M}, \tag{35}$$

where $\dot{M}$ is the constant mass flux through the disk and

$$\nu \Sigma \frac{d\Omega}{dR} = \Sigma V_R \Omega + \frac{C}{2\pi R^3} \tag{36}$$

where C is a constant.

In order to determine C we note the following, the angular velocity is nearly Keplerian ($\Omega \sim R^{-3/2}$) almost throughout the entire disk. However, if the accreting white dwarf is non-magnetic, then near its surface, the angular velocity in the disk has to join smoothly to the angular velocity of the white dwarf, Ω_{WD}. Therefore, for thin disks, there exists a (usually assumed) narrow boundary layer (of dimensions typically much smaller than R_{WD}), within which Ω diverges from the Keplerian value [19], (Fig. 8). From Fig. 8 we see that there necessarily exists a point, R_m, at which $d\Omega/dR = 0$. From eq. (36) we find that $C = \dot{M} R_m^2 \Omega_m$, namely, C represents the angular momentum influx. Because of the fact that the boundary layer is normally assumed to be narrow, it has been common practice to take R_m to be at the white dwarf surface, R_{WD}, so that $C = \dot{M}(GM_{WD}R_{WD})^{1/2}$. In this case, eq. (36) assumes the form (for a non-rotating white dwarf)

$$\nu \Sigma = \frac{\dot{M}}{3\pi}\left[1 - \left(\frac{R_{WD}}{R}\right)^{1/2}\right]. \tag{37}$$

From eqs. (35) and (37) we find that $V_R \simeq -(3\nu/2R)$. It should be noted, however, that accurate solutions for the boundary layer may certainly require other values of C (and therefore a direct coupling between the disk structure and the boundary layer). Furthermore, two dimensional hydrodynamic calculations of the flow in the boundary layer [20] show quite complex, time dependent flow patterns which depend strongly on the magnitude of the viscosity (Fig. 9). In particular, it was found that if the kinematic viscosity in the boundary layer is taken to be proportional to the local speed of sound times the pressure scale height (see section 6), then the flow becomes unstable and time dependent. The hydrodynamic calculations carried out so far represent only a few dynamical timescales and thus their results should be treated with caution.

I should also note that the discussion of the boundary layer assumed the white dwarf to be non-magnetic. However, many CVs are known to contain strongly magnetized white dwarfs (*e.g.*, [21]). In this case, the accretion disk cannot be expected to extend inward all the way to the surface of the accreting white dwarf [22]. Rather, once the magnetic energy density exceeds the energy density in the disk (see section 6), the magnetic field disrupts the disk and stars to control the gas flow [23]. In a subclass of the CVs called polars (or AM Her stars), the magnetic field is sufficiently strong (of the order of a few $\times 10^7$ G) to prevent a substantial disk from forming at all. Instead, in this case the accreted material probably follows field lines.

The rate of energy dissipation in the disk (due to the effective viscosity) is [24]

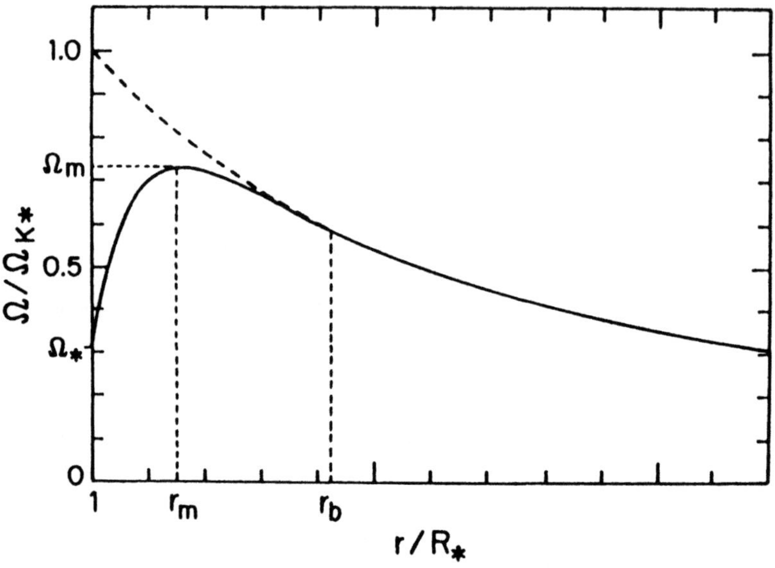

Fig. 8. A schematic representation of the angular velocity as a function of radial distance, defining the boundary layer (from [19]).

$$D(R) = \nu\Sigma\left(R\frac{d\Omega}{dR}\right)^2 = \frac{3GM_{WD}\dot{M}}{4\pi R^3}\left[1-\left(\frac{R_{WD}}{R}\right)^{1/2}\right] \tag{38}$$

(where we have used eq. (37)). Equation (38) reveals, as expected, that the rate of dissipation is determined by the rate of release of gravitational energy (once a mass flux $\dot{M}$ has been established). The total disk luminosity can be found by integration

$$L_{\mathrm{DISK}} \simeq \int_{R_{WD}}^{\infty} D(R)\cdot 2\pi R dR = \frac{1}{2}\frac{GM_{WD}\dot{M}}{R_{WD}}. \tag{39}$$

Since this is only one half of the total release rate of available potential energy, the other half, (which is still retained by the material just outside the boundary layer as kinetic energy), has to be dissipated in the boundary layer (if a boundary layer exists).

More generally, it can be shown [20] that if we write the angular velocity at R_m as $\Omega(R_m) = \beta\Omega_{Kep}(R_m)$, then $L_{DISK} = (1-\frac{1}{2}\beta^2)L_{acc}$, where L_{acc} is the rate of release of potential energy. For the rate of energy dissipation in the boundary layer one finds $L_{BL} = L_{DISK}[1-\Omega_{WD}/\Omega_{Kep}(R_{WD})]^2$, where Ω_{WD} is the rotation angular velocity of the white dwarf.

The calculation of the spectrum of an accretion disk requires a solution of the vertical radiative transfer equation at each radius [25], [26] (to first order radial radiative transport can be neglected since the gradients are smaller than the vertical ones by a factor $H/R \ll 1$).

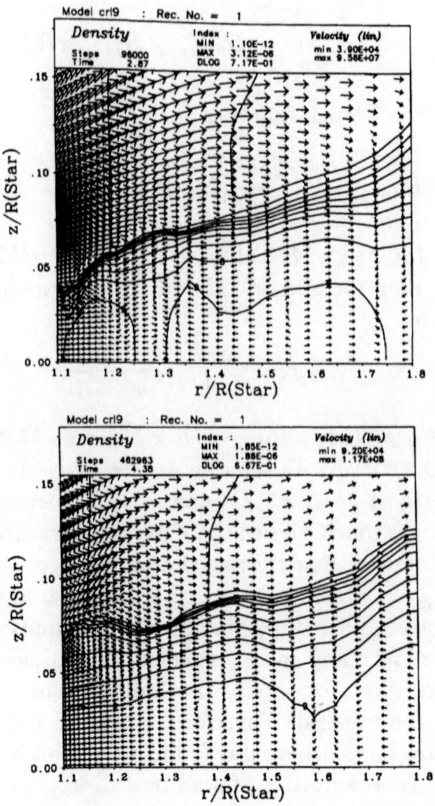

Fig. 9. Density contours and the velocity field in the boundary layer region. The unit of time is the Keplerian period at the stellar surface. Large time variations are observed (from [20]).

As a zeroth order approximation it is frequently assumed that the disk is optically thick and it radiates like a black body (an assumption which is certainly incorrect for many systems). In this case, the energy balance equation can be written as

$$\frac{1}{2}D(R) = \sigma T_{eff}^4, \tag{40}$$

where T_{eff} is the effective temperature and $D(R)$ is given by eq. (38) (the factor 1/2 accounts for radiation from one side of the disk). From eqs. (38) and (40) we find

$$T_{eff} = \left\{ \frac{3GM_{WD}\dot{M}}{8\pi R^3 \sigma} \left[1 - \left(\frac{R_{WD}}{R} \right)^{1/2} \right] \right\}^{1/4}, \tag{41}$$

which for $R \gg R_{WD}$ gives $T_{eff} \propto R^{-3/4}$. The spectrum of the disk (under the assumption of black body radiation) is thus given by

$$S_\nu \propto \int_{R_{WD}}^{R_{disk}} B_\nu(T_{eff}(R)) \cdot 2\pi R dR \tag{42}$$

where $B_\nu(T_{eff})$ is the Planck function

$$B_\nu(T_{eff}) = \frac{2h\nu^3}{c^2}[\exp(h\nu/kT_{eff}) - 1]^{-1}. \tag{43}$$

If we define $T_* = (3GM_{WD}\dot{M}/8\pi R_{WD}^3\sigma)^{1/4}, x = h\nu/kT_{eff}(R)$ (with $x_{out} = h\nu/kT_{eff}(R_{disk})$), then for $\nu \ll kT_*/h$ (which corresponds to $R \gg R_{WD}$), we can write eq. (42) as

$$S_\nu \propto \nu^{1/3} \int_0^{x_{out}} \frac{x^{5/3}dx}{e^x - 1}. \tag{44}$$

Note that for $M_{WD} = 1M_\odot$, $R_{WD} \simeq 5 \times 10^8$ cm, $\dot{M} \simeq 10^{16}$ g s^{-1} (typical for CVs), $T_* \simeq 6.9 \times 10^4$ K. As long as we are interested in frequencies satisfying $kT_{eff}(R_{disk})/h \ll \nu \ll kT_*/h$, (namely, frequencies that correspond to temperatures that are much higher than the temperature at the outer edge of the disk and much lower than that at the inner edge) we can take $x_{out} \to \infty$ and obtain: $S_\nu \propto \nu^{1/3}$. Note, however, that many assumptions went into the derivation of the last result. In particular, the assumption of black body radiation is not satisfied in many cases (especially at quiescence) and the range of relevant frequencies does not satisfy the limits imposed on the frequency for the calculation of the integral. Therefore, it should not be surprising that in many systems accretion disks are not found to exhibit a $\nu^{1/3}$ spectrum (which is considered to be representative of standard disks).

Recently, several serious attempts were made to calculate disk spectra by solving numerically detailed structure and radiative transfer equations [25], [26]. The results of some of these calculations are presented in Figs. 10–12. A good agreement has been obtained with the observations of some nova-like variables, assuming certain system parameters. As can be seen, the emergent spectrum is quite sensitively dependent on the assumed viscosity. In principle, therefore, there exists the potential of using disk spectra as diagnostics for a determination of the disk viscosity (see section 6).

6 Angular Momentum Transport Mechanisms

In spite of the fact that accretion disks have been the subject of extremely intensive research during the past twenty years, the mechanism by which angular momentum is transported outwards in the accretion disk is still unclear.

It has been immediately realized that standard molecular viscosity is many orders of magnitude too low to generate the required mass flux. This can be appreciated by examining the Reynolds number, which to an order of magnitude is given by

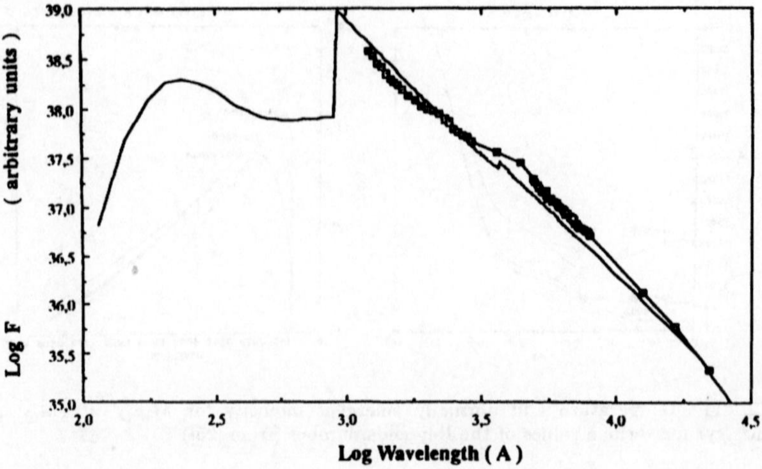

Fig. 10. Comparison between the predicted energy distribution (full line) and the observed one (squares) for RW Sex. The model assumed $\dot{M} = 10^{18}$ g s^{-1}, $M_{WD} = 0.8M_{\odot}$, $\alpha = 0.5$ and an outer disk radius of $15R_{WD}$ (from [26]).

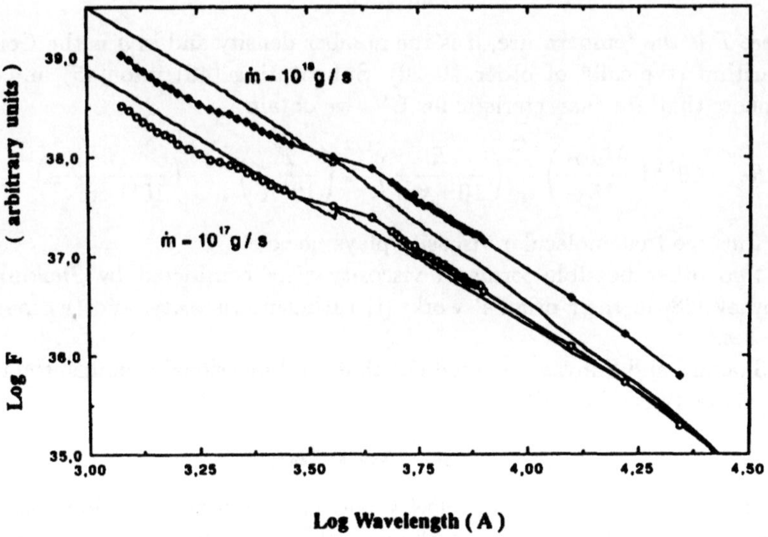

Fig. 11. Comparison of the observed energy distribution of RW Sex with theoretical calculations for two different accretion rates (from [26]).

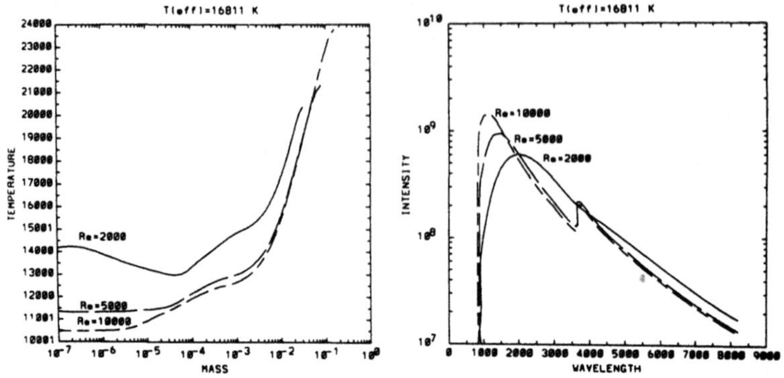

Fig. 12. The temperature and normally emergent intensity for $M_{WD} = 1M_{\odot}$, $\dot{M} = 10^{-11}M_{\odot}/\text{yr}$ for various values of the Reynolds number (from [25]).

$$Re \simeq \frac{v_{\phi}R}{\nu_{mol}}, \tag{45}$$

where ν_{mol} is the molecular kinematic viscosity. Viscous processes are totally unimportant if $Re \gg 1$. To an order of magnitude ν_{mol} is given by [27]

$$\nu_{mol} \simeq \frac{10^{10}}{\ln \Lambda}\frac{T^{5/2}}{n} \text{ cm}^2 \text{ s}^{-1} \tag{46}$$

where T is the temperature, n is the number density and $\ln \Lambda$ is the Coulomb logarithm (typically of order 10–20). Substituting (46) into (45) and using numbers that are characteristic for CVs we obtain

$$Re \sim 10^{14}\left(\frac{M_{WD}}{M_{\odot}}\right)^{1/2}\left(\frac{R}{10^{10}\text{ cm}}\right)^{1/2}\left(\frac{T}{10^4 K}\right)^{-5/2}\left(\frac{n}{10^{15}\text{ cm}^{-3}}\right). \tag{47}$$

We thus see that molecular viscosity plays no role.

Two other possible forms of viscosity were considered by Shakura and Sunyaev [28] in their original work: (i) turbulent viscosity, and (ii) magnetic stresses.

Shakura and Sunyaev pointed out that on dimensional grounds, the turbulent viscosity can be written as

$$\nu_{turb} \sim v_{turb} \cdot \ell_{eddy} \tag{48}$$

where v_{turb} is the turbulent velocity and ℓ_{eddy} is a typical eddy size. Since necessarily $\ell_{eddy} \lesssim H$ and the turbulent velocity can be scaled with the speed of sound, eq. (48) can be written in the form

$$\nu_{turb} \simeq \alpha_{ss}C_S H \tag{49}$$

where α_{ss} is a dimensionless number expressing the magnitude of the viscosity (the viscous force per unit area thus becomes $\rho\nu R d\Omega/dR \simeq \alpha_{ss}p$). Shakura and

Sunyaev argued that v_{turb} cannot exceed the speed of sound since if $v_{turb} > C_S$, shocks are generated, which in turn heat up the gas, increasing the speed of sound. Thus, in the "α-prescription" of Shakura and Sunyaev it was assumed that $\alpha_{ss} \lesssim 1$ (and it is normally assumed that $\alpha_{ss} = \text{const.}$).

Shakura and Sunyaev also considered the possibility that magnetic stresses operate as the mechanism for angular momentum transport, pointing out that the disk shear provides a natural environment for dynamo generation of magnetic fields.

Before I discuss the Shakura-Sunyaev prescription for the treatment of the effects of the magnetic field stress, I review a few basic concepts concerning the magnetic field.

From Maxwell's equations (neglecting the displacement current) it can be easily shown that the magnetic field satisfies the equation

$$\frac{\partial \boldsymbol{B}}{\partial t} = \nabla \wedge (\boldsymbol{V} \wedge \boldsymbol{B}) + \frac{c^2}{4\pi\sigma_E}\nabla^2 \boldsymbol{B} \tag{50}$$

where σ_E is the electric conductivity. For many astrophysical situations, the conductivity (for a fully or partially ionized plasma) and the length scales involved are very large and thus, the second (diffusion) term on the RHS of eq. (50) can be neglected. This has the consequence that the fluid and the magnetic field move together, or, the magnetic field lines can be thought of as being frozen into the fluid. The second thing to note is the fact that the magnetic field exerts a stress on the plasma. The force per unit volume (which is a consequence of the Lorentz force on each moving charge) is

$$\boldsymbol{f} = \frac{1}{c}\boldsymbol{j} \times \boldsymbol{B}, \tag{51}$$

where $\boldsymbol{j}$ is the current density. Using Maxwell's equation (neglecting the displacement current)

$$\nabla \wedge \boldsymbol{B} = \frac{4\pi}{c}\boldsymbol{j}, \tag{52}$$

we obtain (after using some vector identities)

$$\boldsymbol{f} = -\nabla\left(\frac{B^2}{8\pi}\right) + \frac{1}{4\pi}(\boldsymbol{B}\cdot\nabla)\boldsymbol{B}. \tag{53}$$

We therefore see (*e.g.*, compare with the Euler equation) that the term $B^2/8\pi$ plays the role of an isotropic magnetic pressure. The second term on the RHS of eq. (53) acts like a magnetic tension along field lines. Magnetic forces dominate the fluid motion if the magnetic pressure is larger than the gas pressure ($\sim \rho C_S^2$) and the ram pressure ($\sim \rho v^2$). The Alfven speed (at which magnetic waves propagate in a plasma) is defined by $V_A = (B^2/4\pi\rho)^{1/2}$.

We can now imagine that if the disk contains horizontal loops of magnetic field, these loops will be convected by the fluid and a mean Maxwell shear stress will develop of order (eq. 53) $\langle B_R B_\phi/4\pi\rangle$ (where B_R and B_ϕ are the radial and azimuthal field components respectively). The prescription of Shakura and

Sunyaev went as follows: they pointed out that the magnetic stress has to satisfy

$$\left\langle \frac{B_R B_\phi}{4\pi} \right\rangle \lesssim \left\langle \frac{B^2}{4\pi} \right\rangle \lesssim p, \tag{54}$$

where the last inequality is based on the following argument. It is expected that the toroidal field (which is generated by the shear, see 6.3 below) cannot grow indefinitely. Its strength is supposed to be limited by buoyant escape from the disk and reconnections, both of which are expected to increase significantly when the magnetic pressure becomes of the order of the total pressure (eq. 53). On the basis of eq. (54), Shakura and Sunyaev therefore proposed that the magnetic stress can be expressed as $\alpha_{ss}p$ (with $\alpha_{ss} \lesssim 1$), exactly in the same form as the turbulent viscosity stress (eq. (49) and the discussion following it).

It is important to understand that until recently (see 6.3 below) it has never been demonstrated that the accretion flow is unstable, so as to produce turbulence. Similarly, until very recently almost all the work concerning dynamo generation of magnetic fields *assumed* a pre-existing turbulent flow. Therefore, the Shakura-Sunyaev α-prescription should be regarded as nothing more than a convenient dimensional analysis formulation of the (unsolved) angular momentum transport problem.

Recently, several attempts have been made to propose more detailed theoretical models for angular momentum transport mechanisms. I shall now review briefly a few of these models.

6.1 Hydromagnetic Winds

For this mechanism to operate, the disk should be threaded by (or generate by itself) open magnetic field lines. The field should be reasonably ordered and in equipartition with the thermal energy density in the disk. Because of the fact that the magnetic field is frozen in the fluid, the field lines at a given radius R in the disk, can be thought of as orbiting with the local angular velocity. Gas elements are forced to corotate with the lines for as long as their kinetic energy density does not exceed the magnetic energy density in the poloidal field (this condition is satisfied up to the Alfven radius $R_A(R)$). Therefore, the gas is centrifugally accelerated, essentially all the way to $R_A(R)$, much in the same way that a bead is accelerated on a rotating wire. It has therefore been suggested that angular momentum is being carried away from the disk by a disk wind [29] [30]. Strictly speaking, therefore, hydromagnetic winds provide a mechanism for angular momentum removal rather than transport. The angular momentum loss rate is given by

$$\dot{J}_W \simeq \dot{M}_W \Omega R_A^2. \tag{55}$$

Equating this angular momentum loss rate to the rate at which angular momentum has to be transported in the disk, in order to produce a mass transfer rate $\dot{M}_d$ in the disk, gives

$$\dot{M}_W \simeq \dot{M}_d \left(\frac{R}{R_A}\right)^2 . \tag{56}$$

Thus, for $R_A/R \sim 10$, only one percent of the transferred mass has to be lost in the wind in order to remove the necessary angular momentum. Winds with an average rate of $\dot{M}_W \sim 10^{-11} M_\odot/\text{yr}$ or more are indeed observed in CVs [31], and according to the above arguments, such winds could (in principle at least) drive mass transfer rates of $\dot{M}_d \sim 10^{-9} M_\odot/\text{yr}$, which are typical for CVs.

There exists, however, a strong argument which suggests that while hydromagnetic winds may play some role, ***they do not provide the dominant angular momentum transport mechanism in CVs.*** The argument is related to the size (the outer radius) of the accretion disk. As explained in Section 4, the stream material which flows from L_1 tends to form a ring at the circularization radius R_c (eq. 27), which is typically smaller than the white dwarf's Roche lobe radius by a factor 2–3. It is only through the action of some form of viscosity that the outer disk expands all the way to the tidal radius (which is roughly given by the last non-intersecting particle orbit in the restricted three body problem). If angular momentum would have been indeed directly removed from the disk by hydromagnetic winds, we would have expected to observe disk radii that are equal to R_c essentially at all times (including at outburst, see section 8). Instead, as noted in section 4, disk radii which are close to the tidal radius are often observed. This means that some other form of "viscosity" must be operative. Furthermore, the time behaviour of the disk radius during dwarf nova eruptions (section 8.4.1) is *consistent* with the action of viscosity and not with hydromagnetic winds. Finally, the observations of the winds from CVs [31] indicate that the wind originates from the inner disk, or possibly the boundary layer, while the implication of the hydromagnetic wind scenario (if the wind is to remove angular momentum) is that the outer disk should also be an origin of outflow.

6.2 Coherent Spiral Shock Waves

The idea that spiral waves can transport angular momentum in disks was explored quite extensively in the galactic dynamics context [32], [33] and to a lesser extent in the case of protostellar disks [34]. Recently, it has been suggested that this mechanism could also operate in CV disks [35], [36], the spiral shock waves being tidally induced (by the secondary star). I will now give a brief and highly simplified description of how this mechanism is supposed to work.

Suppose we look at some disturbance at the outer disk edge. As the wave that is generated propagates inward, it is being wound up by the differentially rotating disk into a trailing spiral pattern. If we neglect the orbital frequency at the outer edge (because it is much lower), we obtain that the wave frequency (in the comoving frame) for azimuthal wavenumber m is

$$\sigma = \omega - m\Omega \approx -m\Omega . \tag{57}$$

A conserved quantity (along the wave path, with no dissipation present) of the wave is the action [37], given by

$$S_W = \frac{1}{2} \int \frac{\rho V_W^2}{\sigma} dV \tag{58}$$

where V_W is the velocity amplitude of the wave and the integration is carried out over the volume of the wave packet. The angular momentum of the wave, $j = mS_W$, is also conserved [38]. From (57) and (58) we find that the angular momentum is negative

$$j \simeq -\frac{1}{2} \int \frac{\rho V_W^2}{\Omega} dV. \tag{59}$$

Namely, the presence of the wave reduces the total angular momentum of the disk. The idea is that the spiral wave steepens into a shock as the wave propagates inward. As energy is dissipated in the shock, the negative angular momentum is transferred to the disk gas, thus reducing the angular momentum of the gas and allowing it to accrete.

The increase in the amplitude of the wave as it propagates inwards can be shown crudely in the following way. If we denote by ΔR the radial extent of the wave packet, then locally $\Delta R \propto C_S$ (since the front and back of the wave move at a speed proportional to C_S). The volume of the wavepacket is given by $V = 2\pi R \Delta R \cdot 2H$. From eq. (59) we therefore obtain (j is conserved)

$$V_W \propto \left(\frac{\Omega}{\rho R H C_S} \right)^{1/2} . \tag{60}$$

Now, using the standard steady disk algebraic equations (28), (34), (37), (41) and (49), in combination with an equation of state for an ideal gas and Kramers' opacity law, it is easy to solve for the structure of the standard disk and show that $H \propto R^{9/8}$, $\rho \propto R^{-15/8}$ and $C_S \propto R^{-3/8}$ (see *e.g.*, [28]). From (60) we therefore obtain that $V_W \propto R^{-11/16}$, so that the wave is indeed expected to increase inward.

Numerical calculations of two-dimensional disks demonstrated the formation of tidally induced, two-armed spiral shocks (Fig. 13). The spiral shocks persisted for the entire duration of the calculation, which spanned 15–20 rotation periods. The accretion rate that was obtained as a result of the angular momentum transport by the spiral shocks corresponded to an effective viscosity parameter $\alpha_{eff} \sim 0.1$.

In spite of the fact that the results described above appear superficially to be quite promising, there exist a few arguments which suggest that the coherent spiral shocks mechanism for angular momentum transport in CV disks is questionable. The main problems that this mechanism encounters are:

6.2.1 On the Theoretical Side.

(1) The disks that were calculated were unrealistically hot (and therefore unrealistically thick geometrically), with a temperature of the order of $0.1T_V$, where T_V is the virial temperature $T_V = GM_{WD}/R_D\mathcal{R}$ ($\mathcal{R}$ is the gas constant). This has the effect of helping sustain the spiral structure because

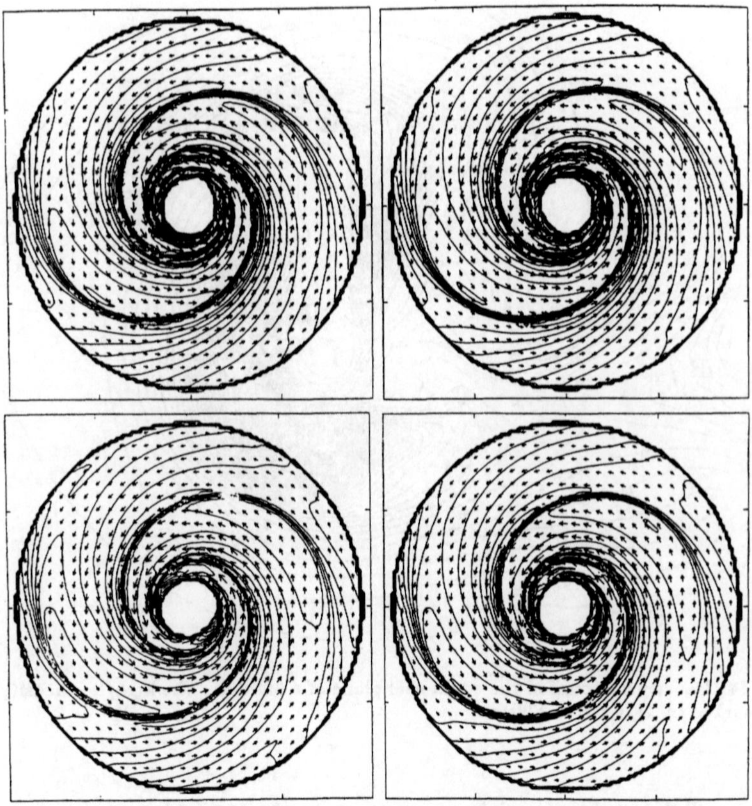

Fig. 13. Tidally induced patterns in 2D disks evolved for $0.2P_{orb}$. The adiabatic exponent increases (while the temperature of the initial model decreases) from 1.1 (upper left) through 1.2 and 1.67 to 2.0. The initial temperature of the hottest disk is $0.11T_v$ (from [36]).

waves can propagate faster between the inner and outer edges of the disk (C_S is larger).

(2) It is not clear at all if the two-armed spirals form in three-dimensions (3D). The only 3D hydrodynamical calculation existing so far [39] produced some traces of spiral waves (Fig. 14) but nothing as coherent and pronounced as in the 2D calculations.

(3) In a study of wave propagation inside a disk in which the density and temperature were stratified in the vertical direction [40], it was found that the temperature stratification causes refraction in such a way that radially moving waves are deflected in the vertical direction (the p-modes). An examination of waves propagating in the vertical (stratified) direction showed that for waves with large azimuthal wavenumbers, energy can be lost in the vertical propagation, while for waves with small wavenumbers, reflection can take place before a shock is formed. As a consequence of both of these effects, the indication is that the waves cannot survive for

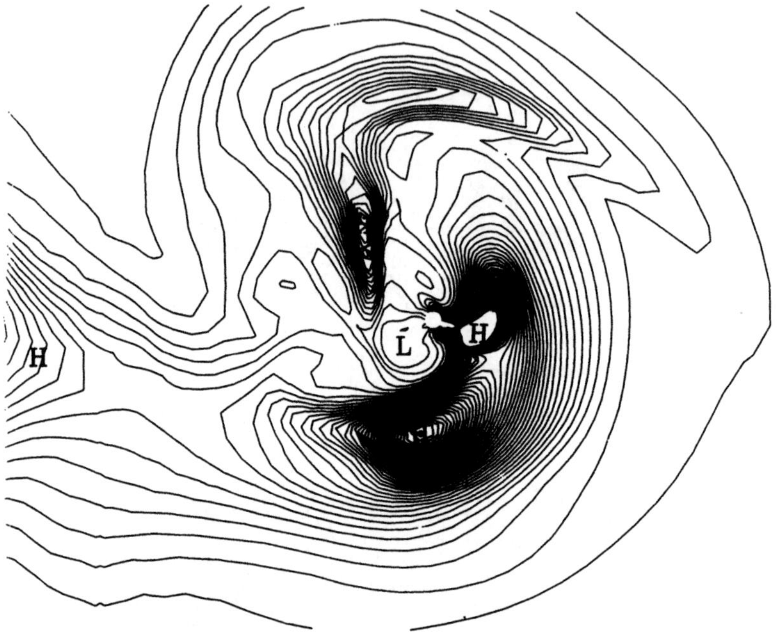

Fig. 14. Density contours in the equatorial plane of a 3D disk. The high/low density regions are marked by H/L (from [39]).

a sufficiently long radial extent, to make this a viable angular momentum transport mechanism.

6.2.2 On the Observational Side.

(1) If the spiral pattern extends all the way from the disk center to the surface then it should manifest itself in the observed emission lines. Under such conditions the pattern should be clearly observable by Doppler tomography techniques [41]. No such pattern was ever observed in CV disks. It has also been shown (theoretically) that a higher temperature spiral pattern should reveal itself in UV eclipse light curves [42] (this has not been investigated observationally yet).

(2) The numerical calculations of the spiral shock mechanism obtained for the accretion timescale $\tau_{acc} \propto \Omega^{-1} q^{-1}$. Using standard disk models it can be shown [43] that this predicts a dependence of the effective viscosity parameter α_{eff} on the mass ratio which is roughly $\alpha_{eff} \propto q$ (the physical reason for this can be understood by the fact that the spiral wave is tidally induced by the secondary). Using the mean recurrence times of dwarf nova eruptions for 20 systems and the disk instability model (see Section 8), Livio and Spruit [43] have shown (with obvious uncertainties) that the observations are consistent with an α_{eff} which is almost independent of q

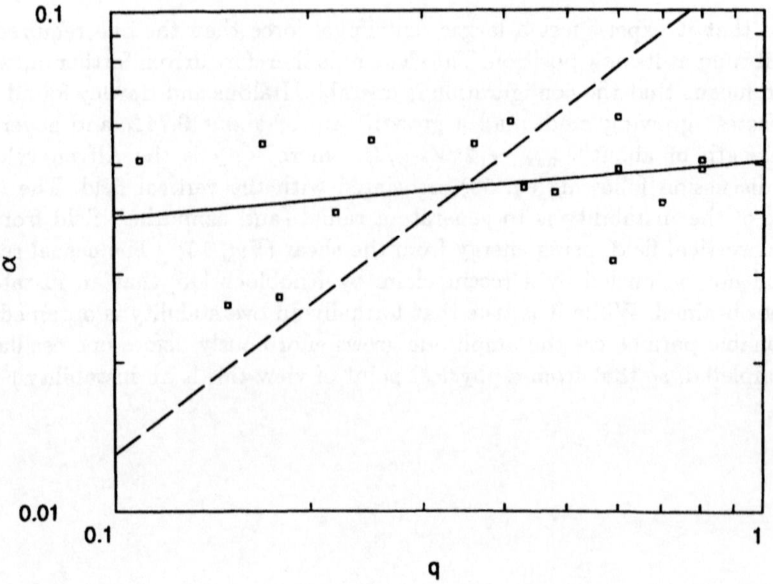

Fig. 15. Viscosity parameters derived from the recurrence timescales of dwarf novae (squares) and a least squares fit (solid line). Broken line gives the dependence of the effective viscosity of spiral shock accretion, as extrapolated from numerical results (from [43]).

(Fig. 15), in contradiction to the prediction of the spiral shock mechanism (at least in its simple form).

6.3 Instabilities and Magnetic Dynamos

Before I discuss specific models involving magnetic dynamos, I would like to describe briefly an important instability, the significance of which has been recently recognized by Balbus and Hawley [44], [45]. These authors have shown that accretion disks that have a weak poloidal field (magnetic energy density smaller than the thermal energy density) and are differentially rotating so that $d\Omega/dR < 0$, are subject to a dynamical axisymmetric shearing instability.

The physical nature of the instability can be understood in the following heuristic way.

Suppose we have a disk in which $d\Omega/dR < 0$, which is threaded by a vertical magnetic field. When the magnetic field is frozen in the fluid it resists both shearing motions and outward displacements which bend the field lines and stretch them. Consequently, a fluid element that is displaced radially outward experiences two forces exerted by the magnetic field, a restoring force and one which tries to enforce corotation. The restoring force is weak for perturbations of a relatively long wavelength. At the same time, the element is forced (by the magnetic field) to rotate too fast for its new environment (because $d\Omega/dR <$

0), so that it experiences a larger centrifugal force than the one required for equilibrium at its new position. The element is therefore driven farther outward, which means that the configuration is unstable. Balbus and Hawley found that the fastest growing mode had a growth rate of about 0.74Ω and a vertical wavelength of about $\lambda_{\max} \sim 2\pi V_{A,Z}/\Omega$, where $V_{A,Z}$ is the Alfven velocity (see discussion following eq. 53) associated with the vertical field. The main effect of the instability is to generate a radial (and azimuthal) field from an initial vertical field, using energy from the shear (Fig. 16). (The casual reader should not be misled by a recent claim by Knobloch [45] that an instability is not obtained. While it is true that formally an overstability is obtained, for reasonable parameters the amplitude grows enormously before one oscillation is completed, so that from a physical point of view this is an instability.)

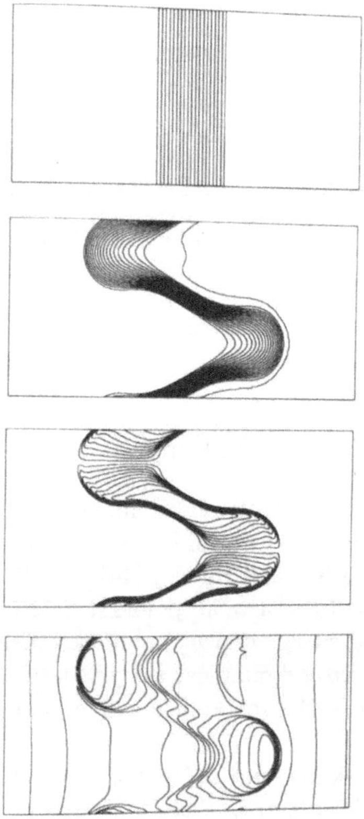

Fig. 16. Contour plots of: (a) the initial poloidal magnetic field, (b) the poloidal magnetic field, (c) toroidal field, (d) angular momentum (from Hawley and Balbus [45]).

Another instability that may play a role in the complete picture of angular momentum transport in disks (see below) is the Parker instability [46].

Very briefly, the physical nature of this instability can be explained as follows. Suppose we have a disk of gas which is threaded by a horizontal field, with hydrostatic equilibrium prevailing in the z (vertical) direction. Clearly, the field is confined by the gravitational force on the gas, the gas itself being supported somewhat by the field (the field is frozen in the fluid). Suppose now that we introduce a perturbation which causes the field lines to undulate slightly (Fig. 17). The gas will tend to slide downward to the lower portions. At the same time, the raised parts of the field lines, now experience less downward pull and therefore they will expand upward even further, leading to an instability. Detailed calculations [47] have shown that the main effect of the Parker instability in a disk is to convert an horizontal magnetic field to a vertical one (Fig. 18).

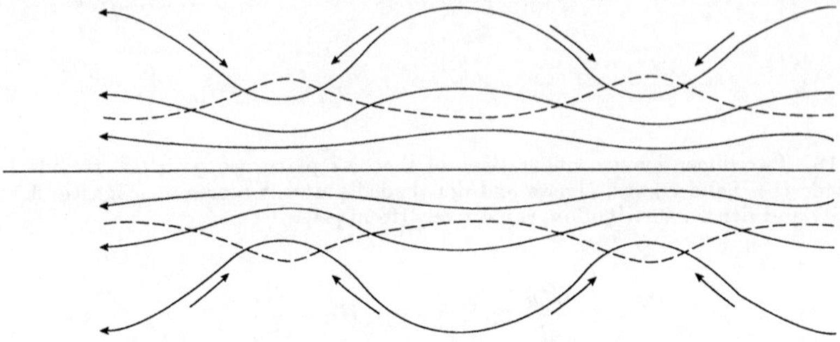

Fig. 17. A schematic presentation of the Parker instability (see text).

Recently, Tout and Pringle [48] suggested a simple physical model which uses the Balbus-Hawley and the Parker instability to operate a magnetic dynamo in the accretion disk and thereby provide an effective viscosity. The proposed mechanism is self-consistent in the sense that it does not depend on an assumed pre-existing turbulence in the disk.

The way in which the dynamo is supposed to operate is the following. Let us follow the time development of the three field components. The azimuthal field, B_ϕ, is generated from the radial field B_R by shear. At the same time the dominant loss mechanism for B_ϕ is assumed to be the Parker instability. Therefore, the equation for B_ϕ takes the approximate form

$$\frac{dB_\phi}{dt} = \frac{3}{2}\Omega B_R - \frac{B_\phi}{\tau_p} \tag{61}$$

where τ_p is the growth timescale for the Parker instability.

The radial field is assumed to be generated from the vertical component B_Z by the Balbus-Hawley instability and it suffers losses by the Parker instability, therefore the relevant equation reads

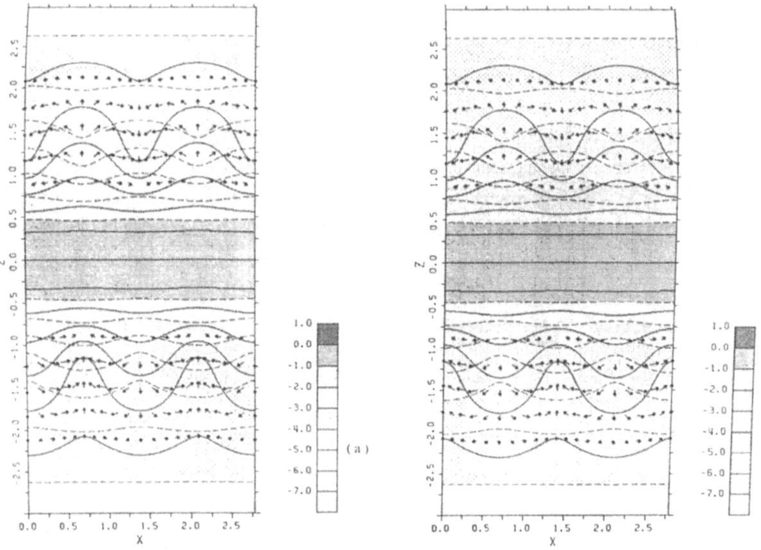

Fig. 18. Two-dimensional configurations of the disk perturbed with the f-mode (a) and $1h$-mode (b). Solid curves, arrows and dot-shaded patterns represent magnetic field lines, velocity and density distribution, respectively (from [47]).

$$\frac{dB_R}{dt} \simeq K(V_{AZ})\Omega B_Z - \frac{B_R}{\tau_P} \tag{62}$$

where $K(V_{AZ})$ is a function of the Alfven speed which can obtain different values in different regimes of V_{AZ}, according to the properties of the Balbus-Hawley instability. In particular, Balbus and Hawley found that for $V_{AZ}/C_S > \sqrt{6}/\pi$ the instability is suppressed (see also Papaloizou and Szuszkiewicz, [45]).

The vertical component B_Z is assumed to be generated by the Parker instability while the main flux loss mechanism is assumed to be reconnection. Namely, patches of B_Z of opposite sign reconnect, leading to loss of flux above the disk surface. So the approximate equation for B_Z reads

$$\frac{dB_Z}{dt} \simeq \frac{B_\phi}{\tau_P} + \frac{B_R}{\tau_P} - \frac{B_Z}{\tau_{rec}} \tag{63}$$

where τ_{rec} is the reconnection timescale. Tout and Pringle solved the set of approximate equations (61)–(63) and showed that a finite magnetic field configuration can be obtained (although non-stationary), with a time-averaged effective viscosity parameter of $\alpha_{ss} \sim 0.4$.

While this magnetic dynamo model is clearly highly simplified (in particular it does not solve the real MHD equations), and certainly cannot be considered to be the final word on the topic, it does represent an attempt to propose a model that is entirely based on known physical processes. As such, this model can be regarded as a very important step towards our understanding of disk viscosity.

A different model, for an internal waves-driven dynamo was proposed by Vishniac and co-workers.

Firstly, Vishniac and Diamond [49] pointed out the existence of a class of slightly non-axisymmetric, low frequency, internal waves in accretion disks. The restoring force for these internal waves (which are in fact g-modes) is provided by the (stable) positive gradient of the specific angular momentum (Rayleigh's criterion for dynamical stability $d(R^2\Omega)/dR > 0$ is satisfied in Keplerian disks). Secondly, Vishniac *et al.* [50] proposed that these waves can drive an $\alpha - \omega$ dynamo effect, by creating a radial field from the azimuthal field component. The azimuthal component was assumed to be generated by the shear (from the radial component) as described above. Finally, Vishniac and Diamond [51] examined the magnetic shearing instability (the Balbus-Hawley instability) and concluded that the net effect of this instability will be to generate a strong MHD turbulent cascade, characterized by a length scale of order V_A/Ω (see above) and a turnover time of order Ω^{-1}.

As a consequence of the picture described above, Vishniac and Diamond concluded that the magnetic dynamo will saturate when the rate of loss of magnetic flux due to diffusion of field lines (which in itself is a consequence of small scale turbulence generated by the magnetic shearing instability) will be equal to the rate of growth. They further claimed that the Parker instability is likely to be suppressed by the MHD turbulence (or that in any case buoyancy is irrelevant for the estimates of the rate of angular momentum transport).

The model of Vishniac and Diamond certainly has some interesting elements in it and it represents an attempt to present a "theory" for the angular momentum transport mechanism. However, the ingredients of the model in terms of physical processes are less transparent (at least to the present author) than in the model of Tout and Pringle. In particular, it has not been demonstrated at all that the internal waves will indeed play the role envisaged for them by Vishniac and Diamond. Even the ability of p-modes and g-modes to transport angular momentum over substantial distances is very questionable (see *e.g.*, discussion in 6.2.1), let alone their ability to complete the dynamo cycle. More rigorous time-dependent calculations and perhaps some numerical MHD work (of the type of the work of Geertsema [52], which demonstrated the development of MHD turbulence) will be required to clarify the question of magnetic dynamos.

At any event, the recently proposed models give hope that the elusive angular momentum transport mechanism in accretion disks will soon be discovered.

7. Accretion from a Stellar Wind—The "Flip-Flop" Instability

In addition to accreting from Roche lobe overflow, stars in binary systems can also accrete from the stellar wind emitted by their companions. This process is important in particular for: (1) massive x-ray binaries, in which neutron stars or black holes accrete from the winds of OB supergiant or Be star companions [53], [54]; (2) symbiotic binaries and some barium stars, in which white dwarfs accrete from the winds of cool giants [55], [56]; and (3) ζ Aur binaries, in which main sequence stars accrete from the winds of cool companions [57].

Here I will discuss only very briefly the general process of accretion onto a compact object from a stellar wind and I will concentrate on some recent work related to the stability of the accretion flow.

7.1 The Mass Accretion Rate

The general problem of accretion by a compact object from an infinite medium received considerable attention over the last 50 years.

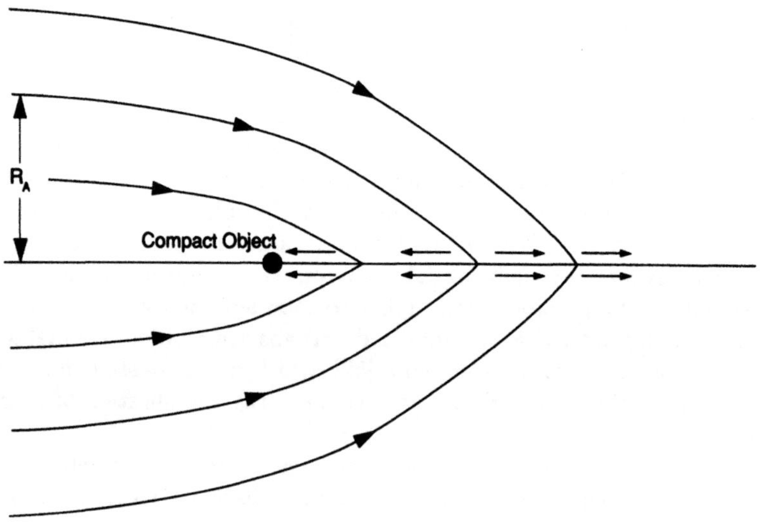

Fig. 19. Bondi-Hoyle accretion onto a compact object (see text).

In the original Hoyle and Lyttleton [58] scenario, the particles of the medium were assumed to follow free (Keplerian) paths far from the accreting object and to interact inelastically only on the down-stream axis (Fig. 19). It has been assumed that these inelastic collisions lead to a cancellation of the momentum component transverse to the accretion line, leaving only the parallel component unchanged. This picture leads to a maximal impact parameter, the "accretion radius", R_A, for which, following the interaction, the material will be accreted (having a velocity smaller than the local escape velocity). On the basis of energy arguments we can deduce an expression for the accretion radius

$$R_A = \frac{2GM}{V^2} \tag{64}$$

where M is the mass of the compact object and V is the velocity at a large distance upstream. Thus, all the material flowing into a cylinder with a radius equal to the accretion radius, should be accreted according to this scenario. This gives an accretion rate

$$\dot{M}_{HL} = \pi R_A^2 \rho_\infty V = \frac{4\pi (GM)^2 \rho_\infty}{V^3} \tag{65}$$

where ρ_∞ is the density of the gas at a large distance upstream. Bondi and Hoyle [59] replaced the singular (in density) accretion line obtained in the Hoyle-Lyttleton picture by a more physical, finite density accretion cone. They found, still for *a velocity dominated flow* (namely, where pressure effects can be neglected) an accretion rate of

$$\dot{M}_{BH} = \frac{4\pi\alpha (GM)^2 \rho_\infty}{V^3} \tag{66}$$

where $1/2 \lesssim \alpha \lesssim 1$ is a parameter which is in fact indeterminate by the theory (which depends on the initial conditions). Bondi [60] considered the *pressure dominated* case, namely, spherical accretion from an initially static medium. He found an accretion rate of

$$\dot{M}_B = 4\pi\lambda(\gamma) R_B^2 \rho_\infty C_\infty = \frac{4\pi\lambda(\gamma)(GM)^2 \rho_\infty}{C_\infty^3} \tag{67}$$

where C_∞ is the speed of sound at infinity and $\lambda(\gamma)$ is a parameter which depends on the specific heats ratio and assumes values in the range 0.25–1.12. Bondi also suggested an interpolation formula between the limiting cases described by expressions (66) and (67), which can be written as (I introduce a factor 2 difference from Bondi's formula since this was found to agree better with more recent numerical calculations)

$$\dot{M}_B \simeq \frac{4\pi (GM)^2 \rho_\infty}{V^3} \frac{\mu^3}{(1+\mu^2)^{3/2}} \tag{68}$$

where μ is the Mach number at infinity.

The realization that the problem of accretion from an infinite medium is very relevant for compact objects accreting from stellar winds, has renewed the interest in Bondi-Hoyle accretion. Much of the more recent work was numerical, and it concentrated on the use of multi-dimensional hydrodynamic codes [61]–[63]. The relatively recent two dimensional, fluid dynamical calculations of Shima *et al.* [64] gave for the accretion rate

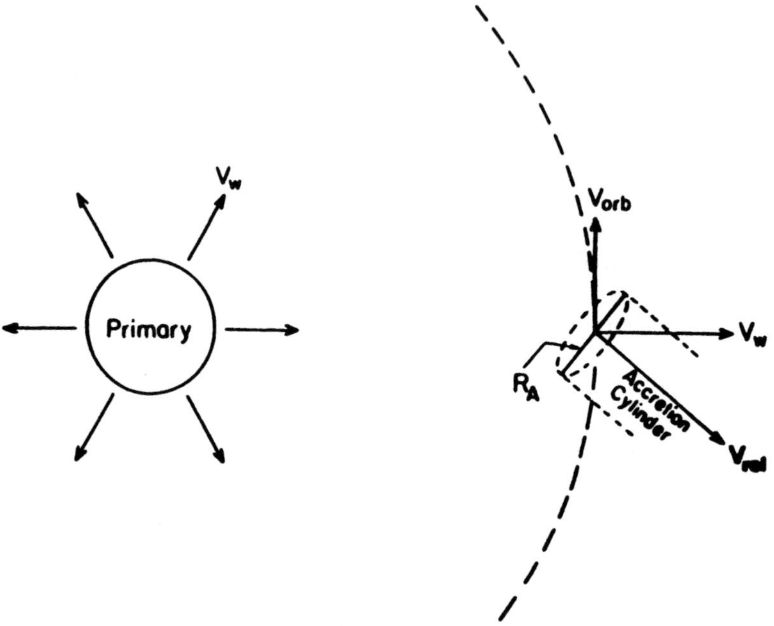

Fig. 20. Accretion by a compact object from the wind of a companion.

$$\dot{M} = F(\gamma, \mu) \cdot \frac{4\pi(GM)^2 \rho_\infty}{V^3} = F(\gamma, \mu)\dot{M}_{HL} \tag{69}$$

where the average (over Mach number) value of $F(\gamma, \mu)$ was about 0.75 for $\gamma = 5/3$ and about 1.0 for $\gamma \lesssim 4/3$. We therefore see that the Bondi-Hoyle theory (as expressed by eq. 68) gives the accretion rate with a quite good accuracy.

7.2 Accretion of Angular Momentum

One of the problems associated with accretion from a stellar wind is the fact that the medium is not homogeneous. It is clear that even in the absence of clumpiness in the wind, the mere existence of a density (or velocity) gradient results in a difference in the density (or velocity) of the material entering the accretion cylinder's cross section (points of the cross section that are closer to the star that emits the wind will receive denser material, see Fig. 20). If the accretion cylinder were to remain unchanged (from the homogeneous case), this would result in a net deposition of angular momentum into it. Because of the fact that no basic theory for the inhomogeneous case exists (for some early attempts see [65], [66]), a number of authors simply *assumed that all the angular momentum deposited into the accretion cylinder is accreted by the compact object* ([67]–[69]).

We can estimate the rate of accretion of angular momentum on the basis of this assumption [69]. In Fig. 20, the relative velocity between the wind and the neutron star at a point that is at a distance r from the mass losing star's center is $v_{rel}(r) = (v_w^2(r)+v_{orb}^2)^{1/2}$, where $v_w(r)$ is the wind velocity and $v_{orb} = 2\pi a/P_{orb}$. The angle between v_w and v_{rel} is given by $\tan\alpha = v_{orb}/v_{rel}$. If we choose coordinates y and z across the cylinder's cross section (with the origin at the center), so that z denotes the distance from the orbital plane and y measures the lever arm with respect to the compact object, then the rate at which angular momentum crosses an area element $dydz$ is $d\tau = \rho(r)v_{rel}^2(r)ydydz$. Since in most cases $R_A \ll a$, we can expand $\rho(r)$ and $v_{rel}(r)$ about the center of the accretion cylinder's cross section (located at $y = 0$, $z = 0$, $r = a$), to first order

$$\begin{aligned} \rho(r) &\simeq \rho(a) - \rho'(a)y\sin\alpha \\ v_{rel}(r) &\simeq v_{rel}(a) - v'_{rel}(a)y\sin\alpha \end{aligned} \tag{70}$$

From the continuity equation for the wind we have

$$\rho'(a) = -\frac{2\rho(a)}{a} - \frac{\rho(a)}{v_w(a)}v'_w(a) \tag{71}$$

and after substituting for $v'_{rel}(a)$ we obtain from (70)

$$\begin{aligned} \rho(r) &\simeq \rho(a)\left\{1 + \left[\frac{2}{a} + \frac{v'_w(a)}{v_w(a)}\right] y\sin\alpha\right\} \\ v_{rel}(r) &\simeq v_{rel}(a)\left\{1 - \frac{v'_w(a)}{v_w(a)}y\sin\alpha + \left[\frac{v_{orb}}{v_{rel}(a)}\right]^2\left[\frac{v'_w(a)}{v_w(a)} - \frac{1}{a}\right] y\sin\alpha\right\}. \end{aligned} \tag{72}$$

Substituting into the rate of accretion of angular momentum we obtain

$$d\tau = \rho(a)v_{rel}^2(a)\left\{1 + \left[\frac{2}{a} + \frac{v'_w(a)}{v_w(a)}\right] y\sin\beta\right\} ydydz \tag{73}$$

where we have denoted $\sin\beta = v_{orb}/v_{rel}(a)$ and kept only terms to order $\sin\beta$. Expression (73) must now be integrated over the accretion cylinder's cross section, the boundary of which is given by

$$y^2 + z^2 = \left(\frac{2GM}{v_{rel}^2(a)}\right)^2 \left[1 + 4\frac{v'_w(a)}{v_w(a)}y\sin\beta\right]. \tag{74}$$

The limits of integration are thus from $z = -R_A \equiv -2GM/v_{rel}^2(a)$ to $z = R_A$ and from $y = -(R_A^2 - z^2)^{1/2} + 2R_A^2[v'_w(a)/v_w(a)]\sin\beta$ to $y \simeq +(R_A^2 - z^2)^{1/2} + 2R_A^2[v'_w(a)/v_w(a)]\sin\beta$. Carrying out the integration and dividing by the mass accretion rate $\dot{M} \simeq \pi R_A^2\rho(a)v_{rel}(a)$ we obtain for the specific angular momentum of the accreted matter

$$j \simeq \frac{1}{2}\Omega_{orb}R_A^2\eta \tag{75}$$

where

$$\eta = 1 + \frac{7}{2} a \frac{v_w'(a)}{v_w(a)}. \qquad (76)$$

Davies and Pringle [70] were the first to note, that in the context of the original Hoyle-Lyttleton picture, particles from both sides have to cancel their momentum component transverse to the accretion line in order to be accreted at all. They suggested that a similar thing should occur in the inhomogeneous case, with the accretion line being somewhat displaced towards the lower density side. They pointed out, therefore, that *the conditions imposed on the material in order to be accreted are in conflict with the idea that angular momentum can be accreted at all* (matter flowing along a line towards a point has no angular momentum). Using a highly simplified, *two dimensional* treatment they have shown that the net rate of accretion of angular momentum was indeed zero. Soker and Livio [71] used a perturbation approach to demonstrate that the accreted angular momentum can be somewhat different from zero in the three-dimensional case.

The first numerical calculations of this problem [72], [73] obtained a a rate of accretion of angular momentum that was considerably lower (by about a factor 5) than the one obtained from eq. (75). It is important to note however, that these early calculations used a particle description of the hydrodynamics and suffered from a relatively poor resolution and a high numerical viscosity. The large viscosity was able to suppress instabilities and produce an artificial steady state. These results therefore represent at best *average* values.

Besides the theoretical interest from the purely hydrodynamical point of view, the importance of the rate of accretion of angular momentum lies in particular in two observable phenomena: (i) The spin-up of the accreting compact object (in particular neutron stars in X-ray binaries) and (ii) the possibility of forming an accretion disk around wind-fed accretors. For a disk to form, the specific angular momentum of the accreted material must be sufficient to allow it to enter a Keplerian circular orbit around the magnetosphere (for magnetized neutron stars) or around the white dwarf (for non-magnetized white dwarfs accreting from the winds of cool giants). This can be expressed by the condition

$$j > (GM_{ns}R_{mag})^{1/2} \quad \text{for neutron stars}$$

$$\text{or} \qquad (77)$$

$$j > (GM_{WD}R_{WD})^{1/2} \quad \text{for white dwarfs.}$$

Typically, conditions (77) are difficult to satisfy. The radius of the magnetosphere can be estimated as follows, the ram pressure of the infalling gas $\left(\rho v_{ff}^2 \sim \frac{2GM\rho}{R_{mag}}\right)$ must be equal to the magnetic pressure ($B_{mag}^2/8\pi$). For an assumed dipole field ($B_{mag} = B_{ns}R_{ns}^3/R_{mag}^3$) we obtain

$$R_{mag} \simeq \left[\frac{B_{ns}^2 R_{ns}^6}{4(GM_{ns})^{1/2}\dot{M}}\right]^{2/7}. \qquad (78)$$

Now, it can be shown that the first of conditions (77) implies (using eq. 78) a relative velocity between the neutron star and the wind, satisfying

$$V_{rel} \lesssim 4\times 10^7 \left(\frac{\xi}{0.2}\right)^{1/4} \mu_{30}^{-\frac{1}{14}} \left(\frac{M_{ns}}{M_\odot}\right)^{5/14} \left(\frac{P_orb}{1 \text{ day}}\right)^{-1/4} \left(\frac{R_{ns}}{10^6 \text{ cm}}\right)^{1/28}$$
$$\left(\frac{L_x}{10^{37} \text{ erg s}^{-1}}\right)^{1/28} \text{cm s}^{-1} \tag{79}$$

where ξ is the factor by which the rate of accretion of angular momentum is reduced (if indeed it is reduced), μ_{30} is the neutron star's magnetic moment (in units of 10^{30} Gauss cm^3) and L_x is the neutron star's x-ray luminosity. Such relative velocities are extremely low since characteristic wind velocities are of order 1000–2500 km s^{-1} [74]. Thus, taken at face value the early results on the accretion of angular momentum would seem to imply that accretion disks do not form in wind-fed systems.

A new phenomenon was introduced into the picture when higher resolution hydrodynamical calculations were performed. Matsuda, Inoue and Sawada [75] found that the flow did not reach a steady state in some cases. Instead, it exhibited a "flip-flop" behavior, in which the shock cone oscillated from side to side, accompanied by short periods of quasi-disk formation. The same type of unsteady behavior was found also in other 2D calculations [76], [77]. Fig. 21, shows a typical flip-flop motion in which the wake oscillates from side to side. The time behavior of the corresponding specific angular momentum at the inner boundary (around the compact object) is shown in Fig. 22. What can be seen is that instead of staying at the relatively low, steady state average value, the specific angular momentum oscillates with a large amplitude (of about 40 times the average value), changing its sign quasi periodically. For the particular parameters of this calculation disk formation occurred for $|j| \gtrsim 0.14 R_A V$. Thus, a temporary disk was formed (for a duration of about $4R_A/V$), changing its sense of rotation with the corresponding j sign reversals. The following phenomenon was noted: when the wake begins to swing to the opposite direction, the specific angular momentum starts decreasing (due to the disk being fed by angular momentum of the opposite sign), this has the effect of emptying the disk material onto the accreting object, resulting in a flare type behavior. The time behavior of the resulting accretion rate is shown in Fig. 23.

7.3 The Stability of Wind Accretion

The two main questions that emerged at this point were the following: (1) Does the development of the flip-flop instability (in 2D) *require* the presence of a density or velocity gradient or is the instability an intrinsic property of Bondi-Hoyle accretion flows (at least in two-dimensions)? (2) Is the instability an artifact of the use of two-dimensions (in the numerical calculations) to describe a flow which is really three-dimensional, or does the instability appear also in three-dimensional (3D) calculations? (in which case it is more likely that it represents a genuine property of realistic flows).

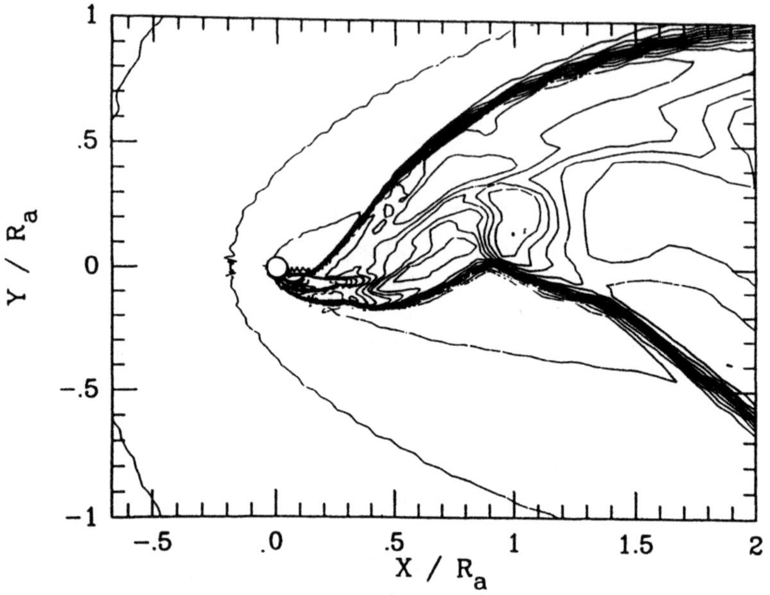

Fig. 21. The density distribution for a flow with a small velocity gradient (from [76]).

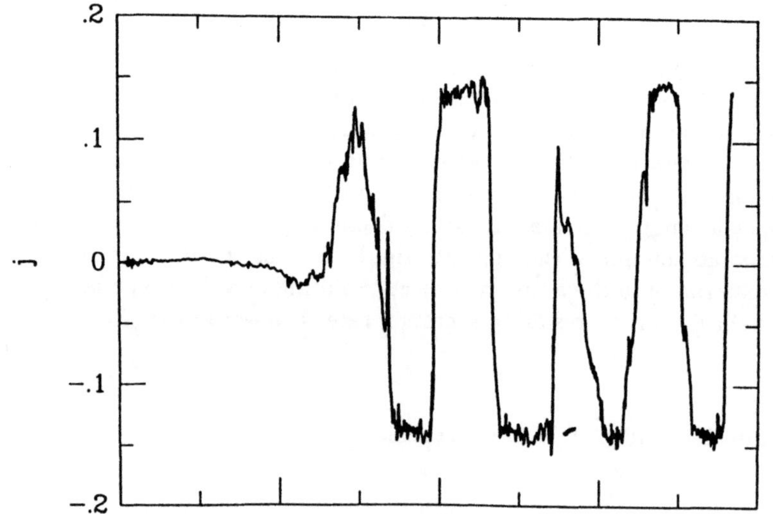

Fig. 22. The specific angular momentum (in units of $R_A V$) at the inner boundary as a function of time (from [76]).

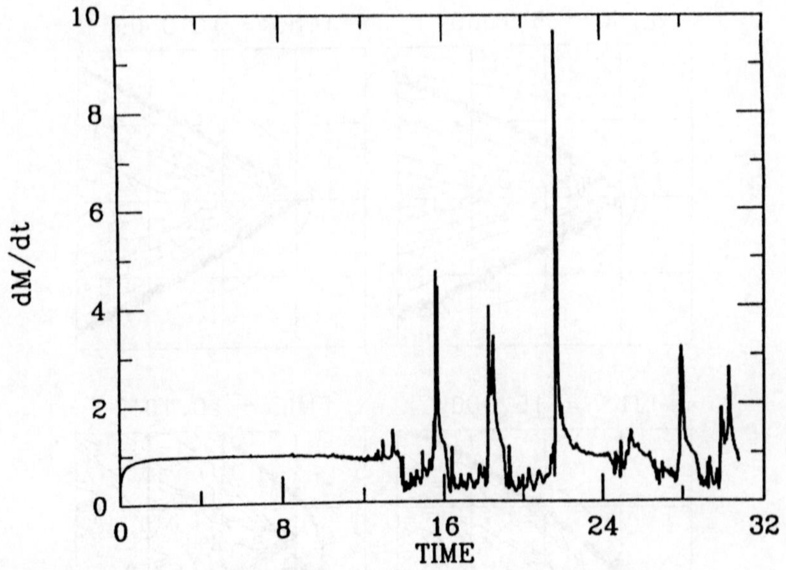

Fig. 23. The mass accretion rate in units of the rate deposited into the symmetric accretion cylinder (from [76]).

In order to answer the first question, Matsuda *et al.* [78], [79] performed a series of calculations in 2D, in which the medium was taken to be homogeneous at infinity. Their results can be summarized as follows:

(1) In all the cases in which the size of the accreting object was taken to be sufficiently small ($R_A/16$, where R_A is the accretion radius) and for a specific heats ratio $\gamma \geq 4/3$, the flip-flop instability appeared and eventually took the form of a violent, aperiodic oscillation. In all of these cases, the violent oscillation was preceded by the formation of a wide, dome-shaped shock, on the upstream side of the accreting object (Fig. 24).

(2) In isothermal calculations, the flip-flop oscillation was quite smooth and quasi-periodic (Fig. 25). In a calculation in which the size of the accreting object was suddenly increased (from $R_A/16$ to $5R_A/16$), after the increase, the oscillation also became quasi-periodic (Fig. 26). The flows which showed a quasi-periodic behaviour were characterized by relatively narrow shock-cones.

(3) When the size of the accreting object was taken to be large ($5R_A/16$) throughout the run, the flow was found to be stable for the duration of the run ($t = 45$, in units of R_A/C_s where C_s is the speed of sound at infinity). For an intermediate size ($3R_A/16$) the oscillation was very weak and quasi-periodic.

(4) Essentially all the unstable flows were accompanied by the shedding of pairs of vortices.

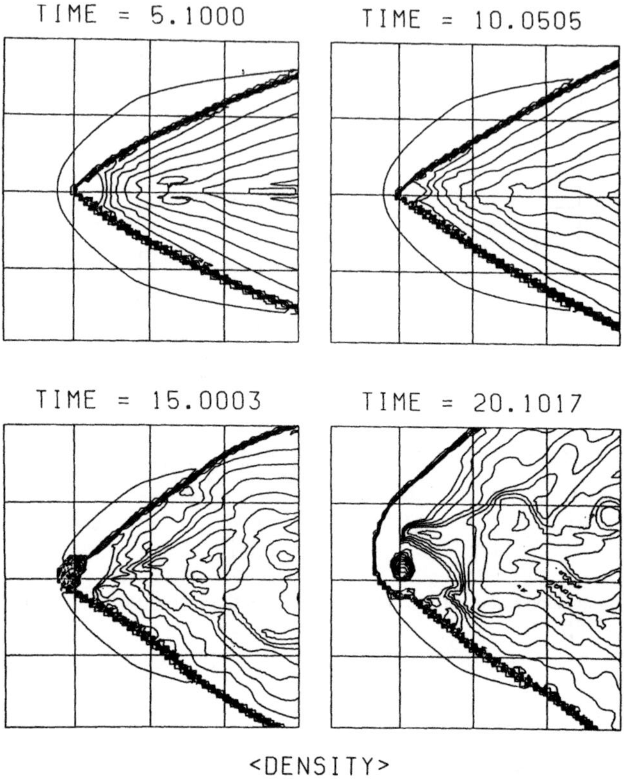

Fig. 24. Density contours in a two-dimensional run, showing the formation of a dome-shaped shock on the upstream side (from [79]).

(5) The instability always started by radial, "pumping" modes, which appeared as pulses in the density and velocity field along the accretion axis.

Analytical studies of Bondi-Hoyle flows treated the stability of two entities: the "accretion line" (on which matter from both sides collides) and the "shock cone" (which results from the supersonic flow).

Analyses of the stability of the accretion line (which treated the linear regime and neglected pressure) demonstrated that [80]–[82]:

i. The accretion line is unstable against radial modes (both in 2D and in 3D).
ii. In 2D, the accretion line is also unstable against tangential modes.

A simplified analysis of the stability of the shock-cone showed that [83]:

i. The shock cone is likely to be unstable to transverse deflections when the shock opening angle is large.
ii. For shock opening angles which are not extremely large, the flow is stabilized when the size of the accreting object exceeds some critical value.
iii. The instability starts near the accreting object.

The picture of the nature of the instability *in 2D* that seems to emerge from all of these studies is the following:

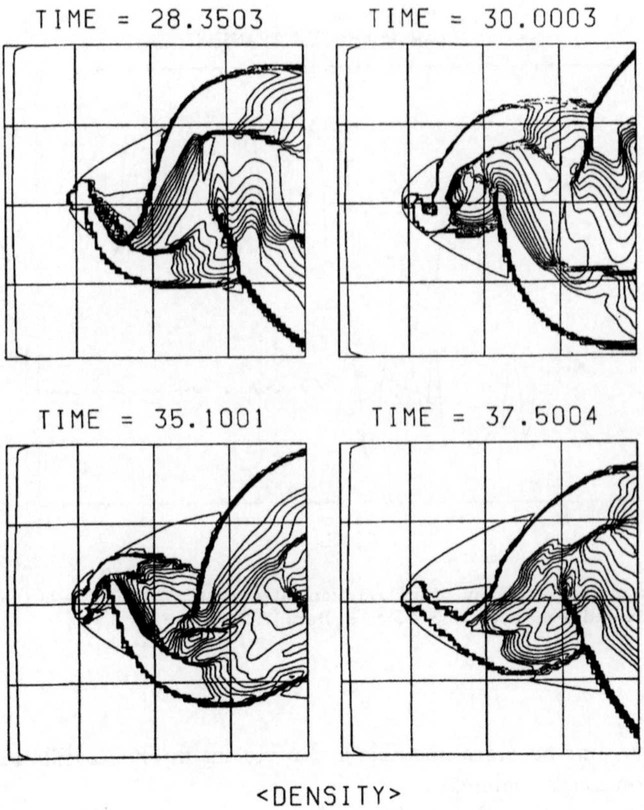

Fig. 25. Density contours in a nearly isothermal calculation (from [79]).

1. Radial modes of the accretion line instability are excited first (with the possible exception of cases in which the accreting object is very large).
2. The radial modes generate a "pumping-type" accretion process from the down-stream side, by causing large variations in the density and velocity along the accretion line.
3. The subsequent evolution depends on whether the shock-cone becomes unstable or not. The following possibilities seem to emerge: (a) The shock-cone remains relatively narrow. This happens for low values of $\gamma(\gamma \sim 1)$ or when the accreting object is sufficiently large, so as to be able to accrete directly the mass inflow pulses which result from the radial modes. In this case the shock-cone remains stable and the flow is dominated by long wavelength tangential modes, which develop into a smooth, quasi-periodic oscillation. (b) The shock-cone opening angle widens. This happens for large values of $\gamma(\gamma \gtrsim 4/3)$ and for small sizes of the accreting object. The cause for the widening can either be the formation of a quasi-rigid atmosphere which forces material upstream, or the accretion or some angular momentum (which results in some type of circular motion). In this case,

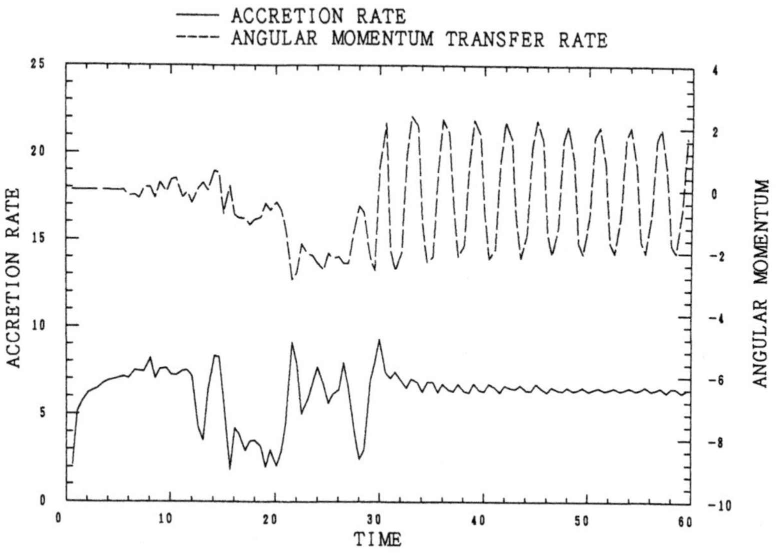

Fig. 26. Mass and angular momentum accretion rates. At $t = 30$ the size of the accreting object was increased from 1×1 to 5×5 cells (from [79]).

the shock-cone becomes unstable and a violent flip-flop oscillation ensues in the non-linear regime.

In addition, we note the shedding of pairs of vortices behind the accreting object, in a somewhat similar manner to Karman vortex shedding in 2D incompressible flows past a circular cylinder (Fig. 27).

In order to answer the second question posed at the beginning of this section, namely, does the instability appear also in 3D? Ishii *et al.* [84] performed a few *preliminary* calculations in 3D. Their results can be summarized as follows:

1. For adiabatic flows, the formation of a bow shock which is detached from the accreting object was observed (unlike the 2D case in which the shock was attached to the body).
2. Radial "pumping" modes developed.
3. Vortex rings-type structures were formed.
4. Distortions did appear in the shock-cone, accompanied by some side to side motion, but no violent flip-flop of the entire shock was observed. This was true for both adiabatic and isothermal flows.

In particular, the shock-cone was found to contain the accretion axis at all times. The preliminary results seem therefore to indicate that the flip-flop instability takes a less violet form in 3D.

However, it should be remembered that in addition to some real differences in the physics (*e.g.*, the absence of tangential modes in 3D and the fact that the 3D shock is more robust against deflections), the resolution of the calculations in 3D is necessarily lower than in 2D. This results in a higher numerical viscosity

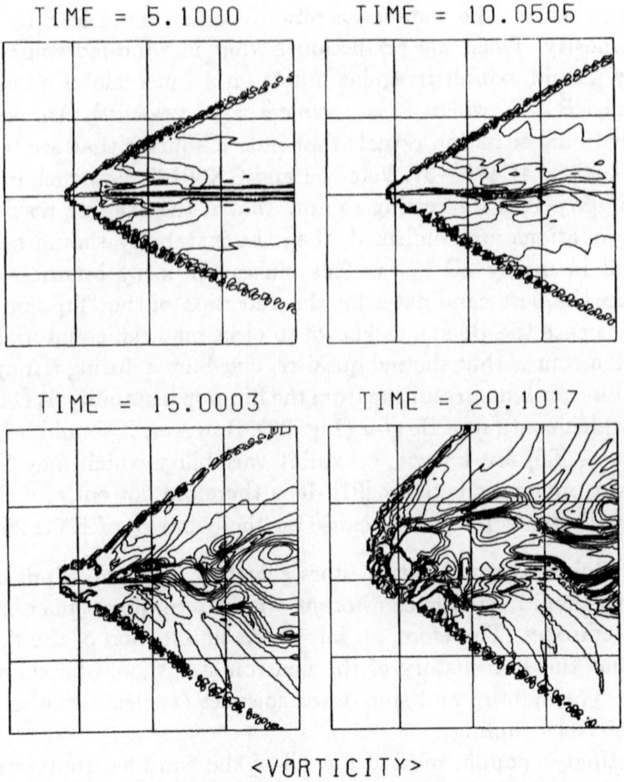

Fig. 27. Dimensionless rotation contours. Rotations above and below the accretion line start by having opposite sign (from [79]).

and a larger damping of potential instabilities. Thus, at present, we cannot determine with certainty the nature of the flow in 3D.

7.4 Possible Observational Implications

Assuming that the flip-flop instability indeed appears in real flows, it can have some extremely interesting observational implications for systems in which compact objects accrete from stellar winds and in particular for massive x-ray binaries. I shall now discuss some of these implications.

7.4.1 Variability in the X-ray Luminosity. During the oscillations of the shock-cone, the compact object (a neutron star or a black hole in the case of massive x-ray binaries) accretes material with a relatively high specific angular momentum of opposite signs, as the shock flip-flops from side to side. This leads to the following phenomenon: a temporary accretion disk forms, and as it starts to be fed by angular momentum of the opposite sign, the disk collapses onto the compact object. This is followed by the formation of a disk rotating in the opposite sense. The sudden accretion events should lead to flares in the x-ray luminosity. Thus, the prediction is that in wind-fed sources, the x-ray luminosity should exhibit irregular bursts on a timescale of minutes to hours (a few times R_A/V, where V is a typical wind velocity). An examination of the observational situation reveals that indeed sources that are believed to be wind-fed, such as 4U1700–37, Vela X-1 and GX301-2 show such irregular x-ray bursts [85]–[87]. It is interesting to note that if the existing results of the numerical simulations are confirmed, then the instability should be much more pronounced in nearly 2D flows. This makes the x-ray binaries with Be star companions the best candidates for the detection of the "flip-flop" instability, since in this case the Be star is known to eject material equatorially.

Indeed, a source that showed quasi-regular bursts during a major outburst, that were interpreted as resulting from the flip-flop instability, is EXO 2030+375 [88], [89] which contains a Be star (Fig. 28). However, it should be remembered that Be stars [90] are known to exhibit variability which may be related to pulsations or magnetic activity [91]. It is therefore not entirely clear whether flip-flop oscillations are the real cause for the behavior of EXO 2030+375.

7.4.2. The Spin History. Neutron stars accreting from the winds of early type companions should experience a torque with alternating signs during the disk formation episodes. Therefore, an important implication of the flip-flop instability is that the spin history of the neutron star should be characterized by shot-noise type spin-up and spin-down episodes (neglecting other effects such as crust-interior coupling).

Interestingly enough, most if not all of the wind-fed sources show indeed an alternating spin-up, spin-down behavior, while the neutron stars in systems that are fed by Roche lobe overflow show a relatively smooth spin-up history (Fig. 29). It is in fact possible to determine which systems are wind-fed simply from an examination of their spin histories (compare for example, Vela X-1 and GX 301-2 which are wind-fed to SMC X-1 and Her X-1 which are fed by Roche lobe overflow). In particular, the angular frequency of Vela X-1 shows variations at a rate of $|\dot{\Omega}/\Omega| = (5.8 \pm 1.4) \times 10^{-3}$ yr^{-1}, with the sign of the derivative reversing on timescales as short as the resolution of the data [92].

7.5 Summary of the Present Status and Remaining Problems

Based on the numerical and analytical work described in the previous sections we note the following preliminary conclusions:

1. In two-dimensions, Bondi-Hoyle accretion flows appear to be non-steady. They exhibit a "flip-flop" instability even in the homogeneous case.
2. More high-resolution calculations are needed in order to determine the possible existence and nature of the instability in three-dimensions.
3. The instability (if it indeed occurs in realistic flows) can have important implications for the possibility to form accretion disks in wind-fed sources and for the spin history of the neutron stars in massive x-ray binaries. In particular, temporary disks of an alternating sense of rotation should form.

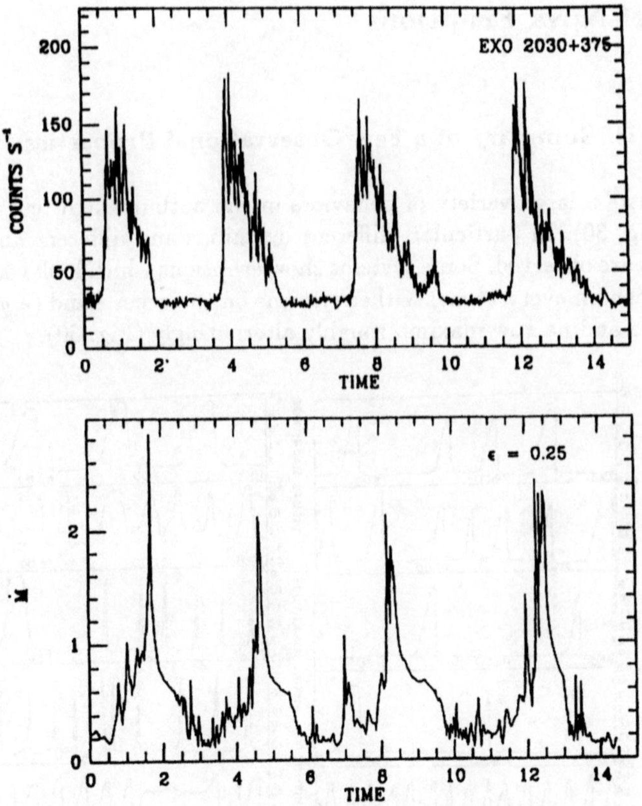

Fig. 28. The count rate in the 1–10 KeV energy range (upper panel) and the mass accretion rate obtained in numerical calculations (lower panel), (from [89]).

4. Observations of massive x-ray binaries are consistent with the predictions based on the existence of the flip-flop instability. In particular, these sources show: (i) irregular flares in their x-ray luminosity and (ii) a random walk-type spin-up, spin-down behavior. X-ray binaries containing Be stars may be the best candidates for exhibiting the flip-flop instability.

   It should be noted, however, that the explanation of the x-ray luminosity behaviour and the spin history in terms of the "flip-flop" instability is not unique. Both of these phenomena could be the consequence of the wind being clumpy.
5. An interesting question is related to Cyg X-1. The compact object in this system (a black hole candidate) is thought to be a wind accretor [90]. Yet, its x-ray light curve is quite smooth [93] and the system exhibits some signatures of a relatively cool accretion disk [94]. The question is: why doesn't this system show the same intensity variations as the other wind-fed sources? This leads to the following interesting possibilities: (1) either the black hole is in fact fed by Roche lob overflow, or (2) the flip-flop in-

8 Dwarf Nova Eruptions

8.1 A brief Summary of a Few Observational Properties

There exists a large variety of behaviors in the outburst light curves of dwarf novae (Fig. 30). In particular, different durations and different amplitudes of outbursts are observed. Some systems show eruptions which look very similar to one another (in every event). Other systems on the other hand (*e.g.*, SS Cyg), show wide and narrow maxima roughly alternatingly (Fig. 30).

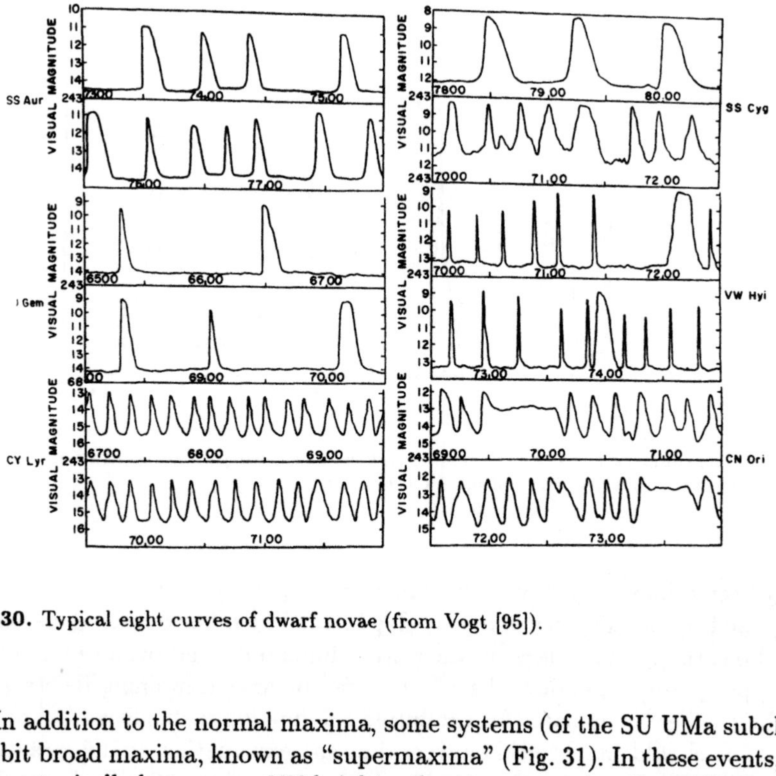

Fig. 30. Typical eight curves of dwarf novae (from Vogt [95]).

In addition to the normal maxima, some systems (of the SU UMa subclass) exhibit broad maxima, known as "supermaxima" (Fig. 31). In these events, the system typically becomes $\sim 0\overset{m}{.}6$ brighter than in a normal outburst (for a short time) and then stays at about normal maximum brightness for 3–5 times longer than in a normal maximum. Another subclass of objects (the Z Cam subclass) exhibits standstills. These are events which always follow a normal outburst, and in which the system remains at about one magnitude below maximum light for periods of time which range from a few days to a few years. These standstills are eventually followed by a return to minimum light.

Typical normal outbursts can have an amplitude of about 4 mag (like in the well studied system VW Hyi). The outbursts last for 4–5 days and recur about every 25–30 days (in VW Hyi).

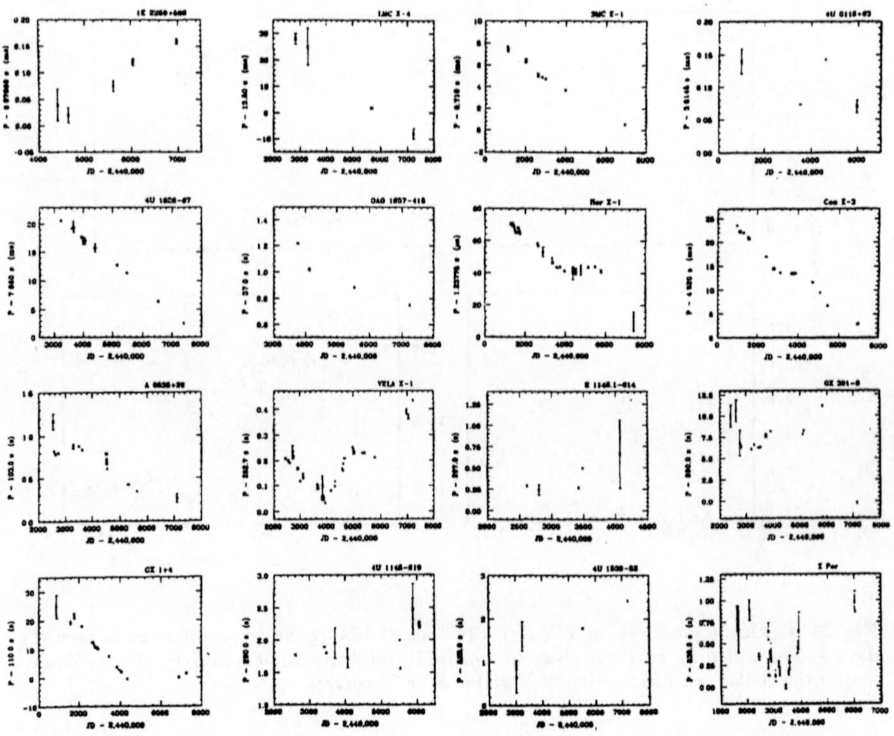

Fig. 29. The pulse period histories for 16 x-ray pulsars (adapted from Nagase [92]).

stability is not operative in this case. The second possibility is particularly intriguing. I note that it is much easier for an accretion disk to form around a black hole than around a magnetized neutron star, because the innermost stable orbit is much smaller in radius than a typical magnetospheric radius and therefore the average specific angular momentum can be much smaller (eq. 77). If a relatively stable disk forms around the black hole and spreads due to viscosity, its presence may suppress the development of the flip-flop instability. If this explanation is correct (which is not clear at all presently), this suggests the exciting possibility that the presence of black holes (rather than magnetized neutron stars) in massive x-ray binaries can be detected simply via the absence of the intensity variations associated with the flip-flop instability.

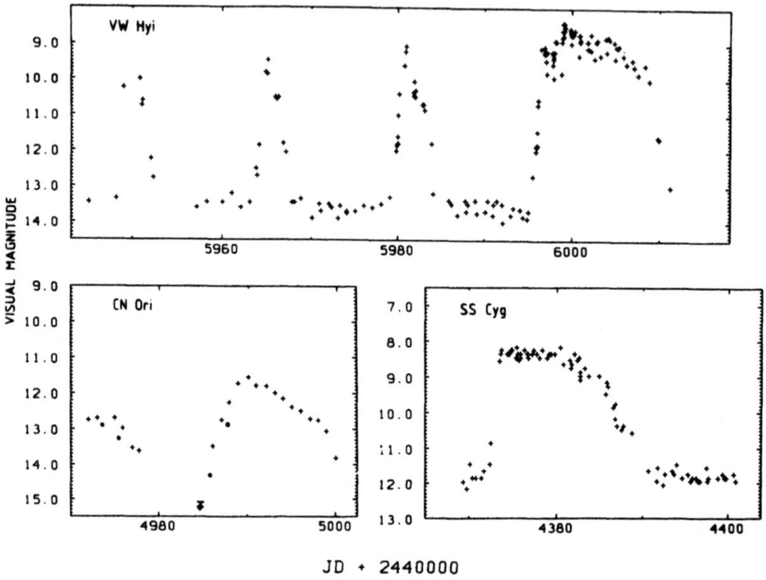

Fig. 31. Optical light curves of VW Hyi, CN Ori and SS Cyg. Supermaxima can be seen. The data was obtained by amateur observers in the Royal Astronomical Society of New Zealand and in the American Association of Variable Star Observers.

For dwarf novae in outburst, observations indicate effective temperatures in the disk of the order of 20000K [95]. For typical disk radii, this yields luminosities at maximum ($L_{disk} \sim 2\pi R^2_{disk}\sigma T^4_{eff}$) of order $15 L_\odot$. The total energy in an outburst is therefore of the order of a few times 10^{40} ergs.

A few other characteristics of the outbursts will be described in relation to the comparison between theory and observations. In the present lectures I will not discuss superoutbursts and standstills. A description of some of the recent work that is related to superoutbursts can be found in [96], [97] and references therein.

8.2 Accretion Disk Timescales

Before we examine models for dwarf nova eruptions it is important to realize that accretion disks are characterized by several timescales. The shortest timescale is the dynamical timescale, which is also the timescale on which vertical hydrostatic equilibrium is established (eq. 34)

$$\tau_{dyn} \sim \Omega^{-1} \simeq \frac{H}{C_S}. \tag{80}$$

A second timescale is the thermal timescale, which is of the order of the heat content divided by the dissipation rate (both calculated per unit area)

$$\tau_{therm} \sim \frac{\Sigma C_S^2}{D(R)} \sim \frac{\Sigma C_S^2}{\nu \Sigma \Omega^2} \sim \frac{1}{\alpha_{ss}} \tau_{dyn} \tag{81}$$

where we have used the fact that $\nu \sim \alpha C_S H$ and eq. (38). The third timescale is the viscous timescale, which is also the radial inflow timescale (see the expression for V_R following eq. 37)

$$\tau_{vis} \sim \frac{R}{V_R} \sim \frac{R^2}{\nu} \sim \frac{1}{\alpha_{ss}} \left(\frac{R}{H}\right)^2 \tau_{dyn}. \tag{82}$$

From eqs. (80)–(82) we see that for thin disks and if $\alpha_{ss} \lesssim 1$ then we have $\tau_{vis} > \tau_{therm} > \tau_{dyn}$.

8.3 Models for the Eruptions

The two main competing models that have been proposed for the cause of dwarf nova eruptions are:
(1) A mass transfer instability.
(2) A disk instability.

The basic idea behind both of these models is that the outburst energy is derived from accretion. Indeed, realizing that the total eruption energy is of the order of $E \sim 10^{40}$ ergs (8.1 above), only an amount of mass $\Delta M_{acc} \gtrsim (E R_{WD}/G M_{WD}) \gtrsim 2 \times 10^{-11} M_\odot$ has to be accreted in order to account for the energetics. Furthermore, for typical CV parameters the kinematic viscosity (in the context of the standard model) is of order $\nu \sim 10^{13}$ cm^2 s^{-1} (eq. 49). We can crudely estimate the mass of a steady state accretion disk by $M_{disk} \sim \dot{M} \tau_{vis}$, which for $R \sim 10^{10}$ cm, $\dot{M} \sim 5 \times 10^{16}$ g s^{-1} gives $M_{disk} \sim 10^{-10} M_\odot$. Thus, it seems that (at least energetically speaking) the accretion of an amount of mass that is a fraction of the disk mass is sufficient to power an outburst. In order to produce the observed variation in the brightness of the system, it is necessary of course, to have a variation in the accretion rate. Indeed, both the mass transfer instability model and the disk instability model agree that dwarf nova outbursts are caused by a change in $\dot{M}_{acc}$. The two models, however, differ in their suggestions for the cause of the change. The mass transfer instability model suggests that the variation is caused by a change in the mass transfer rate from the secondary star. The disk instability model assumes that the mass transfer rate is constant but that the accretion disk undergoes an instability which changes the effective viscosity, thus causing the accretion rate to change.

I shall now examine briefly the theoretical basis for each model.

8.3.1 The Mass Transfer Instability Model. It should be realized that as the cataclysmic binary evolves, secular changes in the mass transfer rate can occur as a result of angular momentum losses from the system (see Section 11). These changes, however, occur on a much longer timescale ($\sim 10^9$ yr). In order to explain dwarf nova outbursts, a mechanism which can produce variations on a timescale of months or less is required. Bath [98], [99], suggested that the outer layers of the secondary stars in CVs are dynamically unstable, because of the presence of the hydrogen and helium ionization zones below the stellar surface.

He pointed out that because of the fact that the secondaries are low mass main sequence stars, their effective temperatures are quite low (a few thousand K). Consequently, the hydrogen and helium ionization zones lie below the surface (and the surface layers are convective). Bath suggested that the lobe-filling star manages to tap the recombination energy in the ionization zones into inducing a dynamical instability. The idea is that if

$$\epsilon_{rec} \gtrsim \frac{GM_2}{R_2^2} h \tag{83}$$

where ϵ_{rec} is the recombination energy per gram and h is the depth of the ionization zone, then mass can be lifted from depth h and pushed through the Lagrangian point L_1. Since for hydrogen $\epsilon_{rec} \sim 10^{13}$ ergs g^{-1}, eq. (83) gives $h/R_2 \lesssim 5 \times 10^{-3}(R_2/R_\odot)(M_2/M_\odot)^{-1}$. If operative, such an instability could cause the mass transfer to increase on the dynamical timescale of the secondary (typically about an hour, $(GM_2/R_2^3)^{-1/2}$). Numerical calculations [100] suggested that under favorable conditions (ionization zone very close to the surface), an instability is indeed obtained. Bath further suggested that the quiescence timescale is determined by the thermal timescale of the envelope layers $\tau_{therm} \simeq \mathcal{R}\bar{T}\delta M/\delta L$, where δM is the mass of the envelope region in which a luminosity deficit δL was generated and $\bar{T}$ is the region's average temperature ($\mathcal{R}$ is the gas constant). For typical values, τ_{therm} is of the order of a month.

Even before we examine possible observational consequences of this model, it is important to note that from a purely theoretical point of view it is not clear how the model would work for most CVs. The point is, that many CVs contain secondaries with masses $M_2 < 0.5M_\odot$. For such stars, the convective envelope is so deep (for $M_2 < 0.3M_\odot$ the star is fully convective) that it is difficult to see how the ionization zones could drive the instability envisaged by Bath.

Another possibility (in principle at least) would be that the mass transfer rate is modulated by magnetic activity in the secondary's surface layers. Such an activity can in fact be expected, because of the fact that the envelope is convective and the stars are forced to corotate. Thus, dynamo generation of magnetic fields is a likely outcome. However, observations seem to indicate that the magnetic cycles tend to exhibit a longer timescale [101], [102], of the order of years to tens of years, which is inadequate for dwarf nova eruptions.

In spite of the theoretical difficulties for the mass transfer instability model described above, this model should not be discarded a priori. The main reason is that mass transfer variations which are still unexplained are observed in a number of CVs (*e.g.,* TT Ari, MV Lyr) and in Algol systems. Furthermore, changes in the accretion rate are observed in AM Her [103], which is believed to have no disk because of the white dwarf's strong magnetic field (so that the variations certainly cannot originate from the disk). Therefore, the real test for this model at present should be observational rather than theoretical (see section 8.4).

8.3.2 The Disk Instability Model. The time evolution of the accretion disk is governed by eq. (31) (which I repeat here for clarity)

$$\frac{\partial \Sigma}{\partial t} = \frac{3}{R}\frac{\partial}{\partial R}\left[R^{1/2}\frac{\partial}{\partial R}\left(\nu \Sigma R^{1/2}\right)\right]$$

and an equation describing the thermal evolution [105]

$$\frac{1}{2}\left[\Sigma C_V \frac{dT_c}{dt} - (\Gamma_3 - 1)C_V T_c \frac{d\Sigma}{dt}\right] = \frac{9\nu\Sigma\Omega^2}{8} - \sigma T_{eff}^4 - \frac{H}{R}\frac{\partial}{\partial R}(RF_R) \quad (84)$$

where T_c is the mid-plane temperature, Γ_3 is the adiabatic exponent, F_R is the radial radiative flux and the time derivatives are Lagrangian derivatives ($d/dt = \partial/\partial t + V_R \partial/\partial R$). In thermal equilibrium (and if we neglect the radial flux, assuming that the gradients in the Z direction are much larger), eq. (84) becomes

$$\sigma T_{eff}^4 = \frac{9\nu\Sigma\Omega^2}{8}. \quad (85)$$

Using eq. (85), the vertical hydrostatic equilibrium, the relation between viscosity and pressure and the relation between the central and effective temperatures (assuming only radiative transfer, κ is the opacity)

$$T_c \simeq T_{eff}\left(\frac{3}{8}\kappa\Sigma\right)^{1/4}, \quad (86)$$

we can construct a local model for the disk structure. In particular, the functional relation between Σ and $f \equiv \nu\Sigma = f(\Sigma)$ (or equivalently T_{eff}, eq. (85), or $\dot{M}$, eq. (37)) can be established [104], [105].

As it turns out, the function $f(\Sigma)$ exhibits an interesting property. Because of the strong temperature dependence of the (realistic) opacity in the temperature range which corresponds to partial ionization of hydrogen and helium, a branch of $f(\Sigma)$ is obtained, on which $df/d\Sigma < 0$. In general, the solutions of the structure equations show that the function $f(\Sigma)$ is multi-valued (see Fig. 32). For a given interval in surface density, equilibrium solutions exist in which the disk is either hot and ionized or cool and mostly neutral. It is easy to convince ourselves that the hot and cool branches (with $d\log T_{eff}/d\log\Sigma > 0$) are thermally stable and therefore represent valid equilibria. Indeed, on these branches if the surface density increases (thereby increasing the rate of dissipative energy generation) then the effective temperature increases too (thereby increasing the rate at which energy is radiated away). The middle branch (with $d\log T_{eff}/d\log\Sigma < 0$) is however *thermally unstable*. (Incidentally, this branch is also viscously unstable since it represents a diffusion equation with a negative diffusion coefficient, eq. (31)). This means that if the mass transfer from the secondary star has a value corresponding to the unstable branch, no stationary solution is possible. Instead, the particular region in the disk to which the S curve corresponds, will undergo a limit cycle (as indicated by the arrows in Fig. 32) as follows. Suppose the region was originally on the cool branch,

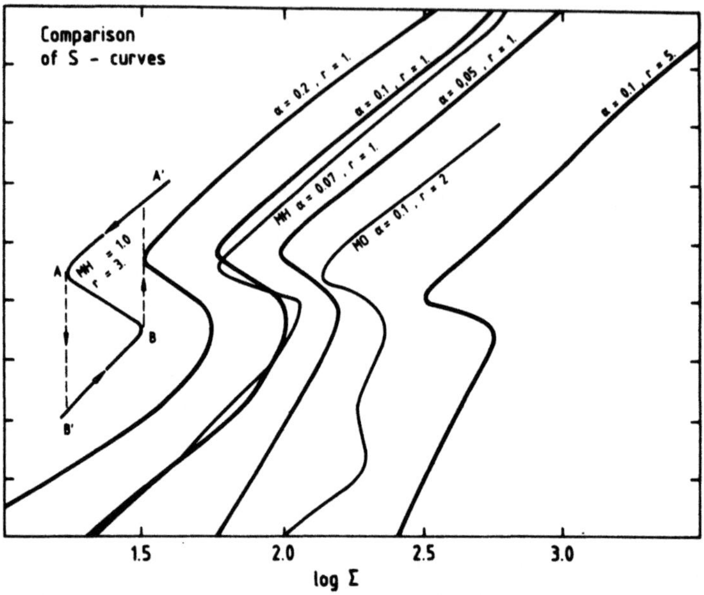

Fig. 32. Curves obtained from the solution of vertical structure equations. The units of r are 10^{10} cm. Results of several authors ([104], [105]) are shown. The limit cycle that is obtained is indicated on the small figure on the left (from [109]).

this means that the mass transfer rate through the disk is lower than the rate at which mass is supplied by the secondary. Consequently, the surface density will increase (on the viscous timescale) and the disk region will move up along the cool branch until it reaches the critical surface density Σ_B. At this point the region will jump to the only available stable equilibrium, on the upper hot branch. On the hot branch, however, the region is transferring the mass at a faster rate than it is being supplied, therefore the surface density will decrease (along the hot branch) until Σ_A is reached, at which point a jump to the lower branch will occur. In order to obtain a coherent outburst from the entire disk, the information of the upward and downward transitions has to be communicated to at least a substantial fraction of the disk. This occurs via transition waves. Once a region in the disk exceeds the critical surface density Σ_B and moves to the hot branch, a heating wave starts to propagate, bringing the entire disk region with surface densities between Σ_A and Σ_B to the hot branch (which results in an outburst). Similarly, a cooling wave brings the disk to the cool branch, thus ending the outburst. The light curves that are obtained from this limit cycle behavior resemble (at least qualitatively) dwarf nova light curves (Fig. 33), ***provided that it is assumed (at least in some cases) that the viscosity parameter*** α ***is larger by at least a factor of 3 on the hot branch than its value on the cool branch*** $(\alpha_h/\alpha_c \gtrsim 3)$.

The following thing should be noted. In the disk instability model, the outburst can start at different parts of the disk [106]. If the accretion rate is

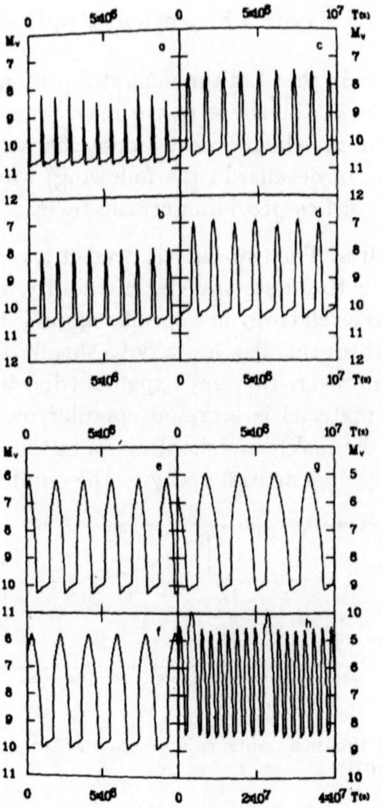

Fig. 33. Theoretical outburst light curves. In (a)–(d) $\dot{M} = 9\times10^{-12}$, 1.8×10^{-11}, 3.6×10^{-11}, 7.2×10^{-11}. In (e)–(h) $\dot{M} = 1.8\times10^{-10}$, 3.6×10^{-10}, 7.2×10^{-10}, 1.44×10^{-9} (from Lin *et al.* [105]).

low (but still sufficient to drive the instability), then the mass has time to diffuse inwards in the accretion disk (even for relatively low viscosity) and the critical density Σ_B (Fig. 32) will be first exceeded in the inner part of the disk (where the critical Σ_B is lowest). Thus, for low transfer rates from the secondary, the outburst will start at the inner edge of the disk. On the other hand, at high transfer rates (but not too high to put it on the upper stable branch), the critical surface density may be reached first in the outer disk (since the inward diffusion is limited by the viscous timescale) and thus, the outburst can start at the outer edge. It is also true, of course, that when the outbursts start at a given edge, the frequency of the outbursts increases with increasing $\dot{M}$ (since the critical density is reached sooner).

8.4 Comparison of a Few Theoretical Predictions with Observations

Because of the freedom that both models still presently have, in choosing some of the parameters (*e.g.*, the shape of the mass transfer burst, α_h/α_c etc.) it is important to determine which observational characteristics can really test the models, irrespective of details. In the following, I examine a few such characteristics which I regard as providing crucial tests.

8.4.1 The Disk Radius. The two models predict a very different behavior of the disk radius [107]. In the mass transfer instability, the arrival of an increased mass flux which has a relatively low specific angular momentum (corresponding to R_c, eq. (27)), causes the disk initially to shrink (towards R_c). Following the initial shrinkage, however, the disk expands (due to viscosity and due to the fact that as more material is accreted, angular momentum is transported to the outer parts of the disk), overshooting its initial equilibrium radius. This is followed by a nearly exponential decay to the equilibrium value (Fig. 34).

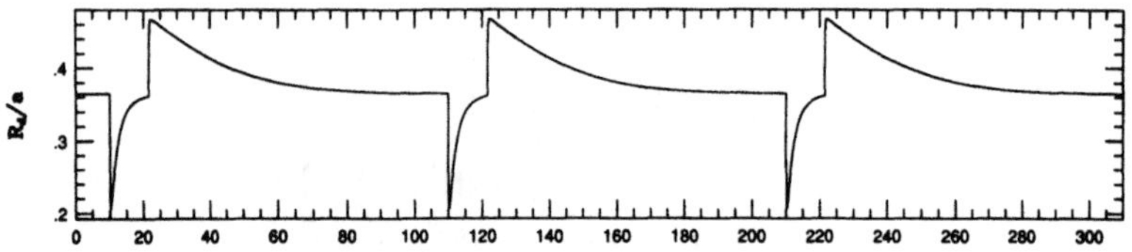

Fig. 34. Behaviour of the disk radius as a function of time for a mass transfer burst (from Ichikawa and Osaki [107]).

In the case of the disk instability model, the disk immediately starts expanding. This is due to the fact that as the disk becomes unstable, matter starts to diffuse inwards. Since angular momentum has to be conserved, the angular momentum of the inward moving material is transferred to the outer parts of the disk, which consequently expands. This is then followed by an exponential decay (Fig. 35).

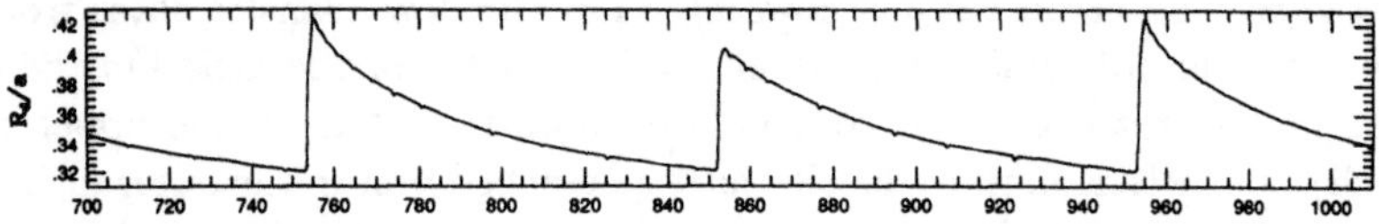

Fig. 35. Behaviour of the disk radius as a function of time for a disk instability model (from Ichikawa and Osaki [107]).

Since the two models predict a very different behavior, a careful monitoring of the disk radius through an eruption can determine which model agrees with the observations. Unfortunately this is not an easy observational task. Disk radii are obtained from eclipse observations of the bright spot that is generated at the point of impact of the stream from L_1 onto the disk edge. However, the exact time of occurrence of dwarf nova outbursts cannot be predicted in advance and consequently there exist only very few radius measurements that are covering outbursts. The most convincing case so far comes from observations of U Gem (see also observations of Z Cha by O'Donoghue) [108] (Fig. 36). In this case, *a clear increase in the radius is seen at outburst (in contradiction to the prediction of the mass transfer instability model).* In spite of the fact that the observations of U Gem do not start before the outburst, it appears that they start sufficiently close to the beginning of the eruption (when the system was brighter only by a few tenths of a magnitude than the quiescence level) to be able to rule out the occurrence of a disk shrinkage.

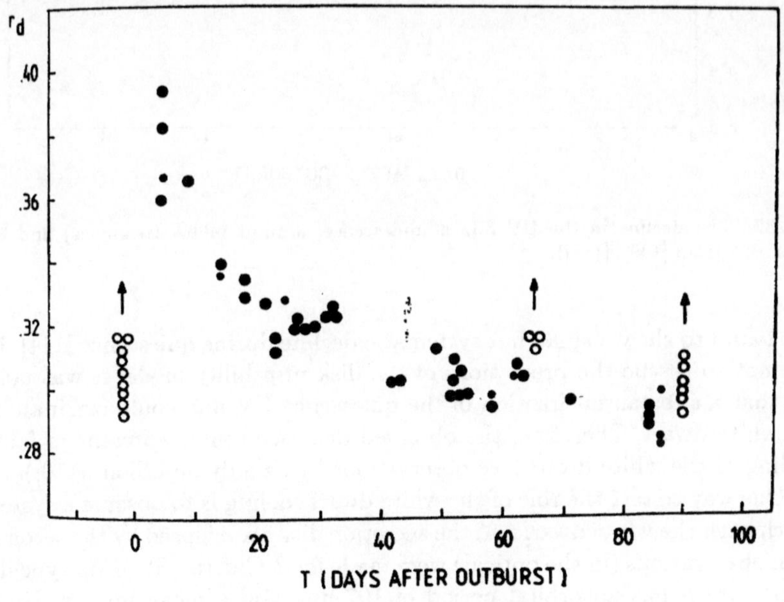

Fig. 36. The time behaviour of the disk radius in U Gem during outburst. Small symbols are of lower weight. Open circles are from eclipses observed at the outburst onset (from [108]).

8.4.2 UV, X-ray and Optical Observations Between Outbursts. An examination of Fig. (32) reveals that the disk instability model predicts that between outbursts the UV and x-ray fluxes should rise. This is due to the continuous build-up of the surface density that precedes the next outburst and the expected increase in the radiation from the boundary layer.

Clearly, the mass transfer instability predicts the UV and x-ray fluxes to decline slightly (or stay constant) between outbursts, as the mass transfer rate from the secondary returns to its normal, low state.

Observations in the UV of WX Hyi [109] and VW Hyi [110] find the UV flux to decline between the outbursts (see Fig. 37). The x-ray flux of VW Hyi was

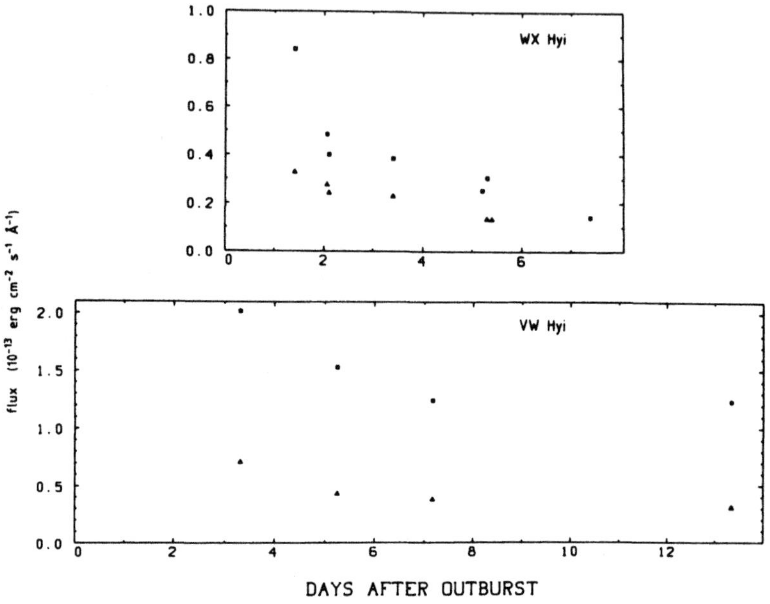

Fig. 37. The decline in the UV flux at quiescence, around 1460Å (triangles) and 2880Å (squares) (from [109], [110]).

also found to show a small but systematic decline during quiescence [111]. In an attempt to rescue the predictions of the disk instability model it was pointed out that a substantial fraction of the quiescence UV flux could originate from the white dwarf. Therefore, the observed decrease could represent in fact the cooling of the white dwarf (see observations by Szkody and Sion, [110]).

One way to test the role of the white dwarf cooling is to observe a system in which both the white dwarf and the accretion disk are eclipsed by the secondary. Such observations (in the optical) were made for Z Cha (an SU UMa type dwarf nova), which has an orbital period of 107 min and a mean interval between normal outbursts of 80 d (280 d between superoutbursts). These observations showed no evidence for a secular variation in the flux of the accretion disk and the white dwarf during quiescence [112]. Similar results were found for VW Hyi.

Thus, *the observations of the UV and x-ray fluxes between outbursts present a serious difficulty to the disk instability model in its standard formulation.*

8.4.3 The UV Delay. Observations of a few dwarf novae indicate that in outburst, the rise of the optical flux precedes the rise of the UV flux by sometimes more than half a day.

A detailed comparison of VW Hyi, which has very fast eruptions and CN Ori, which has quite slow eruptions [113], revealed that the optical clearly precedes the UV in VW Hyi, while the two fluxes rise almost together in CN Ori.

These observations actually agree quite well with expectations based on the mass transfer instability. In that case, if the increase in the mass transfer rate occurs on a timescale which is short compared to the viscous timescale, then the optical flux (which is generated mostly in the outer parts of the disk) is expected to rise first. The UV flux is expected to be delayed until viscosity allows the material to diffuse to the inner disk regions. If on the other hand, the mass transfer enhancement occurs on a timescale that is longer than the viscous timescale (corresponding to slower outbursts), it is expected that the luminosity from all the disk parts will increase more or less simultaneously, thus producing UV and optical fluxes which rise together.

The UV delay observations of CN Ori and VW Hyi, however, present a serious difficulty for the disk instability model in its standard formulation. The expectation from this model is that for fast eruptions, in which the heating transition waves propagate rapidly through the disk, the UV and optical fluxes will rise simultaneously, while for slow outbursts a delay between the optical and UV fluxes will be observed. These expectations are contrary to the observations.

Several modifications to the standard disk instability model were suggested in order to resolve this difficulty. Most of these suggestions attempted to find ways to delay the transition of the disk to the upper hot branch, by making it linger for a longer time around temperature 6000K (near the critical density Σ_B). One such suggestion relied on an increase in the thermal timescale (due to heat being deposited into ionization) and on radial heat transport [114]. Another work proposed that due to the disappearance of convection around 6000K [115], a reduction is expected in the effective viscosity (if the viscosity is at all related to the presence of convection). Since this extends the S curve to larger values of Σ around 6000K, it is expected, according to this idea, that the disk will spend a longer time around 6000K, until the higher Σ is reached.

Both of these rather contrived suggestions are not convincing, in particular in view of the fact that the observations of VW Hyi strongly indicate that the disk has undergone the transition to the hot branch (in that the optical light is near maximum) before the UV flux even started to increase.

8.4.4 The Bright Spot Luminosity. At the point of impact of the stream onto the disk, a bright spot is formed. This is a consequence of the conversion of the stream's kinetic energy to thermal energy in the shocked material at the impact area [116].

The bright spot in eclipsing systems is eclipsed separately from the white dwarf and thus, its luminosity can be determined. If in addition, it is assumed that the bright spot radiates like a black body then the rate of mass transfer into it can be approximately determined by

$$\dot{M}_{BS} \simeq \beta r_S^2 \frac{2\sigma T_{BS}^4}{[V_r^2 + (V_\theta - V_{kep})^2]}, \tag{87}$$

where r_S is the stream's radius, V_r and V_θ are the components of the stream's velocity, V_{kep} is the (Keplerian) velocity of the disk's edge and $\beta = 1$ or 2 depending on whether the spot emits from one or two sides. It is clear that the mass transfer instability model predicts an increase in the bright spot's luminosity in outburst, as the mass transfer rate increases by more than an order of magnitude.

In the disk instability model a slight decrease may in fact be expected, as the disk radius expends at the onset of the outburst, making the potential well into which the stream material falls shallower.

Observations of the bright spot during the rise to maximum of VW Hyi [117], CN Ori [118] and OY Car [119] reveal a very small (less than a factor 3) increase in the spot's luminosity. While formally speaking, this seems to contradict the predictions of both models, it is probably fair to say that due to the uncertainties involved in the observations, ***these observations present a much more serious difficulty for the mass transfer instability model.***

Attempts were made to suggest, that due to stream penetration into the disk, not all of the stream's kinetic energy is in fact released in the bright spot and therefore, that a more modest increase in the bright spot's luminosity should be observed [120]. However, more detailed calculations of the stream-disk interaction, showed (with obvious uncertainties) that while it is true that stream penetration moderates somewhat the expected increase in the spot's luminosity, the expected brightening is still quite substantial and in conflict with observations [121].

8.5 Assessment of the Viability of the Two Standard Models

From the previous section it becomes clear that both standard eruption models (at least in their simplest form) encounter quite serious difficulties.

In the present section I attempt to present (what will obviously be a very biased view of) where we stand with respect to these two models and suggest possible alternatives.

In my opinion, the difficulties confronting the mass transfer instability model at this point appear to be quite fatal for this model. On the theoretical side, there has never really been a sound physical basis for this model to work with realistic secondary stars in CVs (as explained in section 8.3.1). On the observational side, the observations of the behavior of the disk radius in particular and of the bright spot luminosity (to a lesser extent, see section 8.4) make it very difficult for this model to be acceptable. More continuous observations of the behaviour of the disk radius before and during an outburst are strongly encouraged.

Concerning the disk instability model, the general theoretical basis is relatively sound but many serious problems remain. In particular:

(1) The assumption that $\alpha_h/\alpha_c > 3$ appears quite ad hoc at this stage. However, until the nature of the angular momentum transport mechanism is clarified, it can be justifiably argued that this assumption merely reflects our ignorance of the transport mechanism. Some (relatively vague at this stage) suggestions have in fact already been made, of how this requirement could result from the magnetic dynamo models presented in section 6.3.

(2) The observations of the UV delay appear to present a very serious problem for the standard disk instability model.

(3) The UV, x-ray and optical observations between outbursts present a very serious problem to the standard disk instability model.

The two main questions at present therefore appears to be: (i) Can some modifications be made to the standard disk instability model so as to resolve problems (2) and (3) above (assuming that solving (1) will require a better understanding of the angular momentum transport)? and (ii) Is there an alternative model to the two standard models which does not encounter the above difficulties?

Before attempting to answer these questions I would like to note the following point. Both the UV delay problem and the problem of decreasing UV and x-ray fluxes at quiescence would (in principle) be solved, if it turned out that in quiescence the disk lacks its innermost part (the hottest disk region and the boundary layer), so that it is actually more like a ring.

In outburst the ring would start to expand inwards (as the viscosity increases), reaching the white dwarf surface (this will clearly produce a UV delay, since the UV is emitted mostly by the hot, inner disk regions). Following the outburst, as the inner part is accreted, the ring will retreat to its original radial extent (which will result in decreasing UV and x-ray fluxes).

An attempt to introduce a modification into the standard model along the lines of the above picture has been recently made by Meyer [122]. He suggested that a coronal "syphon flow" evaporates matter from the inner disk, which then accretes onto the white dwarf. According to this idea, the coronal mass flow leads to a complete evaporation of the innermost part of the disk during quiescence (on a timescale of a few days), thus solving the existing problems as explained above. While this model could work in principle, it seems (at least to the present author) somewhat contrived. A possible observational test would be to determine whether indeed disks lack their inner parts, using Doppler tomography techniques. I should also note that for Meyer's scenario to work, the effective viscosity parameter α has to be quite high ($\alpha \simeq 1/3$) in the corona, otherwise the coronal matter would not accrete (because it still possesses angular momentum).

I should also note that another way (in principle at least) for a hole to be generated at the disk center (thus providing a possible solution to the UV delay and decreasing UV flux problems) is if the white dwarf has a sufficiently strong magnetic field (like that of an intermediate polar, $\sim 10^6$ G) to disrupt the disk. The material in this case will follow field lines from the inner edge of the disrupted disk to be accreted onto the white dwarf's polar caps [23]. This is expected to produce x-rays, as the (radially) inflowing material is shocked near the white dwarf surface. In this case, however, the prediction would be that a modulation of the x-rays with the white dwarf spin period should be observed, as in the case of DQ Her and GK Per (unless the rotation axis of the white dwarf happens to be aligned with its magnetic axis). The absence of such a modulation in dwarf novae can therefore be regarded as evidence against this type of scenario. The presence of a hole at the center of a disrupted disk should also affect (in any scenario) somewhat the burst duration and the interval between outbursts [123].

To conclude, while the disk instability model seems to provide a good basis for the explanation of dwarf nova eruptions, quite a few questions remain to be answered. The possible role of a relatively weak surface magnetic field of the white dwarf has to be explored.

9 The Formation of Binaries Involving a Compact Component—Some Aspects of Common Envelope Evolution

A stage in which the two components of a binary system are revolving about their center of mass while being engulfed by a common gaseous envelope is defined as a "common envelope" (CE) phase. Usually the term refers to a situation in which the envelope is not corotating with the binary system. As a consequence of the velocity difference between the binary components and the CE, they experience a drag force which works to reduce the binary separation. The formation of CVs very probably involved a CE phase. This allows us to understand how a binary involving a white dwarf can have an orbital separation of the order of a solar radius, while the radius which the white dwarf's precursor (an AGB star) must have achieved was orders of magnitude larger.

In fact, many of the binaries which involve at least one compact component and which have orbital periods of less than a few days probably experienced a CE phase in their evolution.

Common envelope evolution is often the result of *a dynamical mass transfer event* (another cause, secularly unstable tidal dissipation, will not be discussed here). Such events happen when at least one (and in many circumstances both) of the following two conditions occur:

(a) Mass is being transferred from the more massive to the less massive component.

(b) The mass losing star has a deep convective envelope (e.g. in the giant or AGB phase).

These conditions lead to a situation in which the mass losing primary expands as a result of mass loss (or at least is unable to contract as rapidly as its Roche lobe, see section 3), and thus an unstable mass transfer process ensues [11]. The high mass transfer rate (often exceeding the Eddington limit), drives the secondary star out of thermal equilibrium and causes it to expand rapidly, thus resulting in a CE configuration [124]–[127].

Attempts to follow CE evolution numerically were usually forced to make many simplifying assumptions because of two major difficulties: (i) the problem is intrinsically three-dimensional, (ii) the problem involves hydrodynamic and thermodynamic processes which occur on a wide range of both length scales and timescales. Spherically symmetric calculations [128], [129] demonstrated the importance of determining *the efficiency with which orbital energy is deposited into envelope ejection.* Two-dimensional and three-dimensional calculations showed that *mass ejection takes place preferentially in the orbital plane* [130], [131]. I shall now discuss briefly some of the results of these calculations, emphasizing the physics involved.

The principal effect which the CE is supposed to produce is a reduction in the binary separation, accompanied (at least in some cases) by the ejection of the common envelope. It is thus clear that the most important physical parameter which determines the final outcome of CE evolution is the efficiency of energy deposition into mass ejection [131], [132]. This can be parametrized in terms of

$$\alpha_{CE} \equiv \frac{\Delta E_{bind}}{\Delta E_{orb}}, \tag{88}$$

where ΔE_{bind} is the binding energy (gravitational minus thermal) of the ejected material and ΔE_{orb} is the change in the binary's orbital energy (between the beginning and the end of the spiralling-in process).

The two major processes that can act to reduce α_{CE} are:

(a) *Efficient energy transport in the CE.*

(b) *Non-spherical (local or axisymmetric) effects.*

The first effect is important in situations in which the timescale for energy transport in the CE is short compared to the orbital decay timescale. Under such conditions, the energy deposited by the spiralling-in process is transported to the surface and radiated away, without causing significant mass motions. This can be expected to be the case in the outer layers of evolved AGB star-type configurations or in the case that the energy deposition rate constitutes only a small perturbation to the local luminosity. The latter situation arises, for example, in the case that a brown dwarf (or very low mass star) companion spirals into an AGB star [133]–[135]. Efficient energy transport was found to suppress mass ejection also in spherically symmetric calculations of CE evolution [128], [129]. However, in this case, the result was probably produced artificially by the fact that the frictional energy was deposited into an entire spherical shell of a relatively large mass, rather than locally or into a torus. The second process ((b) above) which can reduce α_{CE} is related to the fact that multi-dimensional calculations revealed that mass is ejected primarily in the orbital plane. This has the effect that a smaller amount of mass is given more energy than it needs to escape, rather than each mass element getting exactly its binding energy, thus reducing the efficiency of mass ejection. The results of such a spiralling-in process can be seen in Fig. 38, taken from the three-dimensional calculation of Livio and Soker [131], which shows the importance of local energy deposition in producing mass ejection in the orbital plane. Similarly, Fig. 39, taken from Taam and Bodenheimer [130], shows that mass ejection takes place in a narrow region (half angle of order 12°) around the equatorial plane. As a consequence of these non-spherical effects, values of α_{CE} in the range 0.3–0.6 were obtained.

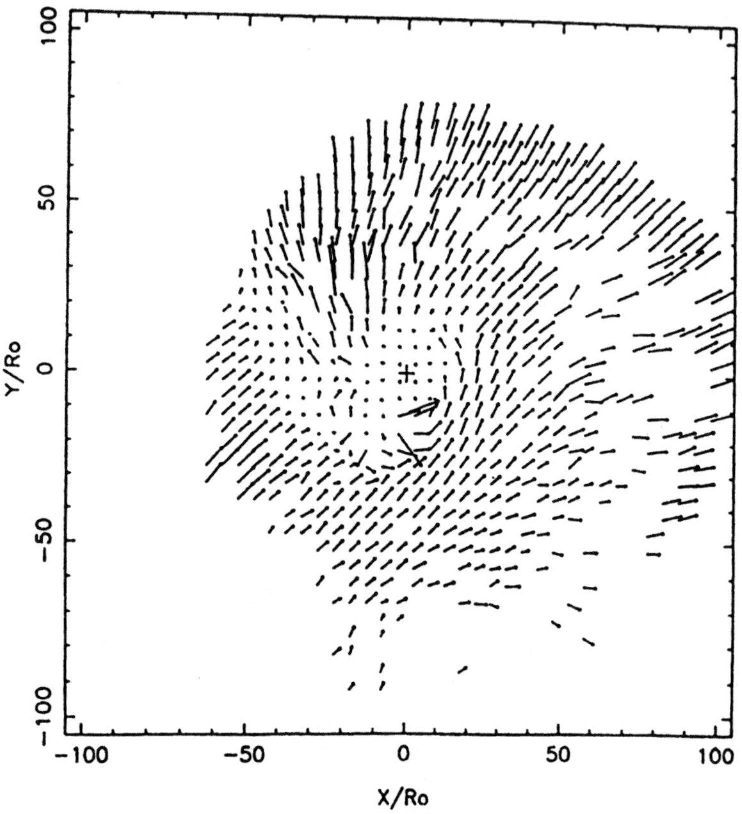

Fig. 38. The velocity field during spiralling-in, in the orbital plane, in the rest frame of the giant's core (cross). Heavy arrow denotes the secondary (from [131]).

If we consider additional sources of energy (other than the orbital energy) which can help in inducing mass ejection, but at the same time continue to use formally the same definition of α_{CE} as in eq. (88), then there are a number of processes that can (in principle at least) increase the value of α_{CE} [136]. Here I will consider only two such processes: (1) spin-up of the envelope, (2) recombination energy. Concerning the first process, onc can define a parameter

$$\gamma_{CE} \equiv \frac{\tau_{spin-up}}{\tau_{decay}} \tag{89}$$

where $\tau_{spin-up}$ is the timescale on which the spiralling-in binary can spin-up the interior CE, and τ_{decay} is the orbital decay timescale. If a situation arises in which $\gamma_{CE} \lesssim 1$, the orbital decay is slowed down significantly, since the drag is reduced. Under such conditions, it may happen that an AGB star will eject the envelope in a similar manner to that of single stars, without a large reduction in the binary separation. Low values for γ_{CE} can be expected in highly evolved AGB type configurations (low envelope mass) and for relatively massive secondaries [131]. Concerning the recombination energy ((2) above), clearly if this

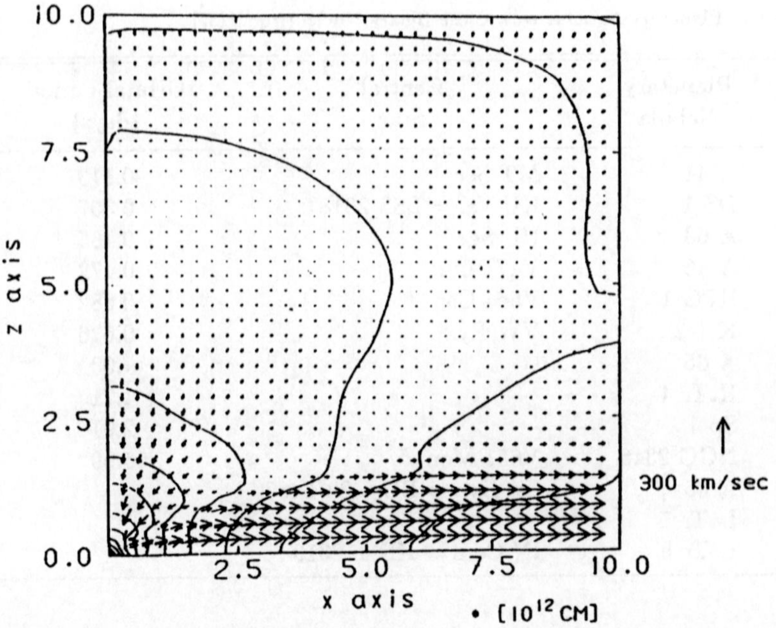

Fig. 39. The velocity field and density distribution in the common envelope. The z axis is perpendicular to the orbital plane (from [130]).

energy in the hydrogen and helium ionization zones can be deposited into inducing mass motions, then envelope ejection will be greatly facilitated (this has been originally suggested for planetary nebulae ejection by single stars, [138]).

While CE evolution is now widely recognized as an essential phase in the formation of cataclysmic variables, of double WD systems, of binary pulsars and of at least some of the x-ray binaries, no really direct observational evidence for the existence of this phase in ordinary binary star evolution exists (the only exception, which is however different in some respects, is provided by classical novae in outburst, see section 10.5). ***The closest to direct evidence is the existence of planetary nebulae (PNe) with close binary nuclei.*** In this case, it seems quite clear that we are observing systems which recently (because of the short lifetime of the nebulae) emerged from the CE phase, the binary having spiralled-in while ejecting the nebula. Table 1, taken from Bond and Livio [139], gives an updated list of planetary nebulae with close binary central stars (for references see Bond and Livio). Because these objects constitute the best evidence we have for the occurrence of a CE phase in normal evolution, it is important to use them to test CE evolution theory (recently, the exciting possibility that the carbon star V Hya is presently in a CE configuration has been suggested by Kahane *et al.* [137]). There are at least three ways in which such tests can be attempted:

(1) To use the properties of the binary central stars in conjunction with stellar evolution theory in an attempt to determine α_{CE}.

Table 1. Planetary Nebulae with Close-Binary Nuclei (from [139])

Planetary Nebula	Central Star	Orbital Period (days)
A 41	MT Ser	0.113
DS 1	KU Vel = LSS 2018	0.357
A 63	UU Sge	0.465
A 46	V477 Lyr	0.472
HFG 1	V664 Cas	0.582
K 1-2	VW Pyx	0.676
A 65	...	1.00
HtTr 4	...	1.71
Sp 1	...	2.91
NGC 2346	V651 Mon	15.99
A 35	Lw Hya = BD −22°3467	
LoTr 1	...	
LoTr 5	IN Com = HD 112313	

(2) To use CE theory in conjunction with assumptions about the distributions in primary mass, separation, and mass ratio of zero age binaries, in order to determine the distributions of these parameters in the expected binary central stars and compare those with the observed objects.

(3) To use CE theory in conjunction with the "interacting winds" process for the shaping of PNe and compare the obtained PN morphologies with the observed ones.

The first method ((1) above) was used by Iben and Tutukov [140] in an attempt to construct an evolutionary path for V651 Mon, LSS 2018, MT Ser and UU Sge. Iben and Tutukov obtained values of α_{CE} in the range 0.8–1.0. It should be noted, however, that at present there exist large uncertainties both in the evolutionary paths and in the observations and these do not allow any firm conclusions at this stage.

The second method ((2) above) was used by de Kool [141]. A typical distribution of primary masses, secondary masses and orbital periods is shown in Fig. 40. These are supposed to represent post CE binaries with orbital periods less than 1 day (starting with an initial distribution of 10^5 zero age binaries). These particular distributions were calculated based on the assumption that the initial mass ratio distribution is obtained by picking both stars individually from the same initial mass function.

Unfortunately, the number of systems in Table 1 is still too small (and the masses are in fact known for two systems at best) to allow any meaningful comparison with the calculated distributions. Nevertheless, the method provides a potentially important tool for the future, when better statistics will become available.

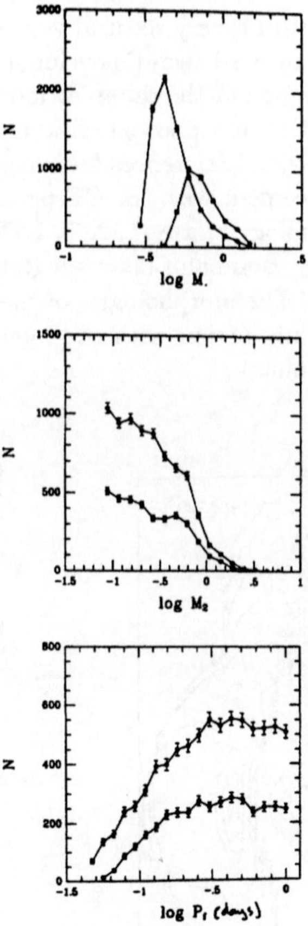

Fig. 40. The distribution of primary masses (a), secondary masses (b) and orbital periods (c) of post common-envelope binaries with $P < 1$ day. Filled circles are for $\alpha_{CE} = 1$, open for $\alpha_{CE} = 0.1$ (from [141].

The third method ((3) above) was used by Bond and Livio [139] in a way which I shall now describe briefly. In the "interacting winds" model [143], after most of the central star's envelope has been ejected (in a slowly moving wind), the hot nucleus that is exposed emits a fast (but low density) wind. The fast wind runs into the slowly moving material, shocks it, and produces a snowplow effect. Balick, and Icke and co-workers [142]–[144] proposed a simple morphological scheme, in which PNe are divided into three basic types: round, elliptical and butterfly. Balick used the interacting winds model to suggest, that the physical parameter which determines which of three types will be formed is the "density contrast" between the equatorial and polar directions (namely, the ratio of the respective densities). A moderate contrast was supposed to result in an elliptical morphology, while a high contrast in a butterfly. Soker and Livio [145] noted that the CE phase in the case of PNe with binary central stars, produces a density contrast in a very natural way, as explained above. They, and Icke and co-workers then used two-dimensional hydrodynamic calculations to demonstrate the production of the different morphologies (Figs. 41, 42). Bond and Livio presented CCD and photographic imagery for nine of the thirteen PNe in Table 1 and provided references for images of the remaining four. They have shown that in agreement with the CE predictions, HtTr 4 and NGC 2346 show a butterfly morphology and A41, A65, LoTr1, DS1, A63 and LoTr5 show an elliptical morphology. Bond and Livio suggested that the round nebula SP 1 is seen close to pole-on. The morphologies of the other PNe were probably significantly modified by interaction with the interstellar medium (except K 1-2 which is extremely peculiar).

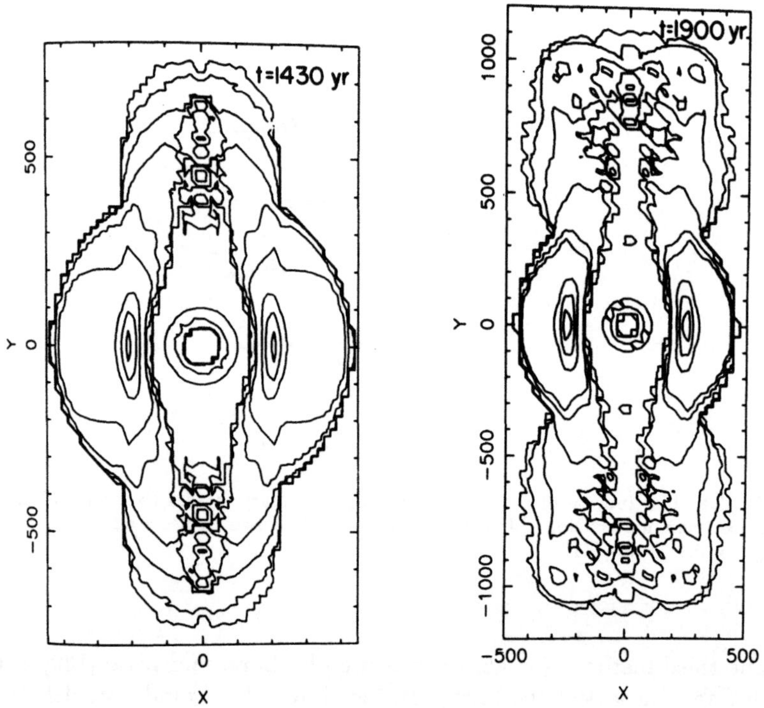

Fig. 41. Density contours obtained from interacting winds in the case of a density contrast between the equatorial (horizontal) and polar directions (from [145]).

To summarize this section, observations of PNe with close binary central stars provide us at the moment with one of the best tools for testing the theory of CE evolution (the use of classical novae as another important tool is described in section 10.5).

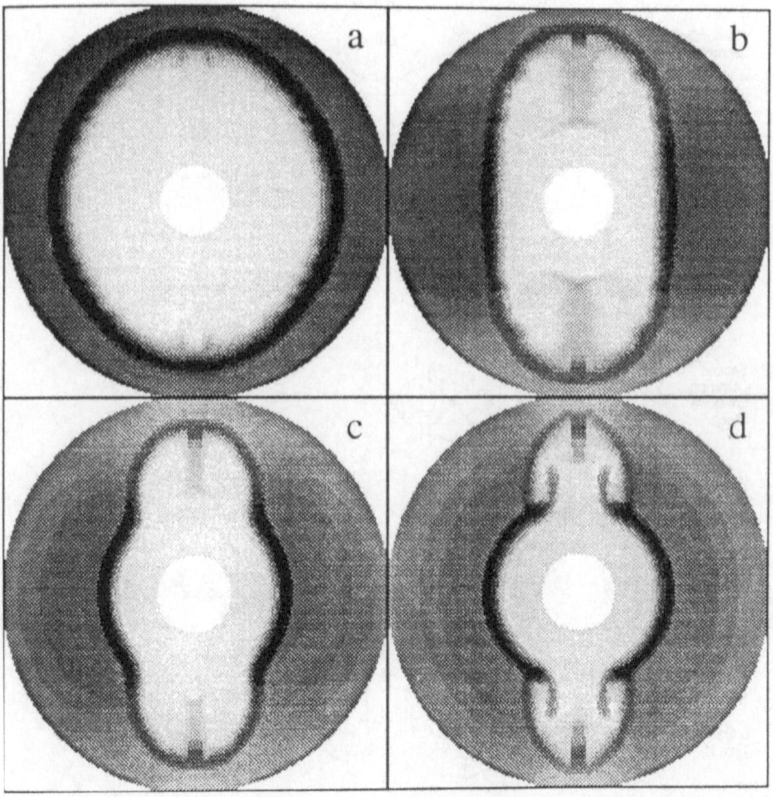

Fig. 42. Density distribution, with darker shades representing higher values. The equatorial plane is horizontal. The formation of different morphologies is clear (from Mellema *et al.* [144]).

9.1 The Coalescence of White Dwarfs and Type I Supernovae

One of the scenarios for Type Ia supernovae (SN Ia) involves the coalescence of two white dwarfs [146], [147]. In one of the possible paths of this scenario, a binary system with initial masses in the range 5–9 $M_{\odot}$, and an initial separation in the range 70–1500 $R_{\odot}$, evolves through one or two CE phases, to form a double white dwarf (WD) binary. The emission of gravitational radiation (see section 11.1.1) can then decrease the separation of the double WD system, leading to a merger in less than Hubble time if the initial orbital period of this system was less than $\sim$ 10 hr. Since the total mass of the two WDs can exceed the Chandrasekhar limit in some cases, some violent outcome of the merger process or at least the formation of a neutron star seems almost inevitable. It has been speculated that a SN Ia explosion is a possible outcome. A typical representation of the entire scenario is shown in Fig. 43 (taken from [146]), it involves two CE phases (a scenario starting with stars of almost equal masses, which involves only one CE phase, was described by Webbink [147]). We can es-

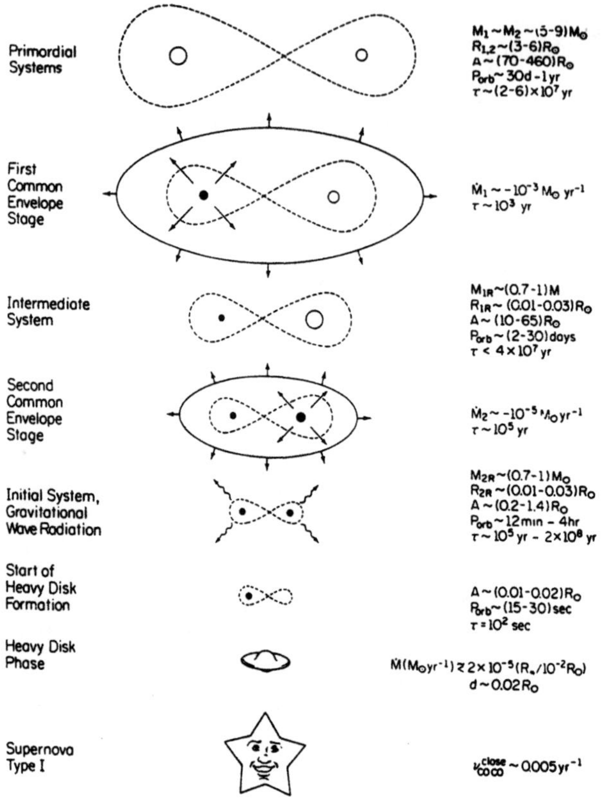

Fig. 43. A suggested scenario leading to Type Ia supernovae (from [146]).

timate very crudely the final separation after the two CE phases approximating the binding and orbital energies in eq. (88) as

$$\frac{GM_1^2}{a_i} \simeq \alpha_{CE} \frac{GM_{1R}M_2}{a_f^{(1)}} \tag{90}$$

where a_i, $a_f^{(1)}$ are the initial and final separations, (following one CE phase) respectively, and M_{1R} is the mass of the core remnant of the primary star. The final separation (after two CE phases) can then be expressed as

$$a_f \simeq 0.5 R_\odot \left(\frac{\alpha_{CE}^{(1)}}{0.4} \right) \left(\frac{\alpha_{CE}^{(2)}}{0.4} \right) \left(\frac{M_{1R}}{M_\odot} \right)^2 \left(\frac{M_{2R}}{M_\odot} \right) . \tag{91}$$

The timescale on which coalescence can be expected to occur (due to gravitational wave radiation) is of order

$$\tau_{GR} \simeq 6 \times 10^8 \text{yr} \left(\frac{M_{1R}}{M_\odot} \right)^{-1} \left(\frac{M_{2R}}{M_\odot} \right)^{-1} \left(\frac{M_{1R} + M_{2R}}{M_\odot} \right)^{-1} \left(\frac{a_f}{R_\odot} \right)^4 . \tag{92}$$

It should be noted right away that an obvious prediction of the coalescence scenario is that there should exist a population of close binary white dwarf systems. This prediction has received some support from the discovery that L 870-2 is a double line spectroscopic binary white dwarf system, with an orbital period of $P = 1.55578$ days [148] and 0957–666 is also a double WD system, with a suggested period of 1.17–1.18 days [149]. Both of these systems, however, will not produce SN Ia (too long periods and too small masses). Other potential double WD candidates, showing variability, were discovered [150] but never confirmed. Thus, while it seems that ***the existence of a population of moderately close double WD systems has been established, it is not clear yet whether this population can account for SN Ia statistics.*** In particular, the null results obtained by Robinson and Shafter [151] from a sample of 44 WDs (for orbital periods between 3 hr and 30 s) the null results of Foss *et al.* [152] from a sample of 25 white dwarfs (for orbital periods between 3 and 10 hr) and the fact that only one candidate was identified so far in a sample of about 100 WDs [149], begin to place meaningful constraints.

On the theoretical side, it has been shown that once the less massive WD fills its Roche lobe, a dynamical mass transfer process ensues, on a timescale of order

$$\tau_{ins.} \sim P_{orb}^{1/3} \tau_{GR}^{2/3}. \tag{93}$$

This is followed by a short CE phase, lasting about [155]

$$\tau_{decay} \sim f(q) \frac{M_2}{\rho(a) a^3} P_{orb} \tag{94}$$

where $\rho(a)$ is the local density in the CE and $f(q)$ is some function of the mass ratio. For a WD binary with $M_1 = 1.2 M_\odot$, $M_2 = 0.9 M_\odot$ and $P_{orb} = 29$ sec, Mochkovitch and Livio [153] found that a CE forms in about 0.18 yrs and the CE phase lasts for about 0.03 yrs. In the next phase, calculations by Benz *et al.* [154] show that the secondary WD is completely dissipated within a few orbital periods, to form a thick disk around the primary (see Fig. 44). Mochkovitch and Livio [153], [155] constructed post-coalescence primary-massive disk configurations, using the self-consistent field method [156]. They have shown that *the subsequent evolution of these configurations depends crucially on the rate of transport of angular momentum through the disk.*

Mochkovitch and Livio considered different mechanisms for angular momentum transport and removal and found that turbulent viscosity and hydromagnetic winds (if operative, see section 6) can act (in principle at least) very efficiently to remove angular momentum from the disk. An even more efficient mechanism can be non-axisymmetric unstable modes. Such modes were recently found to transport angular momentum in self-gravitating disks on a timescale of ten orbital periods [157]. If moderately efficient mechanisms indeed operate, then they may lead to conditions at the time of the explosion that are rather different from the standard ones, in which carbon is ignited at the center of a 1.4 $M_\odot$ WD [158]. In fact, off-center carbon ignition in non-degenerate material will almost certainly occur [155]. The subsequent evolution may resemble

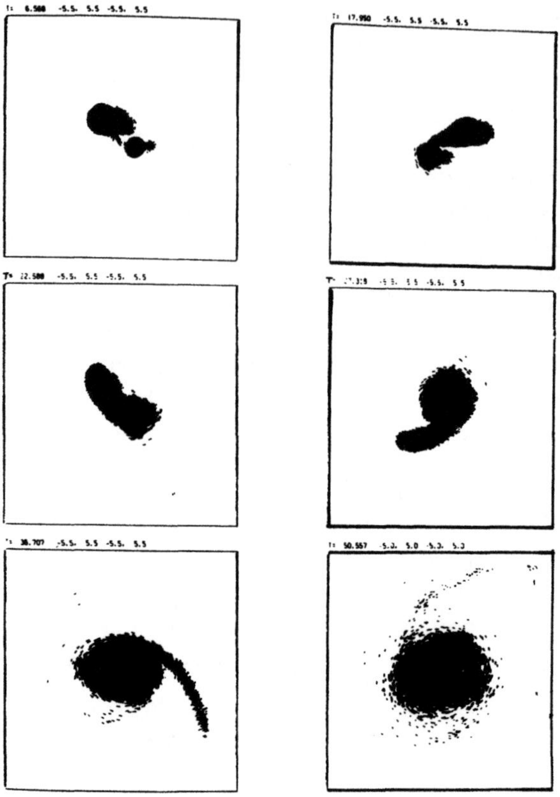

Fig. 44. The time evolution of the WD merger process. The velocities are projected onto the orbital plane (from [154]).

the one calculated by Kawai *et al.* [159] in which the effective accretion rate onto the primary is controlled by carbon burning. This can result in accretion rates $\dot{M} \gtrsim 10^{-5} M_\odot/\mathrm{yr}$, accompanied by a possible transformation of the WD to O-Ne-Mg composition (by the inward moving burning front, [160]). Such an object may eventually produce a SN by the "collapse-bounce" mechanism [275]. If angular momentum removal is *extremely* efficient (as the existence of non-axisymmetric unstable modes seems to imply), an immediate collapse will probably ensue, followed perhaps by the simultaneous ignition of large parts of the WD [161], [162]. In any case, we see that the uncertainties involved in the mechanisms for removal of angular momentum prevent us, at present, from being able to predict the final outcome.

10 Elements in the Theory of Classical Nova Outbursts

10.1 A Brief Summary of a Few Observational Properties

Novae represent dramatic explosions, second only to supernovae in terms of energetics. By definition, a *classical nova* is an object which was seen in outburst only once. Furthermore, until now, no system that has undergone a nova explosion was studied extensively prior to the outburst. There exists a small class (presently containing 8 objects) of *recurrent novae* which were observed to undergo more than one nova-type outburst. I will discuss this class briefly in section 11.2.1 (see [163], [164] and references therein for recent works on recurrent novae).

All classical novae show a very rapid rise (taking of the order of 1–3 days) to maximum light, with a typical amplitude (or range) of 10 magnitudes or more (Fig. 45) (19 mag from the long-term pre-outburst magnitude in the case of V1500 Cyg). This is followed by a decline (*in the optical*) and an eventual return to the pre-outburst magnitude. The timescale of the decline shows a great variety of behaviours and this has led Payne-Gaposchkin [165] to the definition of the nova *speed class*. Her definitions are given in Table 2 (the time (in days) it takes the nova to decline by two magnitudes from maximum light is usually denoted by t_2 and dv/dt is the rate of decline).

Table 2. Classification of Light Curves According to Speed Class

Speed Class	Definition	Rate of Decline mag/day
Very fast	Fall of 2 mag in 10 days or less	>0.20
Fast	Fall of 2 mag in 11 to 25 days	0.18 to 0.08
Moderately fast	Fall of 2 mag in 26 to 80 days	0.07 to 0.025
Slow	Fall of 2 mag in 81 to 150 days	0.024 to 0.013
Very slow	Fall of 2 mag in 151 to 250 days	0.013 to 0.008

Very broadly, optical nova light curves are divided into a few stages which are also characterizing the spectral evolution. Such a "typical" light curve is presented in Fig. 46.

The initial rise takes typically 1–3 days and is often missed. One notable exception is V1500 Cyg, in which this phase was covered [166]–[168]. The pre-maximum halt occurs in many novae and lasts from hours (for fast novae) to days (for slow novae). In the transition region (Fig. 46), some novae exhibit a minimum (in the optical) which can last a few months, and which is followed by a return to the continuation of the decline light curve (examples are T Aur and DQ Her). Other novae (*e.g.*, GK Per, V603 Aql), show large-scale (about 1 mag) oscillations during the transition phase.

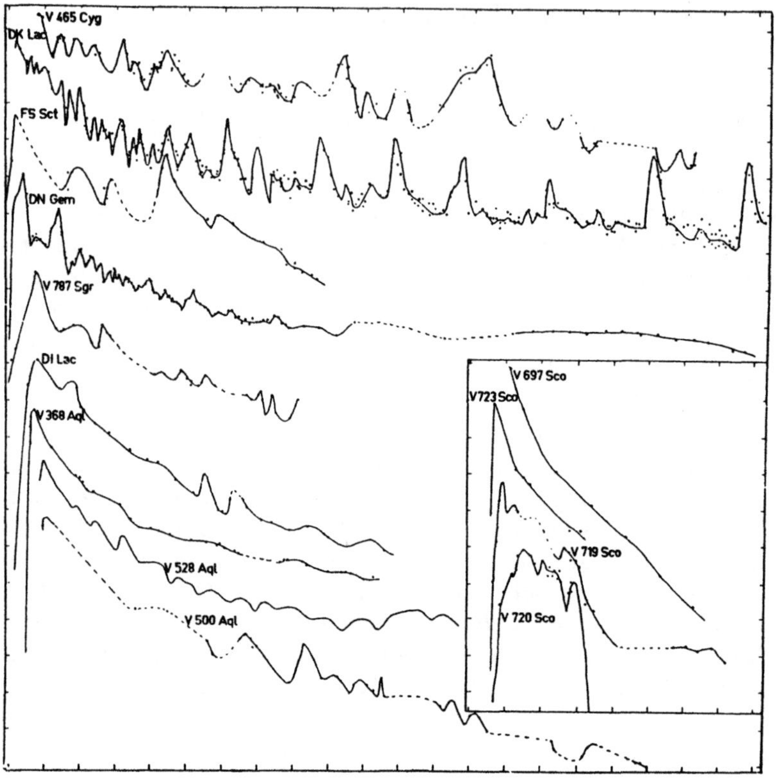

Fig. 45. Light curves of several novae in outburst (from [165]). Intervals are of one magnitude and ten days.

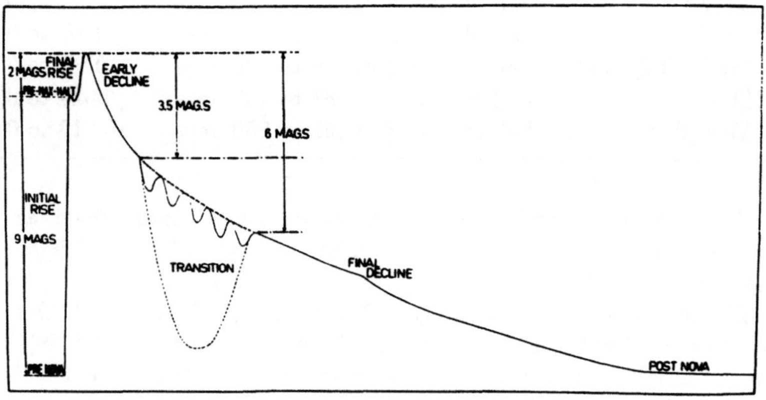

Fig. 46. A typical morphology of a nova light curve (from Warner [165]).

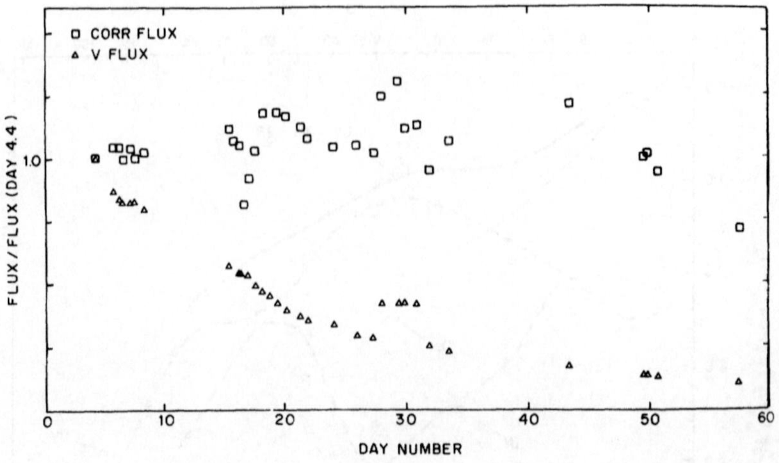

Fig. 47. The total flux of FH Ser between 1550 and 5460 Å and the flux in V (arbitrarily normalized to unity on day 4.4) (from [169]).

One of the most important discoveries of UV observations of novae, was the fact that in FH Ser (nova of 1970), while the optical brightness was declining by an order of magnitude during the first 60 days, the total (UV + optical) flux remained nearly constant [169] (Fig. 47).

Furthermore, during the transition phase, in which the optical dropped by about 5 magnitudes, an infrared excess developed [170], so that the (infrared + optical) flux remained the same as during the early decline. The variation of the different fluxes of V1668 Cyg is shown in Fig. 48 [171].

It is thus by now established that most novae experience *a constant bolometric luminosity phase.*

Two other observations which have important consequences for nova theory concern the masses of the ejecta and their composition. The determination of ejecta masses is not easy, because the interpretation of the emission lines in the phase in which the spectrum (in the optical) starts to evolve towards that of a planetary nebula (the "nebular stage"), is complicated by uncertainties about the "filling factor" (spatial inhomogeneities). Typical ejecta masses range from about $10^{-5} M_\odot$ [172] to about $10^{-3} M_\odot$ for HR Del (I do not include recurrent novae in this discussion, they have typically smaller ejecta masses of 10^{-6}–$10^{-7} M_\odot$).

Abundance determinations in the ejected material are also very uncertain but all of the most reliable ones seem to indicate large overabundances (over solar values) of heavy elements (see Table 3). The average mass fraction of heavy elements, Z, is about $\langle Z \rangle \simeq 0.38$ (compared to the solar value of $Z_\odot \sim 0.02$). In particular, it is interesting to note that in about one third of the novae for which more or less reliable abundance determinations exist, large overabundances of Ne and Mg have been detected (see Table 3, these are called "Neon Novae").

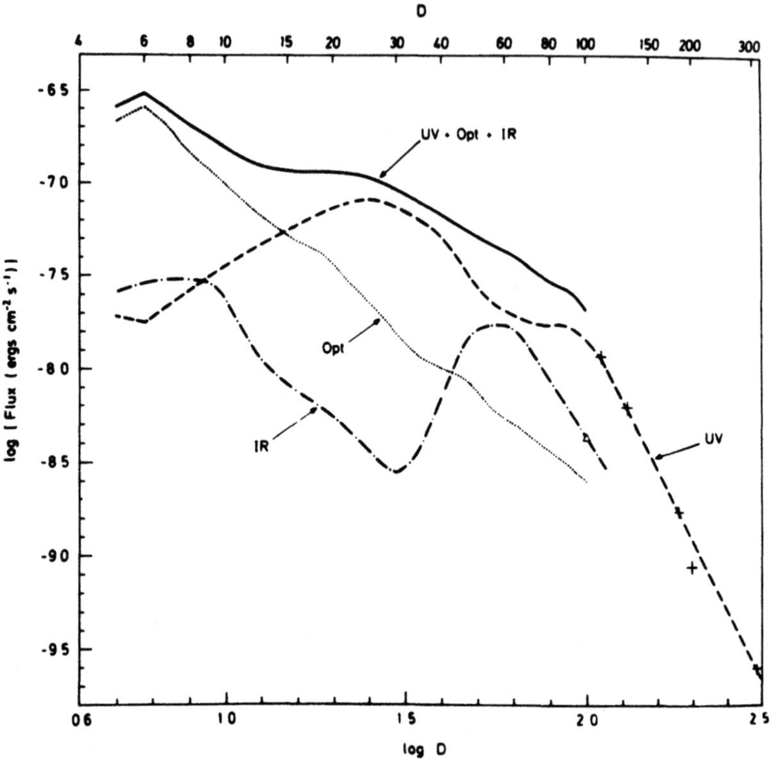

Fig. 48. Fluxes of V1668 Cyg in the uv (from 1140 to 3290 Å), optical (3920 to 12000 Å) and infrared (1.2 to 19.5 μm) (from [171]).

Another property of novae which has to be mentioned is the existence of a relation between the peak luminosity and the rate of decline of novae (the magnitude at maximum vs. rate of decline or MMRD relation) [173]. This relation is presented in Fig. 49. For the central portion $1.0 \lesssim \log t_3 \lesssim 2.0$ (where t_3 is the time it takes the nova to decline three magnitudes from its magnitude at maximum M_B), the M_B–t_3 relation is nearly linear [174],

$$M_B = -10\overset{m}{.}67(\pm 0\overset{m}{.}30) + 1.80(\pm 0\overset{m}{.}20) \log t_3 . \tag{95}$$

Another empirical relation is the result

$$\langle M_{15} \rangle = \text{const.}, \tag{96}$$

namely, all novae (irrespective of speed class) have almost the same absolute magnitude 15 days past maximum light. Both of these relations ((95) and (96)) make novae potential extragalactic distance indicators (see section 10.6).

From the point of view of understanding the physics of nova systems, one of the most significant discoveries concerning novae has been the realization that nova systems are close, interacting binaries, in which a normal (in most cases

Table 3. Heavy Element Abundances in Classical Novae (from [193])

Object	Year of Outburst	Mass Fractions											
		H	He	C	N	O	Ne	Na	Mg	Al	Si	S	Fe
RR Pic	1925	0.53	0.43	0.0039	0.022	0.0058	0.011						
HR Del	1967	0.45	0.48	0.027	0.047	0.0030							
T Aur	1891	0.47	0.40	0.079	0.051								
PW Vul	1984	0.69	0.25	0.0033	0.049	0.014							
V1500 Cyg	1975	0.49	0.21	0.070	0.075	0.13	0.023						
V1668 Cyg	1978	0.45	0.23	0.047	0.14	0.13	0.0068						
V693 CrA	1981	0.29	0.32	0.046	0.080	0.12	0.17	0.0016	0.0076	0.0043	0.0022		
GQ Mus	1983	0.27	0.32	0.016	0.19	0.19	0.0034		0.0014	0.00056	0.0028	0.0016	0.00047
DQ Her	1934	0.34	0.095	0.045	0.23	0.29							
V1370 Aql	1982	0.053	0.088	0.035	0.14	0.051	0.52		0.0067		0.0018	0.10	0.0045

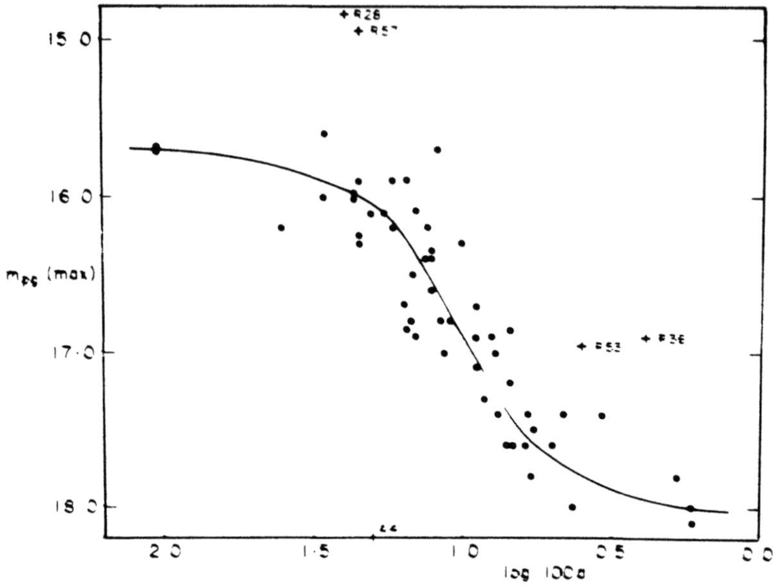

Fig. 49. The relation between the magnitude at maximum and the rate of decline (from van den Bergh [173]).

low mass main sequence) star is transferring mass onto a white dwarf [175], at a rate of 10^{-11}–$10^{-8} M_\odot$/yr.

10.2 The Thermonuclear Runaway Model

It is by now commonly accepted that nova outbursts represent thermonuclear events on the surface of accreting white dwarfs. As hydrogen accumulates on the surface of the white dwarf, the density and pressure at the base of the accreted envelope increase, until the nuclear reactions are ignited. Detailed hydrodynamic and semi-analytical calculations have shown [176]–[178] that the strength of the outburst is largely determined by the pressure at the base of the accreted layer

$$P = \frac{GM_{WD} \Delta M_{acc}}{4\pi R_{WD}^4}. \qquad (97)$$

Strong outbursts are obtained once a critical value of order $P_{crit} \sim 2 \times 10^{19}$ dynes cm^{-2} is exceeded. The reason that a runaway is obtained (rather than steady burning) is the fact that at the high densities (and in some sense low temperatures, see eq. (98) below) that are obtained on a white dwarf, the nuclear reactions are ignited under degenerate conditions. In degenerate matter, the pressure does not depend on the temperature to first order. This means that the normal safety valve that operates, say, in the ignition of nuclear reactions at the centers of stellar cores does not operate. Namely, as the reactions are ignited, the temperature increases with no concomitant increase in the pressure.

Therefore, immediate expansion (which would have lowered the temperature) does not occur. Instead, the energy generation rate, which is extremely sensitively dependent on the temperature, increases further and a thermonuclear runaway (TNR) ensues. Whether or not a degenerate TNR occurs is determined by how the temperature in the ignited shell compares with respect to the Fermi temperature (which determines the degeneracy condition),

$$T_F \simeq 4 \times 10^7 \mathrm{K} \left(\frac{\rho}{10^4 \mathrm{~g~cm^{-3}}} \right)^{2/3} . \tag{98}$$

For $T > T_F$ the degeneracy is lifted and expansion occurs. However, the shell can react by expansion only on the dynamical timescale $\tau_{dyn} \sim C_S/g$ (where g is the surface gravity). If the nuclear reactions are sufficiently rapid, so that $\tau_{nuc} \ll \tau_{dyn}$, then the temperature may rise significantly (and with it the rate of energy generation ϵ_{nuc}) before the shell has time to respond. For typical white dwarf parameters $\tau_{dyn} \sim 1$ sec, while τ_{nuc} can be much shorter than 1 sec if the abundances of CNO nuclei are enhanced (in which case energy generation rates of order $\epsilon_{nuc} \sim 10^{17}$ ergs $\mathrm{g^{-1}~s^{-1}}$ can be achieved, see below). Therefore, very strong TNRs are obtained at relatively low accretion rates (which result in a small amount of compressional heating), onto massive white dwarfs (large pressure build up) of relatively low luminosity (and therefore low temperature, see section 10.6.4 for a discussion of the effects of the different parameters). All of these physical parameters lead to ignition under highly degenerate conditions. As we shall see later, a low accretion rate also allows for various mixing processes to mix white dwarf material into the accreted layer, thereby increasing the CNO concentration.

10.2.1 The Nuclear Reactions Network. The energy in nova outbursts is generated via the hot CNO cycle [179]–[181]. A schematic representation of the reactions involved is shown in Fig. 50. The changes in the abundances of the different nuclear species with time are normally calculated using a nuclear reactions network of the form [182]–[184] (for each nucleus, with the sums over all possible reactions)

$$\frac{dY_i}{dt} = -\sum_j Y_i Y_j \rho N_A \langle \sigma v \rangle_{i,j} + \sum_{k,l} Y_k Y_l \rho N_A \langle \sigma v \rangle_{k,l} \tag{99}$$

where Y_i is the number fraction of nucleus i and N_A is Avogadro's number. The average of the product of the cross section for the reaction under consideration and the relative velocity enters because the reaction rate is given by the flux times the total cross section $r = \sigma(v) N_i \cdot v N_j$, and this has to be integrated over the relative velocity spectrum $\phi(v)$. Numerical simulations of nova explosions presently involve many dozens of nuclei and the energy generation rate is calculated in conditions that are far from equilibrium. These calculations revealed that the β^+ unstable nuclei ^{13}N, ^{14}O, ^{15}O and ^{17}F play a crucial role in the outburst. The β^+ decay times of these nuclei are 863 sec, 102 sec, 176 sec and 92 sec respectively. In the high temperatures that are obtained in the burning

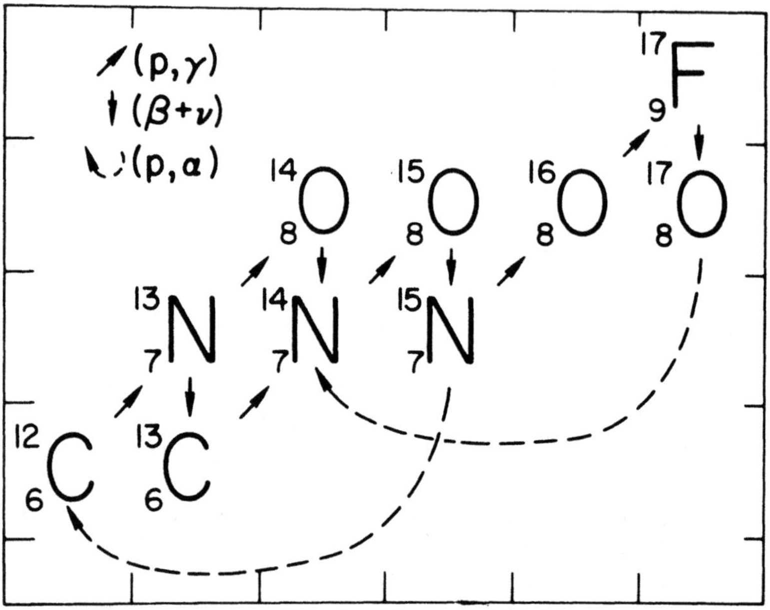

Fig. 50. A schematic representation for the nuclear reactions network of the CNO cycle in the temperature range 10^7–4×10^8 K (from Starrfield [176]).

shell ($\sim 10^8$ K), the lifetime of the nuclei against proton captures becomes shorter than the decay times. This has the consequence that the β^+ decays become the bottle-neck of the reactions, with the β^+ unstable nuclei becoming the most abundant CNO nuclei in the shell. Furthermore, as the temperature in the shell rises, the entire envelope becomes convective. Because of the fact that the convective turnover timescale ($\sim 10^2$ sec) is comparable (or shorter) to the decay times, a large fraction of the β^+ unstable nuclei can be transported to the very surface layers (up to the outermost $10^{-8} M_\odot$ or so).

The decay of these nuclei near the surface deposits large amounts of energy (in excess of $\sim 10^{46}$ ergs) which is ultimately responsible for the envelope expansion.

10.2.2 The Evolution of the Light Curve. We are now in the position to describe the gross features of the development of the nova light curve on the basis of the TNR model.

The early phases of nuclear burning are dominated by the PP chain (which is relatively temperature insensitive $\sim T^4$) and therefore changes are relatively slow. The temperatures starts to rise quickly once the CNO reactions start to dominate (the energy generation rate is then proportional to T^{18}). At a temperature of $\sim 8 \times 10^7$ K, the convection region reaches essentially all the way to the surface and the surface layers start expanding as a result of the energy deposition by the β^+ unstable nuclei. In spite of the enormous luminosity at this stage ($\sim 10^5 L_\odot$), it is mostly radiated in the EUV because of the relatively

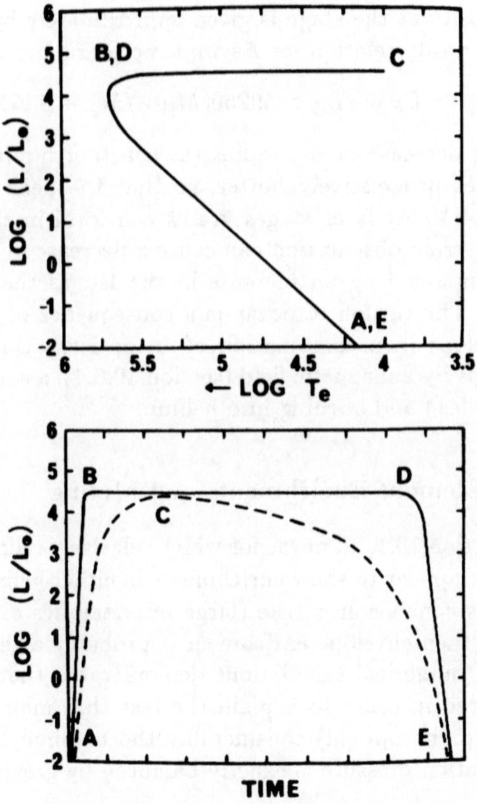

Fig. 51. The evolution of a relatively slow nova through the outburst (see text). The bolometric (solid curve) and visual (dashed) curve are also indicated (from [185]).

small surface area (the expansion velocity is still low) and concomitant high temperature ($\sim 10^5$–10^6 K). At the visual maximum novae reach temperatures of order 10^4 K and since the luminosity at that stage exceeds $10^4 L_\odot$, this implies photospheric radii of order 10^{12}–10^{13} cm.

Therefore, significant envelope expansion must occur prior to the visual maximum, typically on a timescale of a few days (only for the very slow novae like HR Del, RR Tel, the visual maximum was reached after more than 100 days). The numerical calculations indeed produce such an expansion, mainly as a consequence of the energy deposited by the radioactive decays. In the next stage, the nova enters the constant bolometric luminosity phase. The theoretical explanation for this phase is quite simple [185]. In fact, the nova follows quite closely the evolution of the central stars of planetary nebulae. This fact is not surprising since the luminosity of both novae and the pre-white dwarf central stars at this stage is powered by thin shell burning on top of a degenerate core. The white dwarf moves along the path CD (Fig. 51) in the H-R diagram, as the radius of the remnant decreases at constant bolometric luminosity.

The luminosity at the stage is given approximately by the Paczynski [186] core mass-luminosity relation for Asymptotic Giant Branch stars

$$L_{PAC}/L_{\odot} \simeq 59250(M_{WD}/M_{\odot} - 0.522). \tag{100}$$

Because of the decrease in the radius (at constant luminosity), the radiating surface becomes progressively hotter, so that the peak in the luminosity is shifted to the UV. At later stages grains can form in the ejecta [187]–[189]. The increasing grain obscuration can cause a decrease in the visual light curve which is accompanied by an increase in the IR, as the grains reradiate the absorbed flux. The turnoff of novae is a consequence of the loss of the entire hydrogen envelope in a combination of an optically-thick wind phase [190], assisted perhaps by a magnetic field (section 10.6.4), a common envelope phase (see section 10.5.1) and burning into helium.

10.3 Heavy Element Enrichments and Mixing

As noted in section 10.1, all novae for which relatively reliable abundance determinations exist appear to show enrichments in either helium or heavy elements or both. If this conclusion is true (large uncertainties exist in abundance determinations), then envelope enrichment is probably a general phenomenon in novae. In fact, numerical calculations demonstrated that enrichment in CNO nuclei is required in order to explain the fact that many novae at maximum light exceed (in the optical) considerably the nominal Eddington luminosity (at which radiation pressure is exactly balanced by gravity)

$$L_{EDD} \simeq \frac{4\pi c G M_{WD}}{\kappa_{e.s.}} \simeq 3.8 \times 10^4 L_{\odot} \left(\frac{M_{WD}}{M_{\odot}}\right), \tag{101}$$

where $\kappa_{e.s.}$ is the electron scattering opacity. The basic physical reason for this requirement is the fact that the ratio of the nuclear energy release per gram (obtained from proton captures on every available CNO nucleus) to the binding energy per gram of envelope material is given by

$$\frac{\epsilon_{nuc}}{\epsilon_{bind}} \simeq 0.01 \left(\frac{n_{CNO}}{n_{CNO}(\odot)}\right) \left(\frac{M_{WD}}{M_{\odot}}\right)^{-1} \left(\frac{R_{WD}}{5 \times 10^8 \text{ cm}}\right). \tag{102}$$

Here n_{CNO} and $n_{CNO}(\odot)$ are the number fractions of CNO nuclei in the envelope matter and solar matter, respectively. From eq. (102) it is clear that in order to obtain a truly dynamical event (associated with super-Eddington luminosities), a significant enhancement of the CNO concentration is required.

The main question therefore becomes, how do these observed high concentrations arise. The first thing to note is that it is very unlikely that such extreme values of Z have resulted from mass transfer from the companion star. The second possibility (in principle at least), is that the heavy elements are generated by nuclear burning. However, at the temperatures obtained typically in nova calculations $\sim 1.5 - 3 \times 10^8$ K, the controlling CNO cycle reaction sequences act to rearrange existing CNO isotope abundances and not to increase the total number of CNO nuclei. Helium shell flashes can in principle generate carbon and oxygen (if helium accumulates at all on the surface of the white dwarf). However, helium flashes cannot explain the enrichments in neon and magnesium observed in about one third of all novae (Table 3). At very high temperatures ($T \gtrsim 3.5 \times 10^8$ K), breakout of the CNO cycle hydrogen burning cycles via $^{15}\mathrm{O}(\alpha,\gamma)\ ^{19}\mathrm{Ne}(p,\gamma)\ ^{20}\mathrm{Na}$ can occur [191], [192], yielding neon and heavier nuclei. Such high temperatures are normally not obtained, however, in nova calculations (except on white dwarfs near the Chandrasekhar mass). In any case, the observed enrichments in CNO elements cannot be explained in this manner.

These arguments have led to the general belief that the envelope enrichment occurs via dredge-up of material from the underlying white dwarf (and the presence of Ne, Na, Mg and Al enrichments is therefore considered as evidence for the existence of a significant population of ONeMg white dwarfs). (I would like to express my personal opinion that in view of the uncertainties that still exist both in abundance determinations and in nova calculations, I do not think that the dredge-up hypothesis has been proven beyond doubt).

The question is therefore, what is the mechanism that is responsible for mixing between the accreted envelope and the white dwarf core.

10.4 Mixing Mechanisms

I shall now review the various mixing mechanisms that have been proposed (see also [193]). These can be grouped into two classes: (a) mechanisms in which most of the mixing occurs prior to the thermonuclear runaway, and (b) mechanisms in which the mixing takes place only during the TNR. It is important to note that *convection plays a crucial role in all the mechanisms.* I begin by discussing the mechanisms of the first type, for which the mixing occurs prior to the TNR.

10.4.1 Diffusion Induced Convection. The mechanism of diffusion induced convection, first discussed by Prialnik and Kovetz [194], operates in the following way. During the accretion phase, small amounts of hydrogen diffuse across the boundary between the white dwarf core and the hydrogen-rich accreted envelope, where a sharp composition gradient is assumed. The hydrogen that penetrates the core (assumed to be of CO or ONeMg composition) ignites in a high Z environment. The development of convection in this region, subsequently mixes CO (or ONeMg)-rich material into the envelope.

A similar mechanism has been proposed for the explanation of the surface compositions of white dwarfs [195].

The main points demonstrated by the calculations of Kovetz and Prialnik are the following: (i) the level of enrichment of envelope matter by white dwarf material (the enriched fraction) depends strongly on the accretion rate, and (ii) the enriched fraction is relatively weakly dependent on the white dwarf's mass (or luminosity). The first point is very easy to understand, for low accretion rates the interval between outbursts (the time required to build the critical

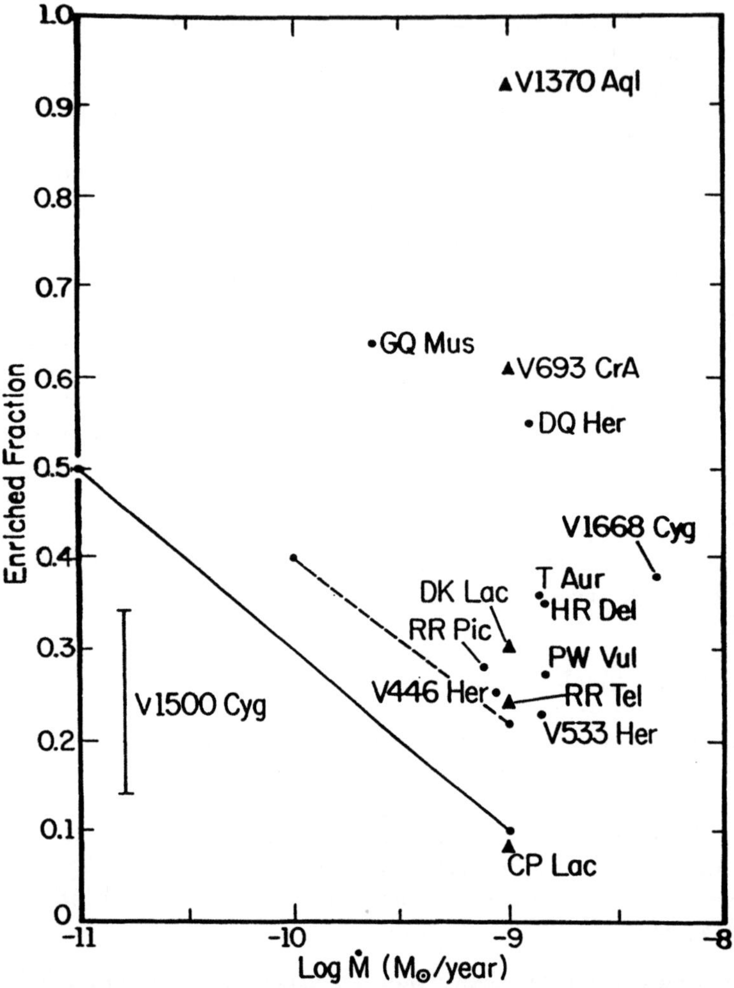

Fig. 52. The fraction of white dwarf matter mixed into the envelope, on the basis of abundances in nova ejecta as a function of the accretion rate (see text). The lines represent diffusion calculations [194] (from [193]).

pressure to trigger a TNR) becomes longer, thus allowing diffusion more time to operate. The level of enrichment achieved is not exactly inversely proportional to $\dot{M}$, since the partial pressure forces during the diffusion process also depend on $\dot{M}$. Nevertheless, the level of enrichment does decrease significantly with increasing $\dot{M}$ (see Figure 52). The relatively weak dependence on the white dwarf mass is a consequence of the fact that the ratio of the accreted mass Δm_{acc} to the depth (in mass) of the heavy element rich shell in which the TNR occurs, is found numerically to be almost independent of the white dwarf mass, M_{WD}. This point may require further study.

The main strengths of the diffusion induced convection mechanism are the following:

(1) It provides a self consistent picture, which relates the accretion phase to the TNR and which is calculable (and thus testable) at some level even in the context of the existing one-dimensional numerical codes (apart from uncertainties in the diffusion coefficients and possible effects of shear mixing, see below).

(2) It is consistent with the fact that even novae with strong magnetic fields, such as V1500 Cyg, show a significant enrichment in heavy elements (see Table 3), since the diffusion process is not strongly effected by the presence of a magnetic field.

(3) It is consistent with the fact that the recurrent nova T Pyx, the outbursts of which may be caused by a TNR (see section 11.2.1), does not show enrichments in heavy elements. For this mechanism, the absence of such enrichments would be a direct consequence of the short recurrence times of the outbursts ($\sim$ tens of years), which do not allow diffusion to take place (note however that recurrent novae may not represent TNRs and therefore this point may be irrelevant).

The fact that the main prediction of diffusion induced convection is the dependence of the level of enrichment on $\dot{M}$, led Livio and Truran [193] to attempt the following test. For the CNe for which abundance determinations exist, they calculated the accretion rates obtained from magnetic braking or gravitational radiation (where appropriate, see section 11). This is supposed to represent the average accretion rate during the inter-burst interval. It is important to note that accretion rates deduced from observations of CNe up to a few tens of years following the outburst are probably considerably higher than the secular mean (due to irradiation of the secondary by the hot white dwarf) and thus should not be used here. Figure 52 presents the observed enrichments as a function of the accretion rates. The orbital periods of V1370 Aql, V693 CrA, DK Lac, RR Tel and CP Lac are not known; these objects were arbitrarily placed at $\dot{M} \simeq 10^{-9} M_{\odot}/\mathrm{yr}$ (which is comparable to many of the determined accretion rates). The accretion rate for V1500 Cyg was estimated on the basis of known constraints on its pre-outburst brightness. Also shown in the figure are the results of two calculations of Kovetz and Prialnik (the points were arbitrarily connected by straight lines). Two features become immediately apparent from the figure: (1) *there appears to be no correlation between the observed enrichments and* $\dot{M}$, and (2) *diffusion induced convection generally predicts somewhat lower enrichments than are observed.*

In view of the large uncertainties, both in the accretion rates and in the abundance determinations, no firm conclusion can be drawn from these apparent discrepancies. Taken at face value, however, the disagreement may mean either that the diffusion coefficients that have been used are incorrect (there are considerable uncertainties associated, for example, with the inclusion of screening effects), or that diffusion induced convection is not the dominant mechanism (at least in some cases) that produces the enrichments.

10.4.2 Shear Mixing. The basic idea underlying the shear mixing mechanism is very simple [196]–[198]. In non-magnetic (or very weakly magnetic) systems, the accretion disk extends all the way to the white dwarf surface. For material to actually be accreted, *at least a fraction of the angular momentum must be transferred to the white dwarf* (some angular momentum may be carried away by material flowing outward from the interaction region [20]). This leads on a short (dynamical) timescale to the formation of a boundary layer (see section 5). On the longer (accretion) timescale, this angular momentum must be transported to the interior, so that differential rotation is inevitably generated. *The presence of strong differential rotation is often associated with various hydrodynamical instabilities.* For example, the surface layers may be subjected to the Kelvin Helmholtz instability, which in the presence of a stabilizing gravitational stratification is expressed by the Richardson stability criterion

$$R_i \equiv \frac{N^2}{\left(r\frac{\partial\Omega}{\partial r}\right)^2} \gtrsim \frac{1}{4}, \tag{103}$$

where N is the Brunt-Väisälä frequency

$$N^2 = \frac{1}{\rho}\frac{dp}{dr}\left(\frac{1}{\rho}\frac{d\rho}{dr} - \frac{1}{\Gamma_1 p}\frac{dp}{dr}\right). \tag{104}$$

The Brunt-Väisälä frequency measures the stabilizing effect of buoyancy and thus the Richardson criterion expresses the condition that the work that has to be done against gravity should exceed the energy deposited by the shear. Hydrodynamical instabilities can generate turbulence which, in turn, can both redistribute angular momentum and can cause elemental mixing. Once small amounts of hydrogen are mixed into the core, ignition will occur and mixing by convection can proceed in a similar manner to that associated with the diffusion mechanism discussed previously.

Estimates based on the baroclinic instability [198] lead to dilution factors of the accreted material of the order of the observed ones under optimum conditions, $\Delta \equiv X_o/X \lesssim 5$, where X_o and X are the hydrogen mass fractions in the accreted material and in the mixed layer.

The main points in favor of the shear mixing mechanism are the following:

(1) In view of the large specific angular momentum of the material at the accretion disk-white dwarf boundary (if the boundary exists), it is difficult to see how at least some degree of shear mixing can be avoided. We should note perhaps that if shear mixing operates, it tends to smooth out the composition discontinuity, thereby reducing the potential effects of diffusion.

(2) The degree of enrichment is not strongly dependent on the accretion rate, a feature which is entirely consistent with the observations (Figure 52).

There are several difficulties presently associated with the shear mixing mechanism:

(i) To date, no self-consistent formalism for an actual calculation of the mixing process has been developed. This is, of course, a direct consequence of the fact that what is involved is a three dimensional, non-linear process which is driven by hydrodynamical instabilities. A quantitative comparison with observations is therefore essentially impossible.

(ii) Shear mixing in its simple form (disk-white dwarf interaction) is not expected to operate for the strongly magnetized white dwarfs, since in those cases the disk is either disrupted or even non-existent (the flow is dominated by the magnetic field). However, the AM Her object V1500 Cyg shows a heavy element abundance [199] of $Z = 0.15 - 0.34$. While a different form of shear mixing may operate in these systems, when the material accumulated at the magnetic poles breaks the confinement and spreads over the entire surface [200], this possible alternative mechanism has not been explored at all. Thus, the enrichments observed in V1500 Cyg may indicate that a different mixing mechanism is dominant (at least in magnetic systems).

I shall now discuss the mechanisms of the second type, for which mixing takes place during the TNR.

10.4.3 Convective Overshoot Induced Flame Propagation. The mechanism of mixing driven by convective overshoot was proposed by Woosley [201]. He used the convective overshoot prescription, in which nonconvective spatial zones that are immediately adjacent to convective regions are also slowly mixed, roughly on about ten times the radiative diffusion timescale (without energy transport). The actual diffusion coefficient that he employed was the minimum of (i) the value obtained from the convective velocity, when a 1% superadiabatic temperature gradient was assumed, and (ii) 0.1 times the radiative diffusion coefficient.

When this procedure was employed *during the TNR*, when the hydrogen rich shell became fully convective, small amounts of hydrogen were found to be mixed into the CO core. Since the base of this mixed layer had a temperature comparable to that of the burning shell, every proton mixed inwards was immediately captured. The associated release of energy, led eventually to the linkage of the substrate by means of a diffusive wave into the white dwarf core. In the numerical study of Woolsey, it was found that $4 \times 10^{-6} M_\odot$ of carbon and oxygen were dredged up in this manner during the TNR (the accreted envelope mass was $\Delta m_{acc} \simeq 2.7 \times 10^{-5} M_\odot$).

The main successes of the convective overshoot induced flame propagation model can be considered to be the following:

(1) It relates the mixing process directly to the TNR, which is the cause of the explosion and thus should certainly occur in all novae (including those possessing strong magnetic fields).

(2) It does not appear to depend (certainly not strongly) on the accretion rate, in agreement with observations.

(3) Its efficiency may be related to the efficiency of convection, which in turn may be related to the strength of the TNR. This may explain why enrichments are higher for the more massive ONeMg white dwarfs, since the outbursts are stronger for such systems.

There are two main difficulties encountered by this model:

(i) There are large uncertainties associated with the operation of the convective overshoot mechanism. In particular, a two dimensional hydrodynamical study of the ignition process, which did not assume mixing length convection, did not obtain any mixing between the hydrogen rich layer and core material [202]. This was a direct consequence of the steep gradients at the core surface. On the other hand, a recent 2D calculation did result in significant mixing [203], when the strength of the outburst has been artificially increased by assuming a large fraction of carbon in the accreted material. Additional multi-dimensional calculations will be required to settle this question and a more careful treatment of the initial conditions.

(ii) The fact that the recurrent novae U Sco, T Pyx and V394 CrA do not appear to show enrichments in heavy elements may be regarded as being inconsistent with the convective overshoot mechanism if the outbursts of these systems are indeed caused by TNRs. Since the mixing in this case is associated with the TNR itself, and is not a function of the details of the accretion process, the recurrent novae should experience comparable levels of enrichment. If future observations will confirm that enrichments do not occur in recurrent novae, this would mean one of two things: either the outbursts of these systems are not caused by TNRs or the convective overshoot induced mixing mechanism does not operate (at least for these systems).

10.4.4 Convection Induced Shear Mixing. Kutter and Sparks [204] proposed a mechanism of convection induced shear mixing, which effectively combines the action of shear mixing with the action of convection during the TNR. Kutter and Sparks suggested specifically the following sequence of events. The shear between the accretion disk and the white dwarf produces (by a Kelvin-Helmholtz instability) an accretion belt which rotates at near Keplerian velocities. In a particular calculation this situation was achieved after 7080 years. At this point shear mixing ceased. Kutter and Sparks suggested that accretion continues (with no further mixing occurring) until the TNR. During the TNR, convection transports angular momentum from the outer (rapidly rotating) layers inwards, causing the inner layers to rotate faster. This was assumed to lead to a new shear instability between the bottom layers of the convection zone and the white dwarf layers immediately below, thus resulting in mixing.

It should be noted, however, that because of the fact that the proposed mechanism of convection induced shear mixing employs physical processes both from mechanisms which operate before the TNR (shear mixing) and from processes that operate during the TNR (convection), it can encounter difficulties associated with both types of mechanisms. For example, the convection induced shear mixing model will find both magnetic novae and recurrent novae (if they are powered by TNRs) difficult to explain (see the discussions of shear mixing and of convective overshoot induced flame propagation).

This survey of the different mixing mechanisms which have been proposed to explain the substantial levels of heavy element abundances in nova envelopes reveals that, at present, all the proposed mechanisms encounter some difficulties. Diffusion induced convection and convective overshoot induced flame propagation are presently the only mechanisms for which there exist relatively well defined prescriptions for their calculation. However, even in these cases, considerable uncertainties exist in the values of the relevant physical parameters.

There exist several critical observations, which can possibly help in determining the relative importance of the different mechanisms:

(1) Abundance determinations for the ejecta of those recurrent novae which are believed to be powered by thermonuclear runaways (in particular T Pyx, which may erupt any moment now) [205]. For these cases, it is extremely important to determine whether enrichments in heavy elements do exist. If the ejecta of all such recurrent novae will be found to be characterized by matter of solar composition, this would present serious difficulties for the convective overshoot and convection induced shear mixing models (or it could mean that recurrent novae do not represent TNR events). Conversely, if the ejecta of the recurrent novae will be found to be systematically enriched in heavy elements, this would deal a fatal blow to the diffusion model.

(2) The identification of other magnetic novae and abundance determinations in these objects. High enrichments in these objects, for which the disk should be disrupted or non-existent, would present serious difficulties for the shear mixing and convection induced shear mixing models.

I must conclude this section noting that the mixing mechanism in classical novae has yet to be unambiguously identified. Abundance data provide an extremely useful tool towards obtaining an understanding of this mechanism and, thereby, an understanding of a large number of non-linear processes associated with accretion and thermonuclear runaways in classical nova systems.

10.5 The Common Envelope Phase in Novae

Until now we have treated the physics of the outburst neglecting the fact that the white dwarf is a member of a close binary system. The questions that now arise are, what possible effects can the presence of a close companion have on the nature of the outburst and on the other hand, what are the effects of the outburst on the binary system.

In order to answer these questions we must examine the properties of a *common envelope* phase.

In the case of novae, the common envelope configuration is somewhat different from the one discussed in section 9 because the envelope is of a very low mass.

The potential importance of the common envelope (CE) phase in novae was originally recognized by MacDonald [206]–[208]. More recently, detailed calculations of this phase were carried out [209], [210].

The first thing to point out is the fact that the following properties characterize most known nova systems:

(a) The orbital periods of most nova systems are of the order of a few hours; this implies orbital separations of order (see eq. (3))

$$a \simeq 1.1 R_\odot \left(\frac{M_{WD}+M_2}{M_\odot}\right)^{1/3} \left(\frac{P_b}{3hr}\right)^{2/3} . \tag{105}$$

(b) Novae at maximum light often exceed the Eddington luminosity (eq. (101)).
(c) Novae at maximum are characterized by effective temperatures $T_e \lesssim 10^4$ K.

The combination of properties (b) and (c) implies photospheric radii of $R_{phot} \gtrsim 4\times10^{12}$ cm. An examination of typical separations as given by eq. (105) reveals that *in most classical nova systems the secondary star is necessarily engulfed in the expanding nova envelope.* Furthermore, the visual light curves of some novae (*e.g.*, DQ Her, RR Pic, T Aur) indicate that these systems may remain in a CE configuration for a few months. This raises the question: what are the effects (if any) of the CE phase on the nova system characteristics? This question is extremely important, because nova systems provide us for the moment with almost *the only opportunity to observe a system during and immediately after the CE phase.*

There are at least three potential effects:
(1) Enhanced mass loss.
(2) Shaping of the nova shell.
(3) Effects on the binary separation.

I shall now discuss each of these effects separately.

10.5.1 Enhanced Mass Loss As the secondary moves inside the expanded envelope it deposits energy due to drag at a rate

$$L_{drag} \simeq \pi R_A^2 \rho (v - v_e)^3 \tag{106}$$

where ρ is the local density, $v - v_e$ is the relative velocity between the secondary and the envelope and R_A is the Bondi accretion radius (eq. (64), (68)). The drag energy can (in principle at least) contribute to the process of mass loss from the nova envelope. In Fig. 53 a comparison is presented between nova models with and without drag energy deposition [211]. In this particular case, the effect of the CE was to speed-up envelope exhaustion. In general, drag energy deposition is expected to be more important for slow novae, occurring on less massive white dwarfs, because in this case the envelope mass is larger (*e.g.*, eq. 97) and therefore the density in the envelope is higher.

It is not clear, however, whether the inclusion of *all* the effects of the companion will result in a shortening of the nova decay phase. This uncertainty is a consequence of the fact that the secondary's gravity may suppress somewhat the occurrence of the optically thick wind phase, which is thought to be responsible for much of the mass loss [212].

Thus, the net result of all the effects of the presence of the secondary on the duration of the mass-loss phase have still to be clarified in the context of full calculations which will use updated opacities (see *e.g.*, [213]).

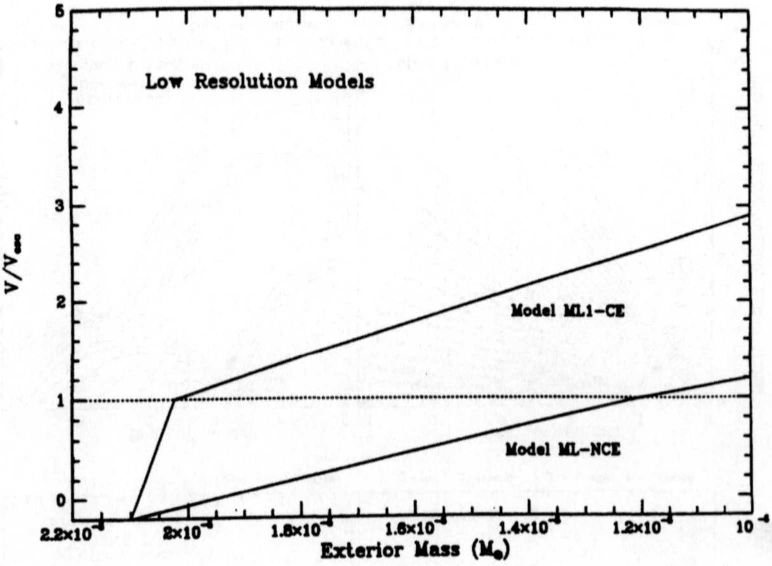

Fig. 53. The velocity in the envelope of a nova model with (MLI-CE) and without (ML-NCE) drag energy deposition (from [211]).

10.5.2 Shaping of the Nova Shell. Two dimensional calculations of the CE phase in novae show, that like in normal CEs, mass loss which results from orbital energy deposition occurs preferentially in the orbital plane (Fig. 54).

Thus, a (mild) density contrast can be generated between the equatorial (orbital plane) and polar directions. In the late phases of the nova, when the hot core is exposed it emits a fast, dilute wind which can catch up with the more slowly moving material and "snowplow" into it. Because of the density contrast, an elliptical (prolate) morphology can be expected for some nova shells (because of the fact that the envelope is of low mass, the density contrast is never extreme). This is true in particular for systems in which the binary experienced a relatively long CE phase, namely, relatively slow novae. This expectation is confirmed in the case of DQ Her (Fig. 55). Other prolate nova shells are also indicated [214]. In some cases, however, the shape of the nova shell is sufficiently complex that a simple interpretation cannot be given. It should be noted that other effects may also be responsible for the shaping of nova shells. In particular, magnetic fields may play a role (see section 10.6.4.3).

10.5.3 Effects on the Binary Separation. The nova outburst can affect the binary separation in two ways:

(1) An orbital separation increase is obtained due to mass loss from the system.

(2) An orbital separation decrease can result from frictional angular momentum losses during the CE phase.

An important question is: which of these two effects dominates?

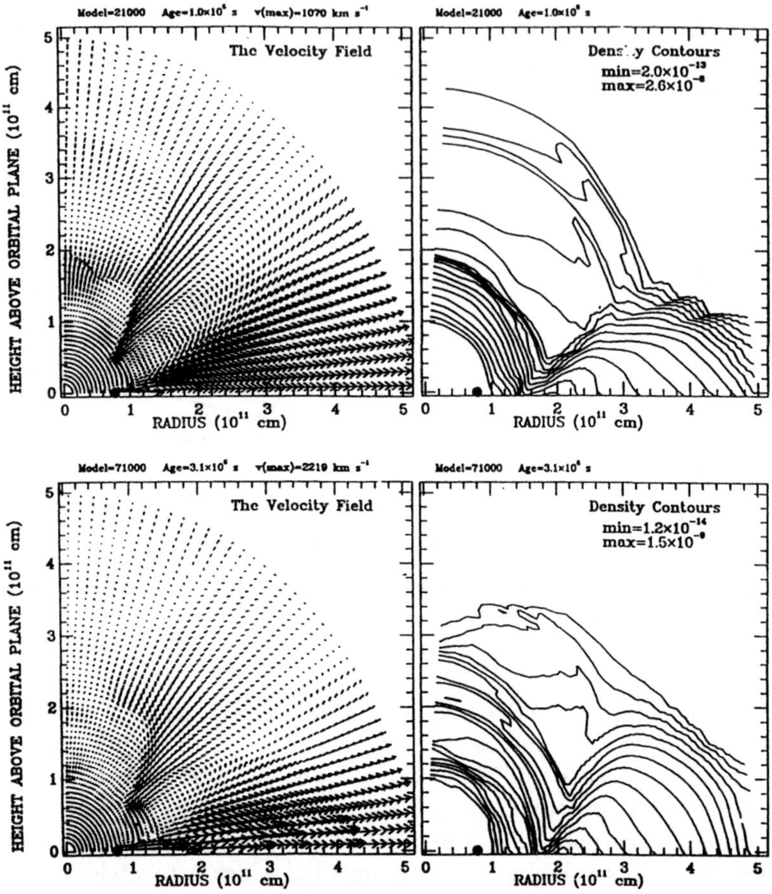

Fig. 54. Time evolution of the velocity and density in a two-dimensional common envelope calculation of novae (from Shankar [211]).

It is relatively easy to estimate the separation increase as a result of mass loss. The angular momentum of the system, J, is related to the binary separation through

$$a = \frac{J^2(M_{WD} + M_2)}{G M_{WD}^2 M_2^2}. \tag{107}$$

Differentiating this gives

$$\frac{da}{a} = \frac{2dJ}{J} - \left(\frac{1+2q}{1+q}\right)\frac{dM_{WD}}{M_{WD}} - \left(\frac{2+q}{1+q}\right)\frac{dM_2}{M_2}. \tag{108}$$

If we assume now that the white dwarf ejects an amount of mass ΔM_{WD}, which has the specific angular momentum of the white dwarf, then the fractional angular momentum carried away is

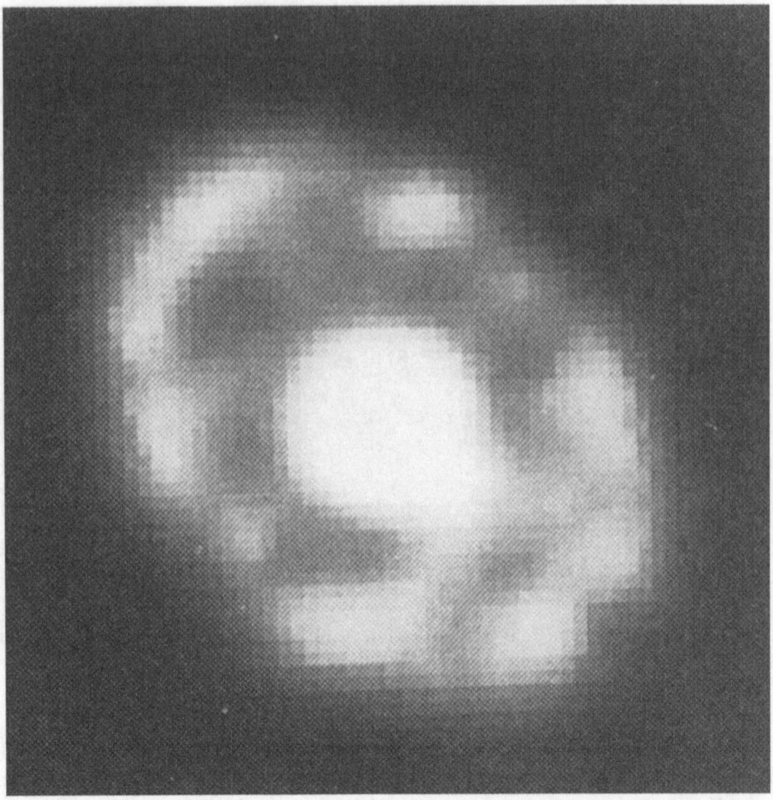

Fig. 55. The DQ Her nova shell (from Barden and Wade [214]).

$$\frac{\Delta J_{WD}}{J} = \frac{-q}{1+q}\frac{\Delta M_{WD}}{M_{WD}}. \tag{109}$$

Combining eq. (108) with (109) gives (assuming that the secondary does not capture any of the ejected mass)

$$\frac{\Delta a}{a} \simeq \frac{\Delta M_{WD}}{M_{WD}}\frac{1}{1+q}. \tag{110}$$

It is more difficult to calculate the separation decrease as a result of the drag.

Realizing that the secondary's accretion radius turns out to be very close in value to its geometrical radius, we can write the drag force as $F_D \simeq \pi R_2^2 \rho v_{rel}^2$ (where v_{rel} is the relative velocity between the secondary and the expanding envelope). The rate of frictional angular momentum loss can then be shown to be

$$\frac{\dot{J}_{friction}}{J} = -A(q)\frac{(1+k_1^2)^{1/2}}{k_1}\frac{\dot{M}_{WD}(a)}{M_{WD}} \tag{111}$$

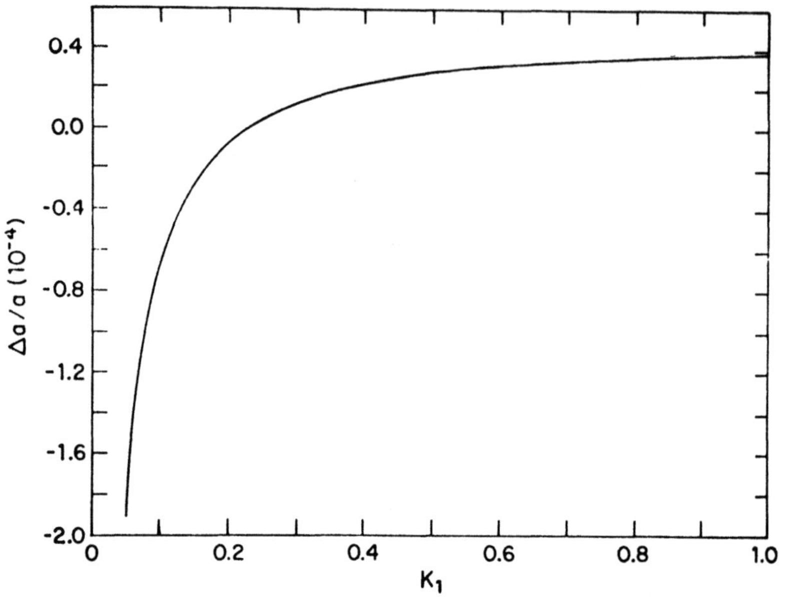

Fig. 56. The fractional change in the separation resulting from a nova explosion, as a function of v_w/v_{sec} (see text, from [215]).

where $k_1 = V_W/V_{sec}$ is the ratio of the expansion velocity (at the secondary's position) to the orbital velocity of the secondary (in the white dwarf's frame). Here, $\dot{M}_{WD}(a)$ denotes the rate at which mass flows past the orbit of the secondary star. The function $A(q)$ results from the fact that the radius of the secondary was taken to be equal to its Roche lobe radius (eq. 5) and thus

$$A(q) = \frac{1+q}{4q}\left[\frac{0.49q^{2/3}}{0.6q^{2/3} + \ln(1+q^{1/3})}\right]^2 . \tag{112}$$

Equation (111) has to be integrated over the duration of the CE phase in order to find the separation decrease. We can note, however, the following qualitative property. It turns out that k_1 increases with increasing orbital separation (because the expanding envelope accelerates as it moves outwards). Consequently, the *separation decrease* (which results from the CE phase) is a monotonically decreasing function of the orbital period. Combining the effects of the separation increase and decrease we therefore realize that there exists such a critical orbital period P_{crit}, that in systems with $P \lesssim P_{crit}$ the separation decreases while in systems with $P \gtrsim P_{crit}$ the separation increases (see Fig. 56). In the calculations of Livio *et al.* [215] this critical period was found to be somewhat shorter than 8 hrs. Unfortunately, so far there is only one system, BT Mon, for which the orbital change as a result of the outburst has been measured. Its period was found to increase [216]. This is consistent with the theoretical prediction since the orbital period of BT Mon is 8 hrs.

Livio [205] predicted a net separation decrease of order $\Delta a/a \sim -5 \times 10^{-6}$ for the recurrent nova T Pyx (assuming its outbursts are similar to those of classical novae). This prediction could in principle be tested in the near future, since the next outburst of T Pyx is already overdue. Unfortunately, recent observations [217] found the pre-outburst orbital period of the system to be insufficiently coherent to allow its determination with the required accuracy (10^{-6}). In view of the potential importance of the measurement of the period for the theory of the CE phase I still urge observers to attempt a more accurate determination of the period of this system (if possible) before the upcoming outburst.

10.6 The Effects of Various Physical Parameters on the Nature of the Outburst and the Use of Classical Novae as Distance Indicators

As explained in section 10.1, nova exhibit a correlation between their absolute (photographic) magnitude at maximum and the rate of decline (expressed in terms of the time it takes the nova to drop by 3 or 2 magnitudes from maximum, eq. 95). This makes novae (in principle at least) good candidates for distance calibration observations [218]–[221], because novae at maximum are brighter (in the optical) than Cepheids and novae are significantly more frequent than supernovae.

In spite of several important attempts [222], [223], the $M_B - t_3$ relation remained until recently essentially unexplained (in terms of basic system parameters) from a theoretical point of view. In the following, I describe briefly the (approximate) theoretical explanation that has recently been suggested [224].

10.6.1 The Absolute Magnitude at Maximum Light. It is by now a well established fact that many novae at maximum light experience a super-Eddington phase [225]. This is exemplified by the fact that 12 out of the 15 systems considered by de Vaucouleurs [218] reached luminosities brighter than magnitude −7.1, which corresponds to the Eddington luminosity for a Chandrasekhar mass white dwarf. Absolute magnitudes at maximum range from about $M_B \sim -6$ (*e.g.*, T Aur, Nova Doradus 1948) to about $M_B \sim -9.5$ (*e.g.*, V1500 Cyg, CP Pup).

Numerical nova model calculations show that the maximum luminosity can be expressed as

$$(L^{\max}/L_\odot) = \beta(M_{\rm WD}, L_{\rm WD}, B, \dot{M}, X_i)(L_{\rm EDD}/L_\odot) \tag{113}$$

where $M_{\rm WD}, L_{\rm WD}$ and B are the white dwarf's mass, intrinsic luminosity and surface magnetic field strength, respectively, $\dot{M}$ is the mass accretion rate, X_i denotes the composition of the accreted material and $L_{\rm EDD}$ denotes the Eddington luminosity.

In eq. (113), *β was found to be most sensitively dependent on the mass of the white dwarf*, with the dependence on the other physical parameters being considerably weaker [226]–[228]. I shall therefore, for the moment, neglect the dependence of β on $L_{\rm WD}, B, \dot{M}$ and X_i and I shall discuss briefly this dependence in section 10.6.4.

The main thing to note about the dependence of β on M_{WD}, is that it is a monotonically increasing function of the white dwarf mass. The basic physical reason for this behavior is the fact that due to the increased surface gravity, the pressure at the base of the accreted envelope at the time of the TNR, is higher the higher the white dwarf mass. This results in stronger runaways. Because of the fact that at the same time the mass of the accreted envelope and its total binding energy is smaller for higher white dwarf masses (see section 10.2), the degree by which the maximum luminosity exceeds the Eddington limit also increases with $M_{\rm WD}$. The results of the numerical calculations [194], [229] can be approximated by a relation of the form

$$\beta(M_{\rm WD}) \simeq (M_{\rm WD}/0.6 M_\odot)^3. \tag{114}$$

I should stress again that in relation (114) the dependence on other physical parameters has been neglected (and the relation itself is only approximate).

Using eqs. (113) and (114) and a bolometric correction of $BC = -0.1$, which is appropriate for novae near maximum light (resembling A to early F type supergiants, [230]), we obtain

$$M_B^{\rm max} \simeq -8.3 - 10.0 \log(M_{\rm WD}/M_\odot). \tag{115}$$

Relation (115) is presented in Fig. 57. We should note that in spite of its limitations, eq. (115) (Fig. 57) does represent the observed range of nova maximum luminosities. Great caution should be exercised, however, when attempting to compare Fig. 57 to any individual system. As I noted above, other factors (than the WD mass) influence the luminosity at maximum. Furthermore, white dwarf masses in nova systems are notorious for being poorly determined.

10.6.2 The Timescale to Decline from Maximum by Three Magnitudes, t_3. We now need to establish the physical processes which determine t_3. Most of the calculations which follow the expansion phase in novae [231], [232], indicate that t_3 is determined essentially by the timescale it takes the nova to eject most of its envelope. This suggests that we may write approximately

$$L_{\rm EDD} t_3 \simeq \xi \frac{G M_{\rm WD}}{R_{\rm WD}} M_{\rm e} \tag{116}$$

where $M_{\rm e}$ is the ejected mass (which is approximately equal to the envelope mass, see below) and ξ is a constant to be determined later. In writing eq. (116), use was made of the fact that during the constant bolometric luminosity phase the luminosity of the nova is nearly Eddington. Another indication to the (approximate) validity of eq. (116) is provided by the fact that the energy flux associated with the mass expansion in novae is the order of the Eddington luminosity [233]. From eq. (116) we obtain (ξ' is a new constant)

$$t_3 = \xi' (M_{\rm e}/R_{\rm WD}). \tag{117}$$

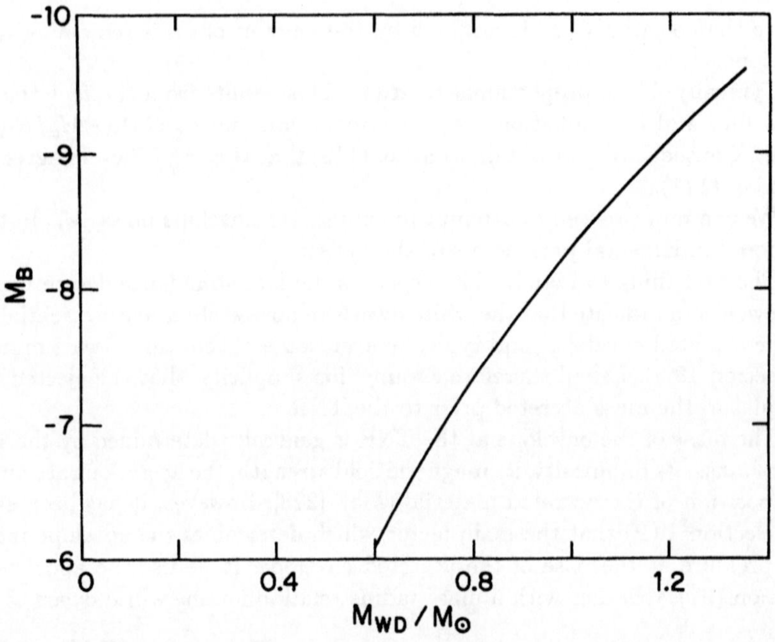

Fig. 57. The absolute magnitude of novae at maximum as a function of the white dwarf mass (from [224]).

We should note that because of the fact that the expansion phase in novae has never been calculated accurately (due to difficulties associated with the radiation-matter interaction in an expanding envelope and the fact that accurate opacities are not available), eq. (116) could (in principle at least) be written with the nuclear luminosity $L_{\rm nuc}$ (the rate of energy production by the nuclear reactions) replacing $L_{\rm EDD}$. However, this would not change the nature of the obtained relation (eq. (117)), since most of the detailed calculations show that the time behaviour of $L_{\rm nuc}$ runs almost exactly parallel to that of the bolometric luminosity $L_{\rm bol}$.

A different derivation for t_3 can be obtained in the following way.

The results of Kato and Hachisu [234] for mass loss in the optically thick wind phase show that the mass loss rate behaves like

$$\dot{M} = \dot{M}_{\rm o}\left(\frac{m}{M_{\rm e}}\right)^{1.5}, \tag{118}$$

where $\dot{M}_o$ is the initial mass flow rate, m is the instantaneous envelope mass and $M_{\rm e}$ is the initial envelope mass. This agrees also with other numerical results [231], which found that $\dot{M} \sim m^{1.6}$. An integration of eq. (118) gives (other power law dependencies would give a similar relation)

$$t_3 = C\left(\frac{M_{\rm e}}{\dot{M}_{\rm o}}\right) \tag{119}$$

where the constant C is determined by the amount of mass remaining in the envelope.

Typically $\dot{M}_o$ is proportional to R_{WD}. This results from the fact that the mass flux and the radiation are in equipartition and so $GM_{WD}\dot{M}_o/R_{WD} \sim L_{EDD}$. Consequently, we obtain from eq. (119) that $t_3 \propto M_e/R_{WD}$ in agreement with eq. (117).

We can now proceed to attempt to express the envelope mass, M_e, in terms of more fundamental parameters of the system.

The first thing to note in this respect is the fact abundance determinations of nova ejecta indicate that the white dwarfs in nova systems eject essentially all of the accreted envelope (and typically even some of the white dwarf material, see secion 10.3). I shall therefore assume, for simplicity, that the ejected mass is equal to the mass accreted prior to the TNR.

The mass of the envelope at the TNR is generally determined by the white dwarf mass, its luminosity, its magnetic field strength, the accretion rate and the composition of the accreted material [235], [228]. However, it has been shown (see section 10.2) that the main factor which determines the envelope mass is the pressure at the base of the accreted envelope. If we use the approximate relation (97), together with a mass-radius relation for the white dwarf

$$R_{WD} = 1.12 \times 10^{-2} R_\odot[(M_{WD}/M_{Ch})^{-2/3} - (M_{WD}/M_{Ch})^{2/3}]^{1/2} \tag{120}$$

we obtain (from eq. (117))

$$t_3 = A\left(\frac{M_{WD}}{M_{Ch}}\right)^{-1}\left[\left(\frac{M_{WD}}{M_{Ch}}\right)^{-2/3} - \left(\frac{M_{WD}}{M_{Ch}}\right)^{2/3}\right]^{3/2} \tag{121}$$

where M_{Ch} is the Chandrasekhar mass and A is a constant that remains to be fixed. If we take for calibration V1500 Cyg, for which $t_3 = 3.6$ days [236] and M_{WD} is estimated to be [237], [238] $1.28 M_\odot$, in order to fix A, we obtain

$$t_3 = 51.3\left(\frac{M_{WD}}{M_{Ch}}\right)^{-1}\left[\left(\frac{M_{WD}}{M_{Ch}}\right)^{-2/3} - \left(\frac{M_{WD}}{M_{Ch}}\right)^{2/3}\right]^{3/2} \text{days.} \tag{122}$$

Relation (122) is presented in Fig. 58. Again I should point out that in obtaining this relation we have neglected the effects of the white dwarf's intrinsic luminosity, its magnetic field and the effect of the accretion rate. Relation (122) produces t_3 values which range from about 0.14 days (for $1.38 M_\odot$ white dwarfs) to about 250 days (for white dwarfs of mass 0.5–$0.6 M_\odot$).

I note that the constant in eq. (121) was fixed on the basis of *one system* (and in fact a quite extreme one). In future work, and when more reliable white dwarf mass determinations in nova systems exist, it will be desirable to determine the value of the constant on the basis of several systems. Because of the effects of the composition of the accreted material it is possible that the value of the constant is different in different galaxies. Better calculations of the expansion phase may actually produce a purely theoretical expression for t_3.

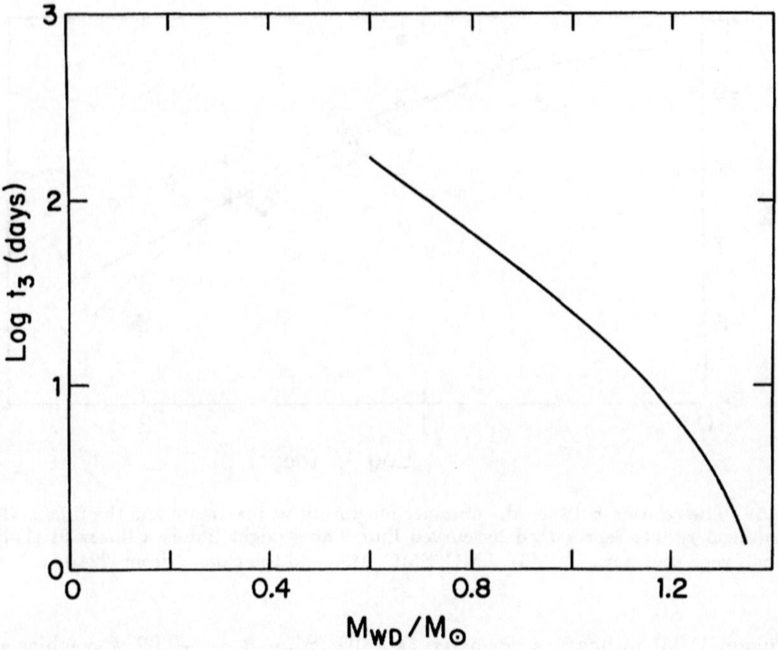

Fig. 58. The time to decline by three magnitudes, t_3, as a function of the white dwarf mass (from [224]).

10.6.3 The $M_B - t_3$ Relation. We are now in the position to obtain the $M_B - t_3$ (MMRD) relation. Combining eqs. (122) and (115) we obtain

$$t_3 = 51.3 \times 10^{\left(\frac{M_B+9.76}{10}\right)} \left[10^{\frac{(M_B+9.76)}{15}} - 10^{\frac{-(M_B+9.76)}{15}}\right]^{3/2} \text{ days.} \qquad (123)$$

Relation (123) is presented in Fig. (59) together with the observational points of Pfau [174].

10.6.4 The Effects of Other Parameters, Limitations of the $M_B - t_3$ Relation. As noted in the previous sections, the derivation of the $M_B - t_3$ relation neglected the effects of a few physical parameters. I shall discuss some of these effects now. This is of importance for the understanding of nova outbursts in general and not just in the context of the $M_B - t_3$ relation.

10.6.4.1 The Effect of the Accretion Rate. The accretion rate affects the value of the envelope mass at the onset of the TNR and thereby the strength of the outburst. If we express the dependence of M_e on $\dot{M}$ as $M_e \propto \dot{M}^{-\beta}$, then the results of Prialnik *et al.* [235] give an average value of $\beta \sim 0.126$ for accretion rates in the range $10^{-11} - 10^{-7} M_\odot/\text{yr}$ onto a $1.25 M_\odot$ white dwarf.

It should be noted, however, that when diffusion is included in the calculation (see section 10.4.1), higher levels of heavy element enrichments are obtained for lower accretion rates. Consequently, the results of calculations with

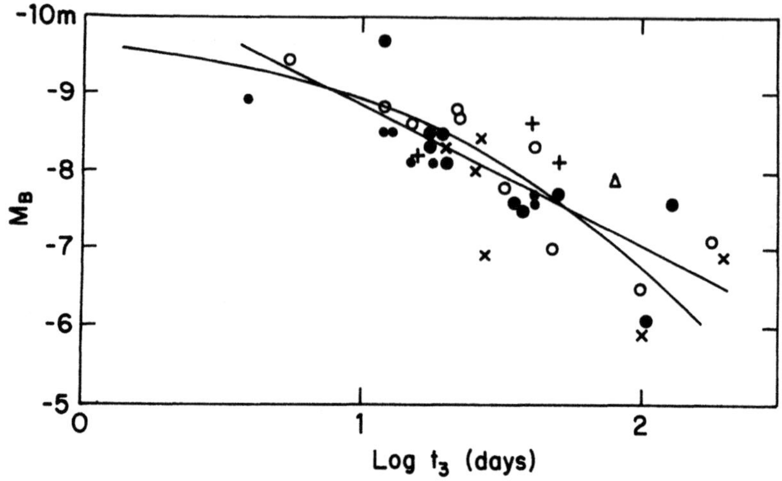

Fig. 59. The relation between the absolute magnitude at maximum and the time of decline. Theoretical results represented by curved line. The straight line is a linear fit [174]. The symbols represent novae in M31, LMC, SMC, M33 and the galaxy (from [224]).

diffusion [194] indicate a negative (small) value $\beta \sim -0.02$, for white dwarf masses of 0.9 and $1.25 M_\odot$ and accretion rates in the range $10^{-11} M_\odot/\text{yr} \le \dot{M} \le 10^{-9} M_\odot/\text{yr}$.

Another effect that the accretion rate can have, is in the amount of compressional heating. For a given white dwarf mass, there exists a critical accretion rate above which no strong outbursts are obtained, because the ignition of the nuclear reactions occurs under only mildly degenerate conditions (due to large compressional heating). The locus of this approximate "nova dud line" in the $M_{WD} - \dot{M}$ plane as obtained on the basis of numerical results is shown in Fig. 60. It is difficult to quantify the net effect of $\dot{M}$ through compressional heating on $L^{\max}$ and t_3. Qualitatively however, the expectation is that for a given white dwarf mass, the higher $\dot{M}$ the lower $L^{\max}$ and the longer t_3.

10.6.4.2 The Effect of the White Dwarf's Intrinsic Luminosity. For a given white dwarf mass and at a given accretion rate, the TNR is initiated sooner on a hotter (higher intrinsic luminosity) white dwarf. Numerical calculations show [194] that if we express the dependence of the envelope mass on the white dwarf's luminosity as $M_e \propto L_{\rm WD}^{-\beta}$, then β is in the range of 0.6–1.0 for different values of $M_{\rm WD}$ and $\dot{M}$. Somewhat smaller values (0.2–0.3) were obtained [232] for accretion onto relatively hot white dwarfs.

The general trend is that the TNR becomes less violent the hotter the white dwarf. This is due to the decrease in the envelope mass and in the degeneracy.

10.6.4.3 The Effect of the White Dwarf's Magnetic Field. Strong surface magnetic fields can influence the TNR development through a number of processes:

(1) Funneling of accreted material onto specific areas on the white dwarf surface [239].

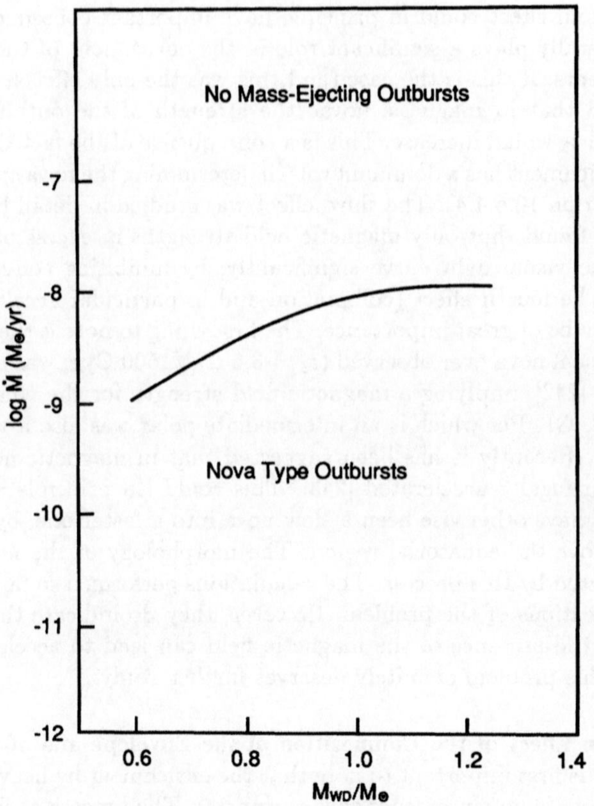

Fig. 60. The "nova dud line" above which no mass ejecting outbursts are obtained in numerical calculations (from [299]).

(2) The disruption of the accretion disk (and thereby the inhibition of shear mixing) [240].
(3) Interference with the development of convection.
(4) Collimation and acceleration of the plasma in the expanding envelope.

The first of these effects is probably not important for our present purposes, since it has been shown [227], that the funneled material is likely to spread on a timescale of

$$\tau \simeq 1.1 \times 10^3 \left(\frac{B}{10^7 G}\right)^{10/7} \left(\frac{M_{\rm WD}}{M_\odot}\right)^{-4/7} \left(\frac{R_{\rm WD}}{5 \times 10^8 \rm cm}\right)^{23/7} \left(\frac{\dot{M}}{10^{-9} M_\odot \rm yr^{-1}}\right)^{-5/7} \rm sec, \tag{124}$$

which is much shorter than the timescale for a TNR to occur (this is the timescale to build a pressure which exceeds the magnetic pressure).

The second effect could in principle have important consequences if shear mixing generally plays a significant role in the enrichment of the envelope by heavy elements. If this is the case (and this was the only effect), then it could be expected that in magnetic novae the strength of the outburst would be reduced and t_3 would increase. This is a consequence of the fact that the heavy element enrichment has a dominant role in determining the nova speed class (see however section 10.6.4.4). The third effect was studied in detail by Livio *et al.* [227]. They found that only magnetic field strengths in excess of $\sim 10^8 G$ can influence the visual light curve significantly, by inhibiting convection during the TNR. The fourth effect (collimation and in particular acceleration of the plasma) may be of great importance. The first thing to note is the fact that the fastest classical nova ever observed ($t_3 = 3.6$ d) V1500 Cyg, was identified as a polar [241], [242], implying a magnetic field strength for the white dwarf of a few $\times 10^7$ G. GK Per which is an intermediate polar was also a very fast nova ($t_3 = 13$ d). Recently it has been suggested that in magnetic novae material can be centrifugally accelerated [243]. This could (in principle at least) turn what would have otherwise been a slow nova into a faster one, by accelerating material above the equatorial region. The morphology of the nova shell may also be affected by this process. The calculations performed so far involve large oversimplifications of the problem. However, they do indicate that at least in some cases the presence of the magnetic field can lead to acceleration of the material. This problem definitely deserves further study.

10.6.4.4 The Effect of the Composition of the Envelope and of the Accreted Material. It is first important to note that the enrichment by heavy elements in nova envelopes is *not* an independent parameter. The process of elemental mixing in novae is not yet fully understood (see section 10.4). However, it is quite clear that whatever the mechanism, the level of enrichment is determined by other system parameters such as the accretion rate, the white dwarf's mass and its magnetic field strength. In principle, therefore, the effects of heavy element abundances on the outburst should not be discussed separately. Nevertheless, in view of the important effect that a high mass fraction of heavy elements, Z, can have (and because the exact dependence of Z on other parameters is not known), I shall briefly summarize it.

As explained in section 10.4, numerical calculations have demonstrated that in order to achieve super-Eddington luminosities, Z has to exceed considerably the solar value of $Z_{\odot} = 0.019$.

It is important to note that a high CNO enrichment can definitely transform what could have otherwise been a slow nova into a fast one. A good example may be provided by DQ Her. The mass of the white dwarf in this system is rather low ($0.66 \pm 0.05 M_{\odot}$) [244], and therefore a very slow nova could have been expected (unless the magnetic field plays a dominant role). However, the enrichment by heavy elements observed in the ejecta is extremely high (Z = 0.56) [240]) and indeed DQ Her is a moderately fast nova ($t_3 = 94^d$).

The question of the composition of the accreted material became relevant with the realization that at least for a subclass of the recurrent novae (if the outbursts of these objects are indeed powered by TNRs), which includes U Sco, V394 CrA and Nova LMC 1990 #2, there exists observational evidence which suggests that the transferred material may be rich in helium [245], [246]. In a series of numerical calculations of nova models involving a $1.38M_{\odot}$ white dwarf (as expected for recurrent novae), for H/He ratios (by mass) of: 3:1,2:1,1:1,1:8, Truran *et al.* [247] have shown that nova outburst-type luminosities are obtained only for a hydrogen mass fraction satisfying $X \gtrsim 0.65$. This can be understood if we realize that the ratio of nuclear energy generated per gram in hydrogen burning to the binding energy per gram is

$$\frac{\epsilon_{\rm nuc}(H\text{burning})}{\epsilon_{\rm bind}} \simeq 6.5X \left(\frac{M_{\rm WD}}{1.38M_{\odot}}\right)^{-1} \left(\frac{R_{\rm WD}}{1.9\times10^{8}\text{cm}}\right) . \tag{125}$$

It is interesting to note, however, that by introducing some artificial modifications to the mixing length and the opacities (the modifications were based nonetheless on a physical picture), Starrfield *et al.* [226] were able to obtain strong outbursts even for $X = 0.2$.

10.6.4.5 Summary of the Various Effects. To summarize the effects of the various parameters, I note that to the extent that present nova codes produce more or less reliable results, the relations derived for t_3 and M_B^{max} should represent approximately average properties of novae. This is a consequence of the fact that the relations are based (with some approximations) on the *main factor* (the white dwarf mass) which determines the outburst properties. However, other physical parameters do have an effect. Therefore, great caution should be exercised in attempts to apply any of the relations to specific systems. In particular, if a given system happens to be characterized by extreme values of one or more of its physical parameters (such as $\dot{M}$ or B), then considerable deviations form the obtained relations might be expected.

Larger deviations from the theoretical $M_B - t_3$ relation may be expected both for very fast and for very slow novae. Specifically, the obtained $M_B^{max} - t_3$ relation does not reproduce the lower bend in the curve, the existence of which seems to be indicated by observations [173] (Fig. 49).

Finally, the expansion phase in novae is still relatively poorly understood. A better understanding of this phase will allow perhaps in the future a more accurate formulation of the expression for t_3.

10.6.5 The t_3 Distribution. The existence of an analytic expression for t_3 (eq. (122)), *allows us for the first time to actually calculate a theoretically expected t_3 distribution in nova systems.* Furthermore, because of the fact that to first order, t_3 is a function only of the white dwarf mass, we can use the results of a recent calculation [248] of the mass distribution in CNe, to obtain the t_3 distribution directly. The resulting distribution is presented in Fig. 61. In order to compare the theoretical distribution to observations, we can use the t_3 values for 120 CNe from Duerbeck [236]. I have excluded from Duerbeck's compilation 19 objects which are almost certainly not CNe (among them are x-ray novae, dwarf novae, symbiotic stars). The observed distribution is also presented in Fig. 61. The main difference between the observed and predicted distributions is in the high fraction of systems with $t_3 \leq 25$ days in the theoretical distribution. This is a direct consequence of the high fraction of white dwarfs with masses $M_{WD} \geq 1.2 M_{\odot}$ in the distribution obtained by Ritter *et al.* [248]. This prominent maximum in outburst frequency at the high end of the mass distribution is in itself the result of a strong selection effect. Other than this difference, the theoretical and observed distributions in t_3 agree quite well, at least qualitatively. It is possible perhaps that some outbursts with very short decline times are missed by observations. The addition of such systems would clearly make the two distributions even more consistent. Another possible cause for a discrepancy is the fact that magnetic fields may play a more important role (than assumed in the derivation) in the determination of the speed class.

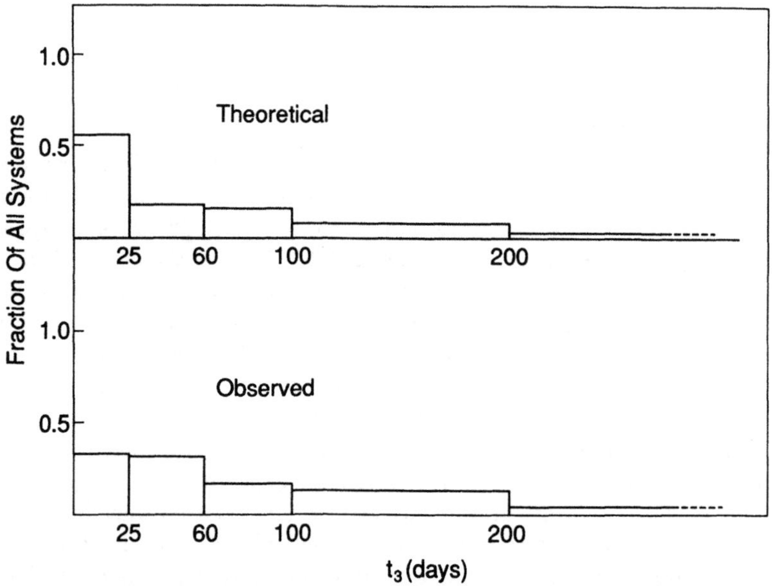

Fig. 61. The theoretically expected and the observed t_3 distributions of novae (from [224]).

10.6.6 Other Possible Calibrations for the Distance Scale Using Novae.

10.6.6.1 The Absolute Magnitude Fifteen Days Following Maximum. Another known characteristic of novae which has been used in calibrations of the extragalactic distance scale, is the fact that 15 days after maximum light, novae appear to have more or less the same mean absolute magnitude $\langle M_{15} \rangle = -5.5 \pm 0.18$ (pg) [218] (see section 10.1).

The reason for this behavior has been partly recognized already by Shara [249]. Because of the fact that during the decline from maximum the nova light curve is almost linear (at least in the early part), the absolute magnitude as a function of time can be described as

$$M_B(t_3,t) = M_B^{\max}(t_3) + \frac{3}{t_3}t, \tag{126}$$

where $M_B^{\max}(t_3)$ is given by the functional dependence eq. (123). Now, because of the fact that the shorter t_3 the brighter $M_B^{\max}$, there exists a time t at which the spread in the values of $M_B(t_3,t)$ is relatively small, the situation is presented in Fig. 62. Differentiating eq. (126) with respect to t_3 gives

$$t(\text{nearly constant} M_B) \simeq \frac{1}{3}t_3^2 \frac{dM_B^{\max}}{dt_3}. \tag{127}$$

If we take the nearly linear part of the $M_B^{\max}(t_3)$ relation then eq. (127) gives $t(\text{nearly constant} M_B) \simeq 0.6t_3$, which is in reasonable agreement with the observations (the average t_3 for novae with $t_3 \lesssim 100$, for which relation (126) is expected to hold is about 39 days).

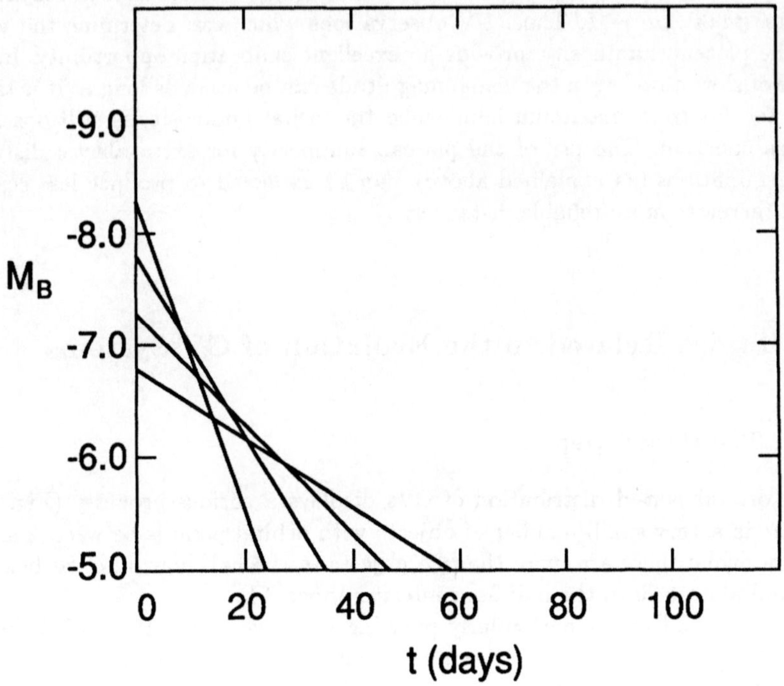

Fig. 62. The approximate time behaviour of the absolute magnitude following maximum for $M_{WD} = 1.0, 0.9, 0.8, 0.7\ M_\odot$ (the brightness at $t = 0$ decreases with the WD mass), (from [224]).

In fact, when the above result is combined with the white dwarf mass distribution obtained by Ritter *et al.* [248], the near constancy of M_{15} gains an additional support. The spread in the M_{15} values is expected to be even narrower than relations (126) and (127) would imply, because of the fact that the distribution of white dwarf masses is quite narrowly peaked, and t_3 was found to depend mainly on the white dwarf mass. Nevertheless, I should emphasize that at least from a theoretical point of view, the M_{15} method of calibration is not more secure than the $M_B - t_3$ relation, since they contain essentially the same physical input.

10.6.6.2 The Plateau Luminosity. Following a brief super-Eddington phase, essentially all novae settle into a plateau of constant bolometric luminosity (see section 10.1). As explained in section 10.2.2, the plateau luminosity is given quite accurately by the Paczynski core mass-luminosity relation (for $M_{\rm WD} \geq 0.7 M_{\odot}$). If we combine this fact with the prediction by Ritter *et al.* [248], that about 30% of all novae have white dwarf masses in the range 0.8–1.2 $M_{\odot}$, we obtain an excellent potential use of novae as standard candles. The idea is to avoid the fastest novae (which may exceed the plateau value considerably). For moderately fast to slow (but not very slow) novae (which probably correspond to $M_{\rm WD}$ in the above range), ***the total range in magnitude is less than*** $\Delta m \sim 1$. Thus, UV observations which can determine the value of the plateau luminosity, provide an excellent calibration opportunity. In the case of slow novae even the visual magnitude can be used, as long as it is taken not too far from maximum light when the visual luminosity is still observed to be constant. The use of the plateau luminosity for extragalactic distance determinations (as explained above), can be expected to produce less scatter and therefore more reliable distances.

11 Topics Related to the Evolution of CV Systems

11.1 The Period Gap

The orbital period distribution of CVs displays a curious property (Fig. 63). There is a very small number of objects with orbital periods between 2 and 3 hours, while there are more than 25 objects in the half hour interval below 2 hr and almost 20 in the half hour interval above 3 hr.

The period gap is particularly puzzling when we realize that it is believed that CV systems undergo a secular evolution in which the systems' orbital period is continuously decreasing as a result of angular momentum losses.

We will now examine briefly the question of why such angular momentum losses are required and what are the suggested loss mechanisms (see also [250] for a comprehensive discussion).

11.1.1 Angular Momentum Loss Mechanisms. As explained in section 2, a star can fill its Roche lobe as a result of its own nuclear evolution. This situation however, cannot happen in CVs, because the secondaries are low mass stars, so that their main sequence lifetime in many cases exceeds a Hubble time. Therefore, in order for the secondary to be able to fill its Roche lobe, the system must evolve in such a way that the Roche lobe contracts towards the

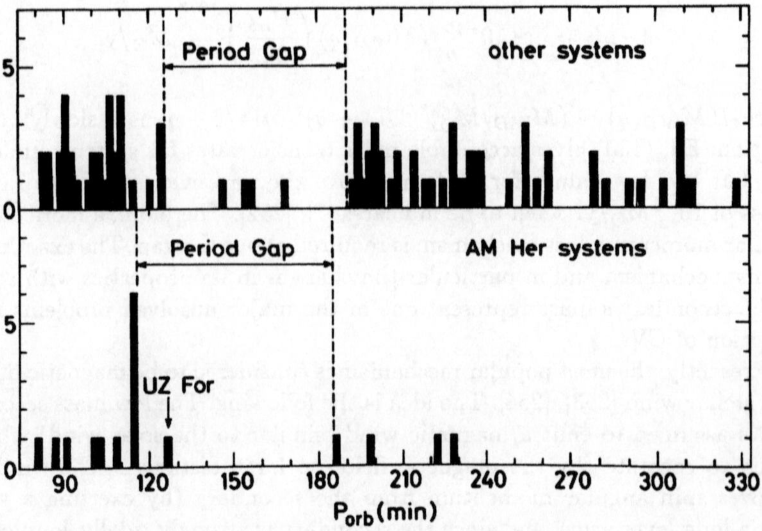

Fig. 63. The Period histogram for magnetic AM Her systems (lower frame) and non AM Her systems (upper frame) (from [251]).

secondary's radius. If we use eq. (107) for the total angular momentum of the system and eq. 6 (for $q \lesssim 0.5$) for the Roche lobe radius we obtain by differentiating (under the assumption that no mass is lost from the system: $\dot{M}_{WD} = -\dot{M}_2$)

$$\frac{\dot{R}_{RL}}{R_{RL}} = \frac{2\dot{J}}{J} + \left(\frac{5}{6} - q\right)\frac{2(-\dot{M}_2)}{M_2}. \tag{128}$$

Since we are dealing typically with $q < \frac{5}{6}$ (and $\dot{M}_2 < 0$), we find that if we want the Roche lobe to contract *the system must lose angular momentum.* Remembering that the secondaries in CVs roughly obey eqs. (7) and (8), we find from eq. (128) that $(-\dot{M}_2)/M_2 \sim (-\dot{J})/J \sim (-\dot{P}_{orb})/P_{orb}$. Note that in order to obtain $|P/\dot{P}| \lesssim 10^{10}$ yrs (if we believe that the distribution in Fig. 63 had to result from evolution during the galaxy's lifetime), we need $|\dot{M}_2| \gtrsim 10^{-11} M_\odot$/yr (which is consistent with most observations of CVs, although large uncertainties exist in the determination of $\dot{M}_2$).

The question now is therefore, what angular momentum loss mechanisms are available.

For the short period systems (under the gap) gravitational radiation (GR) losses seem to be quite adequate [251] and the quadrupole general relatively formula gives

$$\left(\frac{\dot{J}}{J}\right)_{GR} = -\frac{32}{5}\frac{G^3}{c^5}\frac{M_{WD}M_2(M_{WD}+M_2)}{a^4}. \tag{129}$$

If we use eq. (6) (to eliminate a), the fact that the secondary fills its lobe and eqs. (6) and (7) we get

$$(-\dot{M}_2)_{GR} \simeq 10^{-10} f(M_{WD}, q) \left(\frac{P_{orb}}{2hr}\right)^{-2/3} M_\odot/\text{yr} \tag{130}$$

where $f(M_{WD}, q) \sim (M_{WD}/M_\odot)^{2/3}/(1+q)^{1/3}/(4/3-q)$ is a slowly varying function. Eq. (130) gives acceptable mass transfer rates for systems under the gap, but too low values for systems above the gap, where transfer rates in excess of $10^{-9} M_\odot/\text{yr}$ seem to be indicates [7], [252]. Therefore, a more efficient angular momentum loss mechanism is required above the gap. The exact nature of this mechanism and in particular the changes in its properties with respect to the secondary's mass represent one of the major unsolved problems in the evolution of CVs.

Presently, the most popular mechanism is considered to be magnetic braking by a stellar wind [253], [254]. The idea is the following. The low mass secondary star is assumed to emit a magnetic wind (similar to the solar wind) which is forced to corotate (by the magnetic field) to large distances. The wind thus removes spin angular momentum from the secondary (by exerting a torque with a long lever arm), and since the secondary is strongly tidally coupled (see sections 1 and 2) and synchronously rotating with the orbit, this has the effect of removing angular momentum from the orbit. The expression normally used for the magnetic braking torque is based on the empirical finding [255] that the equatorial velocity of G stars behaves with time like $V_{EQ} \propto t^{-1/2}$. Verbunt and Zwaan [253] showed that such a behavior represents a solution to the angular momentum equation $(I/R(dV_{EQ}/dt) = G_{MB}$, where I is the star's moment of inertia and G_{MB} is the magnetic braking torque) if the torque is given by

$$G_{MB} \simeq -5 \times 10^{-29} f^{-2} k^2 M_2 R_2^4 \Omega_2^3 \tag{131}$$

where k is the radius of gyration, Ω_2 is the star's angular velocity and f is a parameter of order unity. Verbunt and Zwaan [253] and Rappaport *et al.* [256], suggested that this torque can be used for the rate of angular momentum losses of CVs (using the fact that the secondary is synchronously rotating). Again, using eqs. (131) and (7) we obtain for the magnetic braking driven mass transfer rate

$$(-\dot{M})_{MB} \simeq 6 \times 10^{-10} \left(\frac{P_{orb}}{3hr}\right)^{5/3} M_\odot/\text{yr}. \tag{132}$$

A somewhat different formulation of the problem by Mestel and Spruit [253] gives a weaker dependence on the orbital period than the one implied by eq. (132).

In a recent work, Tout and Pringle [257] attempted to formulate an approximate self-consistent model for the spin-down rate of fully (or mostly) convective stars. In their model, convectively driven shear coupled with the rotation generate a magnetic dynamo. The energy losses from the dynamo are assumed to give rise to a magnetically driven (and controlled) stellar wind which results in angular momentum loss and spin-down. Tout and Pringle obtained from their model an angular momentum loss rate of

$$\dot{J}_{wind} \simeq 2.2 \times 10^{35} \left(\frac{\delta}{0.3}\right)^{3/2} \left(\frac{\gamma}{10^{-2}}\right) \left(\frac{\eta}{30}\right)^{-2/3} \left(\frac{L_2}{10^{-2}L_\odot}\right)^{2/3} \times \left(\frac{M_2}{0.3M_\odot}\right)^{1/3} \left(\frac{R_2}{0.3R_\odot}\right)^{2/3} \text{dyne cm}, \tag{133}$$

where δ is the ratio between the star's angular velocity and its break-up angular velocity, γ is a measure of the efficiency of the dynamo regeneration term and $\eta \sim 3R_2/l_c$, where l_c is the mixing length. Expression (133) gives a torque which is of a comparable order of magnitude to the one obtained from the Rappaport *et al.* prescription, but the fact that the torque in the Tout and Pringle scenario exists even for fully convective stars is in conflict with at least some of the proposed causes for the occurrence of the period gap (see below).

11.1.2 The Cause for the Period Gap. Since it is difficult to think of an observational selection effect that will prevent the discovery of systems in the period gap [258], the existence of the gap seems to imply one of two possibilities: (i) CVs do not evolve across the gap but rather, once systems above the gap reach the period of ~3 hr, they reverse the direction of their evolution and evolve to longer periods. An explanation of this type is believed to be the cause for the cut-off in the orbital period distribution at about 80 min (I shall not discuss this property here, but see Paczynski and Sienkiewicz [259] and Rappaport *et al.* [256] for a complete description). (ii) The second possibility is that somehow the angular momentum loss mechanism (which drives the mass transfer) stops to operate at about $P \sim 3$ hr [260]. (A third possibility which perhaps has not been sufficiently explored is that for some reason the CV systems evolve across the gap relatively rapdily).

While suggestions of how possibility (i) might materialize have been made [254], [261], this possibility is considered to be unattractive (this does not mean of course that it is necessarily incorrect), especially because it implies that the CV populations above and below the gap evolve differently and independently. Furthermore, this possibility requires a separate explanation for the gap edge at ~ 2 hr.

Before we examine possible causes for a decline in the effectiveness of angular momentum losses by magnetic braking at the upper edge of the gap, we should note the following. The mass transfer timescale implied from eq. (132) near the gap edge at $P_{orb} = 3$ hr is $\tau_{\dot{M}} = M_2/\dot{M}_2 \simeq 5 \times 10^8$ yr. At the same time, the timescale on which the secondary star can adjust its radius thermally is given by the Kelvin-Helmoltz timescale

$$\tau_{KH} = \frac{GM_2^2}{R_2L_2} \simeq 3 \times 10^7 \left(\frac{M_2}{M_\odot}\right)^2 \left(\frac{R_2}{R_\odot}\right)^{-1} \left(\frac{L_2}{L_\odot}\right)^{-1} \text{yr}. \tag{134}$$

For low mass main sequence stars $L_2 \propto M_2^{3.5}$, which gives at $P_{orb} \sim 3$ hr (using eqs. (7) and (8)) $\tau_{KH} \simeq 5 \times 10^8$ yr $\simeq \tau_{\dot{M}}$. Once $\tau_{\dot{M}} \lesssim \tau_{KH}$ it means that the secondary star is forced to transfer mass at a rate which does not allow it to shrink to its thermal equilibrium radius. Therefore, the secondary's radius $R_2(= R_{RL})$ will be larger than the main sequence radius $R_{MS}(M_2)$. Suppose now that at $P_{orb} \simeq 3$ hr the magnetic braking is indeed becoming significantly less effective. In this case, the relevant mass transfer timescale will become the one based on gravitational radiation, which is of order 4×10^9 yr at $P_{orb} = 3$ hr (eq. 130) and thus *much longer* than τ_{KH}. The secondary will therefore shrink (on the timescale τ_{KH}) to its main sequence radius, which will make it detach from the Roche lobe, thus stopping the mass transfer process. Such a system will not be identified as a CV. The Roche lobe radius, however, will continue to shrink on the gravitational radiation timescale and eventually will reach the $R_{MS}(M_2)$ value, thus resuming mass transfer at a shorter orbital period. Thus, this picture predicts the generation of a period gap [256], [259]. From the above discussion it is clear that any angular momentum loss mechanism that can produce similar mass transfer rates will also produce a period gap if switches-off at $P_{orb} \sim 3$ hr. Detailed calculations have shown [256], [260], [262] that in order to produce a gap that is in agreement with the observations $\dot{J}$ must be reduced on a timescale that is much shorter than τ_{KH} at the upper edge of the gap. Furthermore, a rather narrow range in the values of $\dot{J}$ above the gap will produce a gap of the observed width.

The question that now arises is: assuming that magnetic braking is indeed the angular momentum loss mechanism, what can cause it to switch-off at $P_{orb} \simeq 3$ hr.

It was noticed [263] that at around $P_{orb} \simeq 3$ hr the secondary stars become fully convective (this happens for $M_2 \lesssim 0.3 M_\odot$, see eq. (8)). It was therefore speculated that the vanishing of the radiative core either reduces the secondary's magnetic field by interfering with the dynamo process (assumed by some models to require anchoring in the radiative core), or drastically reduces the secondary's wind (again perhaps by interfering with the dynamo processes which are supposed to power the wind). If this speculation is correct, one would expect to see a dramatic reduction in the magnetic activity of stars of spectral type later than M5.

Unfortunately, the observational situation is somewhat ambiguous. For example, while it is true that the x-ray emission from stars later than M5 shows maximum luminosities that are factors of 10–100 less than those of M stars earlier than M5 (*e.g.*, [264], [265]), it is not clear whether this represents a gradual or an abrupt decline. Similarly, there are observations of Ca II and Hα emission which suggest that magnetic activity continues to later spectral types than M5 [266], [267]. As a consequence of the above (and other similar) observations it would be safe to say that there exists no compelling evidence for an abrupt cessation of the magnetic activity. Nevertheless, it has been claimed that there exists observational evidence on the sun which suggests that the dynamo process cannot be simply distributed throughout the convection zone. In particular, measurements of p-mode frequencies [268], [269], seem to imply a differential rotation in the solar convective zone, which is incompatible with the one required by the standard turbulent mean field dynamo models. It was therefore suggested [270] that there exists a layer of nearly toroidal sections of field which does need the radiative core to rest on, with vertical sections connecting to the surface. The above observations and theoretical arguments led Taam and Spruit [271] to propose that the transition to fully convective secondaries is accompanied not by an abrupt switch-off of the magnetic activity, but rather by a rearrangement of the surface field. In particular, the field was assumed to change from one that is regularly ordered (a low-order multipole) to one that is highly tangled (a high-order multipole). Taam and Spruit showed that the effectiveness of magnetic braking can decrease substantially by such a rearrangement, due to the fact that the number of open field lines which reach the Alfven radius is significantly reduced. The angular momentum loss rate (at an orbital period of 3 hr) was found to decrease by three orders of magnitude between a dipole field and a limiting high multipole order field. This loss in stellar wind efficiency did not have to be accompanied by an abrupt decline in the magnetic activity. Thus, the Taam and Spruit scenario achieves the requirements which seem to be imposed by the existence of the period gap.

I should note, however, that from a purely theoretical point of view there is nothing that is absolutely compelling about the interrupted magnetic braking model. The problem of the period gap certainly deserves more work.

11.2 Can Cataclysmic Variables Produce Accretion Induced Collapses or Type I Supernovae?

A large number of recent works have examined the possibility that white dwarfs can grow by accretion to the Chandrasekhar limit, thus yielding either an accretion induced collapse (AIC) [272], or a Type Ia supernova. In particular Wheeler [273] argued that cataclysmic variables can evolve to an explosive endpoint, in contradiction to earlier conclusions by MacDonald [274].

Here, I do not wish to discuss in detail the question of whether an AIC or a Type Ia SN is the more likely product (see [275] for a discussion of this point). Rather, I will concentrate on the general question of whether the white dwarf's mass in CVs can grow to the Chandrasekhar mass. This is an interesting problem both from the point of view of CV evolution and because of the possible relation of CVs to other classes of objects like low mass x-ray binaries.

Since RNe will turn out to be the prime candidates for the growth of the white dwarf mass, I shall discuss this class of objects in some detail (see also [163], [164]).

11.2.1 Recurrent Novae.

11.2.1.1 General Properties. The number of objects in the class of RNe has increased significantly recently due to the discovery of three new RNe. Presently, the class consists of 8 objects: T CrB, RS Oph, V745 Sco, V3890 Sgr, U Sco, V394 CrA, Nova LMC 1990 No. 2, and T Pyx. Other objects previously classified as RNe were rejected by Webbink *et al.* [163]. The years of the recorded outbursts and the orbital periods (when known) for the RNe are given in Table 4.

In spite of the fact that the class includes a small number of objects, RNe seem to fall into three different groups. The first group consists of T CrB,

RS Oph, V745 Sco, and V3890 Sgr. These systems all seem to contain late type giants.

The second distinctive group of RNe consists of U Sco, V394 CrA, and, probably, Nova LMC 1990 No. 2. These are extremely fast novae ($t_3 \sim$ 5–10 days) and their spectra are characterized by the fact that the narrow He II λ4686 emission line is much stronger than Hβ [245], [246], [276]. The third group of RNe presently consists only of T Pyx, which is quite unique in its rather slow light curve development ($t_3 = 88^d$; [236]) and short orbital period [217].

Table 4. The Recurrent Novae

System	Years of Recoded Outbursts	Orbital Period
T Pyxidis	1890, 1902, 1920, 1944, 1966	1.75 hr
U Scorpii	1863, 1906, 1936, 1979, 1987	1.2344 day
V394 CrA	1949, 1987	0.7577 day
Nova LMC 1990 #2	1968, 1990	
T Coronae Borealis	1866, 1946	227.53 day
RS Ophiuchi	1898, 1933, 1958, 1967, 1985	230 day (uncertain)
V745 Scorpii	1937, 1989	
V3890 Sagittarii	1962, 1990	

11.2.1.2 The Outburst Mechanism. Scrutiny of possible outburst mechanisms, based on the observation data that was available in 1986, led Webbink *et al.* [163] to propose that the outbursts of U Sco (and therefore, by similarity, perhaps of V394 CrA and of Nova LMC 1990 No. 2) and of T Pyx are caused by thermonuclear runaways on the surfaces of massive white dwarfs, while they concluded that the outbursts of T CrB and RS Oph (and therefore perhaps also of V745 Sco and V3890 Sgr) are better explained in terms of accretion events onto main sequence stars. I shall mention later some new observational evidence which seems to indicate that the outbursts of T CrB and RS Oph (and therefore probably also of V745 Sco and V3890 Sgr) may also be caused by TNRs on white dwarfs. For the moment, however, I shall concentrate on the "U Sco-like" objects and T Pyx.

A TNR is obtained at the base of the accreted hydrogen layer once the timescale for energy generation by the nuclear reactions becomes shorter than the energy transport timescale. The results of detailed numerical calculations can be quite adequately represented by the expression

$$\frac{\Delta M_{ign}}{M_\odot} \simeq 6.3 \times 10^{-5} \left[\left(\frac{R_{WD}}{0.01 R_\odot}\right)^4 \left(\frac{M_{WD}}{M_\odot}\right)^{-1} \right]^{0.7} . \qquad (135)$$

This expression fits the numerical results sightly better than the expression I used before (eq. 97), but it has a less transparent physical interpretation. In combination with a mass-radius relation for the white dwarf (eq. 120), we can obtain an expression for the recurrence timescale of nova-type outbursts caused by TNRs as

$$\tau_{rec} \simeq 8.66 \times 10^4 \mathrm{yr} \left(\frac{\dot{M}}{10^{-9} M_\odot \mathrm{yr}^{-1}} \right)^{-1} \times$$
$$\left[1.54 \left(\frac{M_{WD}}{M_\odot} \right)^{-7/3} - 2.0 \left(\frac{M_{WD}}{M_\odot} \right)^{-1} + 0.65 \left(\frac{M_{WD}}{M_\odot} \right)^{1/3} \right]^{0.7} . \tag{136}$$

The recurrence timescale as a function of the WD mass is shown in Figure 64, for different mean values of the accretion rate $\dot{M}$. Also shown in Figure 64 are the shortest recurrence timescales of the eight RNe (I ignore for the moment the possibility that the outbursts of some of them may be caused by other mechanisms).

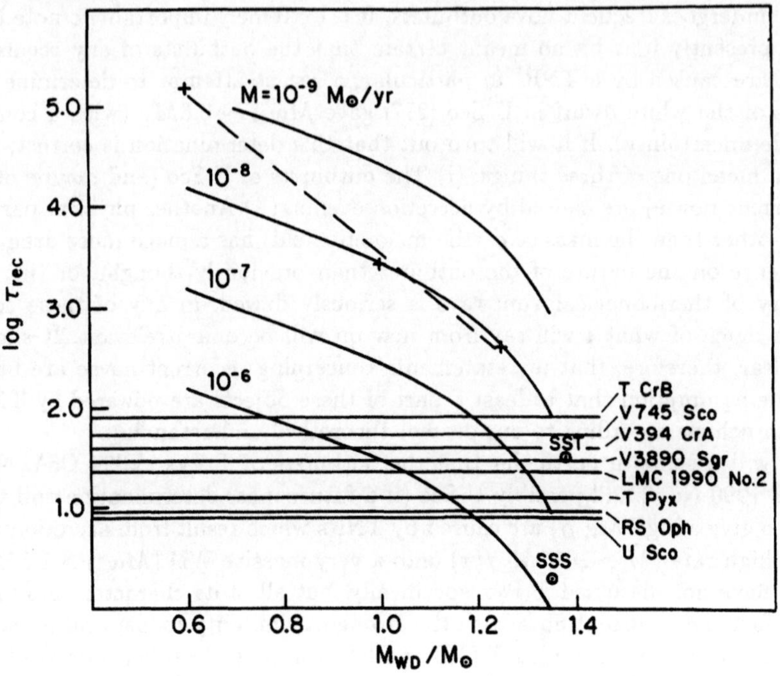

Fig. 64. The recurrence timescale as a function of the white dwarf mass for different accretion rates. Horizontal lines indicate shortest recurrence times for the recurrent novae. The points mark results of numerical calculations. The dashed line is an approximate "nova dud line" (see text) (from [295]).

One thing that becomes immediately clear from Figure 64 is the fact that, in order to obtain the extremely short recurrence timescales characterizing RNe, *very high accretion rates are required.* This raises the important question as to whether nova-like outbursts can occur under such conditions (see section

10.6.4.1). The approximate "nova dud line" is therefore also shown in the figure. When the existence of this nova dud line is taken into account we find that, *for the known RNe to be explained in terms of TNRs, these systems must satisfy* $M_{WD} \geq 1.3M_{\odot}$, $\dot{M} \geq 10^{-8}M_{\odot}$/yr. I would like to note here that normally, the fact that a specific model requires such stringent conditions would raise serious doubts about the model. The main reasons why the TNR model continues to be considered at all for RNe are: (1) the enormous success of the TNR model for classical novae and (2) the fact that the required conditions make the outbursts of these objects much more frequent, so that there is a strong selection effect favoring their discovery (see also [248]).

The above conditions make these systems obvious candidates for the growth of the WD mass to the Chandrasekhar limit, since their WD masses may already be quite close to the limit and they accrete at a high rate. In order to determine whether the Chandrasekhar limit can indeed by reached, we must first be able to determine whether the mass of the WD increases or decreases as it undergoes frequent nova outbursts. It is extremely important to note here that presently it is by no means certain that the outbursts of any recurrent nova are caused by a TNR. In particular, a recent attempt to determine the mass of the white dwarf in U Sco [277] gave $M_{WD} \sim 0.6M_{\odot}$ (with a considerable uncertainty). If it will turn out that this determination is correct, this could mean one of three things: (i) The outbursts of U Sco (and maybe other recurrent novae) are caused by accretion events. (ii) Another physical parameter, other than the mass (*e.g.*, the magnetic field) has a much more dramatic influence on the nature of the outburst than previously thought, or (iii) the theory of thermonuclear runaways is seriously flawed. In any of these three cases much of what I will say from now on will become irrelevant. It should be clear, therefore, that my statements concerning recurrent novae are based on the assumption that at least a part of these objects are powered by TNRs which behave according to our present theoretical understanding.

I will assume in particular that the outbursts of T Pyx, V394 CrA, Nova LMC 1990 No. 2 and possibly U Sco (if a future mass determination will turn out to give a high M_{WD}) are caused by TNRs which result from accretion at a very high rate ($\dot{M} > 10^{-8}M_{\odot}$/yr) onto a very massive WD ($M_{WD} > 1.3M_{\odot}$).

I have not discussed T Pyx specifically, but all of its characteristics make it in fact the recurrent nova with the strongest similarity to classical novae.

The situation concerning T CrB and RS Oph (and the other objects of this sub-class, V745 Sco and V3890 Sgr) is more complicated. In fact, based on the observations then available, Webbink [278] and Livio *et al.* [279], suggested that the outbursts of these RNe are caused by accretion events onto main sequence stars. However, a number of recent observations [280] (e.g. of very high velocities in the accretion disk, suggesting a WD accretor), suggest the possibility that the outbursts of T CrB and RS Oph are rather caused by TNRs on the surfaces of WDs.

Since these objects are not "classical" CVs (the contain giants), I shall not discuss them any further here.

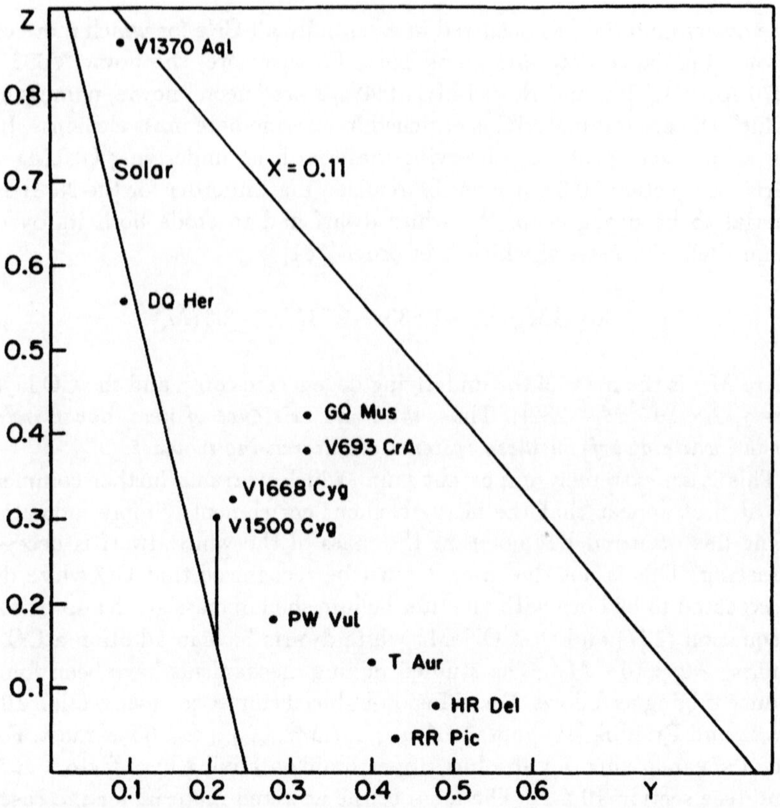

Fig. 65. The location of different novae in the plane of the mass fraction of heavy elements, Z, and helium mass fraction Y (see text, from [295]).

11.2.2 Does the Mass of the White Dwarf Grow in CVs? Clearly the most important question to be considered in our attempt to determine whether the white dwarfs in CVs can reach the Chandrasekhar limit is the following: Can the white dwarfs grow in mass? Let us proceed to address this question from both the observational and the theoretical points of view.

11.2.2.1 The Implications of Observed Abundances in Novae Ejecta. The abundance determinations for the ejecta of novae are crucial to the question of the evolution of the white dwarf mass. As explained in section 10.1, essentially all novae show enrichments. Figure 65 shows the location of all the systems in the mass fractions (Z, Y) plane. Also shown in the figure are two lines indicating the expected Z–Y relations if the accreted layers are mixed directly with core material (this assumes there to be no residual helium layer on top of the white dwarf). The line marked "Solar" represents a solar composition of the accreted material, while that marked "$X = 0.11$" represents the accretion of material that is very helium rich (as seems to be the case for the RNe of the U Sco subgroup). Figure 65 clearly indicates that a considerable dredge-up of white dwarf material has occurred in essentially all CNe for which more or less reliable abundance determinations exist. Furthermore, the novae V693 CrA, V1370 Aql, QU Vul and Nova LMC 1990 #1 are "neon" novae, namely, novae in which the ejected material is enriched in intermediate mass elements. In this case, we are very probably observing material from underlying ONeMg white dwarfs (see section 10.3). It must be realized that, in order for the Ne rich core material to be dredged-up, the white dwarf had to erode both its overlying helium shell, the mass of which is of order [281]

$$\log \Delta M_{He} \simeq -1.835 + 1.73 M_C - 2.67 M_C^2 \tag{137}$$

(where M_C is the mass of the underlying degenerate core), and the CO layer (of mass $\leq 2 \times 10^{-3} M_\odot$ [282]). Thus, *the mere existence of neon novae indicates that the white dwarfs in these systems are decreasing in mass.*

This is an extremely important point which warrants further comment. It may at first appear that the heavy element enrichments simply indicate that mixing has occurred but not that the mass of the white dwarf is necessarily decreasing. This is not the case. It must be recognized that CO white dwarfs are expected to be born with an outer helium shell of mass given approximately by equation (137) and that ONeMg white dwarfs have in addition a CO layer of mass $\sim 2 \times 10^{-3} M_\odot$. The studied mixing mechanisms have been found to produce mixing as follows: For diffusion-induced convection (see section 10.4.1), Kovetz and Prialnik [194] obtain $\Delta m_{mixed}/\Delta m_{accreted} \simeq 1.65$ at most. For the case of shear mixing, an absolute upper limit on mixing by a factor $\lesssim 5$ was found (see section 10.4.2). The mass of the accreted material for the case *e.g.*, of a $1 M_\odot$ white dwarf is $\sim 8.5 \times 10^{-5} M_\odot$, while the initial mass of the helium layer on the same white dwarf is $\sim 1.7 \times 10^{-3}\ M_\odot$. Therefore, there is no way in which, by mixing alone, a CO enrichment can be achieved. The problem is clearly even more severe for ONeMg white dwarfs, since they have an extra $\sim 2 \times 10^{-3} M_\odot$ of CO material through which mixing should have somehow occurred. Thus, *if* indeed the enrichments result from mixing (see the doubts I raised in section 10.3), then it appears that *the mass of the WD in CNe typically decreases as a result of nova explosions.*

In a sharp contrast to the situation for CNe, there is no evidence for enrichments in heavy elements in the ejecta of recurrent novae. The nebular shell that is resolved around T Pyx [283] shows approximately solar abundances [284], and abundance analyses of both the quiescent and outburst spectra of U Sco are consistent with normal CNO abundances. While U Sco, V394 CrA and Nova LMC 1990 #2 show an extreme overabundance of helium (He/H $\sim$ 1–2 by number), this appears not to result from dredge-up, since a similar overabundance is observed in quiescence. Thus, the companion is probably transferring helium rich material. It therefore appears likely that the white dwarfs in recurrent nova systems grow in mass. I would like to emphasize, however, that a serious abundance analysis has never been performed on T Pyx. In view of the crucial importance of such an analysis, this should be a high priority item in the observations of the upcoming outburst of this system.

11.2.2.2 Theoretical Considerations.

Dredge-Up and Mixing. The observational situation outlined in 11.2.2.1 above points towards the following picture: white dwarfs in which significant mixing between the accreted material and the core material occurs decrease in mass, while white dwarfs in which no significant mixing occurs increase in mass. The former situation appears to be the case for CNe while the latter characterizes (perhaps) RNe. This suggests that in order to understand the conditions under which the white dwarf can grow in mass, we must have a better understanding of the mixing or dredge-up process (see section 10.4).

Diffusion induced convection has been explored in far more detail than any of the other mechanisms, I shall therefore discuss first its possible implications for the question of the evolution of the white dwarf's mass.

As explained in section 10.4.1, the results of Kovetz and Prialnik [194] show that, if diffusion induced convection is the dominant mixing mechanism, then $\Delta m_{env}/\Delta m_{acc}$ *is almost independent of the white dwarf mass, but depends strongly on the accretion rate.* Here Δm_{env} is the total mass that is affected by the mixing process and Δm_{acc} is the accreted mass. The level of heavy element enrichment is given in this case approximately by (assuming a solar composition in the accreted material) $Z \simeq 1{-}0.98\Delta m_{acc}/\Delta m_{env}$. Clearly, the higher the value of Z, the stronger the TNR and the higher the fraction of Δm_{env} that is ejected. For $Z \gtrsim 0.3$, essentially the entire Δm_{env} is expected to be ejected. An examination of the results of Kovetz and Prialnik reveals that, in order to obtain $Z \sim 0.3$, the accretion rate must satisfy $\dot{M} \leq 3 \times 10^{-10} M_{\odot}/\text{yr}$. Thus, if diffusion induced convection is indeed the dominant mixing mechanism, then the fact that in CNe the white dwarf mass is decreasing while in RNe it is increasing would be entirely attributable to the much higher accretion rate in RNe (the exact numerical value of the "critical" accretion rate is probably quite uncertain). There exists an additional observational point that appears to support (or at least is consistent with) the last conclusion. One could wonder why the neon novae (V693 CrA, V1370 Aql, QU Vul and Nova LMC 1990 #1), which also contain massive white dwarfs (like the RNe), appear to show a mass decrease, while the RNe appear to grow in mass. Based on the above conclusion, the answer could simply be that this is a consequence of the much higher accretion rates in RNe. If this picture is correct it would mean that, independently of the white dwarf's mass, the systems in which the white dwarf grows in mass are those with the high accretion rates.

In view of the uncertainties that still exist concerning all of the mixing mechanisms (see section 10.4), it is important to examine also the possible consequences of other mechanisms than diffusion induced convection.

A second possibility which can be examined in general terms, without specific reference to any particular mixing mechanism, is that *the amount of mixing increases with increasing recurrence time between outbursts.* One can easily imagine that a mixing process that includes perhaps some combination of diffusion and shear mixing could have this general property. For example, in the model of Kippenhahn and Thomas [285], which involves the development of a layer that is marginally stable against the Richardson criterion, the inner boundary of the mixed layer advances inward on the accretion (recurrence) timescale. An examination of Eq. (136) and Figure 64 reveals immediately that, if this possibility correctly represents the mixing process, we would expect the white dwarfs to grow in mass in systems in which both M_{WD} and $\dot{M}$ are high. This again is clearly consistent with the observations of RNe.

Likely Binary Parameters for RNe. Two additional points should be made. First, if we believe that the secular mean accretion rate (above the period gap) is determined by magnetic braking or a similar mechanism (see section 11.1.1), then we can expect a relation of the form $\dot{M} \sim P^{\alpha}$ with α being some positive constant. Observations indicate that such a relation may indeed exist [7]. If this is the case, then we expect RNe to be found preferentially with relatively long orbital periods (compared to other CVs). An examination of the observational situation reveals that indeed the orbital periods of U Sco and V394 CrA (1.23 days and 0.76 days respectively, [286]) are (with the exception of GK Per), the longest known among cataclysmic binaries. Of course T CrB, RS Oph, V745 Sco and V3890 Sgr can have high mean accretion rates due to the presence of a giant. It therefore appears that T Pyx, with an orbital period of about 1.75 hr [217], is presently the only RN with a short orbital period.

Incidentally, since the frequency distribution of zero age cataclysmic binaries as a function of orbital period is expected to be a quite smooth function above $P \sim 6$ hr [287], then if $\dot{M} \sim P^{\alpha}$, we would expect the accretion rate also to behave smoothly. This means that there should exist CV systems with massive WDs and with accretion rates that are intermediate between the ones in the neon novae ($\dot{M} \sim 10^{-10} M_{\odot}$/yr) and the ones in RNe ($\dot{M} > 10^{-8} M_{\odot}$/yr). These systems should produce nova outbursts with intermediate recurrence times ($\tau_{rec} \sim 10^2$–10^3 yrs). It would therefore be of great interest to investigate the possibility that some of the known novae have experienced previously recorded outbursts a few hundred years ago.

The second point that should be mentioned is that the maximum secondary mass that is stable against dynamical or thermal timescale mass loss is related to the white dwarf's mass by $M_2^{max} \simeq q_{crit} M_1$. Thus, the systems with more massive white dwarfs (such as perhaps are the RNe) contain on the average also more massive secondaries. Since the secondary star is the mass reservoir with the help of which the white dwarf's mass can grow, this means that the RNe have this advantage too, in trying to grow the white dwarf to the Chandrasekhar limit.

11.2.2.3 The Composition of the Accreted Material. As noted already, the spectra of both U Sco and V394 CrA in quiescence have a He II λ4686 line that is much stronger than Hβ. Even in outburst, the He II feature is by far the more prominent. The observations seem to indicate a large overabundance of helium, He/$H \sim 2$ by number. An apparent large overabundance of helium was found also in Nova LMC 1990 #2 (He/$H \sim 1$; [245]). Such an overabundance of helium in the accreted material, *if real*, can have important consequences. The accretion of material deficient in hydrogen necessarily leads to a situation in which only a relatively small fraction of the accreted envelope is ejected, due to insufficient energy generation, thus favoring a growth in the mass of the white dwarf. Indeed, detailed numerical models (see section 10.6.4.4) which adopted a mass fraction of hydrogen of $X = 0.11$ predicted an increase in the white dwarf's mass.

It is beyond the scope of these lectures to discuss the evolutionary path which can lead to the transfer of helium rich material. I do note, however, that the secondary stars in U Sco and V394 CrA, both of which have long orbital periods, are evolved. It is interesting also to note that the spectrum of U Sco is very similar to that of some low mass x-ray binaries (*e.g.*, LMC X-2), in the dominance of the He II lines [288], [289]. Consequently, it has in fact been speculated [288] that this may point towards the possibility that low mass x-ray binaries form from recurrent novae via accretion induced collapse. While this type of evidence can be regarded as circumstantial at best, it is certainly true (as pointed out above) that the (massive) white dwarfs in RNe that accrete helium rich material have the best chance of growing in mass.

11.2.2.4 Can the Chandrasekhar Mass be Reached? Even if the conditions are such that the mass of the white dwarf indeed grows, it is not clear that in most cases in CVs the Chandrasekhar mass can be reached. This depends on the accumulation ratio η and on the mass of the secondary. We can define η by

$$\eta \equiv \frac{\Delta m_{growth}}{\Delta m_{trans}} \tag{138}$$

where Δm_{growth} is the mass that remains on the white dwarf and Δm_{trans} is the mass transferred by the secondary. Kato and Hachisu [290] calculated η for a variety of white dwarf masses and accretion rates, assuming that mass loss occurs mainly in an optically thick wind phase. Their calculations neglected the effects of mixing and therefore their values of η should be regarded in some sense as upper limits (for example, they could not obtain negative values of η).

From the results of Kato and Hachisu it becomes clear that, even in the case of the weakest TNRs, it is in general extremely difficult to grow white dwarfs to the Chandrasekhar mass. For example, for $M_{WD} \simeq 1M_\odot$ they obtain $\eta \simeq 0.3$, even for an accretion rate as high as $10^{-7}M_\odot/\text{yr}$. Consequently, for a $1M_\odot$ white dwarf, the secondary (of mass $M_2 \geq 0.4M_\odot/\eta$) needs to have a mass higher than $1.3M_\odot$. Such systems cannot represent CVs experiencing stable mass transfer ($M_2 > M_1$). Furthermore, even if helium does accumulate, the mass may be prevented from growing substantially by the occurrence of helium flashes. Kato, Saio and Hachisu [291] showed that relatively high accumulation ratios are obtained in helium novae only for $\dot{M} \gtrsim 5 \times 10^{-7}M_\odot/\text{yr}$ onto a $1.3M_\odot$ white dwarf.

It therefore appears that only CV systems which contain very massive white dwarfs and for which the mass indeed increases, ***such as possibly the RNe***, have a chance of growing the mass to the Chandrasekhar limit. Furthermore, systems of the U-Sco subclass, which appear to accrete helium rich material at a very high rate, are also able to avoid the mass loss associated with helium flashes.

It should be remembered that all of this is true under the assumption that the outbursts are caused by "standard" TNRs.

11.2.3 Some Statistical Considerations. From the preceding sections, it has become clear that the CV systems in which the WD can be expected to grow to the Chandrasekhar mass are those that contained a very massive WD at their zero age cataclysmic binary (ZACB) stage and in which the mean accretion rate is very high. The exact values of the critical (lower limit) white dwarf mass and critical (lower limit) accretion rate are somewhat uncertain. Based upon the arguments presented in Sections 11.2.1 and 11.2.2, however, it seems that we can conservatively assume that only systems that have $M_{WD} \geq 1.2 M_\odot$ and $\dot{M} \geq 10^{-8} M_\odot \mathrm{yr}^{-1}$ can reach the Chandrasekhar limit. We can now attempt to estimate the rate of birth of such systems. To this goal, the following assumptions are made:

(1) The space density of CVs is of order $n_{CV} \sim 10^{-4}$ pc^{-3}. This number seems to be implied by observations [292].

(2) Mass transfer is assumed to be driven by magnetic breaking, (section 11.1.1) giving a rate which can be approximated by eq. (132).

(3) The fractional birth rate of cataclysmic binaries per unit area of the galactic disk (in different WD mass and orbital period intervals) is given by the results of Politano [287].

Adopting the above assumptions, we obtain a rate for systems to reach the Chandrasekhar limit

$$R \simeq 1.3 \times 10^{-4} \left(\frac{n_{CV}}{10^{-4}\mathrm{pc}^3}\right) \left(\frac{A_{eff}}{10^9 \mathrm{pc}^2}\right) \left(\frac{\Delta M_{av}}{0.2 M_\odot}\right)^{-1} \left(\frac{\dot{M}}{10^{-8} M_\odot \mathrm{yr}^{-1}}\right) \mathrm{yr}^{-1} \tag{139}$$

where ΔM_{av} is the average mass that has to be accreted to reach the Chandrasekhar mass and A_{eff} is the effective galactic area. If we assume instead that the critical WD mass that is required in order for the Chandrasekhar mass to be reached is $1.3 M_\odot$, then the rate is reduced to $R \sim 9.8 \times 10^{-5}$ yr^{-1}. Clearly, if the value of $n_{CV} \simeq 6 \times 10^{-6}$ pc^{-3} [7] is used, then we obtain (for $M_{crit} = 1.2 M_\odot$) $R \sim 7.8 \times 10^{-6}$ yr^{-1}.

We can attempt now to compare the above numbers with the frequencies expected for accretion induced collapses or Type Ia supernovae. First we note that Nomoto and Kondo [275] find that, for $M_{WD} \gtrsim 1.2 M_\odot$ and $\dot{M} \geq 10^{-8} M_\odot$ yr^{-1}, neutron star formation is expected to occur, rather than a Type Ia supernova. This is true both for the case in which the WD is of ONeMg composition (in which case collapse in triggered by electron capture on ^{24}Mg and ^{20}Ne) and for C/O white dwarfs (in which case collapse is expected to ensue as a result of carbon deflagration).

The rate obtained from eq. (139) is consistent with the expected birth rate of LMXB [293]. It should be noted, however, that both numbers are model dependent and that the assumptions made in the two calculations are not entirely independent (and sometimes not even entirely consistent).

On the other hand, the rate of occurrence of SN Ia in the galaxy is estimated to be [294]

$$R_{SN\ Ia} \simeq 3 \times 10^{-3} \left(\frac{H_o}{75 \mathrm{kms}^{-1} \mathrm{Mpc}^{-1}} \right)^2 \left(\frac{L_{Gal,B}}{2 \times 10^{10} L_\odot} \right) \mathrm{yr}^{-1} \tag{140}$$

where H_o is the Hubble constant and L_{Gal} is the galactic luminosity. This is significantly higher than the rate obtained from Eq. (139). In addition, as noted above, the systems which are expected to reach the Chandrasekhar limit based on the present discussion, will very probably produce AICs rather than SNe Ia.

It is important to note that all of this discussion involved specifically only CV systems and not other systems, such as symbiotic stars, which also contain accreting white dwarfs and which may contribute to the production of SNe Ia [295], [296].

12. Critical Observations

The present lectures discussed mostly theoretical aspects of the physics of CVs and to some extent x-ray binaries. There are some critical observations that can be performed that would both test and clarify some of the topics that have been presented. In the following I give a list (definitely far from being complete) of a few such critical observations (not in order of importance).

(1) More continuous observations of the behaviour of the disk radius before and during a dwarf nova outburst are required. These will test both the predictions of the dwarf nova eruption models and mechanisms for angular momentum transport (e.g. the centrifugally driven wind model).

(2) Doppler tomography of accretion disks to determine: (i) whether a spiral pattern can be detected and (ii) if disks often lack their inner parts (due to magnetic fields).

(3) Multi-wavelength observations of disks in order to test fits of theoretical models to the continuum and to attempt to determine the effective viscosity parameter and the accretion rate.

(4) More observations of outflows from disks, in particular: (i) to attempt to determine from which part of the disk the outflow is emanating. (ii) In systems which have material at large distances around them (e.g. T CrB, RS Oph, from the giant's wind), to attempt to determine whether the outflow collimates to something like a jet (which can interact with the wind material). (iii) To determine whether winds are observed in intermediate polars, in which the inner part of the disk is supposed to be truncated.

(5) Determine the spin behaviour of the neutron stars in massive x-ray binaries on short timescales (hours or less) in order to test predictions of the "flip-flop" instability. Similarly, attempt to determine changes in the spin during short x-ray flares.

(6) Discover and determine the binary parameters of close binary nuclei of planetary nebulae, to allow tests of common envelope evolution theory.

(7) Attempts to discover more binary white dwarf systems and determine the parameters of these systems. This will allow both tests of CE evolution and of the merger scenario for Type Ia Supernovae.
(8) Determinations of the masses of the white dwarfs in nova systems. In particular in neon novae, in very fast novae and in recurrent novae. The prediction of the standard theory is that these should be high.
(9) Attempts to discover more magnetic CVs. In particular fast novae should be searched for evidence. This will help elucidate the potential role of the magnetic field.
(10) Attempts to discover outbursts of known nova systems, that occurred in the last century. Such systems should exist.
(11) Abundance determinations of nova ejecta, in particular in recurrent novae (T Pyx the outburst of which is upcoming) and magnetic novae. This will allow tests of the mixing mechanisms.
(12) If still somehow possible, attempt to determine the orbital period of T Pyx to six significant figures. This will provide an important test for CE theory.
(13) Attempts to determine whether changes in the wind properties occur in stars around spectral type M5, to test period-gap scenarios.
(14) Attempts to determine variations in the accretion rate (of amplitude of order 3–30%) in nova systems shortly (about a month) after the outburst. These variations (which are expected due to the small eccentricity introduced in the orbit by the outburst) should occur with a period that is longer by a few percent than the orbital period [297], [215].
(15) It is important to continue to monitor nova systems which did not return to their pre-outburst brightness, to see if a luminosity decrease occurs. This will confirm the effect of irradiation which is assumed in the cyclic evolution-"hibernation" scenario [298], [299]. Systems in this group include FH Ser, V1500 Cyg, CP Pup, HR Del, IV Cep and V1229 Aql.
(16) Images of nova shells (e.g. CCD, speckle) will help determine the shaping processes (common envelope, magnetic field, etc.).

Acknowledgements

The material presented in this series of lectures represents some of my research interests in the last few years. During this time, I have benefited from many helpful discussions with many colleagues. I would like to acknowledge in particular conversations with: Uli Anzer, Howard Bond, France Cordova, Peter Eggleton, Keith Horne, Icko Iben, Andrew King, Don Lamb, Jean-Pierre Lasota, Takuya Matsuda, Friedrich and Emmy Meyer, Ken Nomoto, Dina Prialnik, Oded Regev, Hans Ritter, Mike Shara, Giora Shaviv, Joe Smak, Noam Soker, Warren Sparks, Sumner Starrfield, Paula Szkody, Ron Taam, Jim Truran, Sasha Tutukov, Frank Verbunt, Brian Warner, Ron Webbink, Craig Wheeler and Leva Yungelson. I am especially indebted to Jim Pringle for many invaluable discussions. I would like to thank Ed van den Heuvel and Steve Shore, the co-contributors to the course and this book for the wonderful time we had while giving this course and finally, many thanks to Harry Nussbaumer and Astrid

Orr who made it all possible. The participation of the students contributed to the improvement of these lectures. This work has been supported in part by the Director's Research Fund at the Space Telescope Science Institute, by the Israel Academy of Sciences at the Technion, and by NSF grant AST 89-17442 at the University of Illinois.

References

1. *Interacting Binary Stars*, J. E. Pringle and R. A. Wade (eds.), (Cambridge: Cambridge University Press), (1985).
2. Frank, J., King, A. R. and Raine, D. J., *Accretion Power in Astrophysics*, (Cambridge: Cambridge University Press), (1985).
3. *Accretion Disks in Compact Stellar Systems*, J. C. Wheeler (ed.) (Singapore: World Scientific Publishing Co.) 1992, in press.
4. Shapiro, S. L. and Teukolsky, S., *Black Holes, White Dwarfs and Neutron Stars* (New York: Wiley-Interscience) (1983).
   Cordova, F. A. and Mason, K. O., in *Accretion-Driven Stellar X-ray Sources*, W. H. G. Lewin and E. P. J. van den Heuvel (eds.), (Cambridge: CUP), p. 147 (1983).
5. Eggleton, P. P., *Ap. J.*, **268**, 368 (1983).
   Kopal, Z., *Close Binary Systems* (Chapman & Hall), (1959).
   Kluźniak, W., in *Texas/ESO-CERN Symposium on Relativistic Astrophysics, Cosmology and Fundamental Physics*, J. D. Barrow, L. Mestel and P. A. Thomas (eds.), (New York: The New York Academy of Sciences), Annal. New York Acad. Sci., Vol. 647, 587 (1991).
6. Paczynski, B., *Ann. Rev. Astr. Astrophys.*, **9**, 183 (1971).
7. Patterson, J., *Ap. J. Suppl.*, **54**, 443 (1984).
8. Lacy, C., *Ap. J. Suppl.*, **34**, 479 (1977).
9. Webbink, R. F. in *Interacting Binary Stars*, J. E. Pringle and R. A. Wade (eds.), (Cambridge: CUP), p. 39 (1985).
10. Pringle, J. E. in *Interacting Binary Stars*, J. E. Pringle and R. A. Wade (eds.), (Cambridge: CUP), p. 1 (1985).
11. Paczynski, B. and Sienkiewicz, R., *Acta Astron.*, **22**, 73 (1972).
12. Edwards, D., and Pringle, J. E., *M.N.R.A.S.*, **229**, 383 (1987).
13. Flannery, B. P., *M.N.R.A.S.*, **170**, 325 (1975).
14. Lubow, S. H. and Shu, F. H., *Ap. J.*, **198**, 383 (1975).
15. Plavec, M. and Kratochvil, P., *Bull. Astr. Inst. Czech*, **15**, 165 (1964).
16. Geyer, F., Herold, H. and Ruder, H., in *Accretion Powered Compact Binaries*, C. W. Mauche (ed.), (Cambridge: CUP), p. 308, (1990).
17. Smak, J., *Acta Astron.*, **29**, 325 (1979).
18. White, N. E. and Holt, S. S., *Ap. J.*, **257**, 318 (1982).
19. Tylenda, R., *Acta Astron.*, **31**, 127 & 267 (1981)
   Regev, O., in *Structure and Emission Properties of Accretion Disks*, C. Bertout *et al.* (eds.), (Editions Frontieres), p. 311, (1991).
20. Kley, W., *Astron. Astrophys.*, **247**, 95 (1991).
21. Schmidt, G. D. in *White Dwarfs*, G. Wegner (ed.), (Berlin: Springer-Verlag), p. 305, (1989).
22. Lamb, D. Q. and Patterson, J., in *Cataclysmic Variables and Related Objects*, IAU Colloq. 72, M. Livio and G. Shaviv (eds.) (Dordrecht: Reidel), p. 229, (1983).
   Ghosh, P. and Lamb, F. K., *Ap. J.*, **232**, 259 (1979).
   Ghosh, P. and Lamb, F. K., *Ap. J.*, **234**, 296 (1979).
23. Hameury, J.-M., King, A. R. and Lasota, J.-P., *M.N.R.A.S.*, **218**, 695 (1986).
24. Pringle, J. E., *Ann. Rev. Astr. Astrophys.*, **19**, 137 (1981).
25. Hubeny, I., in *Theory of Accretion Disks*, F. Meyer *et al.* (eds.), (Dordrecht: Kluwer Academic Publishers), p. 445, (1989).
26. Shaviv, G. and Wehrse, R., *Astron. Astrophys.*, **251**, 117 (1991).

27. Spitzer, L., Physics of Fully Ionized Gases, (New York: Wiley), (1962).
28. Shakura, N. I. and Sunyaev, R. A., *Astron. Astrophys.*, **24**, 337 (1973).
29. Blandford, R. D. and Payne, D. G., *M.N.R.A.S.*, **199**, 883 (1982).
30. Cannizzo, J. K. and Pudritz, R. E., *Ap. J.*, **327**, 840 (1988).
Pudritz, R. E. and Norman, C. A., *Ap. J.*, **301**, 571 (1986).
31. Drew, J. E., in *Structure and Emission Properties of Accretion Disks*, C. Bertout *et al.* (eds.) (Edition Frontieres), p. 331, (1991).
32. Binney, J. and Tremaine, S., *Galactic Dynamics* (Princeton: Princeton University Press), (1987).
33. Lin, C. C. and Shu, F. H., *Proc. Nat. Acad. Sci.*, **55**, 229 (1966).
34. Lin, D. N. C. and Papaloizou, J. C. B., *Ap. J.*, **309**, 846 (1986).
35. Spruit, H. C. in *Theory of Accretion Disks*, F. Meyer *et al.* (eds.), (Dordrecht: Kluwer Academic Publishers), p. 325 (1989).
36. Rozyczka, M. and Spruit, H. C. in *Theory of Accretion Disks*, F. Meyer *et al.* (eds.) (Dordrecht: Kluwer Academic Publishers), p. 341 (1989).
37. Lighthill, M. J., *Waves in Fluids*, (Cambridge: Cambridge University Press), (1978).
38. Narayan, R., Goldreich, P. and Goodman, J., *M.N.R.A.S.*, **228**, 1 (1987).
39. Sawada, K. and Matsuda, T., *M.N.R.A.S.*, **255**, 17p (1992).
40. Lin, D. N. C., Papaloizou, J. C. B. and Savonije, G. J., *Ap. J.*, **364**, 326 (1990).
Lin, D. N. C., Papaloizou, J. C. B. and Savonije, G. J., *Ap. J.*, **365**, 748 (1990).
41. Marsh, T. R. and Horne, K., *M.N.R.A.S.*, **235**, 269 (1988).
42. Bunk, W., Livio, M. and Verbunt, F., *Astron. Astrophys.*, **232**, 371 (1990).
43. Livio, M. and Spruit, H. C., *Astron. Astrophys.*, **252**, 189 (1991).
44. Chandrasekhar, S., *Proc. Nat. Acad. Sci.*, **46**, 53 (1960).
45. Balbus, S. A. and Hawley, J. F., *Ap. J.*, **376**, 214 (1991).
Hawley, J. F. and Balbus, S. A., *Ap. J.*, **376**, 223 (1991).
Knoblock, E., *M.N.R.A.S.*, **255**, 25p (1992).
Papaloizou, J. and Szuszkiewicz, E., preprint (1992).
46. Parker, E. N., *Ap. J.*, **145**, 811 (1966).
47. Horiuchi, T., Matsumoto, R., Hanawa, T. and Shibata, K., *Pub. Astron. Soc. Japan*, **40**, 147 (1988).
48. Tout, C. A. and Pringle, J. E., preprint (1992).
49. Vishniac, E. T. and Diamond, P., *Ap. J.*, **347**, 435 (1989).
50. Vishniac, E. T., Jin, L., and Diamond, P. *Ap. J.*, **365**, 648 (1990).
51. Vishniac, E. T. and Diamond, P., preprint (1992).
52. Geertsema, G. T., Ph.D. Thesis, Astronomical Institute Utrecht (1992).
53. Henrichs, H. F. in *Accretion Driven Stellar X-ray Sources*, W. H. G. Lewin and E. P. J. van den Heuvel (eds.), (Cambridge: CUP), p. 393, (1983).
54. Rappaport, S. and Joss, P. C. in *Accretion Driven Stellar X-ray Sources*, W. H. G. Lewin and E. P. J. van den Heuvel (eds.), (Cambridge: CUP), p. 1, (1983).
55. Livio, M. and Warner, B., *Observatory*, **104**, 152 (1984).
56. Reimers, D. and Cassatella, A., *Ap. J.*, **297**, 275.
57. Che-Bohnenstengel, A. and Reimers, D., *Astron. Astrophys.*, **156**, 172 (1986).
58. Hoyle, F. and Lyttleton, R. A., *Proc. Camb. Phil. Soc.*, **35**, 405 (1939).
59. Bondi, H. and Hoyle, F., *M.N.R.A.S.*, **104**, 273 (1944).
60. Bondi, H., *M.N.R.A.S.*, **114**, 195 (1952).
61. Hunt, R., *M.N.R.A.S.*, **188**, 83 (1979).
62. Okuda, T., *Pub. Astron. Soc. Japan*, **35**, 235 (1983).
63. Fryxell, B. A., Taam, R. E. and McMillan, S. L. W. *Ap. J.*, **315**, 536 (1987).
64. Shima, E., Matsuda, T., Takeda, H. and Sawada, K., *M.N.R.A.S.*, **217**, 367 (1985).
65. Gething, J. P. D., *M.N.R.A.S.*, **111**, 468 (1951).
66. Dodd, K. N. and McCrea, W. H., *M.N.R.A.S.*, **112**, 205 (1952).
67. Illarionov, A. F. and Sunyaev, R. A., *Astron. Astrophys.*, **39**, 185 (1975).
68. Shapiro, S. L. and Lightman, A. P., *Ap. J.*, **204**, 555 (1976).
69. Wang, Y.-M., *Astron. Astrophys.*, **102**, 36 (1981).
70. Davies, R. E. and Pringle, J. E., *M.N.R.A.S.*, **191**, 599 (1980).
71. Soker, N. and Livio, M., *M.N.R.A.S.*, **211**, 927 (1984).
72. Livio, M., Soker, N., de Kool, M. and Savonije, G. J., *M:N.R.A.S.*, **222**, 235 (1986).
73. Anzer, U., Börner, G. and Monaghan, J. J., *Astron. Astrophys.*, **176**, 235 (1987).

74. White, N. E., in *Interacting Binaries*, P. D. Eggleton and J. E. Pringle (eds.), (Dordrecht: Reidel), p. 249, (1985).
75. Matsuda, T., Inoue, M. and Sawada, K., *M.N.R.A.S.*, **226**, 785 (1987).
76. Taam, R. E. and Fryxell, B. A., *Ap. J.*, **327**, L73 (1988).
77. Fryxell, B. A. and Taam, R. E., *Ap. J.*, **335**, 862 (1988).
78. Matsuda, T., Sekino, N., Sawada, K., Shima, E., Livio, M., Anzer, U. and Börner, G., *Astron. Astrophys.*, **248**, 301 (1991).
79. Matsuda, T., Ishii, T., Sekino, N., Sawada, K., Shima, E., Livio, M. and Anzer, U., *M.N.R.A.S.*, **255**, 183 (1992).
80. Cowie, L. L., *M.N.R.A.S.*, **180**, 491 (1977).
81. Soker, N., *Ap. J.*, **358**, 545 (1990).
82. Soker, N., *Ap. J.*, **376**, 750 (1991).
83. Livio, M., Soker, N., Matsuda, T. and Anzer, U., *M.N.R.A.S.*, **253**, 633 (1991).
84. Ishii, T., Matsuda, T., Shima, E., Livio, M., Anzer, U. and Börner, G. *Ap. J.*, submitted, (1992).
85. Doll, H. and Brinkmann, W., *Astron. Astrophys.*, **173**, 86 (1987).
86. Mason, K. O., Branduardi, G. and Sanford, P. W., *Ap. J.*, **203**, L29 (1976).
87. White, N. E., Kallman, T. R. and Swank, J. H., *Ap. J.*, **269**, 264 (1983).
88. Parmar, A. N., White, N. E., Stella, L., Izzo, C. and Ferri, P., *Ap. J.*, **338**, 359 (1989).
89. Taam, R. E., Fryxell, B. A. and Brown, D. A., *Ap. J.*, **331**, L117 (1988).
90. Rappaport, S. and van den Heuvel, E. P. J., in IAU Symp. 98, *Be Stars*, M. Jaschek and H.-G. Groth (eds.) (Dordrecht: Reidel), p. 327, (1982).
91. Balona, L. A., *M.N.R.A.S.*, **245**, 92 (1990).
92. Nagase, F., *Publ. Astron. Soc. Japan*, **41**, 1 (1989).
Deeter, J. E., Boynton, P. E., Lamb, F. K. and Zylstra, G., *Ap. J.*, **336**, 376 (1989).
Rappaport, S. A. and Joss, P. C., in *Accretion Driven Stellar X-Ray Sources*, W. H. G. Lewin and E. P. J. van den Heuvel (eds.), (Cambridge: CUP), p. 1 (1983).
93. Mason, K. O., Hawkins, F. J., Sanford, P. W., Murdin, P. and Savage, A., *Ap. J.*, **192**, L65 (1974).
94. Done, C., Mulchaney, J. S., Mushotzky, R. F. and Arnaud, K. A., *Ap. J.*. in press (1992).
95. Verbunt, F., in *The Physics of Accretion onto Compact Objects*, K. O. Mason, M. G. Watson and N. E. White (eds.), (Berlin: Springer-Verlag), p. 59, (1986).
Vogt, N. in *Classical Novae*, M. F. Bode and A. Evans (eds.), (Chichester: John Wiley & Sons), p. 225 (1989).
96. Whitehurst, R., in Structure and Emission Properties of Accretion Disks, C. Bertout *et al.* (eds.) (Editions Frontieres), p. 275, (1991).
97. Osaki, Y., in Structure and Emission Properties of Accretion Disks, C. Bertout *et al.* (eds.) (Editions Frontieres), p. 271, (1991).
98. Bath, G. T., *M.N.R.A.S.*, **171**, 311 (1974).
99. Papaloizou, J. and Bath, G. T., *M.N.R.A.S.*, **172**, 339 (1975).
100. Edwards, D. A., *M.N.R.A.S.*, **212**, 623 (1985).
101. Bianchimi, A. in *Physics of Classical Novae*, A. Cassatella and R. Viotti (eds.), (Berlin: Springer-Verlag), p. 13 (1990).
102. Warner, B., *Nature*, **336**, 129 (1988).
103. Liebert, J. and Stockman, H. S., in *Cataclysmic Variables and Low Mass X-ray Binaries*, D. Q. Lamb and J. Patterson (eds.), (Dordrecht: Reidel), p. 151, (1985).
104. Meyer, F. and Meyer-Hofmeister, E., *Astron. Astrophys.*, **104**, L10 (1981).
105. Faulkner, J., Lin, D. N. C. and Papaloizou, J. C. B., *M.N.R.A.S.*, **205**, 359 (1983).
Lin, D. N. C., Papaloizou, J. and Faulkner, J., *M.N.R.A.S.*, **212**, 105 (1985).
Smak, J., *Acta Astron.*, **34**, 161 (1984). Mineshige, S. and Osaki, Y., *Pub. Astron. Soc. Japan*, **35**, 377 (1983).
106. Cannizzo, J. K., Wheeler, J. C. and Polidan, R. S., *Ap. J.*, **301**, 634 (1986).
107. Livio, M. and Verbunt, F., *M.N.R.A.S.*, **232**, 1P (1988).
Ichikawa, S. and Osaki, Y., preprint (1992).
108. Smak, J., *Acta Astron.*, **34**, 93 (1984).
O'Donoghue, D., *M.N.R.A.S.*, **220**, 23p (1986).
109. Hassall, B. J. M., Pringle, J. E. and Verbunt, F. *M.N.R.A.S.*, **216**, 353 (1985).
Verbunt, F. in *The Physics of Accretion onto Compact Objects*, K. O. Mason *et al.* (eds.), (Berlin: Springer), p. 59 (1986).

110. Verbunt, F., Hassall, B. J. M., Pringle, J. E., Warner, B. and Marang, F., *M.N.R.A.S.*, **225**, 113 (1987).
Szkody, P. and Sion, E. M. in *White Dwarfs*, G. Wegner (ed.), (Berlin: Springer-Verlag), p. 92 (1988).
Sion, E. M. and Szkody, P., in *Physics of Classical Novae*, A. Cassatella and R. Viotti (eds.), (Berlin: Springer-Verlag), p. 59 (1990).
111. Van der Woerd, H. and Heise, J., *M.N.R.A.S.*, **225**, 141 (1987).
112. van Amerongen, S., Kuulkers, E. and van Paradijs, J., *M.N.R.A.S.*, **242**, 522 (1990).
113. Pringle, J. E., Verbung, F. and Wade, R. A., *M.N.R.A.S.*, **221**, 169 (1986).
114. Minejhige, S., *Astron. Astrophys.*, **104**, L10 (1988).
115. Duschl, W. J., *Astron. Astrophys.*, **225**, 105 (1989).
116. Livio, M., in *Accretion Disks in Compact Stellar Systems*, J. C. Wheeler (ed.), (World Scientific Publishing Co.), 1992, in press.
117. Haefner, R., Schoembs, R. and Vogt, N., *Astron. Astrophys.*, **77**, 7 (1979).
118. Schoembs, R., *Astron. Astrophys.*, **115**, 190 (1982).
119. Krzeminski, W. and Vogt, N., *Astron. Astrophys.*, **144**, 124 (1985).
120. Bath, G. T., Edwards, A. C. and Mantle, V. J., in IAU Colloq. 72, *Cataclysmic Variables and Related Objects*, M. Livio and G. Saviv (eds.), (Dordrecht: Reidel), p. 55, (1983).
121. Dgani, R., Livio, M. and Soker, N., *Ap. J.*, **336**, 350 (1989).
122. Meyer, F., preprint, 1991.
123. Angelini, L. and Verbunt, F., *M.N.R.A.S.*, **238**, 697 (1989).
124. Paczynski, B., in IAU Symp. 73, *Structure and Evolution of Close Binary Stars*, P. Eggleton, S. Mitton and J. Whelan (eds.), (Dordrecht: Reidel), p. 75 (1976).
125. Ostriker, J. P., paper presented at IAU Symp. 73, (1975).
126. Webbink, R. F. in *Critical Observations vs. Physical Models for Close Binary Systems*, K. C. Leung (ed.) (New York: Gordon and Breach), p. 403 (1988).
127. Livio, M. and Iben, I. Jr., in preparation 1992.
128. Meyer, F. and Meyer-Hofmeister, E., *Astron. Astrophys.*, **78**, 167 (1979).
129. Taam, R. E., Bodenheimer, P. and Ostriker, J. P., *Ap. J.*, **222**, 269 (1978).
130. Taam, R. E. and Bodenheimer, P., *Ap. J.*, **337**, 849 (1989).
131. Livio, M. and Soker, N., *Ap. J.*, **329**, 764 (1988).
132. Tutukov, A. V. and Yungelson, L. R. in IAU Symp. 83, *Mass Loss and the Evolution of O-Type Stars*, P. S. Conti and C. W. H. de Loore (eds.) (Dordrecht: Reidel), p. 415 (1979).
133. Livio, M. and Soker, N., *M.N.R.A.S.*, **208**, 763 (1984).
134. Livio, M. and Soker, N., *M.N.R.A.S.*, **208**, 763 (1984).
135. Soker, N., Horpaz, A. and Livio, M., *M.N.R.A.S.*, **210**, 189 (1984).
136. Livio, M., *Space Sci. Rev.*, **50**, 299 (1989).
137. Kahane, C., Audinos, P., Barnbaum, C. and Morris, M., preprint (1992).
138. Paczynski, B., paper presented at IAU General Assembly, Commis. 35, (1967).
139. Bond, H. E. and Livio, M., *Ap. J.*, **355**, 568 (1990).
140. Iben, I. Jr. and Tutukov, A. V., in IAU Symp. 131, *Planetary Nebulae*, S. Torres-Peimbert (ed.), (Dordrecht: Reidel), p. 505, (1989).
141. De Kool, M., *Ap. J.*, **358**, 189 (1990).
142. Balick, B., *A. J.*, **94**, 671 (1987).
143. Icke, V., *Astron. Astrophys.*, **251**, 369 (1991).
144. Icke, V., Balick, B. and Frank, A., *Astron. Astrophys.*, **253**, 224 (1992).
Mellema, G., Eulderink, F. and Icke, V., *Astron. Astrophys.*, **252**, 718 (1991).
145. Soker, N. and Livio, M. *Ap. J.*, **339**, 268 (1989).
146. Iben, I. Jr. and Tutukov, A. V., *Ap. J. Suppl.*, **54**, 335 (1984).
Tutukov, A. V., Yungelson, L. R. and Iben, I. Jr., *Ap. J.*, **386**, 197 (1992).
147. Webbink, R. F., *Ap. J.*, **277**, 355 (1984).
148. Saffer, R. A., Liebert, J. and Olszewski, W., *Ap. J.*, **334**, 347 (1988).
149. Bragaglia, A., Greggio, L., Renzini, A. and D'Odorico, S., *Ap. J. Lett.*, **365**, L13 (1990).
150. Tytler, D. and Rubenstein, E., in *White Dwarfs*, IAU Colloq. 114, G. Wegner (ed.) (Berlin: Springer-Verlag), p. 524 (1988).
151. Robinson, E. L. and Shafter, A. W., *Ap. J.*, **322**, 296 (1987).
152. Foss, D., Wade, R. A. and Green, R. F., *Ap. J.*, **374**, 281 (1991).
153. Mochkovitch, R. and Livio, M., *Astron. Astrophys.*, **236**, 378 (1990).

154. Benz, W., Bowers, R. L., Cameron, A. G. W. and Press, W. H., *Ap. J.*, **348**, 647 (1990).
155. Mochkovitch, R. and Livio, M., *Astron. Astrophys.*, **209**, 111 (1989).
156. Ostriker, J. P. and Mark, J. W. K., *Ap. J.*, **151**, 1075 (1968).
157. Papaloizou, J. C. B. and Savonije, G. J., *M.N.R.A.S.*, **248**, 353 (1991).
158. Nomoto, K., Thielemann, F. K. and Yokoi, K., *Ap. J.*, **286**, 644 (1984).
159. Kawai, Y., Saio, H. and Nomoto, K., *Ap. J.*, **328**, 207 (1988).
160. Nomoto, K. and Iben, I. Jr., *Ap. J.*, **280**, 282 (1984).
161. Sutherland, P. G. and Wheeler, J. C., *Ap. J.*, **297**, 282 (1984).
162. Iben, I. Jr., *Ap. J.*, **324**, 355 (1988).
163. Webbink, R. F., Livio, M., Truran, J. W. and Orio, M., *Ap. J.*, **314**, 653 (1987).
164. Webbink, R. F., in *The Physics of Classical Novae*, IAU Colloq. 122, A. Cassatella and R. Viotti (eds.) (Berlin: Springer-Verlag), p. 405 (1990).
165. Payne-Gaposchkin, C., *The Galactic Novae* (Amsterdam: North-Holland), (1957). Warner, B., in *Classical Novae*, M. F. Bode and A. Evans (eds.), (Chichester: John Wiley & Sons), p. 1 (1989).
166. Alksne, Y. and Samus, N. N., IAUC 2839 (1975).
167. Samus, N. N., IAUC 2836 (1975).
168. Liller, W., Shao, C. Y., Mayer, B., Garnavich, P. and Harbrecht, R. P. IAUC 2848 (1975).
169. Gallagher, J. S. and Code, A. D., *Ap. J.*, **189**, 303 (1974).
170. Hyland, A. R. and Neugebauer, G., *Ap. J. Lett.*, **160**, L177 (1970).
171. Stickland, D. J., Penn, C. J., Seaton, M. J., Snijders, M. A. J. and Storey, P. J., *M.N.R.A.S.*, **197**, 107 (1981).
172. Anderson, C. M. and Gallagher, J. S., *P.A.S.P.*, **89**, 264 (1977).
173. Arp, H., *A. J.*, **61**, 15 (1956).
Rosino, L., *Ann. d'Ap.*, **27**, 498 (1964).
van den Bergh, S., in *Stars and Stellar Systems Vol. 5: Galaxies and the Universe*, A. Sandage, M. Sandage and J. Kristian (eds.), (Chicago: University of Chicago Press), p. 509, (1975).
174. Pfau, W., *Astron. Astrophys.*, **50**, 113 (1976).
175. Kraft, R. P., *Ap. J.*, **139**, 457 (1964).
176. Starrfield, S., Sparks, W. M. and Truran, J. W. *Ap. J.*, **201**, 136 (1985)
Starrfield, S. in *Classical Novae*, M. F. Bode and A. Evans (eds.), (Chichester: John Wiley & Sons), p. 39 (1989).
177. Fujimoto, M., *Ap. J.*, **257**, 752 (1982).
178. MacDonald, J., *Ap. J.*, **267**, 732 (1983).
179. Starrfield, S., Truran, J. W., Sparks, W. M. and Kutter,G. S., *Ap. J.*, **176**, 169 (1972).
180. Prialnik, D., Shara, M. M. and Shaviv, G., *Astron. Astrophys.*, **62**, 339 (1978).
181. Nariai, K., Nomoto, K. and Sugimoto, D., *Pub. Astron. Soc. Japan*, **32**, 472 (1980).
182. Fowler, W. A., Caughlan, G. R. and Zimmerman, B. A., *Ann. Rev. Astron. Astrophys.*, **5**, 525 (1967).
183. Fowler, W. A., Caughlan, G. R. and Zimmerman, B. A., *Ann. Rev. Astron. Astrophys.*, **13**, 69 (1975).
184. Harris, M. J., Fowler, W. A., Caughlan, G. R. and Zimmerman, B. A., *Ann. Rev. Astron. Astrophys.*, **21**, 165 (1983).
185. Truran, J. W. in *Essays in Nuclear Astrophysics*, C. A. Barnes, D. D. Clayton and D.Schramm (eds.) (Cambridge: CUP), p. 467, (1982).
186. Paczynski, B. 1971, *Acta Astron.*, **21**, 417 (1971).
187. Bode, M. F. and Evans, A. in *Classical Novae*, M. F. Bode and A. Evans (eds.), (Chichester: John Wiley and Sons), p. 163 (1989).
188. McLaughlin, D. B., *Pub. Am. Astr. Soc.*, **8**, 145 (1935).
189. Clayton, D. D. and Wickramasinghe, N. C. *Ap. Sp. Sci.*, **42**, 463 (1976).
190. Kato, M., *Pub. Astron. Soc. Japan*, **37**, 19 (1985).
191. Wallace, R. K. and Woosley, S. E., *Ap. J. Suppl.*, **45**, 389 (1981).
192. Wiescher, M., Görres, J., Thielemann, F.-K. and Ritter, H., *Astron. Astrophys.*, **160**, 56 (1986).
193. Livio, M. and Truran, J. W. in *Florida Workshop on Nonlinear Astrophysical Fluid Dynamics*, J. R. Buchler and S. T. Gottesman (eds.), *Annals New York Academy of Science*, **617**, 126.

194. Prialnik, D. and Kovetz, A., *Ap. J.*, **281**, 367 (1984).
Kovetz, A. and Prialnik, D., *Ap. J.*, **291**, 812 (1985).
195. Michaud, G., Fontaine, G. and Charland, K., *Ap. J.*, **280**, 247 (1984).
196. MacDonald, J., *Ap. J.*, **273**, 289 (1983).
197. Livio, M. and Truran, J. W., *Ap. J.*, **318**, 316 (1987).
198. Fujimoto, M. Y., *Astron. Astrophys.*, **198**, 163 (1988).
199. Ferland, G. J. and Shields, G. A., *Ap. J.*, **226**, 172 (1978).
200. Livio, M., Shankar, A. and Truran, J. W. *Ap. J.*, **325**, 282 (1988).
201. Woosley, S. E., in *Nucleosynthesis and Chemical Evolution*, B. Houck, A. Maeder and G. Magnet (eds.), (Sauverny: Geneva Observatory), p. 1 (198€).
202. Oettl, R., Ph.D. Thesis, Max Planck Institut für Astrophysik, Garching bei München (1988).
203. Shankar, A., Arnett, D. and Fryxell, B. A., *Ap. J. Lett.*, submitted (1992).
204. Kutter, G. S. and Sparks, W. M., *Ap. J.*, **340**, 985 (1989).
205. Livio, M., *Ap. J. Lett.*, **369**, L5 (1991).
206. MacDonald, J., *M.N.R.A.S.*, **191**, 933, (1980).
207. MacDonald, J., *Ap. J.*, **305**, 251 (1986).
208. MacDonald, J., Fujimoto, M. Y. and Truran, J. W., *Ap. J.*, **294**, 263 (1985).
209. Livio, M., Shankar, A., Burkert, A. and Truran, J. W., *Ap. J.*, **356**, 250 (1990).
210. Livio, M., in *Nonisotropic and Variable Outflows from Stars*, A. Nota and C. Leitherer (eds.), (1992), in press.
211. Shankar, A., Livio, M. and Truran, J. W., *Ap. J.*, **374**, 693 (1991).
Shankar, A., Ph.D. Thesis, University of Illinois (1990).
212. Kato, M. and Hachisu, I., *Ap. J.*, **383**, 761 (1991).
213. Kato, M. and Iben, I. Jr., preprint, (1992).
214. Barden, S. C. and Wade, R. A., in *A.S.P. Conference Series 3, Fiber Optics Astronomy*, S. C. Barden (ed.) (San Francisco: Astron. Soc. Pacific), p. 113 (1988).
Wade, R. A., in *Physics of Classical Novae*, IAU Colloq. 122, A. Cassatella and R. Viotti (eds.), (Berlin: Springer-Verlag), p. 179, (1990).
215. Livio, M., Govarie, A. and Ritter, H., *Astron. Astrophys.*, **246**, 84 (1991).
216. Schaefer, B. E. and Patterson, J., *Ap. J.*, **268**, 710 (1983).
217. Schaefer, B. E., Landolt, A. U., Vogt, N., Buckley, D., Warner, B., Walker, A. R. and Bond, H. E., *Ap. J.*, in press (1992).
218. de Vaucouleurs, G., *Ap. J.*, **223**, 351 (1978).
219. Pritchet, C. J. and van den Bergh, S., *Ap. J.*, **318**, 507 (1987).
220. Capaccioli, M., Della Valle, M., D'Onfrio, M. and Rosino, L., *A. J.*, **97**, 1622 (1989).
221. Capaccioli, M., Della Valle, M., D'Onfrio, M. and Rosino, L. *Ap. J.*, **360**, 63 (1990).
222. Hartwick, F. D. A. and Hutchings, J. B., *Ap. J.*, **226**, 203 (1978).
223. Shara, M. M., *Ap. J.*, **243**, 926 (1981).
224. Livio, M., *Ap. J.*, in press (1992).
225. Truran, J. W., Hayes, J., Shankar, A. and Livio, M., in preparation, 1992.
226. Starrfield, S., Sparks, W. M. and Shaviv, G., *Ap. J.*, **325**, L35 (1988).
227. Livio, M., Shankar, A. and Truran, J. W., *Ap. J.*, **330**, 264.
228. Starrfied, S., Sparks, W. M. and Truran, J. W., *Ap. J. Lett.*, **303**, L5 (1986).
229. Shankar, A. and Livio, M., unpublished.
230. Allen, C. W., *Astrophysical Quantities* (London: Athlone Press), (1973).
231. Prialnik, D., *Ap. J.*, **310**, 222 (1986).
232. Prialnik, D. and Kovetz, A., *Ap. J.*, **385**, 665.
233. Bath, G. T. and Shaviv, G., *M.N.R.A.S.*, **175**, 305 (1976).
234. Kato, M. and Hachisu, I., *Ap. J.*, **329**, 808.
235. Prialnik, D., Livio, M., Shaviv, G. and Kovetz, A., *Ap. J.*, **257**, 312 (1982).
236. Duerbeck, H. W., *Sp. Sci. Rev.*, **45**, 1 (1987).
237. Webbink, R. F., in Accretion Powered Compact Binaries, C. W. Mauche (ed.), (Cambridge: CUP), p. 177 (1990).
238. Horne, K. and Schneider, D. P., *Ap. J.*, **343**, 888 (1989).
239. King, A. R., in *Cataclysmic Variables and Related Objects*, M. Livio and G. Shaviv (eds.) (Dordrecht: Reidel), p. 181 (1983).
Wu, K. in the *Viña del Mar Workshop on Cataclysmic Variables*, N. Vogt (ed.) in press, (1992).

240. King, A. R., Mouchet, M. and Lasota, J.-P., in *Structure and Emission Properties of Accretion Disks*, C. Bertout *et al.* (eds.), (Editions Frontieres), p. 213 (1991).
241. Stockman, H. S., Schmidt, G. D. and Lamb, D. Q., *Ap. J.*, **332**, 287 (1988).
242. Kaluzny, J. and Chlebowski, T., *Ap. J.*, **332**, 287 (1988).
243. Orio, M., Trussoni, E. and Ögelman, H., *Astron. Astrophys.*, in press, (1992).
244. Horne, K., Welsh, W. F. and Wade, R. A., in *Accretion Powered Compact Binaries*, C. W. Mauche (ed.), (Cambridge: CUP), p. 383 (1990).
245. Shore, S. N., Sonneborn, G., Starrfield, S. G., Hamay, M., Williams, R. E., Cassatella, A. and Drechsel, H., *Ap. J.*, **370**, 193 (1991).
246. Sekiguchi, K., Stobie, R. S., Buckley, D. H. H. and Caldwell, J. A. R., *M.N.R.A.S.*, **245**, 28p (1990).
247. Truran, J. W., Livio, M., Hayes, J., Starrfield, S. and Sparks, W. M., *Ap. J.*, **324**, 345 (1988).
248. Ritter, H., Politano, M. J., Livio, M. and Webbink, R. F., *Ap. J.*, **376**, 177 (1991).
249. Shara, M. M., *Ap. J.*, **243**, 268 (1981).
250. King, A. R., *Q.J.R.A.S.*, **29**, 1 (1988).
251. Ritter, H. and Kolb, U., *Astron. Astrophys.*, submitted, (1992).
252. Warner, B., *M.N.R.A.S.*, **227**, 23 (1987).
253. Verbunt, F. and Zwaan, C., *Astron. Astrophys.*, **100**, L7 (1981).
Mestel, L. and Spruit, H. C., *M.N.R.A.S.*, **226**, 57 (1987).
254. Eggleton, P. P., in IAU Symp. 73, *Structure and Evolution of Close Binary Systems*, P. P. Eggleton, S. Mitton and J. A. J. Whelan (eds.), (Dordrecht: Reidel), p. 209 (1976).
255. Skumanich, A., *Ap. J.*, **171**, 565 (1972).
Smith, M. A., *P.A.S.P.*, **91**, 737 (1979).
256. Rappaport, S., Verbunt, F. and Joss, P. C., *Ap. J.*, **275**, 713 (1983).
257. Tout, C. A. and Pringle, J. E., *M.N.R.A.S.*, **256**, 269 (1992).
258. Robinson, E. L. in *Cataclysmic Variables and Related Objects*, M. Livio and G. Shaviv (eds.), (Dordrecht: Reidel), p. 1, (1983).
259. Paczynski, B. and Sienkiewicz, R., *Ap.J. Lett.*, **28**, L27 (1981).
260. Spruit, H. C. and Ritter, H., *Astron. Astrophys.*, **124**, 267 (1983).
261. Taam, R. E. and McDermott, P. N., *Ap. J. Lett.*, **319**, L83 (1987).
262. Hameury, J. M., King, A. R., Lasota, J.-P. and Ritter, H. *Ap. J.*, **316**, 275 (1987).
McDermott, P. N. and Taam, R. E., *Ap. J.*, **342**, 1019 (1989).
263. Robinson, E. L., Barker, E. S., Cochran, A. L., Cochran, W. D. and Nather, R. E., *Ap. J.*, **251**, 611 (1981).
264. Golub, L. in IAU Colloq. 71, *Activity in Red Dwarf Stars*, P. B. Byrne and M. Rodono (eds.), (Dordrecht: Reidel), p. 83 (1983).
265. Maggio, A., Sciortino, S., Vaiana, G. S., Mayer, P., Bookbinder, J., Golub, L., Harnden, F. R. Jr., and Rosner, R., *Ap. J.*, **315**, 687 (1987).
266. Soderblom, D. R., in IAU Symp. 102, *Solar and Stellar Magnetic Fields: Origins and Coronal Effects*, J. O. Stenflo (ed.), (Dordrecht: Reidel), p. 439, (1983).
267. Giampapa, M. S. and Liebert, J., *Ap. J.*, **305**, 784 (1986).
268. Christensen-Dalsgaard, J., Gough, D. and Toomre, J., *Science*, **229**, 923 (1985).
269. Libbrecht, K. G., *Nature*, **319**, 753 (1986).
270. Spruit, H. C. and Roberts, B., *Nature*, **304**, 401 (1983).
272. Bailyn, C. and Grindlay, J., *Ap. J.*, **353**, 159, (1990).
Romani, R. W., *Ap. J.*, **357**, 493 (1990).
Canal, R., Isern, J. and Labay, J., *Ann. Rev. Astron. Ap.*, **28**, 183 (1990).
273. Wheeler, J. C., in *Frontiers in Stellar Evolution*, D. L. Lambert (ed.), in press, 1991.
274. MacDonald, J., *Ap. J.*, **283**, 241 (1984).
275. Nomoto, K. and Kondo, Y., *Ap. J.*, **367**, L19 (1991).
276. Hanes, D. A., *M.N.R.A.S.*, **213**, 443 (1985).
277. Johnston, H. M. and Kulkarni, S. R., *Ap. J.*, in press (1992).
278. Webbink, R. F., *Nature*, **262**, 271 (1976).
279. Livio, M., Truran, J. W. and Webbink, R. F. *Ap. J.*, **308**, 736 (1986).
280. Selvelli, P. L., Cassatella, A. and Gilmozzi, R., *Ap. J.* in press (1992).
281. Iben, I. Jr. and Truran, J. W., *Ap. J.*, **220**, 980 (1978).
282. Nomoto, K., *Ap. J.*, **277**, 291 (1984).
283. Shara, M. M., Moffat, A. F. J., Williams, R. E. and Cohen, J. G., *Ap. J.*, **337**, 720 (1989).

284. Williams, R. E., *Ap. J.*, **261**, 170 (1982).
285. Kippenhahn, R. and Thomas, H.-C., *Astron. Astrophys.*, **63**, 625 (1978).
286. Schaefer, B., *Ap. J. Lett.*, **355**, L39 (1990).
287. Politano, M. J., Ph.D. thesis, University of Illinois, (1988).
288. Williams, R. E., Phillips, M. M. and Heathcote, S. A., *Astrophys. Space Sci.*, **131**, 681 (1987).
289. Bradt, H. V. D. and McClintock, J. E., *Ann. Rev. Astron. Ap.*, **21**, 13 (1983).
290. Kato, M. and Hachisu, I., *Ap. J.*, **346**, 424 (1989).
291. Kato, M., Saio, H. and Hachisu, I., *Ap. J.*, **340**, 509 (1989).
292. Shara, M. M., Potter, M., Moffat, A. F. J., Bode, M. and Stephenson, F. R. in *The Physics of Classical Novae*, IAU Colloq. 122, A. Cassatella and R. Viotti (eds.), (Berlin: Springer-Verlag), p. 57 (1990).
293. Kulkarni, S. and Narayan, R., *Ap. J.*, **335**, 755 (1988).
294. Evans, R., van den Bergh, S. and McClure, R. D., *Ap. J.*, **345**, 752 (1989).
295. Livio, M. and Truran, J. W., *Ap. J.*, **389**, 695 (1992).
296. Kenyon, S. J., *The Symbiotic Stars*, (Cambridge: Cambridge University Press), (1986).
297. Edwards, D. A. and Pringle, J. E., *Nature*, **328**, 505 (1987).
298. Shara, M. M., Livio, M., Moffat, A. F. J. and Orio, M., *Ap. J.*, **311**, 163 (1986).
299. Livio, M., in the *Viña del Mar Workshop on Cataclysmic Variables*, N. Vogt (ed.), in press (1992).

Interacting Binaries: Topics In Close Binary Evolution

Edward P.J. van den Heuvel[1,2]

[1] Astronomical Institute, University of Amsterdam, The Netherlands
[2] Center for High Energy Astrophysics (CHEAF), Amsterdam, The Netherlands

1. Summary and Introduction.

The ideas of how binary systems evolve with mass exchange have been largely inspired by the surprising characteristics of Algol-type eclipsing binary systems[1].

Such systems consist of an unevolved main-sequence star, in the case of Algol: a B 8 V star of 3.7 $M_\odot$, together with a less massive subgiant (i.e. more evolved) companion star, in the case of Algol: of spectral type G 8 III (0.8 $M_\odot$).

This situation, with the more evolved star having the smaller mass of the two, is just opposite to what one would expect on evolutionary grounds, as stars of larger mass are expected to live shorter and thus at any time to be in a more advanced stage of evolution than stars of smaller mass.

This is what is called the "Algol-paradox". Crawford [50] was the first one to realize that this paradoxical situation can be explained if one assumes that large-scale mass transfer can take place during the evolution of a binary system: he hypothesized that the subgiant components in Algol systems were originally the more massive components of these systems. As the more massive star evolved faster than its less massive companion it was the first to have evolved away from the main-sequence towards the giant branch. The presence of the close companion, however, prevented such an evolution: when the outer layers of the expanding (sub)giant came under the gravitational influence of the companion, they were captured by that star, causing the latter to increase its mass at the expense of the sub(giant) companion. The (sub)giant transferred so much of its mass that it was finally able to restabilize its interior structure. At that moment it had become the less massive of the two stars.

The first -very courageous- attempt to carry out a real calculation of this type of evolution with mass transfer was by Morton [174]. He demonstrated the correctness of Crawford's conjecture that mass tranfer, once it begins, becomes unstable and continues until the (sub)giant has become the less massive star of the two. In his calculations, however, he still assumed that the orbital period of the system does not change during the mass transfer. He just kept it fixed.

---

[1]Algol-type means here: physically similar to the Algol system, i.e. with one subgiant and one main-sequence component. This is to be distinguished from: "Algol-type light-curve", which is a type of light-curve of an eclipsing binary that can also be shown by systems consisting of two main-sequence components.

This is not correct since, if one assumes the total mass of the system to be conserved (as he did), one also expects the orbital angular momentum to be conserved. This implies that during mass transfer the orbital period and separation change in a well determined way (which is described by Livio, this volume, e.g. in his equation (11), and in our equations (21)-(27).

The first ones to calculate -independently of one another- the evolution of close binaries in this more realistic approach were Paczyński [193] [194] [195] [196] [197] [198], Plavec [209] and Kippenhahn and Weigert [132]. Their work was the foundation for all subsequent works on the evolution of close binary systems.

Furthermore, the discovery of the X-ray binaries in 1971/1972 has given a great stimulus to research in this field. Without the occurrence of extensive mass transfer from the original primary star to the secondary star many of these systems could not have survived the supernova explosion of the primary star, in which the neutron star was formed [321].

The realization of the importance of accretion as an energy source in many types of binary systems, ranging from X-ray binaries to Cataclysmic Variable systems and Symbiotic stars, has, since the early seventies, been a further important source of inspiration for new research on the structure and evolution of close binary systems.

The discovery of many binary pulsars during the last few years, of which several are double neutron stars with extremely narrow and eccentric orbits has clearly demonstrated that even though binaries may lose over 90 per cent of their mass and over 99 per cent of their original orbital angular momentum, still the remnants of the two stars may survive as a (very close) binary system. The amazing properties of these double neutron stars and the many relativistic effects that are measurable in them with unprecedented accuracy (e.g. Taylor [276]; Taylor and Weisberg [278]) have created a new and very fundamental branch of binary astrophysics, which has produced the most accurate measurements of masses of (neutron) stars so far (cf. [364]).

The existence of these systems has demonstrated that a precise knowledge of binary evolution is of vital importance for our understanding of fields of fundamental astrophysics as diverse as: the generation of strongest bursts of gravitational radiation, observable with detectors on Earth, and the synthesis of r-process elements in bulk in the Universe. Both phenomena are expected to be produced by the merging of close double neutron stars, which is expected to occur at a rate of order 10^{-5} yr to 10^{-4} yr in normal spiral galaxies like our own. Such events are expected to be detectable with gravitational radiation detectors on Earth out to 200 *Mpc* distance (Whitcomb [358]), providing an event rate on Earth of order 20/yr (Clark [42]; Schutz [247]; Phinney [207]; Narayan et al. [177], van den Heuvel [316]).

Finally, recently the realization that all type I supernovae, notably also the types Ib and Ic (which together with type Ia form about half of all supernovae events) are most probably related to the evolution of binary systems ([356]; [180]), has once more shown how important knowledge of binary evolution is for an understanding of observational stellar evolutionary processes in general. Indeed, over half -and possible close to 100%- of all stars are found to be members of binary systems ([1]; [111]). Thus, by the sheer abundance of binaries in nature, an understanding of the evolution of binary systems is essential to our understanding of the stellar evolutionary processes in general.

This review is organized as follows.

In section 2 we consider some basic aspects of the celestial mechanics of binary systems, the meaning and limitations of the Roche-lobe concept, the changes in orbital period and orbital separation that may be induced by processes of mass loss and mass transfer in binary sytems. For general background reading on this subject and on the subject of chapter 3 we refer the reader to the excellent monographs of Tsesevitch [287] and Batten [12].

In section 3 we give a classification of the various types of close binary systems, concentrating most of our attention to systems in which at least one component is an evolved star: either a (sub)giant, a white dwarf, a neutron star or a black hole. We describe here in some detail the various types of X-ray binaries and binary radio pulsars.

In section 4 we summarize the evolution of single stars and classify the types of remnants produced by this evolution.

In section 5 the various possible types of evolution of a close binary are described, as a function of initial primary mass, mass ratio and orbital separation. It is assumed here that the evolution proceeds in a 'conservative' way, i.e. that total mass and total orbital angular momentum of the system are conserved during the evolution. A comparison with observed types of systems is presented. These are Algol-type systems, Wolf-Rayet binaries, massive X-ray binaries and -most probably- B-emission systems and runaway OB stars.

In section 6 we consider more exotic types of close binary evolution, so-called 'non-conservative' evolution, in which large losses of mass and of orbital angular momentum during phases of mass transfer are taken into account.

The conditions for the occurrence of 'conservative' vs. 'highly non-conservative' evolution are discussed, and the possible outcome of highly non-conservative types of evolution (common-envelope evolution) is described. The products of this evolution include: the Cataclysmic Variables and Low-Mass X-ray Binaries, double white dwarfs, binary radio pulsars and runaway pulsars.

In order to understand the evolution of the Cataclysmic Variables and Low-Mass X-ray binaries and the final outcome of their evolution, we consider, in section 7, the mechanisms that are driving the mass transfer in Cataclysmic Variable binaries and Low-Mass X-ray binaries.

In section 8, we discuss the evolution of a neutron star, born as a companion to a normal star, in the case of a massive as well as a low-mass companion, taking the possible effects of magnetic field evolution and common envelope evolution into account. The final results of these systems, binary and millisecond pulsars, are discussed.

In section 9 we discuss the incidence of close binaries in the general unevolved (newborn) stellar population, as well as the statistics of initial mass ratios and orbital separations. From this, using the general knowledge of close binary evolution with mass exchange we derive the statistically expected incidence of evolved close binaries in the general stellar population in the galaxy. As an example, it is found from this that one expects that of order 50 % per cent or more of supernova explosions from massive stars occur in bare cores (stars that have lost their hydrogen-rich envelopes in a preceding phase of mass transfer), which seems to confirm recent observational results on type Ib and Ic supernova explosions [175].

Also in this section we present an example of the effects of close binary evolution on the Hertzsprung-Russell diagrams of open star-clusters, by evolving synthetic star clusters with a realistic percentage of close binaries. Observed characteristics of open clusters, such as the presence of blue stragglers and some spread in the main-sequence colours of stars of similar luminosity are a natural outcome of these types of experiments. It is argued that in early-type clusters blue stragglers that have been produced by mass transfer are likely to be B-emission stars. Also the incidence of close double neutron star systems in the Galaxy is discussed.

For general references on interacting binaries we refer to the books of Eggleton and Pringle [71], Tsesevitch [287], Batten[12], de Loore and Doom [65]. For references especially to systems with one or two neutron star components, to the books of: Shapiro & Teukolsky[248], Lewin and van den Heuvel[152], Srinivasan and Radhakrishnan[259], Pacini [192], Helfand and Huang[104], Ögelman and van den Heuvel[187], Kundt[148], Ventura and Pines [334], van den Heuvel and Rappaport [324], and the reviews by Canal et al. [32], Bhattacharya and van den Heuvel [18] and Verbunt [341].

2. Mechanical Aspects of Mass Transfer and Mass Loss in binary systems.

2.1 Shape of the two stars in the Roche approximation.

For the definition and description of the Roche equipotential surfaces I refer to section 2 of Livio's article (this volume). In the Roche approximation it is assumed that the gravitational fields of the two stars can be described by those of two point masses, that the orbits are circular and that the two stars are in synchronous rotation with the orbital motion (so-called co-rotation). Using a co-rotating Cartesian coordinate system centered on star 1, with the X-axis pointed to star 2 and the Y-axis in the orbital plane (angular velocity of rotation Ω) the effective gravitational potential Φ (x,y,z) corrected for centrifugal effects due to the rotation of the frame is then given by equations (4) and (1) of Livio (this volume).

For the shapes of the Roche equipotential surfaces Φ (x,y,z)=const., see figures 1 to 3 of Livio's article.

Since stars tend to be strongly centrally concentrated, the Roche approximation is expected to give a good representation of the potential in the tenuous outer layers of the stars.

In a co-rotating frame the equation of motion of a gas element in the star is given by:

$$\frac{d^2 \vec{r}}{dt^2} + 2\,\vec{\Omega} \times \frac{d\,\vec{r}}{dt} = -\frac{1}{\rho}\nabla P - \nabla\Phi \tag{1}$$

where ρ is the gas density and P the pressure.

Note that the centrifugal acceleration has been included in $\nabla\Phi$. For synchronously rotating stars the left-hand side of eq. (1) vanishes, since the stars are not moving, in the co-rotating frame. This implies that the stars are in hydrostatic equilibrium:

$$\nabla P = -\rho\nabla\Phi \tag{2}$$

Hence surfaces of equal pressure (isobars) coincide with equipotential surfaces. Also, since $\rho = dP(\Phi)/d\Phi$, which depends only on Φ, surfaces of equal density coincide with equipotential surfaces. Hence, in the Roche approximation, the shape of a binary component will be given by that of the Roche equipotential surfaces. Deep inside the Roche lobe (the critical equipotential surface through the first Lagrangian point L_1) these surfaces are practically spherical in shape. Close to the Roche lobe they are, however, strongly deformed and almost pear-shaped (see Livio's figures 1, 2 and 3).

When a star fills its Roche lobe, matter can freely flow along L_1 towards the companion star. As a measure of the size of the Roche lobe one in general uses the so-called critical Roche radius R_c, which is the radius of a sphere that has the same volume as the Roche lobe.

The ratio R_c/a, where a is the orbital radius of the system, depends only on the mass ratio $Q = M_1/M_2$ of the two stars. Convenient approximation formulas for R_c/a as a function of Q are given by equations (5) and (6) of Livio's article in this volume.

2.2 Limitations of the Roche-concept.

It is important to notice that the Roche approximation has several limitations. First of all it holds only for stars that are co-rotating with the orbital motion. This is because the Roche equipotential surfaces are defined in a co-rotating coordinate system.

Stars which are not co-rotating will therefore not have a shape given by the Roche equipotential surfaces. Still, on such stars there will be tidal bulges, approximately directed towards the companion star. An observer on the surface of the non-synchronously rotating star will notice that the stellar surface is periodically lifted up when the tidal bulge passes by (when the companion star passes overhead). This happens with a period

$$P = 2\pi/(\Omega - \Omega_{rot}) \tag{3}$$

where Ω_{rot} is the angular velocity of rotation of the star.

Secondly, even for co-rotating stars, the meaning of the Roche-surfaces outside of the critical one through the L_2 point is often mis-interpreted. Since the whole coordinate system is rotating, the outward directed centrifugal accelerations will increase when one moves away from the center of mass of the system. At the same time the gravitational potential felt from the two stars will decrease outwards. This makes that the Roche-equipotential surfaces reach their maximum heights in the vicinity of the critical surface through L_2 (L_3) as can be clearly seen in figures 2 and 3 of Livio, this volume. Further outwards the surfaces slope down. This might give the impression that matter, once it has overcome the potential peaks near L_2 (and L_3) will move outwards by itself, sliding down along the equipotential surfaces, and leave the system. This is, however, a totally misleading idea.

Matter will only do so if it is kept in co-rotation with the system out to large distances. That can only be achieved if the matter would be sliding along stiff rods fixed in the two co-rotating stars. However, no such stiff rods exist in real stars (possibly magnetic fields might play such a role in some cases).

Let us therefore consider what will really happen when the two stars overflow their Roche lobes and start filling one of the closed common equipotentional surfaces just outside the critical surface through L_1. Here co-rotation is still possible as matter (which surrounds both stars) still belongs to the two stars. If, however, the stars keep swelling, there will come a point that they fill the equipotential surface through the second Langarian point L_2. Matter will now flow out from L_2 and is no longer connected to the two stars: it will form a ring around the system.

If the Roche-model could be extended infinitely beyond L_2, matter would now simply be centrifuged out of the system which is, of course, impossible, as the matter leaving L_2 is no longer forced to co-rotate. The particles flowing out through L_2 into the ring will only feel the attractive gravitation of the two stars, and are therefore, of course, still bound to the system.

Indeed, seen from a large distance, the two stars can be considered as one point mass of mass $M_1 + M_2$. Thus seen from a distance the whole system is deep inside a gravitational potential well and a particle released in the vicinity of the system will fall towards the two stars and will certainly not be centrifuged out of the system.

The above implies that, in fact in figures 2 and 3 of Livio's article (this volume), the parts of the equipotential surfaces beyond the surface through L_2 should be ignored. The real equipotential surfaces beyond L_2 keep sloping upwards when one moves outwards, and not downwards as in the figures. Figure 1 illustrates this more realistic situation schematically.

The above considerations are important if one wishes to study what happens to the orbit of the binary system when matter (with angular momentum) leaves the system, for example through the L_2 point. This will be the topic of the next section.

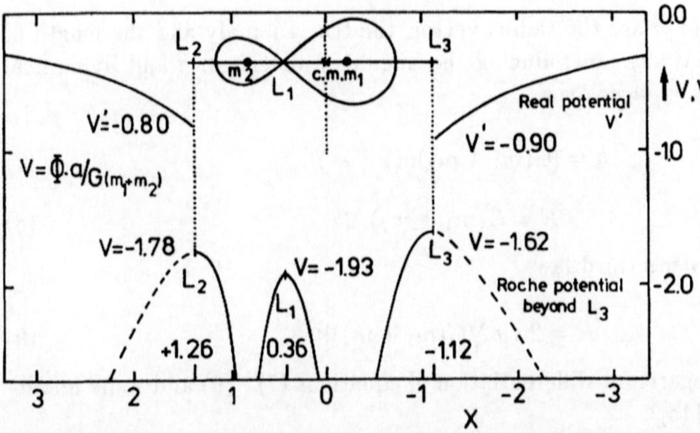

Fig. 1. Roche potential Φ (here represented by $V = \Phi a/G(M_1 + M_2)$ and the real potential (schematically) around a binary system. Real potentials are fully drawn curves. The Roche potential is defined in a rotating coordinate system; the rotational contribution $-\frac{\Omega^2}{2}(x - \frac{a}{2})^2$ becomes more and more negative at increasing distances r from the center of gravity. However, beyond the points L_2, L_3 the Roche potentials have no meaning any more (see text). The real potential V' beyond L_2, L_3 is the sum of the gravitational potentials of the two stars, which are revolving around their common center of gravity ($V' = -a/r$). The system depicted here has $M_2/M_1 = 1/3$.

2.3 Orbital changes due to mass transfer and mass loss in binary systems.

Introduction. In this section we consider effects on the orbits of binaries, produced by the exchange of matter between the components and by the loss of matter and angular momentum form the systems. We consider here some general "modes of mass loss" identified in the important paper of Huang[113], see also Kruszewski[142] and Tsesevitch[287] for a detailed treatment.

General equations. The relative motion in the two-body problem can be described with the aid of two integrals:
1) The dynamical energy ϵ:

$$\epsilon = -G\frac{m_1 m_2}{2a} \tag{4}$$

2) The orbital angular momentum J

$$J^2 = G\frac{m_1^2 m_2^2}{m_1 + m_2} a\,(1 - e^2) \tag{5}$$

(where e is the orbital eccentricity), together with the equation describing the shape of the orbit:

$$R = \frac{a(1 - e^2)}{1 + e\cos(\theta - \omega)} \tag{6}$$

where R, θ and ω are the radius vector, the true anomaly and the length of periastron, respectively. Introducing the angular momentum h and dynamical energy E per unit *reduced* mass

$$h = [G(m_1 + m_2)a(1 - e^2)]^{1/2} \tag{7}$$

$$E = G(m_1 + m_2)/2a \tag{8}$$

and using Kepler's third law:

$$P = 2\pi[a^3/G(m_1 + m_2)]^{1/2} \tag{9}$$

one finds by logaritmic differentiation of equations (7) - (9) and some manipulations (Huang[113]):

$$\frac{da}{a} = \frac{d(m_1 + m_2)}{(m_1 + m_2)} - \frac{dE}{E} \tag{10}$$

$$\frac{dP}{P} = \frac{d(m_1 + m_2)}{m_1 + m_2)} - \frac{3dE}{2E} \tag{11}$$

$$\frac{ede}{1 - e^2} = \frac{d(m_1 + m_2)}{(m_1 + m_2)} - \frac{1}{2}\frac{dE}{E} - \frac{dh}{h} \tag{12}$$

Equations (10), (11) and (12) describe the variations of the orbital elements in terms of the amount of mass lost, $d(m_1 + m_2)$ and of dE and dh. They are of formal nature only. In order to solve them one has to *know* how E and h change during the mass loss, i.e. one should know dE/E and dh/h as a function of $d(m_1 + m_2)/(m_1 + m_2)$. However, there is no general way in which one can know these quantities, because dh/h and dE/E depend on the precise ways in which the mass is ejected from the systems, i.e. on the direction and velocity of the mass ejection. This can be illustrated as follows.

Ejection of an element dm_1 from star 1 results in a change in the *position* and *velocity* of the center of star 1. If an element of mass dm is ejected to a position vector $\vec{r}$ (from the stellar center), star 1 will undergo a displacement $\vec{s}$ given by

$$\vec{s} = -\frac{dm_1}{m_1}\vec{r} \tag{13}$$

Consequently, also R and θ will change. Kruszewski[142] lists the changes of the various orbital elements due to the ejection, e.g.

$$dR = x_1\frac{dm_1}{m_1}, d\theta = \frac{y_1}{R}\frac{dm_1}{m_1} \tag{14}$$

etc. (x,y,z are the coordinates in a co-rotating rectangular system, defined earlier). Thus the *direction* and the velocity of ejection determine the orbital changes. An exact general solution for this problem does not exist since we are dealing with the three-body problem: the two stars and the mass element that is being ejected.

Solutions can, however, for some simple, physically relevant cases be obtained, for which the changes dE/E and dh/h either are known as a function of the mass loss, or can be specified by using some plausible assumptions. Such a specification of the changes in h and E is called a "mode of mass loss" (Huang[113]). We will now consider a few simple and relevant examples of such modes.

"Modes of mass loss'. Huang[113] considered the following three modes of mass loss.

I. *Jeans' mode*: spherically symmetric mass ejection from one component, with very high velocities. This occurs, for example, if one of the components explodes as a supernova. As the ejection velocity of the supernova shell is of order $10^4\ km/s$, some two orders of magnitude larger than the orbital velocities of the two stars, the mass ejection occurs in this case instantaneously in very good approximation. The effects of this instantaneous mass ejection are to drastically change the orbital parameters of the system and if the system remains bound, to impart an extra velocity to the center of mass of the binary ("slingshot effect"). In the case that the system becomes unbound, the two stars will run away with velocities of the order of their original orbital velocities. This mode was studied extensively by Blaauw[22] and Boersma[26] in order to examine whether this may explain the origins of the runaway OB stars (cf. van den Heuvel[314], Stone[263]). This mode, obviously, is of great importance also in the formation of X-ray binaries, for which we know that one of the stars underwent a supernova explosion. For a detailed treatment of the mechanics we refer to the above-mentioned papers by Blaauw and Boersma, to Flannery and Van den Heuvel[79], Wijers et al.[361], and Nomoto and Yamaoka[185], and references therein.

We make the following simplifying assumptions:

- the stars are point masses, so effects of impact of the supernova shell on the companion can be neglected,
- an originally circular orbit, with radius a_o,
- an instantaneous explosion, with spherically symmetric mass ejection.

Due to the mass ejection the total mass of the system is suddenly reduced by an amount $\triangle m$, which causes a sudden reduction in the attraction between the components. Due to the suddenly unbalanced part of the centrifugal accelerations of the orbital motion of the stars, they now will move into a wider and eccentric orbit. Figure 2 explains this situation. The periastron separation in the new orbit equals the orbital radius before the explosion, a_o.

We express the radius vector r connecting the stars, in units of a_o, and define $\mu_f = \frac{M_1^f+M_2^f}{M_1^0+M_2^0}$, where zero and f denote initial and final quantities,

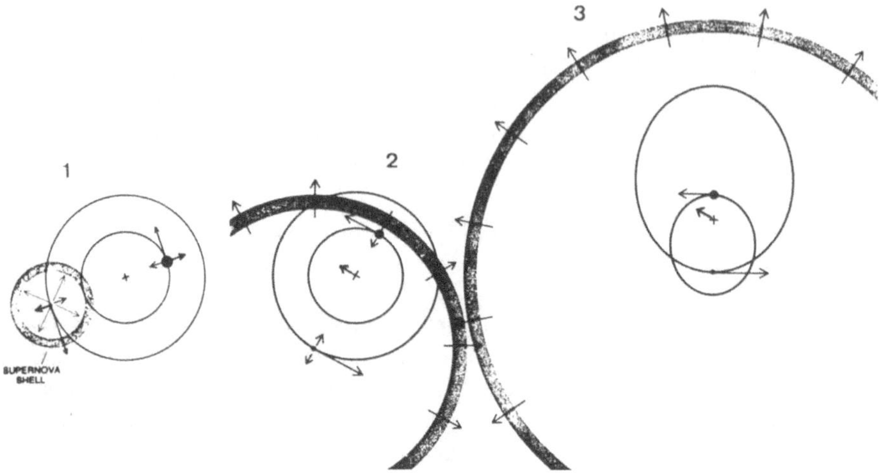

Fig. 2. Illustration of the effects of the explosive mass ejection on binary orbits ("Blaauw effect"). It is assumed that the initial orbits are circular and one star ejects instantaneously a spherical shell of matter which does not interact with the companion. The companion only "notices" the mass loss once the shell has passed it (frame 2): it then suddenly feels less gravitational attraction from the exploding star than before. So, the centrifugal force of its orbital motion is no longer balanced by the other star's gravitational attraction, and the stars start moving into wider, elliptic, orbits. Since the ejected shell has momentum due to the orbital motion of the exploding star, the entire system will get a recoil velocity in the opposite direction (indicated by the arrow at the center of gravity in frames 2 and 3).

respectively (final meaning: after the explosion): $a_f = a^f/a_0$, $p_f = P^f/P_0$. It appears then that the solution of this problem is simple (see e.g. Flannery and van den Heuvel[79]), and that a_f and e_f can be simply expressed in terms of μ_f, as follows.

$$a_f = \mu^f/(2\mu^f - 1) \tag{15}$$

$$e_f = (a_f - 1)/a_f = (1 - \mu_f)/\mu_f \tag{16}$$

This equation for e_f follows from the fact that after the explosion the periastron distance remains $r = 1$. P_f (final orbital period) is given by Kepler's third law: $a_f^3/p_f^2 = \mu_f$, hence

$$p_f = \mu_f/(2\mu_f - 1)^{3/2} \tag{17}$$

Equations (15) - (17) show that for $\mu_f < 0.5$ the system becomes unbound ($e_f > 1$ and $a_f < 0$), hence: the orbit becomes hyperbolic.

This condition is simply due to the virial theorem, which for the initially circular orbit stated: $E_{pot} = -2E_{kin}$. When half of the mass disappears, the potential energy is decreased by a factor of two, while E_{kin} does not change (per unit reduced mass), hence for $\Delta m > \frac{1}{2}(m_1 + m_2)$ one suddenly has $E_{kin} > E_{pot}$, and the system becomes unbound.

The runaway velocity of the center of gravity of a system that remains bound after star 2 with orbital velocity V_2 has suddenly ejected an amount of mass Δm will simply be:

$$V_g = \Delta m . V_2 / (M_1^f + M_2^f) \tag{18}$$

Hence, at most the system can attain a velocity V_2 (for the limiting case $(\Delta m = (M_1^f + M_2^f) = \frac{1}{2}(M_1^0 + M_2^0))$.

II. *Slow mode.* If the ejection velocities are smaller than the escape velocity, the ejected matter will finally be captured by one of the components. Since gas-rings are observed around the more massive components in many Algol-type systems, it seems likely that the accreted matter is first captured in a ring and from there rains down onto the accreting component. In the slow mode it is assumed that no matter will leave the system, i.e. the mass lost by one of the stars is completely captured by the other star. Hence:

$$m_1 + m_2 = const. = m_1^0 + m_2^0 \tag{19}$$

The orbital angular momentum of the system is almost always orders of magnitude higher than the rotational angular momentum. One therefore has in good approximation that in case that no mass is lost from the system the orbital angular momentum is constant, hence according to eq. (5) for circular orbits:

$$\frac{m_1^2 m_2^2}{m_1 + m_2} a = const. \tag{20}$$

which together with eq. (19) yields:

$$\frac{a}{a_0} = (\frac{m_1^0 m_2^0}{m_1 m_2})^2 \tag{21}$$

in which sub- and superscripts zero indicate the initial situation. From eq. (9) then further follows that

$$\frac{P}{P_0} = (\frac{m_1^0 m_2^0}{m_1 m_2})^3 \tag{22}$$

(one derives from eq. (12) that for initially circular orbits in this case $\delta e = 0$; hence, the orbits remain circular during the exchange; cf.Huang[113]).

Differentiation of eq. (22) yields

$$dP/P = -3\, dm_1/m_1 - 3\, dm_2/m_2 \tag{23}$$

and as

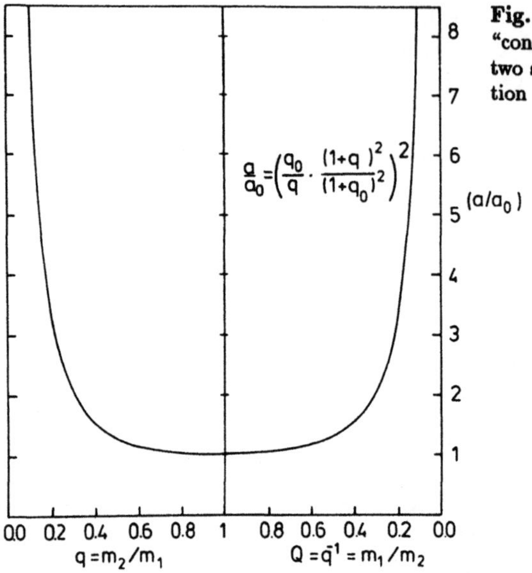

Fig. 3. **Changes in separation during "conservative" mass transfer. When the two stars have the same mass, the separation is at its minimum.**

$$dm_1 = -dm_2 \tag{24}$$

one finds:

$$dP/P = -3dm_1[\frac{1}{m_1} - \frac{1}{m_2}], \tag{25}$$

$$\frac{da}{a} = -2dm_1[\frac{1}{m_1} - \frac{1}{m_2}] \tag{26}$$

Since $dm_1 < 0$, it follows that P decreases if the more massive component is losing mass and increases in the opposite case. Equation (25) can be also be written as:

$$\frac{dP}{dt} = 3P(\frac{m_1 - m_2}{m_1 m_2})\frac{dm_1}{dt} \tag{27}$$

where dm_1/dt is the mass loss per unit time, from star 1.

III. *Intermediate mode.* Here it is supposed that the ejected particles have enough energy to overcome the attraction of the individual components and escape through L_2 (or L_3) to form a ring revolving around the entire system.

If the angular momentum per unit mass of the escaping matter is h_e and the angular momentum per unit mass of the system is h_0, one can rather simply derive that

$$\frac{dh_0}{h_0} = \frac{h_e - h_0}{h_0}\frac{d(m_1 + m_2)}{m_1 + m_2} \tag{28}$$

Hence, since $d(m_1 + m_2) < 0$, one has: $dh_0 < 0$ if $h_e > h_0$.

In order to be stable, the ring must be large compared with the binary system. Hence, in first approximation the particles in the ring can be considered as moving under the gravitational attraction of a mass m_1+m_2. The ring will probably be circular due to the frequent particle collisions. If the ring revolves in the same sense as the binary system at a distance a_e from the centre of mass, the angular momentum per unit mass in the ring will be

$$h_e = (G(m_1 + m_2)a_e)^{1/2} \tag{29}$$

On the other hand from equation (5) it follows that

$$h_0 = \frac{m_1 m_2}{(m_1 + m_2)^2}(G(m_1 + m_2)a)^{1/2} \tag{30}$$

Hence , since $m_1 < m_1 + m_2$, $m_2 < m_1 + m_2$ and $a < a_e$ it follows that $h_0 < h_e$. Consequently, according to eq. (28) $= dh_0 < 0$. This implies in general that due to the formation of the ring a will decrease, which we will demonstrate for a simple case.

Eliminating dE/E from equations(10)-(12) one obtains:

$$\frac{dP}{P} = -2\frac{d(m_1 + m_2)}{(m_1 - m_2)} + \frac{3dh}{h} + \frac{3ede}{1 - e^2} \tag{31}$$

where h is the angular momentum per unit reduced mass. Since by definition $h = J(m_1 + m_2)/m_1 m_2$ one can write by logaritmic differentiation:

$$3dh/h = 3dJ/J + 3d(m_1 + m_2)/(m_1 + m_2) - 3(dm_1/m_1 + dm_2/m_2) \tag{32}$$

Using $dJ = h_e d(m_1 + m_2)$ and

$$J = (G(m_1 + m_2)a(1 - e^2))^{1/2} m_1 m_2/(m_1 + m_2), \tag{33}$$

equation (31) transforms into:

$$dP/P = (1 + 3\gamma)\frac{d(m_1 + m_2)}{m_1 + m_2} - 3(\frac{dm_1}{m_1} + \frac{dm_2}{m_2}) + \frac{3ede}{(1 - e^2)} \tag{34}$$

where

$$\gamma = \frac{(m_1 + m_2)^2}{m_1 m_2}(\frac{a_e}{a(1 - e^2)})^{\frac{1}{2}} = h_e/h_0 \tag{35}$$

Let us now consider, as an *example*, a system with $m_1 = m_2$. As a_e must be larger than the distance of the L_2 point, take $a_e = 1.2a$. We then find $\gamma = 4.4$, and with eq. (34) one finds $dP/P = 4.1d(m_1 + m_2)/m_1$. Since $d(m_1 + m_2) < 0$ one thus finds that the orbital period will rapidly decrease when mass is being lost to a ring around the system.

In addition to these three "Huang modes" we consider two more modes that may be of importance in the case of X-ray binaries, especially in the case of "super-critical" accretion.

IV. *"Isotropic re-emission"*: an element dm_1 lost by star 1 is captured by star 2 and from there ejected isotropically with a large velocity (this might, for example, apply to novae). Figure 4 illustrates this situation. The loss of angular momentum due to the ejection from star 2 is:

$$dJ = \omega r_2^2 dm_1 = (J_2/m_2)dm_1 = \frac{m_1 J}{(m_1+m_2)m_2} dm_1 \qquad (36)$$

where J_2 and r_2 denote the orbital angular momentum and distance to the centre of gravity of the second component. In terms of $Q = m_1/m_2$ one then obtains:

$$dJ/J = QdQ/(1+a) \qquad (37)$$

which yields after integration:

$$J/J_0 + ((1+Q_0)/(1+Q))\, exp(Q-Q_0) \qquad (38)$$

which, after substitution of eq. (5) yields:

$$a/a_0 = ((1+Q_0)/(1+Q))(Q_0/Q)^2 exp2(Q-Q_0) \qquad (39)$$

Similarly, if a fraction α of the exchanged element is isotropically reemitted, and the fraction $(1-\alpha)$ remains on star 2, one obtains:

$$(a/a_0) = \frac{(1+Q_0)}{1+Q_0-\alpha(Q_0-Q)}[1+(1-\alpha)(Q_0-Q)]^{\frac{-2}{1-\alpha}} \qquad (40)$$

(For $\alpha \to 1$ this expression approaches equation (39); for a derivation of equations (39) and (40): see Bhattacharya and van den Heuvel[18]).

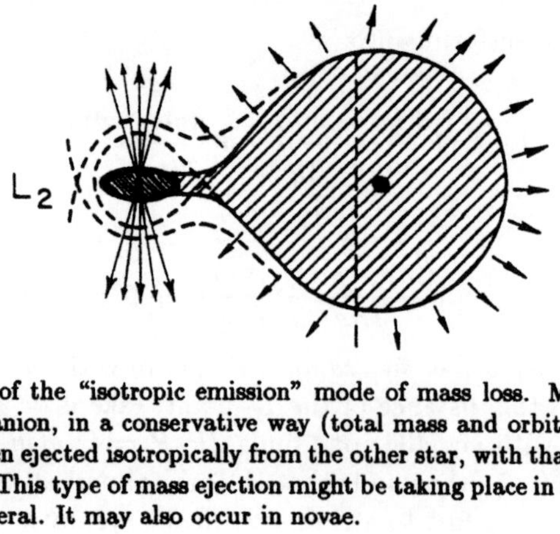

Fig. 4. Illustration of the "isotropic emission" mode of mass loss. Matter flows over from one star to its companion, in a conservative way (total mass and orbital angular momentum conserved), and is then ejected isotropically from the other star, with that star's specific orbital angular momentum. This type of mass ejection might be taking place in SS433, or in high-mass X-ray binaries in general. It may also occur in novae.

V. *Mode V.* In the above modes nothing is specified about the *energy* required to expell the matter. The energy to escape can be supplied either by the radiation field or by the dynamical energy ϵ of the binary system. As to the latter: in the three-body problem one of the bodies can be expelled, at the expense of the energy of the remaining bodies. The element dm_1 will, at the moment of escape, have a mean distance to the center of gravity of the system, of the order of a (as R_{Roche} is of the order of a). Hence, the energy required for its escape is roughly

$$d\epsilon = \frac{G(m_1 + m_2)}{a} dm_1 \tag{41}$$

Let us suppose that fraction $\beta(< 1)$ of this energy is supplied by the dynamical energy of the system, and a fraction $(1 - \beta)$ by the radiation pressure (or other forces). Then, due to the loss of dm_1 from the system, the dynamical energy will decrease by an amount

$$d\epsilon = \beta G(m_1 + m_2) dm_1 / a \tag{42}$$

We assume $m_2 = const.$ Dividing (42) by (4) one obtains:

$$d\epsilon/\epsilon = -2\beta \frac{m_1 + m_2}{m_1 m_2} dm_1 = -2\beta(1 + \frac{1}{Q}) dQ \tag{43}$$

(it is assumed that $m_2 = constant$, as in an X-ray binary). Integration yields (with $Q = m_1/m_2$): $\epsilon/\epsilon_0 = (Q_0/Q)^{2\beta} exp\,(2\beta(Q_0 - Q))$. Substituting ϵ and ϵ_0 from eq. (4) one obtains

$$(a/a_0) = (\frac{Q}{Q_0})^{2\beta+1} exp 2\beta(Q - Q_0) \tag{44}$$

From this follows

$$P/P_0 = ((m_1^0 + m_2)/(m_1 + m_2)^{1/2} (m_1/m_1^0)^{3\beta+1.5} exp(3\beta(m_1 - m_1^0)/(m_2)) \tag{45}$$

The solution with $\beta = 0$ refers to the case that all of the energy required to expell the mass is supplied by the radiation pressure.

Some examples of period changes due to mass transfer or mass loss

A. *Systems with rapid spherically symmetric mass loss by stellar wind.* Apart from the supernova-type mass loss, mentioned above under mode I, it is also possible that a star continuously loses mass in the form of a rapid spherically symmetric stellar wind. A good example of this is the short-period X-ray binary Cygnus X-3, which has an orbital period of 4.8^h which is increasingly rapidly: $\dot{P}/P = 1.66 \times 10^{-6} yr^{-1}$ determined over an 18-yr time interval since 1974.

The system consists of a Wolf-Rayet star and a compact star (the X-ray source, which probably is a neutron star). The Wolf-Rayet spectrum was recently identified in the infrared (van Kerkwijk et al.[328]). Wolf-Rayet stars lose mass at a high rate ($\sim 10^{-5} M_\odot/yr$) at high velocities, $\sim 2000 km/s$ (cf. Abbot and Conti[4]). The orbital changes as a function of the mass loss rate can be computed as follows:

$$J = \frac{m_1 m_2}{(m_1 + m_2)^{1/2}} G a^{1/2} \tag{46}$$

Star 1 is losing mass:

$$dJ = \omega r_1^2 dm_1 = \frac{J dm_1}{Q(1+Q)m_2} = \frac{J dQ}{Q(1+Q)} \tag{47}$$

$(Q = m_1/m_2; m_2 = const.)$

Hence:

$$\frac{dJ}{J} = (\frac{1}{Q} - \frac{1}{1+Q})dQ \tag{48}$$

Also because of (46):

$$\frac{dJ}{J} = \frac{dm_1}{m_1} + \frac{dm_2}{m_2} - \frac{1}{2}\frac{d(m_1+m_2)}{m_1+m_2} + \frac{1}{2}\frac{da}{a} \tag{49}$$

Combination of equations (48) and (49) yields:

$$\frac{da}{a} = -\frac{dQ}{1+Q} = -\frac{dm_1}{m_1+m_2} = -\frac{dm}{m} \tag{50}$$

where $m = m_1 + m_2$.

Equation (50) holds generally for stellar wind mass loss (also if *both* stars would be losing mass by stellar wind). Using Keppler's third law, equation (50) can be written as

$$\frac{dP}{P} = -\frac{2dm}{m} \tag{51}$$

hence:

$$\frac{dm}{dt} = -\frac{1}{2}\frac{m}{P} dP/dt \tag{52}$$

In the case of Cyg X-3 this equation with the above given value of $\dot{P}/P$ yields: $\frac{dm}{dt} = -0.83 \times 10^{-6} m [M_\odot/yr]$.

Assuming $m \simeq 10 M_\odot$, as for a typical Wolf-Rayet star (cf. Abbott & Conti[4]) one finds $\dot{m} \sim -10^{-5} M_\odot/yr$, as is indeed characteristic for a Wolf-Rayet star.

B. *Systems in which the orbits change due to mass transfer.*

Continuous changes of the orbital period are observed in many Algol- and β Lyrae type systems.

Examples:

U Cephei and *β Lyrae.*

The period of the Algol System U Cep has been observed to increase over the past century (Batten[12]). Figure 5 shows that the present period is 2^d. 49302, while the originally determined period of 2^d. 4929005 was used near $E \sim 4000$ (E is the eclipse number). The eclipses in 1969 occurred about 15 hours too late with respect to the 1909 ephemeris. The present orbital period is some 16 seconds longer than in 1909. For a linear increase in period with time, the deviation of the eclipse phase from the one predicted for a fixed orbital period will increase quadratically with time (i.e.(:)E^2), hence the $O-C$ curve will be a parabola. For U Cephei one has, on average $\Delta P/P = 4.3 \times 10^{-9}$ per cycle. Similarly, Kuiper[145] determined for the *β Lyrae* system: $\Delta P/P = 6.10^{-7}$ per cycle (see also Kreiner[140]).

Explanation in terms of mass transfer by means of the "slow mode" (mode(II)):

Equation (27) shows that in this case

$$\frac{dm_1}{dt} = \frac{1}{3P}\frac{dP}{dt}\left(\left(\frac{m_1 m_2}{m_1 - m_2}\right)\right) \tag{53}$$

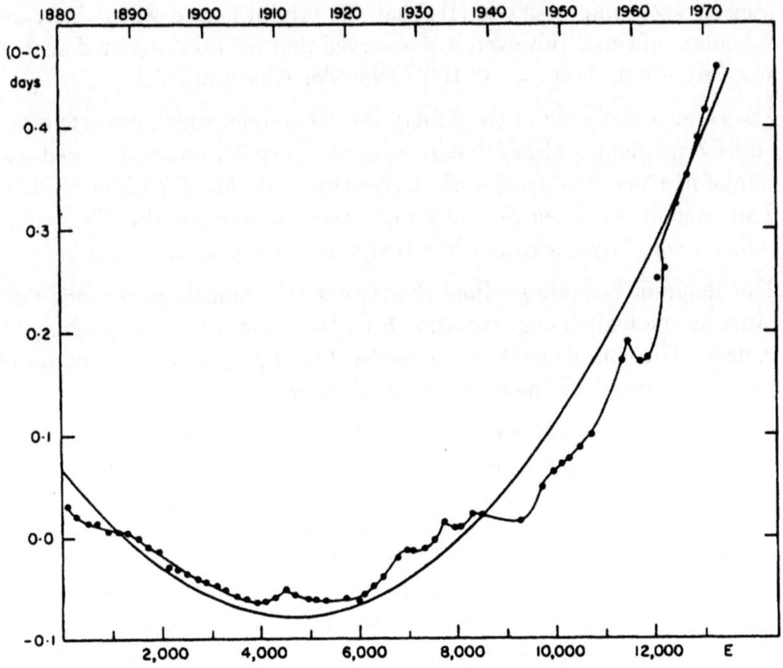

Fig. 5. Residuals of times of minima of U Cephei since its discovery as a variable star. Some individual minima have been omitted in crowded portions of the plot. The adopted period is 2^d.4929005. The observations indicate that the present period is 2^d.49302. The period has increased continually (but not steadily) throughout 90 years of observation (from Batten[12]).

Hence, for given values of m_1, m_2 the measured value of (dP/dt) immediately yields the rate of mass transfer dm_1/dt. For U Cephei, with $m_1 = 2.8M_\odot, m_2 = 4.2M_\odot$, P=2.5 d the observed value of (dP/Pdt) yields $dm_1/dt = 1.76 \times 10^{-6} M_\odot/yr$. For β Lyrae (P=12.9 days), with $m_1 = 2M_\odot$, $m_2 = 12M_\odot$, $\Delta P/P = 6.10^{-7}$ per cycle, one finds $dm_1/dt = 8.14 \times 10^{-5} M_\odot/yr$.

We will consider the causes of these very large rates of mass transfer in section 5.

C. *Systems with orbital change due to other modes of mass loss.* There are several peculiar systems which do not seem to fit to mass loss or transfer by modes (I) or (II), even though mass transfer or rapid mass loss are taking place here.

Examples are the low-mass X-ray binary 4U1820-30 in the globular cluster NGC 6624 and the millisecond-pulsar eclipsing binary PSR 1957+20. 4 U 1820-30 has an orbital period of 11 minutes and clearly must consist of a neutron star (the X-ray source) and a low-mass white dwarf(or helium star) ($M \sim 0.2M_\odot$) which is overflowing its Roche lobe, and provides the mass accretion that powers the X-ray source (see section 3).

As the white dwarf is the less massive component and is transferring mass to the more massive component one would expect in the case of pure mass transfer according to mode (II) that the orbital period would increase in the course of time. However, it is observed that the orbital period decreases at a rate $-7.10^{-5} s/yr = -2.10^{-12}$ (Van der Klis et al.[330]).

The same is observed in the PSR1957+ 20 system, which consists of a 1.6 millisecond radio pulsar, that is eclipsed every 9.2 hours by a red dwarf companion that has a mass not larger than 0.06 $M_\odot$ (Fruchter et al.[84]). This system has been optically identified and the red dwarf companion shows a very large heating effect (see Van Paradijs et al. [337]).

The shape of the radio eclipse shows that the companion is being evaporated by the high-energy radiation from the pulsar that is impinging on its surface. The companion has a cometlike tail of gas, directed away from the pulsar, produced by the evaporation of its envelope.

At large distances (of the order of a lightyear) the system is surrounded by an $H\alpha$-nebula, produced by the matter evaporated from the companion (Kulkarni and Hester[147]). In view of the rapid evaporative mass loss from the companion one would in this case according to mode (I) expect the orbital period to *increase* with time. However, it is observed to *decrease* at a rate $-3.9(\pm 0.6) \times 10^{-11}$ [276]. In both these systems there must be gas leaving the system that carries away higher specific-angular momentum than the mean specific angular momentum of the system. This can be easily achieved by modes such as mode (III): formation of a ring or disk around the system by outflow through L_2 (or L_3) will cause the orbit to rapidly shrink, as demonstrated by the equations of mode III given above. From this ring matter may be lost by being blown away by either the pressure of the X-ray radiation (1820-30) or of the radio pulsar radiation (PSR 1957 + 20).

More general, if matter flows out from L_2 in an Archimedic-spiral type outflow, it will stay behind and will be gravitationally torqued (accelerated "forward" in the rotating frame of the binary) by the revolving binary components, which will cause transfer of orbital angular momentum and energy from the binary to the outflowing stream, by gravitational forces (it is a "three-body" interaction: the binary components losing energy and angular momentum to a "third body": a particle in the stream; see Flannery[77].).

Detailed numerical calculations of the torques exerted by the binary on the outflowing matter in PSR 1957 + 20 indeed show that the binary may lose so much angular momentum that its orbital period decreases (Banit and Shaham [10]; Tavani and Brookshaw[273]).

Figure 6 shows the outflow pattern in PSR 1957+20 as calculated by Tavani and Brookshaw[273]. It should be noticed that, based on the observations of the β Lyrae system Kuiper[145] already concluded that a similar spiral-shaped outflow exists in that system. Also the V Sge system (a

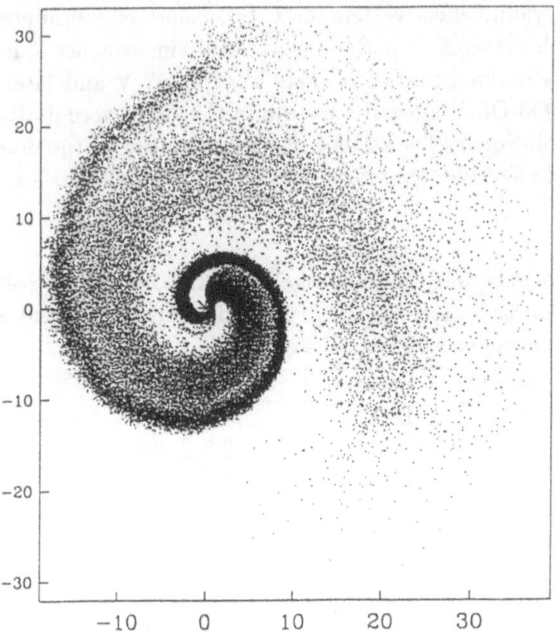

Fig. 6. Outflow pattern from the eclipsing millisecond radio-pulsar binary PSR 1957+20 as calculated by Tavani and Brookshaw[272]. The energy flux from the pulsar causes its small companion star ($M < 0.06M_{\odot}$) to be evaporated. The evaporated matter spirals out in a disk at relatively low velocities, and excerts a torque on the binary system. This causes the 9-hour orbital period of the system to decrease.

β Lyrae like system with $P = 0^d.514$) shows such an outflow pattern (cf. Tsesevitch[287]), and indeed such a pattern is expected on theoretical grounds (Flannery[77]).

In all these cases one therefore expects the effect of the mass transfer mode (mode (II)) to be counteracted in part by angular momentum losses to the outflow.

Whether or not the orbital period will increase or decrease with time will then depend on the balance between these various modes of mass and angular momentum loss, which may vary from system to system. In the cases of U Cephei and β Lyrae clearly mode II dominates whereas in 1820-30 and PSR1957+20 other modes, such as III may dominate.

3. Properties of important classes of interacting binaries

3.1 Introduction

Interacting binaries are almost always spectroscopic or eclipsing binary systems. These are also the systems from which most of our fundamental knowledge on masses and radii of stars has been derived. Popper[215] gives the most reliable data on masses, radii, effective temperatures, colors and luminosities of stars derived from such systems. For the unevolved main-sequence stars (luminosity class V) the thus derived masses of stars of type O7 V and later are listed in table 1. For the O3-O6 V stars the masses in the table were derived form comparisons with evolutionary tracks[166]. For a description of the determination of masses, radii and effective temperatures of binary components see, for example, Tsesevitch[287].

Table 1. Masses of unevolved (main sequence) O7V-G2V stars, as derived from detached spectroscopic binaries that are eclipsing (after Popper[215]). For the O3V-O6V stars the masses were derived from comparison with evolutionary tracks.

Spectral type	Mass ($M_\odot$	Spectral Type	Mass ($M_\odot$)
O3V	100	B6V	4 ± 0.5
O4V	80	B8V	3.5 ± 0.5
O5V	50-60	B9V	3 ± 0.5
O6V	40	A0V	2.4 ± 0.2
O7V	30 ± 5	A2-5V	1.9 ± 0.2
O8V	25 ± 5	A7-9V	1.8 ± 0.2
O9V	20 ± 3	F0V	1.6 ± 0.1
B0V	15 ± 2	F2V	1.5 ± 0.2
B1V	12 ± 1.5	F5V	1.4 ± 0.2
B2V	10 ± 1.5	F6-7V	1.3 ± 0.1
B3-4V	6 ± 1.5	F8V	1.1 ± 0.1
B5V	5 ± 0.5	G2V	1.0 ± 0.1

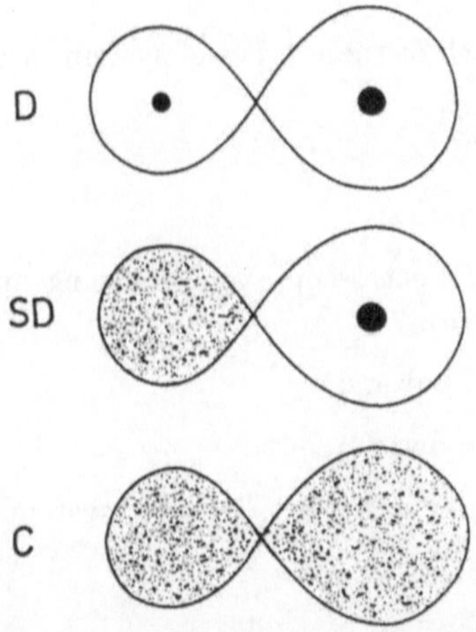

Fig. 7. Classification of close binaries according to the degree in which the components fill their Roche lobes: detached (D), semi-detached (SD) and contact (C).

In this section we will discuss various important types of interaction binaries found in Nature.

The sytems can be classified in various ways, such as:

- by their observational characteristics, or
- by the evolutionary status of their components.

As we are primarily interested here in evolutionary aspects, we will concentrate here on an overview of systems in terms of the second type of classification, which will be given in the following sections. Since, however, the nomenclature of systems is often derived form their observational characteristics, we briefly describe here the much used observational classification in terms of the degree in which one or both components fill their Roche lobes, as depicted in figure 7.

In this classification three categories of systems are distinguished as follows:

I *detached*: neither component fills its critical Roche surface;

II *semi-detached*: one of the two stars fills its critical Roche surface. In short-period systems this is almost always the less massive component of the two, like in classical Algol systems. In wide systems like ζ Aurigae or ε Aurigae it often is the more massive star that is (almost) filling its Roche lobe;

III *contact*: both stars are filling their Roche lobes and have a common outer envelope.

Important groups of each of these types of systems are:

1. *Detached systems*:

   - All visual binaries;
   - Non-evolved spectroscopic and eclipsing binaries (both stars on the main sequence);
   - Double radio pulsars;
   - Double white dwarfs;
   - Pre-cataclysmic variables (detached systems consisting of a red dwarf and a white dwarf).

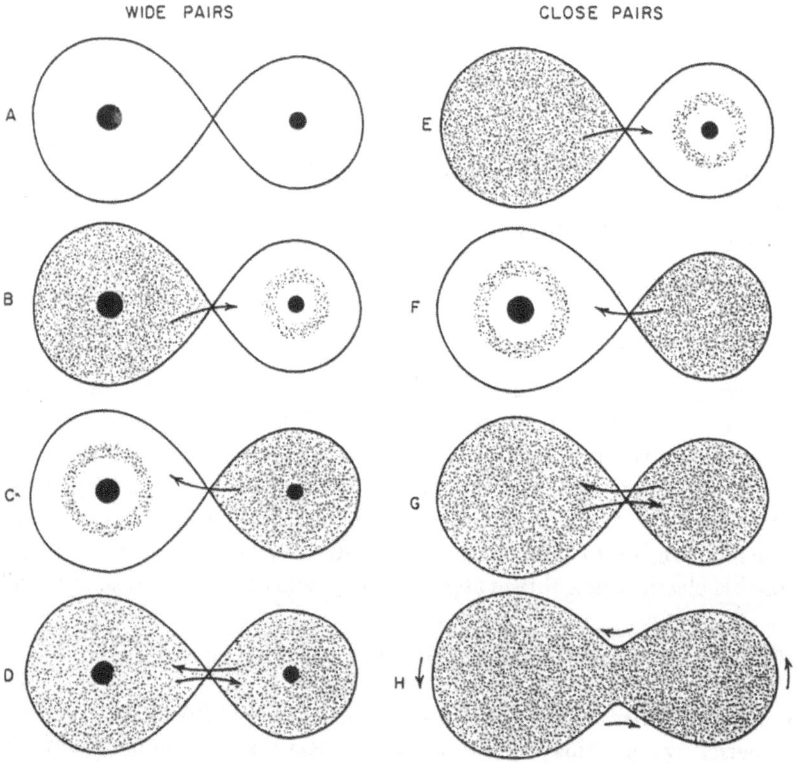

Fig. 8. Schematic theoretical representation of different types of eclipsing binaries (for systems with adopted mass ratio 3:2). *A* represents a detached system. In supergiant systems *B* and *C* the low-density outer atmosphere (chromosphere) of one star fills one of the lobes; the other star has a gaseous ring. System *D* is typical of some supergiant systems: the low-density atmospheres of both stars fill both lobes. In system *E* the more massive star itself fills one of the lobes, the other has a gaseous ring. Such systems are rare. Systems like *F*, on the other hand, are common, e.g.: Algol and U Cep. Systems *G* represents massive contact binaries, H pairs like W Ursa Majoris. Arrows show the direction of the flow of gases between the stars. (Diagram after Struve and Payne-Gaposchkin & Haramundanis [206]).

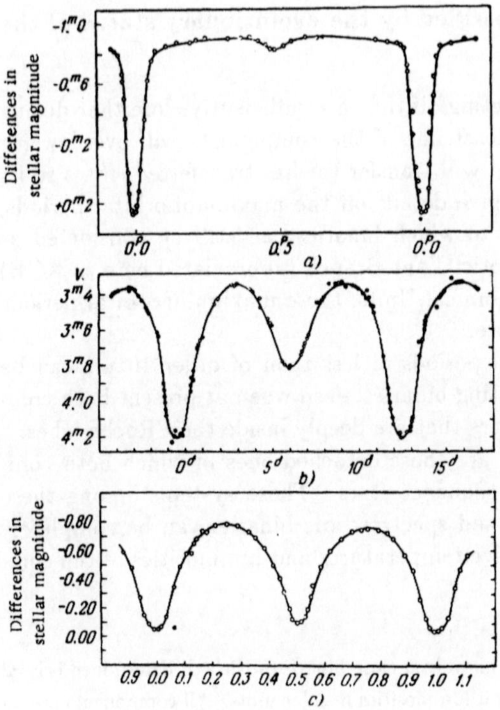

Fig. 9. Some characteristic eclipsing binary lightcurves: a) Algol, b) β Lyrae, c) W Ursa Majoris (in visual light, after Tsesevitch[287]).

2. *Semi-detached (or nearly semi-detached) systems*

   - Algol systems (orbital periods of order days);
   - ζ Aurigae systems (orbital periods of order years);
   - Cataclysmic variables (orbital periods mostly < 0.5 day);
   - X-ray binaries (High Mass and Low Mass);
   - Symbiotic Stars (orbital periods of the order of years).

3. *Contact systems.*

   - W Ursa Majoris binaries (systems of low total mass consisting of two F,G or K dwarfs, orbital periods less than one day);
   - β Lyrae type binaries (high-mass contact systems, consisting of two stars of relatively early type: A, B or O, orbital periods days to weeks, cf. Leung[154]; [155]).

Figure 8 gives a schematic classification of eclipsing systems (after Payne-Gaposchkin and Haramundanis[206]). Figure 9 shows characteristic lightcurves of a semi-detached eclipsing system Algol) and two contact systems (β Lyrae and W Ursa Majoris). For a detailed description of each of these types of systems and of their characteristic lightcurves we refer to the general literature (cf. Batten [12]; Tsesevitch, [287]; Heinz[105], Plavec et al.[210]).

3.2 Interacting binaries as classified by the evolutionary status of the components.

We define here systems as "interacting" if they are sufficiently close that during some stage of their evolution at least one of the components will overflow (or has overflowed) its Roche lobe and will transfer (or has transferred) mass to its companion. In section 5 we will give details on the maximum orbital periods, as a function of stellar mass, up to which binaries can still be considered as interacting. Since on the Asymptotic Giant Branch (abbreviated here as AGB) stars reach radii of several Astronomical Units, these maximum orbital periods are of the order of a decade or more.

Hence all binaries with orbital periods of less than of order 10 yrs can be considered as (potentially) interacting binaries, even when at present both components are still main-sequence stars that are deeply inside their Roche lobes.

The simplest types of systems are those detached ones in which both components are still unevolved main-sequence stars. Those systems among these binaries which are both eclipsing and spectroscopic binaries can be completely solved for the masses, radii, effective temperatures and luminosities of the components.

Table 2. Some characteristic detached unevolved close binaries with well determined orbital parameters (cf. Popper[215]). Masses and luminosities in solar units. All components are on the main sequence (luminosity class V or IV).

Name	Sp.	P(d)	M_1	M_2	log L_1	log L_2
V 382 Cyg	O7.3+O7.7	1.89	26.9(± 2.7)	19.0 (± 2.7)	5.17	4.95
Y Cyg	O9.8+O9.8	3.00	16.7 ± 0.5	16.7 ± 0.5	4.45(± 0.11)	4.45
V 453 Cyg	B 0.4+B 0.7	2.88	14.5 ± 1.2	11.3 ± 1.0	4.69	4.20
α Vir	B1.5 + B4	4.01	10.8 ± 1.3	6.8 ± 0.0	4.33	3.16
CV Vel	B3 + B3	6.89	6.1 ± 0.004	6.00 ± 0.004	3.19	3.19
AG Per	B4 + B5	2.03	4.53 ± 0.07	4.12 ± 0.06	2.89	2.62
RX Her	B9 + A0	1.78	2.75 (± 0.06)	2.33 (± 0.05)	1.79	1.48
V 805 Agl	A5 + F0	2.41	2.06 (± 0.07)	1.60 (± 0.04)	1.26	8.86
CM Lac	A2 + F0	1.60	1.88 (± 0.009)	1.47 ± 0.04	1.10	0.68
CW Eri	F2 + F2	2.73	1.52 ± 0.01	1.28 ± 0.01	0.92	0.57

It is from such systems that the basic data on stellar masses listed in table 1 were derived by Popper[215], to whom we refer for details. The mass-luminosity relation of main sequence stars has been derived from observations of such binaries.

Table 2 lists a few characteristic examples of such detached unevolved binaries (cf. [215]).

Since evolution is the topic of this paper, we now consider the various types of interacting binaries classified according to their evolutionary status.

We divide the systems into the following three categories, which will be dealt within the following sections, as follows:

- Evolved systems with non-degenerate components: section 3.3;
- Evolved systems with one or two white dwarf components: section 3.4;
- Evolved systems with one or two neutron star or black holes components: section 3.5.

3.3 Evolved systems with non-degenerate components

These are either (I) close systems (P < few weeks) in which one of the components has evolved into a subgiant (luminosity class IV or III) and/or shows signs of considerable mass loss, or (II) wide systems in which at least one of the components has evolved into a yellow or red giant or supergiant.

I. Close Systems.

Tables 3, 4 and 5 show some characteristic examples of three varieties of systems of this group. These are: (1) the hot semi-detached systems, (2) the classical Algol systems and (3) the Wolf-Rayet binaries.

The latter type of systems consist of an O-type star together with an -in general- less massive Wolf-Rayet star, which is, however, in a more advanced stage of evolution than its O-type companion. This is evidenced by the fact that the Wolf-Rayet stars in practically all cases show no evidence of hydrogen in

Table 3. Hot evolved (in most cases: semi-detached) systems (Popper[215], De Loore[62], De Greve[52]). Masses, radii and luminosities in solar units.

Name	Sp.	P(d)	M_1	M_2	R_1	R_2	log L_1	log L_2
BD +40°4220	O6f + O7f	6.6	12	42			6.20	6.10
LY Aur	O9.2 + O9.4	4.00	21.7	8.4	6.3	5.3		4.96
AO Cas	O9IV	3.52	19	23	13.9	8.9	5.2	5.2
V Pup	B1 + B2	1.45	17 ± 1	9 ± 1	6.3	5.3	4.20	3.85
β Lyrae	B8I + A0III	12.925	2	12	19.2	13.8		
V356 Sgr	B4 + A2III	8.90	12.1 ± 1.1	4.7 ±1.1	6.0	14.0	3.45	3.00
u Her	B2.5V + B7	2.05	7.3 ±1.0	2.7 ± 0.3	4.8	4.3	3.6	2.65
Z Vul	B3V + A2III	2.455	5.4 ± 0.3	2.3 ± 0.1	4.7	4.5	3.30	2.07

Table 4. Some characteristic classical Algol systems (Popper[215], Popper & Tomkin[216], De Greve[52]).

Name	Sp.	P(d)	M_1	M_2	R_1	R_2	log L_1	log L_2
RY Per	B4V + F5III	6.86	5:	0.8:	3.4:	7	3.0	1.9
RS Vul	B5V + G0III:	4.48	4.5	1.4	4.1	5.6	2.9	1.5
U Cep	B7V + G8III	2.49	4.2	2.8	2.7	4.9	-	-
Algol (=β Persei)	B8V + G8III	2.87	3.7	0.8	3.1	3.2	2.2	0.8
S Cnc	A0V + K2III:	9.48	2.33	0.175	2.2	4.8	1.6	1.1
AS Eri	A3V + K0III:	2.666	1.9	0.2	1.8	2.2	1.2	0.4
RY Aqr	A5V + K1III:	1.97	1.3	0.3:	1.5:	1.7:	0.9	0.2
TW And	A8V + K2III:	4.12	1.8	0.4	2.2	3.4	1.1	0.7

Table 5. Wolf-Rayet binaries: double-lined (after Abbott & Conti [4]).

Star	Spectral type	Period (d.)	M_{WR} *sin* i $M_\odot$	Mass ratio WR/O	M_{WR} ($M_\odot$)	Basis M_o in $M_\odot$
AB6 (SMC)	WN3+O7	6.861	6.4	0.17	6.4	$M_0 \sim 37$
AB5= HD 5980 (SMC)	WN4+O7I	19.266	2.0?	0.3?	8?	See ref.
AB8= Sk 188 (SMC)	WO4+O4V	16.644	1.7	0.27	13.5	$M_0 \sim 50$
γ^2 Vel	WC8+O9I	78.5002	17	0.54	25	$M_0 \sim 35$
HD 90657	WN4+O4-6	8.255	8.4	0.52	11	See ref.
HD94305	WC6+O6-8	18.82	15	0.47	16	$M_0 \sim 35$
HD 94546	WN4+O8V	4.831	2.7	0.43	7	See ref.
HD 97152	WC7+O7V	7.886	3.6	0.59	20	$M_0 \sim 35$
HDE 311884	WN6+O5V	6.94	40	0.84	50	$M_0 \sim 50$
HD 152270	WC7+O5	8.893	1.8	0.36	20	$M_0 \sim 60$
HDE 320102	WN3+O7	8.83	1.8	0.33	11	$M_0 \sim 35$
CV Ser/ HD168206	WC8+O8V	29.707	11	0.48	13	Eclipse
HD 186943	WN4+O9V	9.5548	9-11	0.52	13	$M_0 \sim 25$
HD 190918	WN4.5+O9I	112.8	0.7	0.26	9	$M_0 \sim 35$
V 444 Cyg	WN5+O6	4.21238	9.3	0.40	11	Eclipse
GP Cep/ HD211853	WN6+O8-B0III	6.6884	-	>0.22	10-25	Eclipse
CQ Cep/ HD214419	WN7+O9	1.64	23	1.19	≥ 23	Eclipse

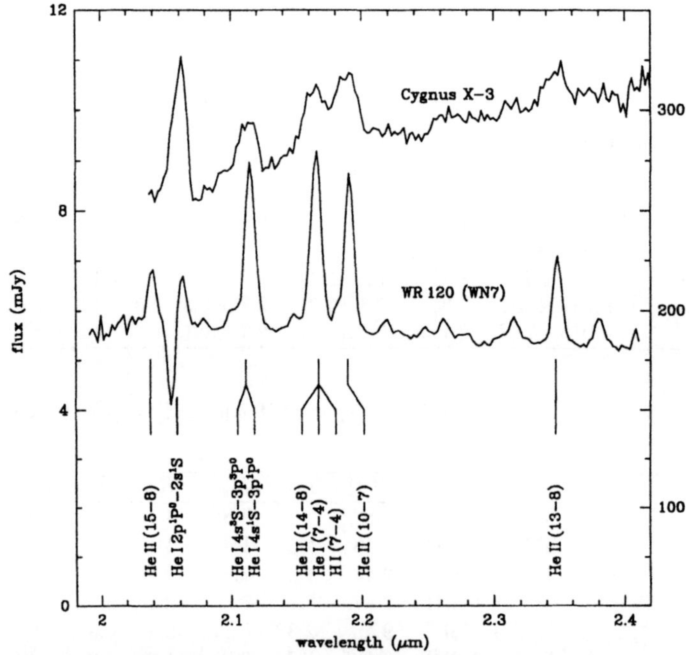

Fig. 10. The spectra of Wolf-Rayet stars are dominated by strong broad emission lines. Hydrogen is weak or absent. The figure shows a portion of the infrared spectra of the star WR 120 together with that of the peculiar X-ray binary Cygnus X-3, van Kerkwijk[329].

their spectra (fig. 10) but, nevertheless, have bolometric luminosities similar to those of their (much) more massive O-type companions. The general consensus is, therefore (see Abbott and Conti[4]; van der Hucht[327]) that Wolf-Rayet stars are the helium cores of the originally more massive components of their binaries. In view of their larger masses their progenitors evolved faster than their (at that time less massive) companions, and now are already in the phase of helium burning while their O-type companions are still in the hydrogen burning phase ([198]; see Chiosi & Maeder[40]; Abbott and Conti[4]). A fourth group of systems, probably related to the WR-binaries, is that of giant-like helium stars of low mass ($\lesssim 1 - 2M_{\odot}$) and rather low effective temperature ($T_e \sim 10^4$ K). These systems, of which ν Sgr is the example, are quite rare. ν Sgr is a spectroscopic binary with P=137,96 days and e=0.06, in which the giant like helium star (M $\sim M_{\odot}$) is transferring mass to a hot companion, possibly also a helium star. See e.g. Habets[99]) for details. Two other examples in this category are KS Per (P=360.47^d, e=0.28) and LSS 4300.

II. Wide Systems. These are the giant and super-giant "Algol Systems", eclipsing binaries with orbital periods of years to decades, in which one of the components is a red giant or super-giant star. Table 6 lists a number of characteristic examples. Several of them are found among the naked-eye stars: ζ and ε Aurigae and 31 and 32 Cygni.

These belong to the most marvellous types of eclipsing binaries, of interest to professional and amateur astronomers alike. In the ζ Aurigae systems, to which also 31 and 32 Cygni belong, the companion is a hot unevolved B-type main-sequence star. The long duration of the eclipses (several months) allows one to carefully monitor the phases of ingress and egress, both spectroscopically and photometrically. Before and during ingress many new absorption lines appear in the spectrum of the B-type star, due to an eclipse by the semi-transparent outer layers (chromosphere and corona) of the red super-giant. Long before the UV- and X-ray detection of chromospheres of red super-giants the study of the spectra during ingress and egress demonstrated the presence of coronae and chromospheres around super-giants and enabled one to study their temperature and density profiles in detail (cf. [362], Ferluga and Hack[76]).

One of the most puzzling systems of this type is ε Aurigae, which has an orbital period of 27.1 years. The nature of the eclipsing object is not known. It is probably a disk with a huge radius (1278 $R_{\odot} \simeq 6$ AU), which eclipses the

Table 6. Giant and Supergiant "Algol" systems (cf. Wilson[362]).

Name	Spectrum	P(d)	M_1	M_2	R_1 ($R_{\odot}$)	R_2 ($R_{\odot}$)	a ($R_{\odot}$)	a(AU)
RZ Cnc	K1III + K4III	21.6	3.20	0.54	10.2	12.2	50.9	0.24
31 Cyg	K3Ib + B3V	3784.3	18	9	174	5.0	3080	14.3
ζ Aur	K4Ib + B8V	972	14	8	190	7.0	1255	5.8
32 Cyg	K5I + B3V	1140.8	22.7	8.1	353	4.0	1447	6.7
VV Cep	M2Iab + B8I	7430	24:	24:	1940	19.4	5830	27.0
ε Aur	?+ A8Ia	9898	42:	28:	1278	716:	8030	37.2

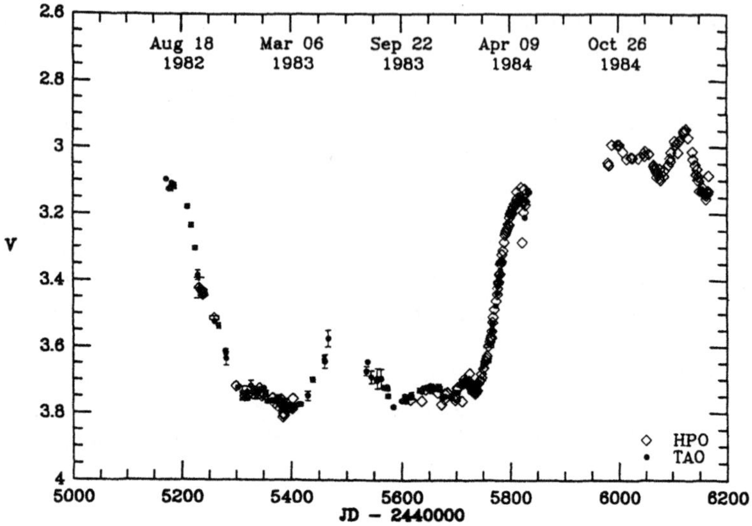

Fig. 11. The visual light curve of the 1982-84 eclipse of Epsilon Aurigae. The cluster of data points near J.D. 2445150 indicate the beginning of the eclipse. Observations contributed by J.L. Hopkins of the Hopkins Phoenix Observatory (HPO) and by S.I. Ingvarsson of the Tjornisland Astronomical Observatory (TAO) are shown.

visible component, an A8Ia super-giant, for almost two years. The system has a semi-major axis as large as that of the orbits of Neptune and Pluto. Figure 11 shows its last eclipse light curve (1982-1984). For a detailed discussion of this system and its possible evolutionary status: see Stencel[262] and Lambert and Sawyer[149]).

3.4 Evolved Systems with one or two white dwarf Component(s).

I. Cataclysmic variable binaries (Novae, Dwarf Novae). These consist of a red dwarf (in most cases: an unevolved main-sequence star) and a white dwarf, the red dwarf is filling its Roche lobe and transferring mass to the white dwarf. The white dwarf is surrounded by an accretion disk which is the main source of light in the system. In general the white dwarf is more massive than its companion. The orbital periods of the systems are in general short, ranging from $\sim$ 80 minutes to 12 hours. The distribution of the orbital periods of CV binaries shows a characteristic "Period Gap" between 2 and 3 hours (see Livio, this volume, fig. 63).

Table 7 lists orbital parameters of a few well-known systems. Complete catalogues of all CV binaries and Low-Mass X-ray binaries are published by Ritter[229] and Webbink[351] and regularly updated. It has been noted that the masses of the white dwarfs in CV-binaries tend to be large, of order $1M_\odot$ or more, whereas most single white dwarfs have masses of only about $0.6M_\odot$ (cf. Law and Ritter[150], Weidemann[355]). There are a few hundred systems now

Table 7. Some characteristic CV binaries (SU=SU UMa-type; DQ=DQ Her-type; AM=AM-Her type; UG= U Gem-type; ZC= Z Cam-type; UX= UX UMa-type) (cf. Webbink[351]; Ritter[229]).

Name	P_{orb} d(h)	Type	Secondary Spectrum	M_{wd} ($M_\odot$)	M_{sec} ($M_\odot$)
WZ Sge	0.0566878455 (1.36050828)	SU	-	1	0.08
V436 Cen	0.06250015 (1.500003600)	SU	-		
EX Hya	0.0628233838 (1,637612112)	DQ	M5.5V		
Z Cha	0.0744991674 (1.782980008)	SU	M5-6V		
AM Her	0.12892737 (3.09425688)	AM	M4-5V		
U Gem	0.1769061911 (4.245748584)			1.18(±0.15)	
DQ Her	0.193620873 (4.646900952)	DQ	M3V		
SS Cyg	0.27512973 (6. 60311520)	UG	K5V	1.39	0.8
Z Cam	0.289840 (6.95616)	ZC	K7V	1.20 (±0.20)	0.85
V 363 Aur	0.32124246 (7.70981940)	UX	K0V		
SY Cuc	0.380 (9.120)	ZC	G9V		
AE Aqr	0.4116580 (9.8797920)	DQ	K5IV	0.94 (±0.1)	
BV Cen	0.6101159 (14.6427816)	UG	G5-8IV-V	0.83(±0.1)	
GK Per	1.996803 (47.923272)	DQ/UG	K2IVp	≥ 0.72	0.25

known. A few CV binaries have orbital periods outside the above-mentioned range. In these systems the companion star of the white dwarf is evolved (a sub-giant, or a giant). Examples are GK Per (Nova Persei 1901) and T Cor Bor (the recurrent nova of 1866 and 1946). GK Per has an orbital period of 1.997 days, $M_{wd} \geq 0.72 M_\odot$, $M_c = 0.25 M_\odot$ (a K2 subgiant). T Cor Bor has P $\sim 227.53^d$; its companion is an M-giant (Kenyon & Garcia[128]). Several other recurrent novae appear to be wide systems like T Cor Bor, for example RS Ophiuchi (P=230^d). Such systems are in general called "symbiotic stars". A number of characteristic symbiotics are listed in table 9.

II. Pre-Cataclysmic Variable Binaries and Double Nuclei of Planetary Nebulae.

These are systems resembling the CV-binaries, but differing from those by the fact that the non-degenerate star is not (yet) filling its Roche-lobe. When

Table 8. Upper part: some characteristic Pre-CV binaries (close detached systems resembling CV binaries). Lower part: some characteristic double nuclei of planetary nebulae (after Webbink[351]).

Name	P_{orb}	Spectrum	Masses ($M_\odot$) *or Plan. Nebula*	Remarks
NN Ser	0.130080094	wd+dMe	-	
PG 0308+096	0.289114	DA4+M2-3V	-	
GK Vir	0.344330809	DA+dM0-6	-	
V 471 Tau (=BD + 16516)	0.521182993	K0V+DA2	0.79+0.80	Hyades
HZ 9	0.56433	DA2+dM4.5e	0.56+0.20	Hyades
EG UMa (=Case 1)	0.66751	DA4 + dM2	$\geq$ 0.4 + 0.36	Hyades
MT Ser	0.1132269	-	Abell 46	
HD 35914	0.165170	O7fp	IC 418	
KV Vel	0.35711296	-	DS 1	
UU Sge	0.46506918	Sd O+K5	Abell 63	
V477 Lyr	0.4717294	O9+K	Abell 46	
LW Hya	0.7653	G8III+sdO	Abell 35	
	2.91		Sp 1	
V651 Mon	16.089	A0III+sdO	NGC2346	

this star expands evolutionary or the orbital separation decreases (e.g. by gravitational radiation losses, see section 7), these systems may turn into CV binaries. This will in most cases take several times 10^9 years.

Closely related to this group are the double nuclei of planetary nebulae, which are clearly an earlier phase of the pre-cataclysmic variables. In the double PN-nuclei the compact stars are genuine nuclei of planetary nebulae: extremely hot degenerate stars of spectral type O that will later on cool to become white dwarfs. Of order 5 to 10 per cent of all PN nuclei are such doubles (see de Kool[56]).

Table 8 lists a number of examples of both categories of objects. Three of the pre-cataclysmics listed belong to the Hyades open cluster.

III. Symbiotic stars.

These form probably a mixed group of systems. They are wide binaries consisting of a red giant or AGB star, probably with $M < 3M_\odot$, together with a hot blue companion. The hot star is accreting mass from its red giant companion, either from a strong stellar wind emitted by the companion, or by Roche lobe overflow. In the former case, since red giant wind velocities are low ($\sim$ 30 km/s) and wind mass loss rates are $\leq 10^{-6}\ M_\odot/yr$, the accretion rates are generally $< 10^{-7} M_\odot/yr$. For Roche lobe overflow accretion rates from $10^{-7} M_\odot/yr$ up to $10^{-3} M_\odot/yr$ can be reached. The red giants typically have luminosities of

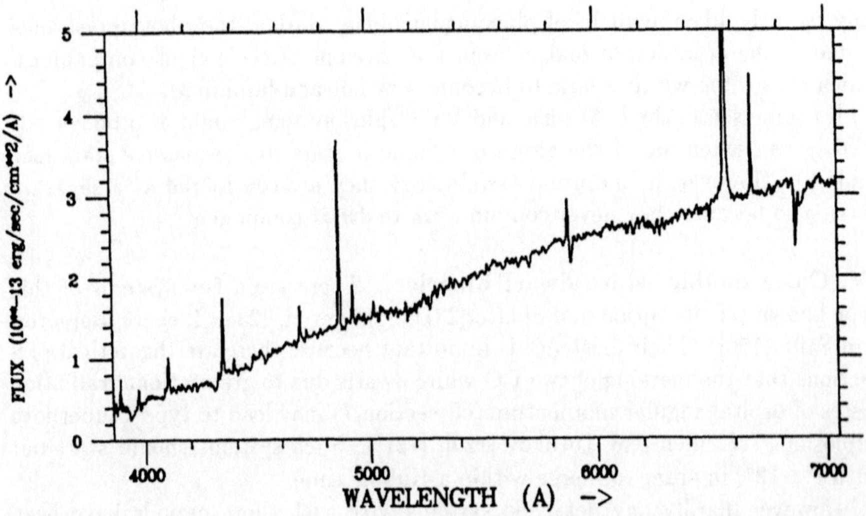

Fig. 12. A low-resolution spectrum of BD-21°38773 observed on February 24, 1988, with the ESO 1.52-m telescope, B&C spectrograph, and a CCD at photometric phase 0.64. Emission lines of H, HeI and He II are superimposed on a reddened G8III spectrum. The truncated Hα line reaches a maximum flux of 13 $\times 10^{-13}$ $erg/(sec\, cm^2 \mathring{A})$ (from Niehues et al. [179]).

Table 9. Some characteristic symbiotic stars and related objects (after Webbink [351]).

Name	P_{orb} (days)	Spectrum	Remark
T CrB	227.53	M4.1 III+sdBe	Recurr. nova 1866, 1946
RS Oph	230	K5.7I-II+OBeq	Recurr. nova
SS Lep	260.0	A0 Veq+M1III	
EG And	481.2	M2.4III+sdOe	
AX Per	681.6	M5.2 III+sdOe	
Z And	756.85	M3.5 III+sdOe	
CI Cyg	855.25	M4.9 III+sdOe	
V1329 Cyg	950.07	WN5+M4II	
CQ Dra	1706	M3IIIa+sdBe	
CH Cyg	5700:	M6.5III+A0Veq	
V 407 Cyg	15700(?)	M+sdBe	

$\sim 100 - 1000 L_{\odot}$. The symbiotic stars are recognized by their spectra which are a combination of the spectrum of a cool giant together with that of a hot blue star (figure 12). Excellent reviews of the properties of these stars can be found in Friedjung & Viotti[83] Kenyon [127]. See also Kenyon and Garcia[128][129], Vogel[345] and Kenyon and Webbink[130].

The hot blue stars are in many cases accreting white dwarfs, in some cases also main-sequence stars (see table 9).

The fact that the white dwarfs in these systems show a hot sdO or sdB spectrum does not necessarily imply that they are young objects that were formed only recently, like the nuclei of planetary nebulae. Rather, their hot appearance is due to the accretion of matter from the envelope of their giant companions, which causes the white dwarfs to become very hot and luminious.

[In some sense, the ζ Aurigae and VV Cephei systems might also be considered as van extension of the group of symbiotic stars to the massive stars (see table 6). However, in the usual terminology they are considered as a separate class, also because they never contain a white dwarf component].

IV. Close double white dwarf binaries. There are a few systems of this type known (cf. Robinson and Shafter[231]; Saffer et al. [238]; Liebert, Bergeron and Saffer[156]). Their existence is important because there are theoretical predictions that the merging of two CO white dwarfs due to gravitational radiation losses of orbital angular momentum (cf. section 7) may lead to type I supernova explosions (cf. Iben and Tutukov [120], [121]). Such systems should start out with $P < 12^h$, in order to merge within a Hubble time.

However, hardly any detached systems with such short periods have been found so far, while they should have been detected (Robinson and Shafter[231]). This is somewhat of a problem for evolutionary models that predict such systems to be quite numerous (Webbink[350]; Iben and Tutukov[120],[121]).

Table 10 lists presently known double white dwarf systems. All of them are consistent with having helium white dwarf components (Liebert et al. [156]).

CO white dwarfs are expected to be of non-DA types, which make up < 25% of all hot white dwarfs ($T_e > 10^4$). The detection of binaries among them is much more difficult than among DA's, because of the much larger pressure broadening in a helium-rich atmosphere. The lack of detected systems of this type is therefore probably largely due to selection effects (cf. Liebert et al.[156]).

Table 10. Close double white dwarf systems (cf. Webbink[351]).

Name	P_{orb} (days)	Spectrum	Remarks
AM CVn (=HZ 29)	0.0121648445 (=17.52 min)	DBp	Semi-detached; two He-white dwarfs
GP Com	0.03231 (=46.53 min.)	DBe	Semi-detached; two He white dwarfs
AA Dor (=LB 3459)	0.2615397198	Sd OB	prob. two hot sub dwarfs (eclipsing)
L 101-26	1.15	DA 2	single-lined binary white dwarf
L 870-2	1.55578	DA7+DA	two DA white dwarfs

3.5 Evolved systems with neutron star and black hole components.

The neutron star binaries fall into two categories: the X-ray binaries and the binary and millisecond radio pulsars. Among the X-ray binaries there are also some strong black hole candidates. We now consider the X-ray binaries and the binary and millisecond pulsars separately.

X-ray binaries.

A. Introduction: The discovery of the neutron-star nature of strong compact galactic X-ray sources. The approximately 100 "bright" galactic X-ray sources with fluxes above 10^{-10} erg cm^{-2} s^{-1} (in the energy range 1–10 keV) above the Earth's atmosphere show a clear concentration towards the galactic center and also towards the galactic plane, indicating that the majority of them belongs to our galaxy (see figure 13). A few strong sources are also found in the two nearby satellite systems of our galaxy, the Large and Small Magellanic Clouds

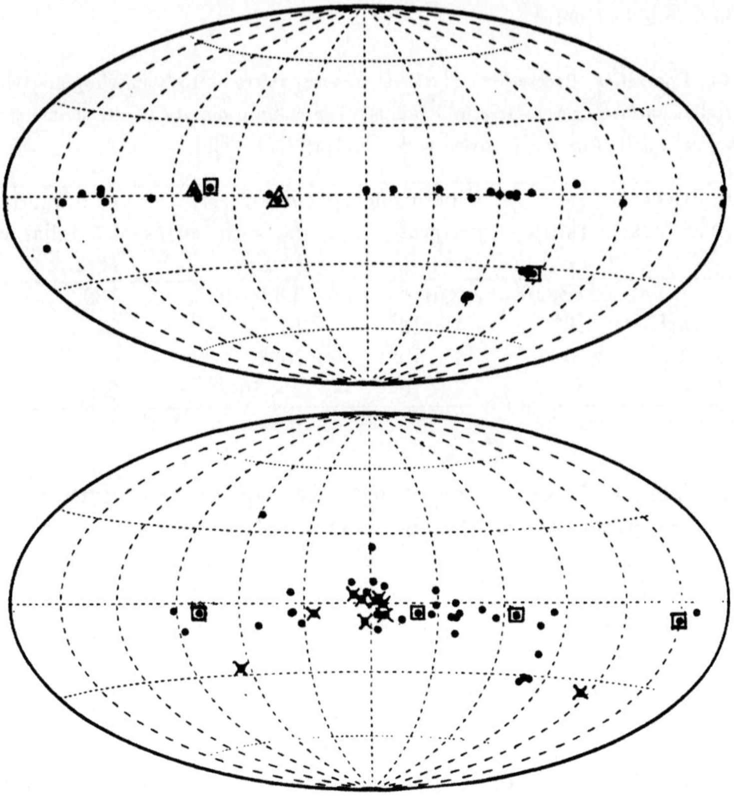

Fig. 13. Sky maps (in galactic coordinates) of the optically identified high-mass X-ray binaries (top panel) and low-mass X-ray binaries (bottom panel); the latter also contains the globular-cluster X-ray sources (indicated by crosses). Black hole candidates are indicated by squares and the "jet"-sources SS433 and Cyg X-3 by triangles (after [338]).

(abbreviated here as LMC and SMC, respectively). For the strong sources around the galactic center as well as those in the Magellanic Clouds a fair estimate of the distance can be given: ~8.5 kpc and ~60 kpc, respectively. Together with the X-ray fluxes measured at the Earth this yields typical source luminosities of 10^{34}–10^{38} erg s^{-1}, some 2.5 to 2.5×10^4 times the total energy output of our Sun, but concentrated in the X-ray part of the spectrum. Also, about a dozen strong sources are found in globular clusters in our galaxy, for which the distances are known. These yield source luminosities of typically 10^{35}–$10^{37.5}$ erg s^{-1}. Zeldovich and Guseynov[370] were the first to suggest that the strong galactic X-ray sources are accreting neutron stars in binary systems.

Slightly earlier it had been pointed out by Salpeter[239] and Zeldovich[369] that, when gas is accreted by a black hole, it is heated so much that a large fraction of its loss of gravitational potential energy can be emitted in the form of electromagnetic radiation before the gas disappears into the hole. Typically an amount of matter m can in this way generate an amount of radiation $(0.1 - 0.4)mc^2$. Table 11 lists the amounts of gravitational potential energy released per unit mass by accretion onto the sun, a white dwarf, a neutron star and a black hole, respectively.

Table 11. Energetics of accretion. The kinetic energy refers to freely infalling matter arriving at the stellar surface or Schwarzschild radius. The column density to the stellar surface for free spherical infall of matter is given by $R^{1/2}L_x/[\pi(2GM)^{3/2}]$.

Stellar object (one solar mass)	R (km)	Kinetic energy per gram (erg)	dm/dt for $L_x = 10^{37} erg/s$ ($M_\odot$/yr)	Column density to stellar surface (g/cm^2)
sun	7×10^5	2×10^{15}	0.8×10^{-4}	190
white dwarf	10^4	1.4×10^{17}	10^{-6}	22
neutron star	10	1.4×10^{20}	10^{-9}	0.7
black hole	3	0.7×10^{21}	2×10^{-10}	0.4

The table also lists the rates of accretion required to generate the typical X-ray luminosities of about 10^{37} erg s^{-1} of the strongest sources in the Galaxy and the Magellanic Clouds. Also the column density towards the stellar surface (or Schwarzschild radius) is listed for the case that the accretion takes place in a spherically symmetric way. The table shows that only for accreting neutron stars and black holes is the column density low enough to allow X-rays to escape, as X-rays are stopped at column densities larger than a few g cm^{-2}. Hence, the strongest galactic X-ray sources cannot be accreting white dwarfs.

The accreting neutron-star binary model was nicely confirmed in 1971 with the discovery (Scheier et al.[246] that the source Centaurus X-3 is regularly pulsing (thus: a neutron star), and is a member of an eclipsing binary system. Its regular X-ray pulsations have a period of 4.84 s, and the regular X-ray eclipses have a duration of 0.488 days, which repeat every 2.087 days. The pulse period shows a sinusoidal doppler modulation with the same 2.087 day period and is in

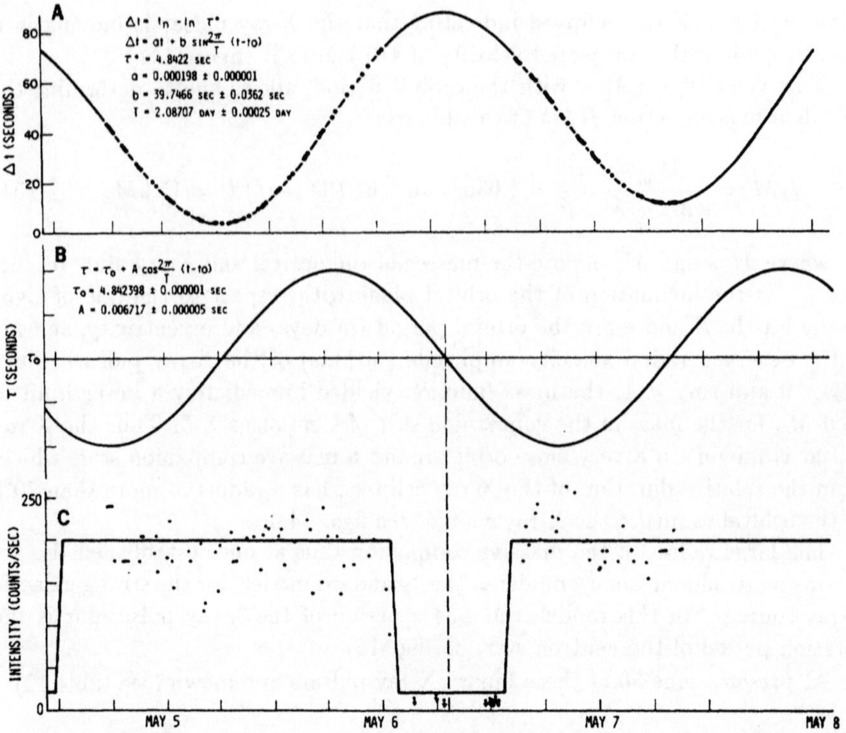

Fig. 14. a.) Bottom: The intensity observed from Cen X-3 (dots) and the light curve prediction for 5-7 May. Top: The difference Δt between the time of occurrence of a pulse and the time predicted for a constant period, plotted as a function of time. A best fit function and the values of the parameters are given. Center: The dependence of the pulsation period τ as derived from the best fit phase function. Note the coincidence of the null points of the period function with the centers of the high and low intensity states (Schreier et al. [246]).

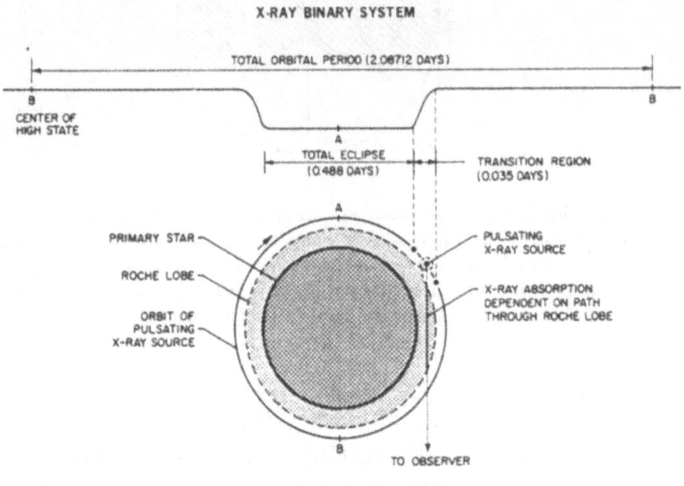

Fig. 14. b.) Schematic representation of the occulting binary X-ray system Centaurus X-3.

phase with the X-ray eclipses, indicating that the X-ray pulsar is moving in a circular orbit with a projected velocity of 415.1 km s^{-1} (figure 14).

This velocity, together with the orbital period, allows one to determine the so-called mass function $f(M)$ of this binary:

$$f(M) = \frac{M_{opt}^3 \sin^3 i}{(M_x + M_{opt})^2} = 1.035 \times 10^{-7} K_x^3 \mathcal{P}(1 - e^2)^{3/2} = 15.6 M_\odot \qquad (54)$$

where M_{opt} and M_x denote the masses of the optical and X-ray star, respectively, i is the inclination of the orbital plane with respect to the line of sight to the Earth, $\mathcal{P}$ and e are the orbital period (in days) and eccentricity, and K_x is the observed radial velocity amplitude (in kms) of the X-ray pulsar. Since $M_x > 0$ and $\sin i < 1$, the mass function yielded immediately a lower limit of 15.6 $M_\odot$ for the mass of the companion star of Centaurus X-3. Thus the X-ray pulsar is moving in a very close orbit around a massive companion star, which, from the relative duration of the X-ray eclipses, has a radius of more than 70% of the orbital radius of the X-ray source (see figs. 14ab).

The large radius of the massive companion thus at once established the accreting neutron-star binary model as the "standard model" for the strong galactic X-ray sources. In this model, the 4.84 s period of the X-ray pulsations is the rotation period of the neutron star (see fig. 15).

At present some 30 of these binary X-ray pulsars are known (see table 12).

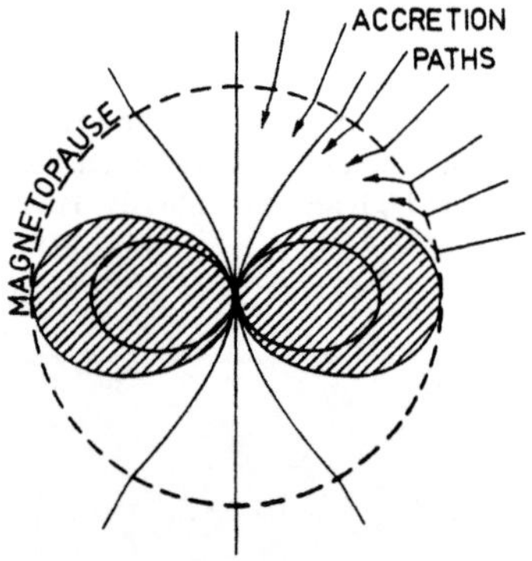

Fig. 15. A simplified dipole magnetosphere around an accreting neutron star. The accreting matter can flow in only near the magnetic poles where the field lines are "open", leading to the formation of two hot X-ray emitting accretion columns, one above each pole. The neutron star rotates around an axis that is inclined with respect to the magnetic axis; this gives rise to the pulsated appearence of the X-ray emission (after Ostriker and Davidson [190]).

B. The High- and Low-mass X-ray Binaries. The X-ray binaries can, broadly speaking, be divided into two groups, the High-Mass X-ray Binaries (HMXBs) and Low-Mass X-ray Binaries (LMXBs) which differ in a number of important characteristics, listed in table 13 and graphically represented in figure 16, as follows.

Table 12. List of confirmed X-ray pulsars (after Ögelman [186] and Nagase[176]).

Name[a]	Rotation period (s)	Orbital period (days)	$a_X \sin i$ (light-seconds)	$f(M)$ $M_\odot$	L_X (erg s^{-1})	Eccentricity	Type[b]
A0538−66	0.069	16.7	–	–	8×10^{38}	0.4	Be
SMC X-1	0.717	3.89	53.46	10.8	5×10^{38}	<0.007	MB
Her X-1	1.24	1.70	13.18	0.9	2×10^{37}	<0.003	LMXB
4U0115+63	3.61	24.3	140.13	5	3×10^{37}	0.34	Be
V0332+53	4.38	34.25	48	0.1	4×10^{35}	0.31	Be
Cen X-3	4.84	2.09	39.79	15.5	5×10^{37}	<0.0008	MB
1E1048.1	6.44	–	–	–	3×10^{34}	–	Be
1E2259+59	6.98	0.03?	< 0.2?	–	5×10^{35}	–	LMXB
4U1627−67	7.68	0.0228	–	–	3×10^{37}	–	LMXB
2S1553−54	9.30	30.6	164	5	–	0.09	Be
LMC X-4	13.5	1.41	26	15	7×10^{38}	<0.02	MB
2S1417−67	17.6	–	–	–	–	–	Be?
GPS1843+01	29.5	–	–	–	–	–	Tr
OAO1653−40	38.2	–	–	–	1×10^{37}	–	TrBe?
EXO2030	41.8	46.5–47.5	–	–	$\approx 1 \times 10^{38}$	–	TrBe
GPS2138+57	66	–	–	–	–	–	Tr
4U1700−37	67.4?	3.4	–	–	3×10^{36}	–	MB
GPS1843−02	94.8	–	–	–	–	–	Tr
A0535+26	104	111	500	20	2×10^{37}	0.3	Be
Sct X-1	111.1	–	–	–	–	–	Tr
GX 1+4	122	304?	–	–	4×10^{37}	–	LMXB
4U1230−61	191	–	–	–	–	–	Tr
GX 304−1	272	133	500	–	5×10^{35}	–	Be
4U0900−40	283	8.96	112	20	2×10^{36}	0.09	MB
4U1145−619	292	188	600	–	3×10^{35}	–	Be
1E1145.1	297	5.65?	–	–	3×10^{36}	–	–
A1118−61	405	–	–	–	5×10^{36}	–	Be
GPS1722−36	414	–	–	–	–	–	Tr
4U1907+09	438	8.38	80	9	4×10^{37}	0.22	MB
4U1538−52	529	3.73	55	13	4×10^{36}	–	MB
GX 301−2	696	41.5	367	31	3×10^{36}	0.47	MB
4U0352+30	835	580	–	–	4×10^{33}	–	Be

[a] The letters in the source names indicate the discovery spacecraft: A: Ariel 5, S: SAS-3, U: UHURU, E: Einstein, GPS: GINGA, EXO: EXOSAT, V: Vela satellites, GX: sources discovered in MIT balloon experiments.

[b] MB: massive binary (standard type), Be: Be system, LMXB: low-mass system, Tr: transient.

i. In the HMXBs, the companion of the X-ray source is a luminous early-type star of spectral type O or B (like in the Cen X-3 system), with a mass typically between 10 and 40 $M_\odot$. In the LMXBs it is a faint star of mass $\lesssim 1.2 M_\odot$. In most cases the stellar spectrum is not even visible here as the light of the systems is dominated by that of the accretion disk around the neutron star.

ii. In space the HMXBs show a strong concentration towards the galactic plane, (but not towards the galactic center) as is typical for a very young stellar population. This fits with the fact that the companion stars in these systems have $M \geq 10 M_\odot$; such stars do not live longer than 2.10^7

Table 13. The two main types of strong Galactic Bianry X-ray Sources.

HMXB	LMXB
-Optical counterparts massive and luminous early type stars, spectrum O and early B; $L_{opt}/L_x > 1$	Faint blue optical counterparts $L_{opt}/L_x < 0.1$
-Concentrated in space towards the galactic plane: young stellar population, age $< 10^7$ years	Concentrated in space towards the galactic center; fairly wide spread around the galactic plane: old stellar population, age (5-15 $\times 10^9$ years
-Type of time variability: Regular X-ray pulsations; no X-ray outbursts	Type of time-variablity: often X-ray outbursts; only in 3 cases regular X-ray pulsations.
- Relatively hard X-ray spectra kT $\gtrsim$ 15 KeV	Softer X-ray spectra: kT $\lesssim$ 10KeV

yrs. Hence, the HMXBs and their neutron stars (or black holes) must have ages less than this value. On the other hand the LXMBs show a strong concentration towards the galactic center, and a much broader distribution around the galactic plane, as is typical for an old stellar population with age of order 10^{10} yr. Indeed, a sizeable number (some ten per cent) of the $\sim$ 130 LMXBs is found in globular star clusters, the oldest objects in our Galaxy (ages $\sim (1.5 - 2) \times 10^{10}$ yrs.).

Figure 13 shows the galactic distribution of the optically identified HMXBs and LMXBs; the globular cluster sources and black-hole X-ray binaries (see below) are indicated separately.

iii. Each group has its own characteristic type of time variability. In most HMXBs the X-ray source is a regular X-ray pulsar indicating that the neutron star has a strong surface magnetic field (surface dipole strength $\sim 10^{11.5} - 10^{13.5}$ G), as is supported by the detection of cyclotron features in the X-ray spectra of many of these sources (cf. Nagase 1989). (Some characteristic pulse profiles are shown in figure 17).

On the other hand, very few LMXBs are regular X-ray pulsars, but many of them show thermonuclear X-ray flashes (so-called type I bursts) of the matter accreted onto the neutron-star surface. As thermonuclear flashes are expected to be suppressed if the field strength exceeds $\sim 10^{11}$G (Lewin

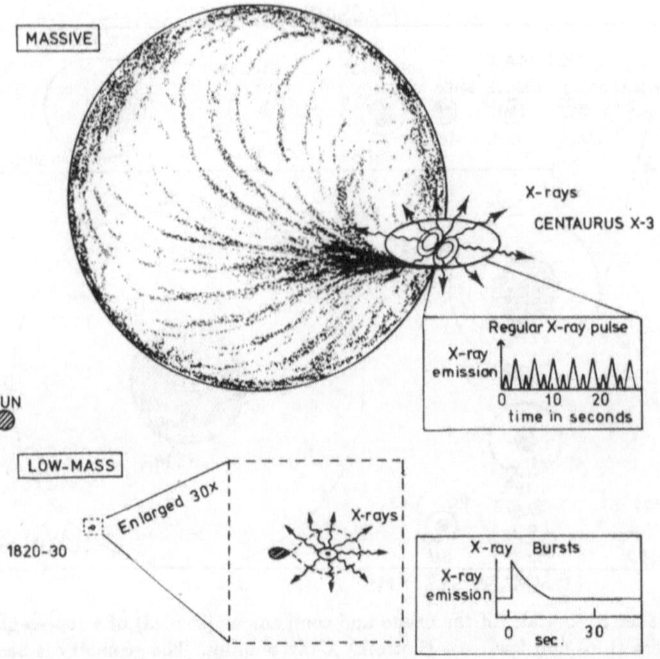

Fig. 16. The two basic types of accreting binary neutron star systems in the Galaxy. The High Mass X-ray Binaries (top) consist of a massive star (8 to 40 $M_{\odot}$) together with a neutron star. In almost all systems the neutron star is regular X-ray pulsar indicating that it has a relatively strong surface dipole magnetic field ($B \geq 10^{11}$ G) which channels the inflowing matter towards the magnetic poles. In the Low Mass X-ray Binaries (lower panel) the neutron star has a companion of mass $\leq 1.2 M_{\odot}$. These sources almost never show regular X-ray pulsations, but many systems show thermonuclear X-ray outbursts. Characteristic representatives of each type of system are Centaurus X-3 and XB 1820-30 respectively, depicted here to scale with the sun. See further the text.

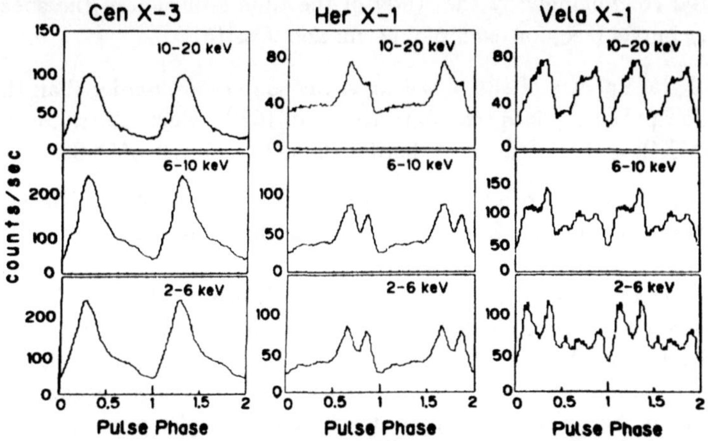

Fig. 17. Sample pulse profiles of three representative pulsating binary X-ray sources in three different energy bands, as observed by Japanese TENMA observatory (after Nagase [176]).

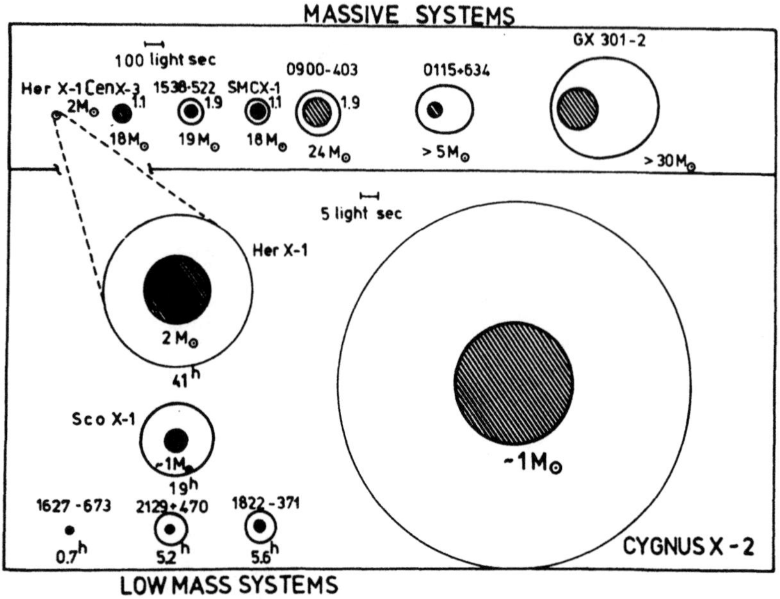

Fig. 18. Sketch, to scale, of the orbits and companions (shaded) of a representative number of high-mass (top) and low-mass (bottom) X-ray binaries. The geometry is based on orbital period and an estimate of the companion mass, derived either from doppler tracking or from its optical characteristics. As indicated in the figure, Hercules X-1 can be considered a member of either group. In the two wide low-mass X-ray binaries the companion is an evolved star (low-mass giant).

and Joss 1983), the surface magnetic fields in the neutron stars in these old sytems are expected to be weaker than this value.

That these burst-systems contain indeed neutron stars has been demonstrated convincingly by the study of the time evolution of the spectra of X-ray bursts (see, for example, Lewin and Joss[151]).

iv. The X-ray spectra of the pulsating X-ray sources are harder than those of the non-pulsating ones (cf. Maraschi et al.[162]). which causes the spectra of HMXBs to be, on average, harder than those of the LMXBs.

Figure 18 shows the orbits of a representative sample of the HMXBs and LMXBs drawn to scale (cf. Bhattacharya & van den Heuvel[18]).

C. Mass determinations of neutron stars in X-ray binaries. Most of the presently known neutron star masses have been obtained from pulsating X-ray sources in eclipsing High Mass X-ray Binaries. For these systems, once the optical component has been identified, the radial velocity orbits of both stars can be determined: that of the pulsar by Doppler-tracking of its pulse period, that of the optical star by measuring the Doppler shifts in its optical spectra obtained at different times.

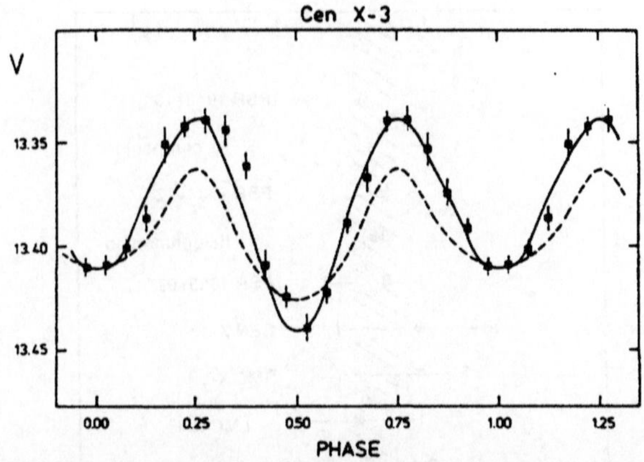

Fig. 19. a.) **Optical lightcurve of the high-mass X-ray binary Cen X-3. Observed data are indicated by squares. The expected purely ellipsoidal light curve is dashed. It has two maxima and minima per orbital period, as explained in figure 19b. The solid curve indicates the theoretical prediction of the effects X-ray heating and the presence of a disk [335].**

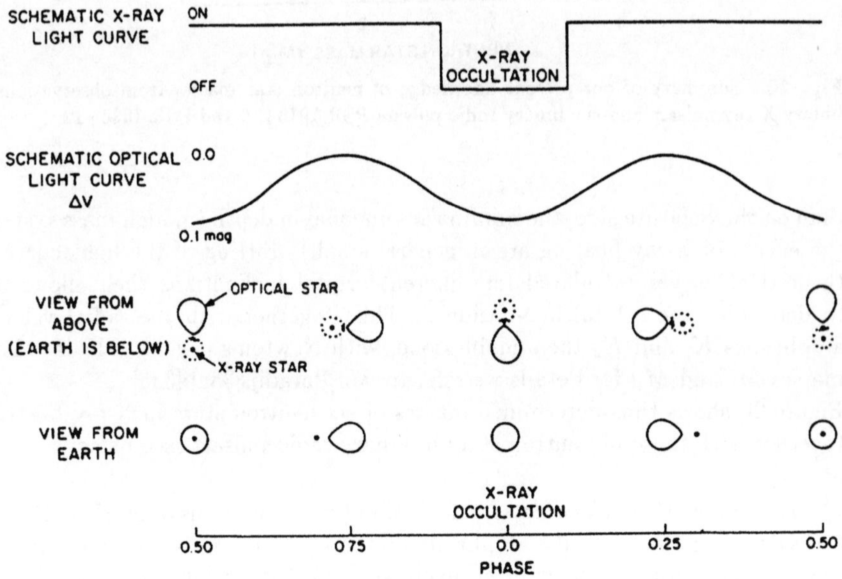

Fig. 19. b.) **The X-ray and visible light curves of an X-ray binary system that contains a supergiant star are shown schematically (top). Also illustrated (bottom) are several aspects of this system as viewed both from above and from Earth [81].**

Important extra information in these systems is obtained from the regular photometric variability of the optical star. Since this star is (nearly) filling its Roche lobe, it is pear-shaped and will show a double-wave optical light curve (two maxima and two minima per orbital revolution) as depicted in figure 19. As the stellar surface on the side of the neutron star is darker ("gravity darkening")

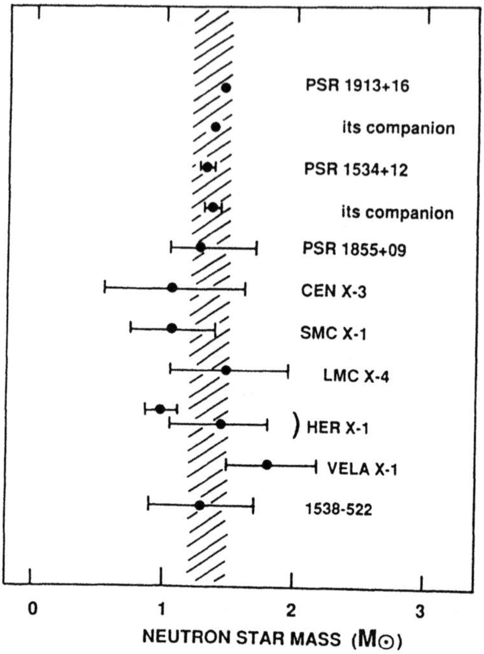

Fig. 20. **Summary of our present knowledge of neutron star masses from observations of binary X-ray pulsars and the binary radio pulsars PSR 1913+16 and PSR 1534+12.**

than on the opposite side, the minima are unequal in depth (in high mass systems the effects of X-ray heating are in general small). Fitting of the light curve to theoretical curves calculated for different orbital inclinations then allows one to determine the orbital inclination i. This, together with the radial velocity amplitudes K_1 and K_2 then enables one, with Newton's law, to calculate both masses M_1 and M_2; for details we refer to van Paradijs[338];[335].
Figure 20 shows thus determined masses of six neutron stars in X-ray binaries, together with those of neutron stars in binary radio pulsars (see below).

D. The Black Hole Candidates. In a number of non-pulsating HMXBs and LMXBs the radial velocity amplitude of the optical star is so large that the mass of the compact star is in all likelyhood $\gtrsim 3.4 M_{\odot}$, which is expected to be the absolute upper limit to the mass of a stable neutron star (Nauenberg and Chapline[178]; Rhoades and Ruffini[228]). One therefore expects the compact objects in these systems to be black holes.

A realistic upper limit to the mass of a neutron star, based on realistic equations of state of neutron-rich matter, is about 2.0 to 2.2 $M_{\odot}$ (cf. Baym[14] and Glendenning[91], Weber and Glendenning[354]).

The X-ray sources in these systems show a number of characteristics that are quite different from those of reguraly pulsating sources and "normal" LMXBs that we know to contain neutron stars (e.g.: the X-ray bursters). The clearest of these characteristics is: an X-ray spectrum that either has a very hard tail, up to energies beyond 100 keV, or is super-soft (softer than observed in the usual LMXBs). Some sources are found to switch between these states (e.g. Cygnus X-1), or to always be in one of these states. It is presently thought that these spectral characteristics are those of an accreting black hole (cf. Tanaka[269]; Mc Clintock[169] and Cowley[47] and references therein).

At present five systems are known with the above-mentioned characteristics. These are the black hole candidates listed in table 14. Figure 21 (after McClintock) illustrates the orbital dimensions of the systems and Figure 22 shows the radial velocity curve of the optical star in one of them, the system of A0620-00).

Table 14. Properties of five Black Hole Binaries.

	Cyg X-1	LMCX-3	A0620-00	GS 2023+338 (V404 Cyg)	GS 1124-68
L_x	2×10^{37}	3×10^{38}	1×10^{38}	2×10^{38}	1×10^{38}
MK type	O9.7Iab	B3V	K5V	KOIII-G9V	K0-4V
d(kpc)	2.5	55	1 (?)	~ 2.7	~ 2.5
m_v	9	17	18	18.4	20
e	0.001 ($\pm$0.01)	0.13 ($\pm$0.05)	0.01 ($\pm$0.01)	0.00	0.00
$K(km/s)$	74.7$\pm$1.0	235 $\pm$11	443$\pm$4	211 $\pm$4	409 $\pm$ 18
$P(d)$	5.6	1.7	0.32	6.47	0.433
$f(M)/M_\odot$	0.25$\pm$0.01	2.3 $\pm$0.3	2.91 $\pm$0.08	6.26 $\pm$0.31	3.07 $\pm$ 0.40

Two of the black hole systems belong to the class of HMXBs: the system of Cygnus X-1, a non-eclipsing HMXB in which the optical star is an O9.7 Iab blue supergiant (the 9th magnitude star HD 226868), a spectroscopic binary with a 5.60 day period and a radial velocity amplitude of 75 km/s. From the absence of X-ray eclipses and the fitting of the double-wave optical lightcurve one finds the orbital inclination to be $i < 70°$. Assuming a (normal) mass of $\gtrsim 20M_\odot$ for the O9.7 Iab supergiant one finds from the mass function: $M_x \gtrsim 6M_\odot$ (see also Dolan and Tapia[70]).

The best-fit solution to all observational data of the Cygnus X-1 system yields (Gies and Bolton[90]): $M_x = 13M_\odot$, $M_c = 33M_\odot$. Nevertheless, the presence of a black hole in this system is still not one hundred per cent certain, since the mass of the O9.7 Iab supergiant has to be assumed in order to find M_x, and purely theoretically speaking this mass may be as low as only $5M_\odot$. Therefore it is still theoretically possible though quite unlikely that the mass of Cyg X-1 is $\lesssim 2.5M_\odot$, and the star is a very heavy neutron star.

For LMC X-3, a 1.7 day binary with a B3V optical star, one obtains $M_x = 9M_\odot$, $M_c = 6M_\odot$ (but also here the value of M_x depends on the assumed value of M_c).

The other three black hole candidates belong to the Low-Mass X-ray binaries: the companion stars, which are filling their Roche-lobes, are main-sequence G-

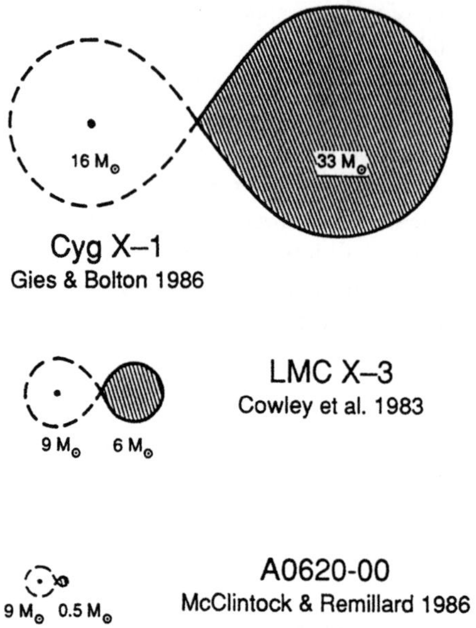

Fig. 21. Schematic sketch, to scale, of plausible models for three dynamical black hole candidates. The optical companions (shaded regions) are shown filling their critical Roche equipotential lobes. The value of the mass function and the masses given in the figure determine the following values of the inclination angle: $i(CygX-1) = 32°$, $i(LMCX-3) = 63°$ and $i(A0620-00) = 45°$. The "Nova Muscae 1991" system closely resembles A0620-00. The fifth system, V 404 Cyg, is in many respects similar to A0620-00 and Nova Mus (see text), (Adapted from Mc Clintock [169]).

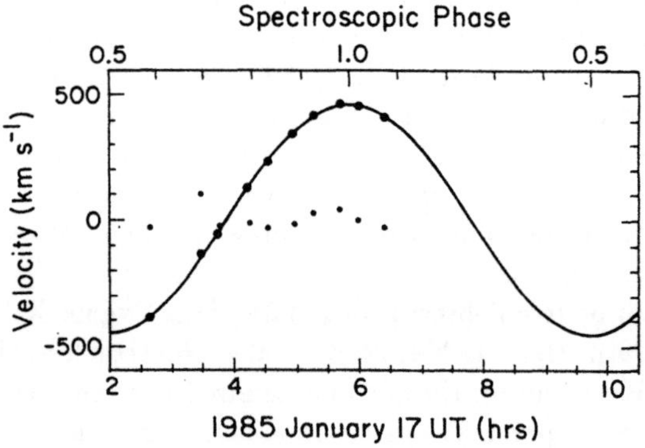

Fig. 22. Radial velocities of the K dwarf in the A0620-00 system (large filled dots) measured by Mc Clintock and Remillard [168] Smooth curve is a fit to a circular orbit using the photometric period. Small filled dots are the residual differences (× 5) between the data and the fitted curves.

or K-dwarf stars. At present there are three such systems known and they are the strongest black hole candidates found so far, since even if one assumes the masses of their optical stars to be zero, the radial velocity curves of these stars still indicate that the masses of their compact companions are $\gtrsim 3M_{\odot}$. Taking into account that their orbital inclinations are $< 90°$, these masses are $\gtrsim 3.5-4.0M_{\odot}$.

All these three systems were discovered as strong X-ray transients ("X-ray novae") which flared up as a strong X-ray source for a duration of several weeks and then faded away. During the X-ray outbursts they also brightened very much optically by 6 magnitudes or more, which allowed their optical identification. During outbursts, the optical spectra are those of a bright accretion disk. After the turning off of the X-ray source, also the optical brightness declined and finally the spectrum of a G- or K-dwarf became visible.

In the first such system discovered, A0620-00 (Nova Monocerotis 1975), the K5V dwarf was found to be spectroscopic binary with an orbital period of 8 hours and a radial velocity amplitude of almost 450 km/s (fig. 22; McClintock and Remillard[168]). Even if the mass of the K-dwarf is zero this already leads to a mass of the compact star $> 3M_{\odot}$. In the systems of GS 2023+338 (V 404 Cyg) and Nova Muscae (GS 1124-68) these values are $\geq 6.5M_{\odot}$ and $\geq 3M_{\odot}$, respectively (Casares et al.[34]; Remillard, Mc Clintock and Baylin[?]. Putting in a realistic mass of $\sim 0.7M_{\odot}$ for the K-dwarf companions in A0620-00 and Nova Muscae, one obtains $M_x > 3.4M_{\odot}$ in all cases.

Table 14 lists the best fit parameters of these five black-hole candidate systems.

E. The High Mass X-ray Binaries in more detail. It appears that the HMXBs fall into two categories which differ in a number of important characteristics:

a. the so-called "standard" HMXBs in which the massive OB-type companion of the X-ray source is close to filling its Roche-lobe, as evidenced from its double-wave optical light curve (see fig. 19), and

b. the B-emission X-ray binaries.

Table 15 lists a number of characteristic examples of both categories and figure 23 depicts the characteristic differences between the two groups (after Rappaport and van den Heuvel[223]; van den Heuvel and Rappaport[323]), which we will discuss.

1. In "standard" systems the companions tend to be evolved (giant, Of or blue supergiant) stars with radii $10-30R_{\odot}$, an optical luminosity L_{opt} $\sim 10^5 L_{\odot}$ and initial mass (derived from their luminosities by using evolutionary tracks) in excess of $20M_{\odot}$. On the other hand the optical components in Be systems are generally unevolved stars of spectral type O9Ve to B2Ve; in some cases they are of luminosity class III or IV ("subgiants"). Such stars have relatively small radii, i.e. $\lesssim 5-10R_{\odot}$, absolute luminosities $< 3.10^4 L_{\odot}$ and masses between about $8M_{\odot}$ and $20M_{\odot}$.

Table 15. The "standard" massive X-ray binaries (top) and some examples of Be X-ray binaries (bottom).

Source	Optical counterpart	Spectral type	Pulse period (s)	Orbital period (d)	Eccentricity
LMCX-4	Sk-Ph	O7 III–V	13.5	1.408	$e = 0.00$
Cen X-3	Krz's star	O6.5 II–III	4.84	2.087	$e = 0.00$
4U 1700–37	HD 153919	O6.5 f	–	3.412	$e \approx 0$
SMC X-1	Sk 160	B0Ib	0.717	3.89	$e = 0.00$
4U 1538–52	* 12	B0Iab	529	3.73	$e = 0.00$
Cyg X-1	HD 226868	O9.7Iab	–	5.60	$e = 0.05$
4U 0900–40	HD 77581	B0.5Ib	283	8.965	$e = 0.09$
GX 301-2	Wra 977	B1.5Ia	696	41.5	$e = 0.47$
4U 0115+63	John's star	Be	3.61	24.3	$e = 0.34$, transient
4U 0352+30	X Per	O9.5(III–V)e	835	580	very weak, steady
A 0535+26	HD 245770	B0Ve	104	111	transient
4U 1145–61	Hen 715	B1Vne	292	188	highly variable
4U 1258–61	MMV star	B2Vne	272	133	highly variable

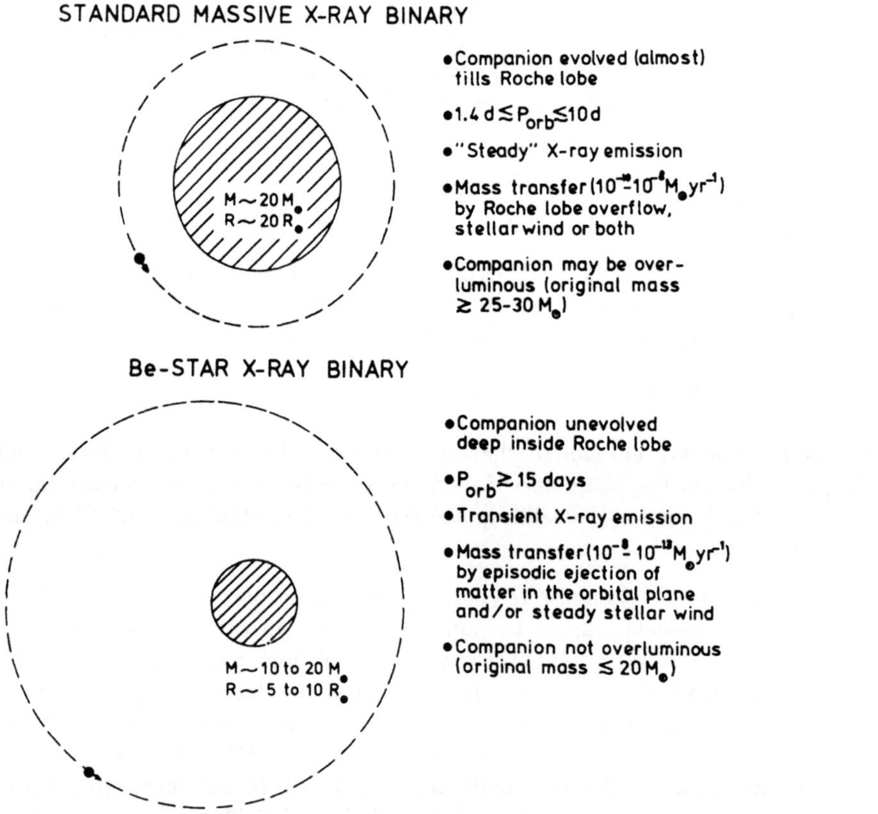

Fig. 23. Differences between the 'standard' and B-emission-type high-mass X-ray binaries. Explanation in the text.

2. The standard HMXBs (with the sole exception of 1223-62=GX 301-2) have orbital periods between 1.4 and 10 days, whereas the Be systems have orbital periods in the range of 15 days to over one year; most of these systems have periods of 3 to 12 months.

3. Characteristics 1 and 2 explain why the "standard" systems in many cases exhibit X-ray eclipses and always show ellipsoidal (double-wave) light variations, indicating that their optical components pratically fill their Roche lobes, and therefore are 'pear-shaped'.

   The mass transfer in these systems is thought to be due beginning atmospheric Roche-lobe overflow or to the strong (enhanced) winds expected from supergiant stars that nearly fill their Roche lobes as will be explained in more detail below, in section F.

   On the other hand, characteristics 1 and 2, together with the absence of X-ray eclipses and periodic (ellipsoidal) light variations in the Be systems, indicate that the Be stars in these systems reside deep inside their Roche lobes. This rules out Roche-lobe overflow or enhanced winds as a source for the mass accretion.

4. The "standard" systems are persistent X-ray sources (i.e. they are always "on"). On the other hand, many of the Be sources are so-called "transient", which means that they may be completely unobservable ("off") for long periods, ranging from months to many years, and they suddenly turn "on" as a strong transient source for a duration of weeks to months. Since Roche-lobe overflow is ruled out here as a source of mass accretion, the mass transfer must be due to the intrinsic mass-losing properties of the Be component, as first argued by Maraschi et al.[161]).

   Be stars show two types of intrinsic mass loss: (I) by a stellar wind (weak), and (II) by mass ejection from their equatorial regions. This latter type of mass loss produces the characteristic emission lines in their spectra. Since Be stars are the most rapidly rotating B stars, the equatorial ejection of the matter is presumably rotationally driven (possibly by the excitation of non-radial oscillations, c.f. Slettebak[253]. The weak winds of Be stars can only generate X-ray luminosities of $\sim 10^{32}$-10^{34} erg/s at most (cf. section F), i.e. of the same order as observed in the nearby weak Be X-ray binaries X Per and γ Cas. On the other hand, the transient outbursts with $L_x = 10^{37} - 10^{38}$ erg/s cannot be explained in this way. They must be due to equatorial outbursts of mass ejection that are characteristic for Be stars, which occur at completely erratic moments, with completely irregular intervals of time. Figure 24 shows as a characteristic example the intensity variations of the strong pulsating transient source V0332+53, which had "on" stages only in 1983/84 and 1973/74 and was "off" for a decade in between. Between 14th of November 1983 and 24th of January 1984 it had three "on" stages of about one week separated by "off" intervals of about 34 days, which together with Doppler-tracking of its pulse period (4.4 s) yielded an orbital period of 34.2 days; its orbit is moderately eccentric

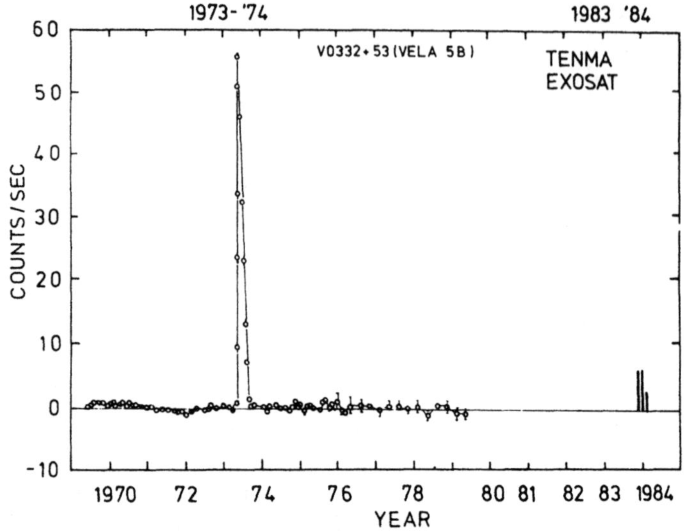

Fig. 24. X-ray intensity versus time for the transient Be/X-ray binary X0332+53. The 1973 outburst was discovered in the Vela satellite database, by Terrell and Priedhorsky [279], after the discovery of the November 1983 outburst by Tanaka et al. [270]. The outbursts in November and December 1983 and January 1984 occurred with approximately 34-day intervals. The intensities of the 1983/84 outbursts are drawn to scale of the 1973 outburst. Further explanation in the text.

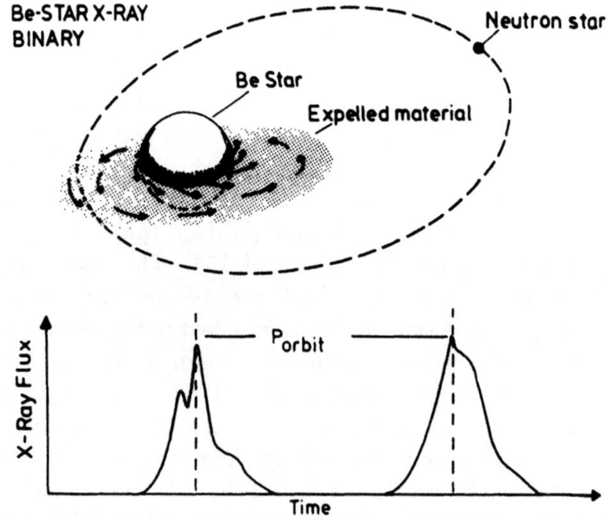

Fig. 25. Schematic model of a Be star X-ray binary system such as X0535+26 and X0332+53. The neutron star moves in a moderately eccentric orbit around the Be star, which is much smaller than its own critical equipotential lobe. The rapidly rotating Be star is temporary surrounded by matter expelled in its equatorial plane. Near its periastron passage the neutron star enters this circumstellar matter and the resultant accretion produces an X-ray outburst lasting several days to weeks.

($e = 0.31$) and its periodic "turn-ons" in in 1983/4 occurred near periastron passage. During these "turn-ons" the infrared brightness of the optical counterpart increased considerably. Since 24 January 1984 the source has not been seen. For a summary of the behaviour of this system we refer to van den Heuvel and Rappaport[323]). Figure 25 gives a schematic picture of how the periodic X-ray flares of a Be/X-ray binary during a phase of equatorial mass ejection from the Be-star are thought to arise.

F. The Low-Mass X-ray binaries in more detail/the globular cluster sources: the oldest neutron stars in the Galaxy. Table 16 lists the known orbital periods of LMXBs, and figure 26 graphically depicts the distribution of these orbital periods, together with that of the Cataclysmic Variable binaries. The CV-binaries are very similar to the LMXBs: the only difference is that in CVs the accreting star is a white dwarf (see table 7). Nevertheless, figure 26 shows that a great difference between the CVs and LMXBs is that the pronounced period-gap in the distribution of orbital periods of CVs (see also Livio, this volume) seems absent for the LMXBs.

A group of LMXBs that requires special attention is that of the Globular Cluster X-ray sources, which are indicated in figure 13. The incidence of Low-Mass X-ray binaries (per unit mass) in globular clusters is about 10^3 times higher than in the Galaxy as a whole (Gursky[98]). With a total mass of $\sim 2.10^7 M_\odot$ the about 120 known globular clusters in the Galaxy contain about 10 per cent of all strong low-mass binary X-ray sources of a Galaxy that has a total mass of $\sim 2.10^{11} M_\odot$. The same is true for M31 (cf. Long and van Speybroeck[157]). This abnormally high incidence of LMXBs in globular clusters is due to a special formation process that can take place in a globular cluster, but not in the galactic disk: binary formation by close encounters -either by tidal capture or by exchange collisions: binary single star interactions which cause one binary component to be exchanged by a passing neutron star (see e.g. the reviews by Verbunt[340]; Bhattacharya and van den Heuvel[18]; Hut[117]). The reason why globular clusters are particularly favourable places for this process is their com-

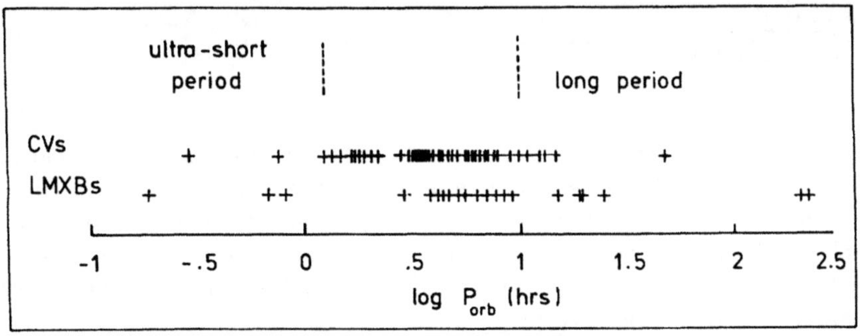

Fig. 26. Distributions of orbital periods of cataclysmic variables (CVs) and low-mass X-ray binaries (LMXBs); data from Ritter [229].

Table 16. The orbital periods of LMXBs (after Parmer and White[205]). The letters B and T in the source name indicate a burst source and a transient nature, respectively. A o indicates that an orbital periodicity is een in the optical band, a x that is seen in the X-ray band.

Source	Period (hr)	Optical	X-ray
XB 1820−303	0.19	–	×
X 1626−673	0.70	o	–
XB 1916−053	0.83	o	×
X 1705−44 [a]	1.30	–	×
MS 1603+26 [a]	1.85	–	×
XB 1323−619	2.9	–	×
XB 1636−536	3.8	o	–
XBT 0748−676	3.8	o	×
XB 1254−690	3.9	o	×
X 1728−169	4.2	o	×
X 1755−338	4.4	o	×
XB 1735−444	4.6	o	–
Cyg X-3	4.8	(IR)	×
XBT 2129+470	5.2	o	×
X 1822−371	5.6	o	×
LMC X-2	6.4	o	×
XB 1658−298	7.2	–	×
XT 0620−003	7.8	o	–
X 2127+119 (M15)	8.5	o	×
X 1957+115	9.3	o	–
X 0547−71 [a]	10.6	–	×
X 1659−487 (G339−4)	14.8	o	–
Cen X-4 [a]	15.1	–	×
Sco X-1	19.2	o	–
X 1813−14 [a]	19.5	–	×
X 1624−490	21	–	×
Cal 83	25	o	–
Her X-1	40.8	o	×
X 0921−630	216	o	×
Cyg X-2	235	o	×

[a] after Ritter [229].

bination of very high star densities (up to $10^5 pc^{-3}$) with low relative velocities (a few tens of km/s). The discovery of many LMXBs in globular star clusters was the first evidence of the presence of a population of old neutron stars in the Galaxy. These neutron stars must date from some 2.10^{10} years ago, from the first few tens of millions of years of the life of the globular clusters, when massive stars were still present in them.

G. The Eddington limit and the maximum accretion rate possible for a compact object. For a star of a given mass M there is a maximum possible luminosity, L_{edd}, above which the star will start blowing away its outer layers by radiation pressure. As this limiting luminosity was first calculated by Eddington around 1920, it is called the "Eddington limit". L_{edd} is defined as the luminosity at which the radiation pressure force on ionized hydrogen plasma near the star balances the gravitational acceleration force exerted by the star on this plasma (see figure 27):

$$\frac{GM}{r^2} = \frac{L\sigma_{el}}{4\pi r^2 m_p c} \tag{55}$$

where r is the distance to the stellar center, m_p is the proton mass and σ_{el} is the Thomson scattering cross section of an electron (the dominant opacity source in an ionized hydrogen plasma of low density). Eq. (55) follows from the fact that a proton is attracted by a force GMm_p/r^2 while the electron associated with the proton feels the outward radiation pressure force $L\sigma_{el}/4\pi r^2 c$. Since the protons move along with electrons (huge Coulomb forces will prevent any charge separation over distances larger than the Debye length) and the electrons have negligible mass, GMm_p/r^2 is also the gravitational force felt by an electron.

Eq. (55) shows that the Eddington luminosity is given by

$$L_{edd} = 4\pi GMm_p c/\sigma_{el} = 10^{4.5}(M/M_\odot)L_\odot \tag{56}$$

where $M_\odot$ and $L_\odot$ are the mass and luminosity of the Sun.

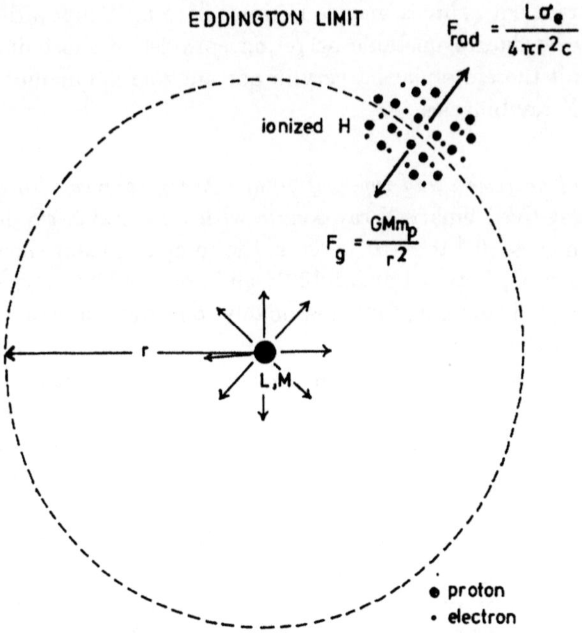

Fig. 27. The Eddington limit is defined as the luminosity at which the radiation pressure on ionized hydrogen gas is just balanced by gravity (explanation in the text).

The Eddington luminosity sets an upper limit to the *accretion luminosity* L_{acc} of an accreting compact object, since for $L_{acc} > L_{edd}$, further accretion of matter will be inhibited by the radiation pressure.

The maximum accretion rate $\dot{M}_{Edd}$ possible for a neutron star (or black hole) of mass M and radius R follows from:

$$L_{acc} = GM\dot{M}/R = 10^{4.5}\left(\frac{L_\odot}{R_6}\right)\left(\frac{M}{M_\odot}\right)\frac{\dot{M}}{1.5\times10^{-8}M_\odot yr^{-1}} \tag{57}$$

Combination of equations (56) and (57) shows that the maximum accretion rate is:

$$\dot{M}_{Edd} = R_6.1.5\times10^{-8}M_\odot/yr \tag{58}$$

where R_6 is the radius of the compact star in units of 10^6 cm(10km). For a black hole, $R = 2GM/c^2$, so $\dot{M}_{Edd}$ is proportional to the mass M of the hole.

Equation (58) shows that for a standard neutron star with R_6= 1 holds: $\dot{M}_{Edd} = 1.5\times10^{-8}M_\odot/yr$.

That the Eddington limit plays indeed a very important practical role in Nature is demonstated by the fact that none of the persistent compact X-ray sources known in our Galaxy, the LMC or the Andromeda Nebula considerably exceeds the Eddington limit for a 1 $M_\odot$ neutron star for extended periods time.

Only occasionally a source may flare up temporarily to a few times L_{Edd}, not more.

Some transient sources may sometimes flare up temporarily to 10^{39} ergs/s, which may be due to unstable accretion episodes of short duration. The Eddington limit therefore plays a very important role in our understanding of the physics of X-ray binaries.

H. Modes of accretion and types of binary X-ray sources. In order to produce a steady long-lived binary X-ray source with a typical X-ray luminosity in the observed range $\sim 10^{34}$ to $\sim 10^{38}$ ergs/s, the companion star should offer a steady accretion rate $\dot{M}_a$ between about 10^{-12} and about $10^{-8} M_\odot$/yr ($\approx \dot{M}_{Edd}$) to the neutron star or black hole for a reasonably long time interval, say, some 10^4 to 10^5 yrs (see below).

If the companion provides continuously more than $\dot{M}_{Edd}$ the source will be choked, as the compact star cannot accept more than $\dot{M}_{Edd}$. The excess inflowing matter will pile up around the source and absorb the X-rays and degrade them to radiation at longer wave-lengths: one will no longer observe an X-ray source but an optical star.

One now may wonder: why do we see in Nature the two major classes of persistent binary X-ray sources: the "standard" HMXBs, in which the companion stars are more massive than about 15 $M_\odot$, and the LMXBs, in which the companion stars are less massive than about 1.2 $M_\odot$: And, furthermore, why do we see the Be/X-ray binaries, which are wide binaries, have companion stars more massive than about 8 $M_\odot$ and are very variable as X-ray sources (transients)? In this respect a most important question is: does the dichotomy between HMXBs and LMXBs have an intrinsic evolutionary origin, i.e.: are two different formation mechanism of neutron stars in binaries operating in Nature, one producing only neutron stars as companions of stars more massive than 8 to 15 $M_\odot$, the other one: as companions of stars less massive than $\sim 1.2 M_\odot$ and are little or no neutron stars produced in the intermediate range?

Or, alternatively, is the virtual absence of persistent binary X-ray sources with companions in the mass range 1.2 $M_\odot$ to 15 $M_\odot$ purely due to the absence of a suitable continuous mode of mass transfer that is able to provide an $\dot{M}_{accr}$ in the range 10^{-12} to $10^{-8} M_\odot/yr$ for these stars? I.e.: do neutron stars exist also as companions to stars in this mass range, but are they unable to become persistent long-lived X-ray sources?

We have already argued in the past (van den Heuvel[305]) that the latter possibility is the most likely one, for the following reasons .

Let us assume that indeed neutron stars are formed as companions to stars of any kind of mass. In which mass ranges do we then expect to see (reasonably) long-lived persistent bright X-ray sources? I.e.: in which mass ranges do we expect companions to lose mass in such a way that the neutron star will have an accretion rate of 10^{-12} to $10^{-8} M_\odot/yr$?

Basically, the companions can lose mass in two ways:

1. By Roche-lobe overflow: the star in this case fills its Roche lobe and transfers mass along the L_1-point to its companion.
2. By a high-velocity stellar wind; in this case the star needs not fill its Roche lobe.

We will deal with these two processes separately.

a. Roche-lobe overflow. In this case the mass transfer takes place at low velocities and practically all the transferred matter will be accreted by the companion. As shown in section 2, if the mass-losing star is more massive than its companion, the orbit of the system and the Roche-lobe shrink. This leads to an unstable type of mass transfer (the star becomes larger than its Roche-lobe and loses mass at an accelerated pace). Therefore rapid mass transfer ensues on a timescale equal to (roughly) the thermal timescale of the mass-losing star (see section 5.1).

$$\tau_{th} = \frac{GM^2}{RL} \simeq 3.10^7 \frac{(M/M_\odot)^2}{(R/R_\odot)(L/L_\odot})) \qquad (59)$$

Using the mass-luminosity and mass-radius relations for main-sequence stars: $L \sim M^{3.5}$; $R \sim M^{0.5}$, one obtains:

$$\tau_{th} = 3.10^7/M^2 yr \qquad (60)$$

(M in solar masses).

As most of the stellar mass ($\sim 0.8M$) will be transferred on this timescale, one therefore has

$$\dot{M}_{Roche} \simeq 3.10^{-8} M^3 [M_\odot/yr] \tag{61}$$

with M in $M_\odot$.

Equation(61) shows that only for $M \lesssim M_\odot$ mass-transfer rates below $\dot{M}_{Edd}$ can be obtained. However, for $M \lesssim M_\odot$, the mass-losing star is no longer more massive than the compact star, and equation (61) is no longer valid, as now the mass transfer is no longer unstable. For companions less massive than the compact star (which has $M_c \sim 1.4 M_\odot$), mass transfer will will cause the orbit to widen, so the slow evolutionary expansion of the star, on a nuclear timescale, will drive the mass transfer. We will for the sake of argument consider this type of transfer here. (We will consider also driving by other causes, such as Gravitational radiation losses, in section 7.). The timescale of the transfer is now: $\tau_{nuc} = 10^{10}(M/L)$ yrs. With $L \sim M^{3.5}$ one obtains $\dot{M} \sim 10^{-10} M^{3.5}[M_\odot/yr]$ (M in $M_\odot$), which holds for $M < M_c$. Assuming for a neutron star $M_c \simeq 1.4 M_\odot$ (Chandrasekhar mass) one observes that, indeed, in systems with companion masses $\lesssim 1.4 M_\odot$, mass transfer rates below $\dot{M}_{Edd}$ can be obtained, in the range $\sim 10^{-10}$ to $10^{-9} M_\odot/yr$. For $M > 1.4 M_\odot$, equation (61) has to be applied and $\dot{M}$ will be $\gtrsim 10^{-7} M_\odot/yr$ and the source will be quenched.

So, Roche-lobe overflow can only produce steady long-lived X-ray sources with $\dot{M} < \dot{M}_{Edd}$ if the if the companion mass is $\lesssim 1.4 M_\odot$.

b. Accretion from a stellar wind Strong stellar winds, with (total) mass-loss rates $\gtrsim 10^{-9} M_\odot/yr$ are found only in stars more massive than about $15 M_\odot$, as depicted schematically in figure 29 (see, e.g. de Jager[54]).

As the outflow velocities observed in these winds are typically of the order $(1-2) \times 10^3$ km/s, the bulk of the wind matter will flow past the neutron star companion without being affected by its presence, as depicted in figure 28. Only a tiny fraction of the matter will be captured. In order to estimate that fraction one defines the accretion radius r_{acc} as follows (Ostriker and Davidson[190]). We consider the wind to consist of free particles. To be captured, the orbit of a passing particle must be deflected by about 90°, such that the particle will collide with particles from the opposite side, also deflected by 90°. The colliding particles will dissipate their transverse velocity components into heat which is radiated away, and then fall to the neutron star and be captured (see figure 28).

Particles deflected over less than 90° will retain part of their wind velocity and flow away -though at somewhat reduced speed.

In order to be deflected over 90° a particle with velocity V_w has to pass the neutron star within a distance r_{acc} given by $V_w = V_{esc}(r_{acc}) = \sqrt{\frac{2GM_n}{r_{acc}}}$.

Hence $r_{acc} = \frac{2GM_n}{V_w^2}$ where M_n is the mass of the neutron star. Thus particles will be captured that pass through a circle of radius r_{acc} around the neutron star, which is itself orbiting at distance a from the center of the companion star (see figure 28). The fraction of the total wind mass loss rate $\dot{M}_w$ from the companion star that is captured by the neutron star will therefore be

$$\dot{M}_{acc} = \frac{\pi r_{acc}^2}{4\pi a^2} \dot{M}_w = \left(\frac{G^2 M_n^2}{V_w^4 a^2} \right) \dot{M}_w = S \dot{M}_w \tag{62}$$

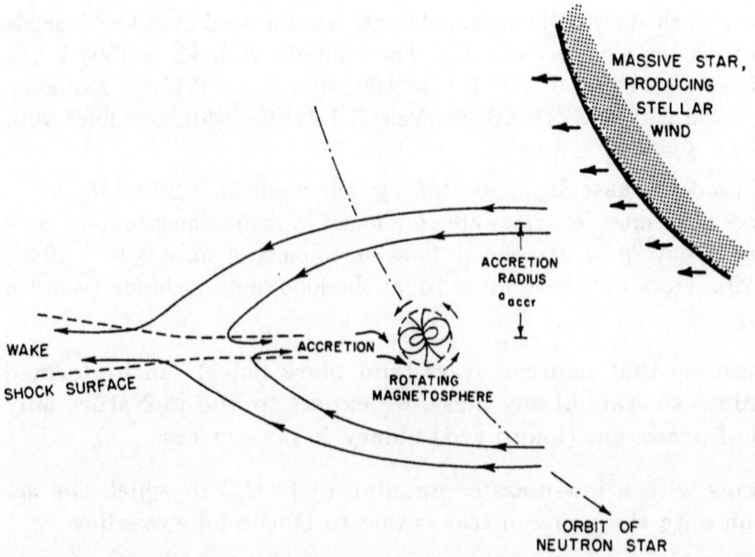

Fig. 28. Streamlines of stellar-wind material, in frame of an accreting neutron star. Relative dimensions are not to scale (from Ostriker and Davidson [190]).

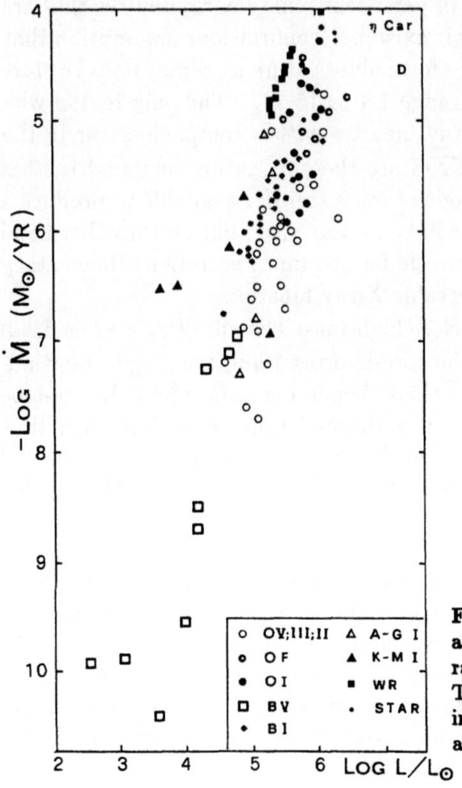

Fig. 29. The relation between mass loss rate and luminosity for stars with known mass loss rate, luminosity, and effective temperature. The meaning of the various symbols is shown in the bottom panel of the figure (from Chiosi and Maeder [40]).

Equation (62) shows that the captured fraction of the wind mass loss depends very strongly on the wind velocity V_w. For example, with $V_w = 1000$ km/s, $M_n = 1.4 M_\odot$ one has $r_{acc} = 0.7 R_\odot$. If V_w=2000 km/s, $r_{acc} = 0.17 R_\odot$. Assuming $a = 50 R_\odot$, as in a typical HMXB like Vela X-1 (4U0900-40), one finds with $V_w = 10^3 km/s$: $S = \frac{\pi r_{acc}^2}{4\pi a^2} = 5.10^{-5}$.

Hence, in order to have $\dot{M}_{acc} \gtrsim 10^{-12} M_\odot/yr$ one needs $\dot{M}_w \gtrsim 2.10^{-8} M_\odot/yr$.

Such large wind mass loss rates are only found in main-sequence stars more massive than about 20-25 $M_\odot$ and in blue supergiants of mass $\gtrsim 15 - 20 M_\odot$ (see figure 29). From the above we arrive at the following conclusion (van den Heuvel[305]):

If we assume that neutron stars (and black holes) can be formed as companions to stars of any mass, we expect to find in Nature only two types of persistent (long-lived) binary X-ray sources:

- **Systems with a low-mass companion ($\lesssim 1.4 M_\odot$) in which the accretion onto the neutron star is due to Roche-lobe overflow.**

- **Systems with a high-mass ($\gtrsim 15 M_\odot$) companion in which the neutron star is accreting from the strong stellar wind of its companion.**

As these are exactly the two types of persistent binary X-ray sources that are found in Nature, we conclude that their existence confirms our assumption that neutron stars (and also black holes) are formed in Nature *as companions to stars of any mass*, hence: also in the mass range 1.4 to 15 $M_\odot$. The only reason why we practically do not observe any X-ray binaries with a companion star in the latter mass range (Her X-1 and LMCX-3 are the only two exceptions) is, that these companions have no suitable mode of mass transfer available to produce a steady long-lived X-ray source: their winds are too weak, but on the other hand by Roche-lobe overflow they would provide far too much accretion. Hence, they are unable to produce persistent observable X-ray binaries.

We will later see section 7, that in close high-mass systems ($P_{orb} \lesssim$ a few days) where the companion star is still in the core-hydrogen burning stage, the wind-accretion may be supplemented by so-called "beginning atmospheric Roche-lobe overflow" which does not take place on a thermal time scale, but on a time scale between the thermal and nuclear one (cf. Savonije[241]). This is expected to be the case in systems like Cen X-3, LMCX-4 and possibly SMCX-1 (cf. Savonije[241]).

c. The third form of mass loss: eruptions from Be stars. We consider only ordinary companion stars, i.e. main-sequence stars and (sub)giants (no peculiar objects such as novae, planetary nebulae, etc.). Apart from the steady forms of mass loss by Roche-lobe overflow in a binary, or by a steady wind in the case of massive stars, there is a third well-known type of mass loss among ordinary stars: the irregular outbursts of equatorial mass loss observed in rapidly rotating B-stars, so-called B-emission stars.

As mentioned in section D, rapidly rotating B-type stars show, at irregular time intervals, outbursts of equatorial mass ejection which produce a rotating ring of gas around the star, giving rise to the sudden appearance of hydrogen emission lines in its spectrum.

Such outbursts of mass ejection are clearly intrinsic to B-star (i.e. are independent of the presence or absence of a companion star), and must be connected to its rapid rotation, as Be stars always have very high rotational velocities (see e.g.: Underhill[292]; Slettebak and Snow[254]). The star can appear as a normal main-sequence B-star for many years and suddenly go through an outburst phase, becoming a Be star for periods ranging from a few weeks to many years (cf. Slettebak and Snow[254]). Others are Be-stars almost permanently, although in practically all systems the strengths of the emission lines is variable.

If such a Be-star has a neutron star companion, this companion will suddenly become a bright X-ray transient when the B-star goes through a B-emission phase and ejects a ring or disk of matter from its equator, part of which is captured by the neutron star (see figure 25).

The fact that indeed also stars that exhibit this third well-known type of mass loss appear to form an important class of X-ray binaries (Be/X-ray binaries are more numerous than "standard" HMXBs like Cen-X-3) is a further confirmation of our hypothesis that neutron stars can, apparently, be born as companions to stars of any possible type and mass.

Only when such companions have a suitable mode of mass ejection available, such a neutron star will appear as an X-ray source to us.

The sample of compact objects in binaries observed on the X-ray sky is therefore, a highly biased sample of the real collection of binaries with compact components in the Galaxy! (van den Heuvel[304], [305]).

I. Spin Evolution as a diagnostic of the mass transfer. Table 12 shows the spin-periods of 32 known X-ray pulsars. A striking fact observed in this table is that, contrary to most known radio pulsars, X-ray pulsars tend to have very long spin periods, of order hundreds of seconds. Spin periods of order of seconds are found only in systems where -from optical and UV observations- there is clear evidence for the presence of accretion disk. These are systems in which (a large part of) the mass transfer is due to Roche-lobe overflow: the low-mass X-ray binaries Her X-1 and 1626-67 and the high mass systems with very narrow orbits Cen X-3, SMCX-1 and LMC X-4. The latter ones are expected to be powered by a combination of stellar wind and beginning atmospheric Roche-lobe overflow, cf. Savonije[241].

The spin-periods of the accreting X-ray pulsars are observed to vary on relatively short time scales. Figure 30 (after Nagase[176]) shows the spin histories of a number of well-known binary X-ray pulsars over almost two decades.

The figure shows that the sources in which an accretion disk is observed (SMC X-1, Cen X-3, U1626-67 and Her X-1) show a more or less continuous decrease of the pulse period on relatively short time scales, of order thousands of years (Cen X-3, SMC X-1) to $\sim 10^5$ yrs (Her X-1).

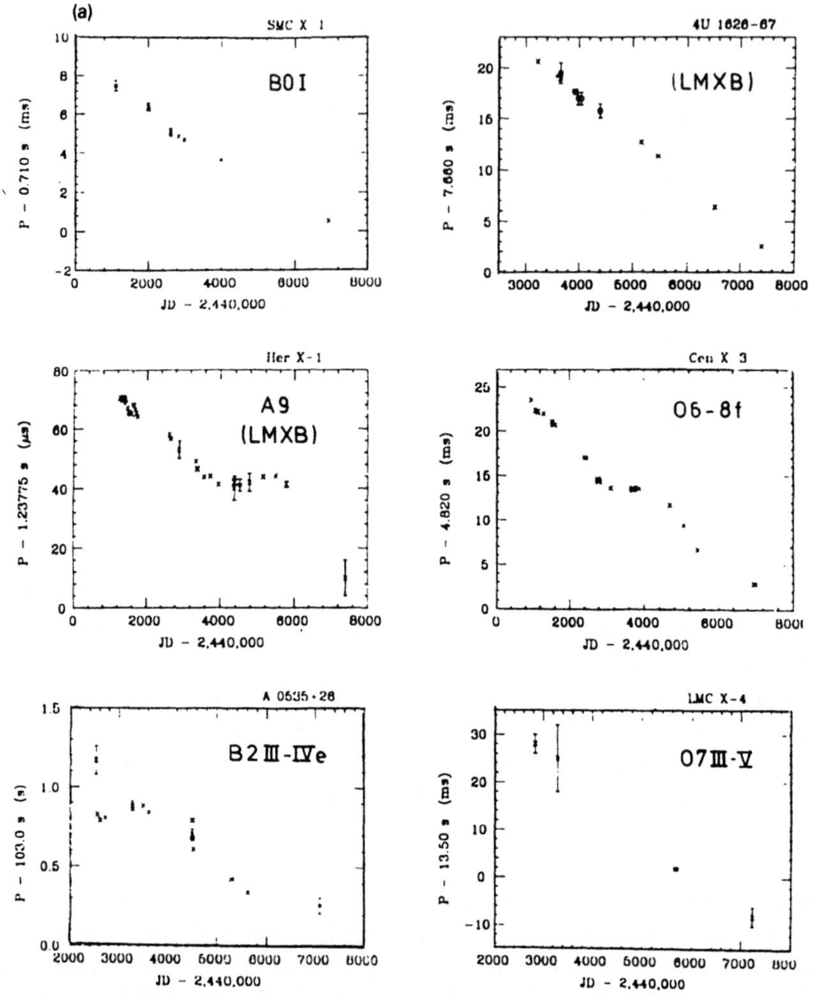

Fig. 30. a.) Pulse period histories of the disk-fed binary X-ray pulsars and one transient X-ray pulsar with a B-emission companion (X0535-26) after Nagase [176]. The disk-fed pulsars have relatively short pulse periods (~ 1 *s*) and show more or less monotonic decrease in pulse period, so-called "spin-up".

This so-called "spin-up" is, of course, due to the angular momentum carried by the accreted disk matter.

On the other hand, sources which are clearly accreting from a wind, such as Vela X-1 (4U0900-40) and 1223-62 (GX 301-2), show erratic variations in the pulse period. We know that these sources are accreting from a wind since their orbits are quite eccentric ($e \sim 0.09$ in Vela X-1 and ~ 0.5 in 1223-62), so the companions cannot be overflowing their Roche-lobes. Also Be-X-ray binaries show alternating spin-ups and spin-downs: during a transient outburst they are often observed to be spinning up, while in quiet episodes between outbursts they are spinning down. We refer to Nagase (1989) for further details.

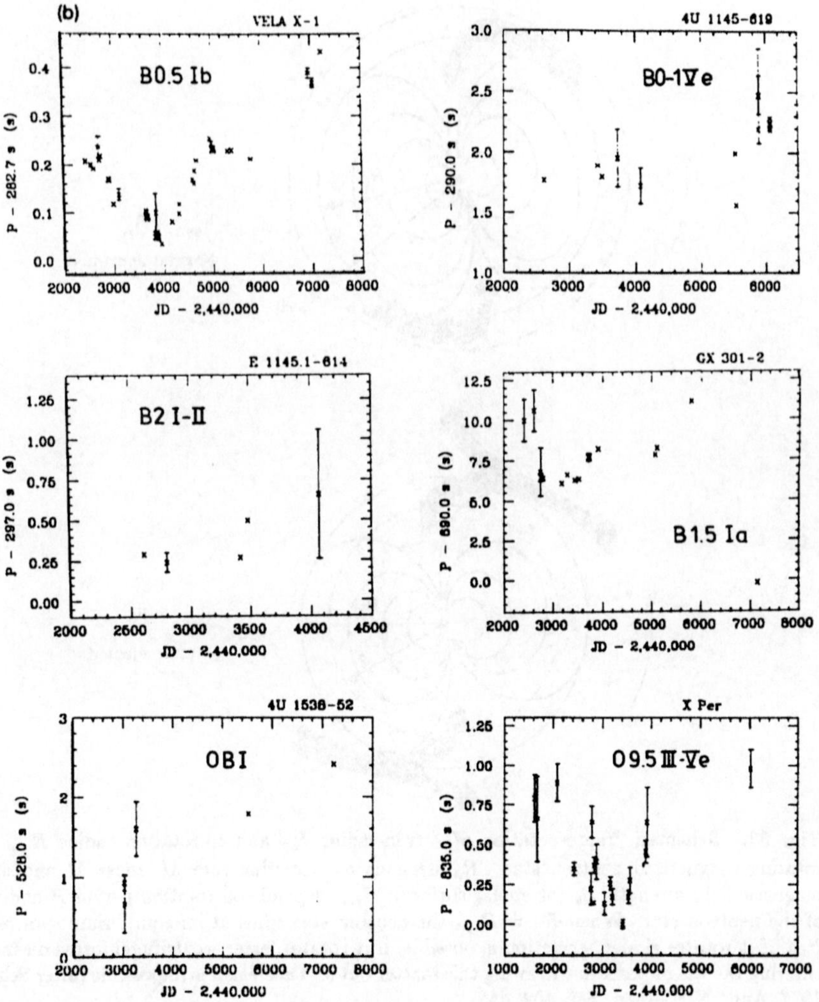

Fig. 30. b.) Pulse-period histories of four wind-fed X-ray pulsars (with early-type supergiant companions) and two X-ray pulsars with B-emission companions. These pulsars have long pulse periods (> 100 *s*) and show erratic pulse-period variations (after Nagase [176]).

J. Equilibrium spin up period/fastest possible spin. The shortest spin period which a magnetic and accreting neutron star can reach for a given surface dipole magnetic field strength B_9 (in units of 10^9 Gauss) and accretion rate $\dot{M}$ is the so-called equilibrium spin period:

$$P_{eq} = (2.4ms) B_9^{6/7} R_6^{16/7} M^{-5/7} (\dot{M}/\dot{M}_{Edd})^{-3/7} \tag{63}$$

(cf. van den Heuvel[307],[313]).

R_6 is the neutron star radius in 10^6 cm, M the neutron star mass in solar units and $\dot{M}_{Edd}$ the Eddington-limit accretion rate. P_{eq} is the spin period at

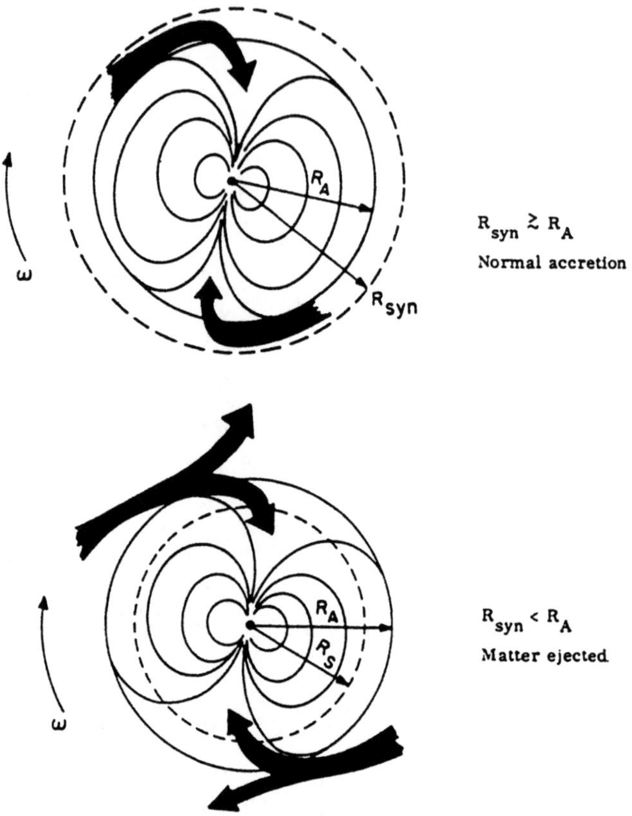

Fig. 31. Schematic representation of Alfvén-radius R_A and co-rotation radius R_{syn} of a rotating magnetized neutron star. R_A depends on accretion rate $\dot{M}$, mass M and dipole magnetic field strength B_9 (at stellar surface); R_{syn} depends on rotation period P and mass of the neutron star. When $R_A = R_{syn}$ the neutron star spins at its equilibrium spin period P_{eq}. If it rotates slower, accretion is possible, if it rotates faster, centrifugal forces on matter entering the magnetoshere will swing this matter out and accretion is impossible (after Schreier 1977, Ann. N.Y. Acad. Sci. 402, 445).

which the matter at the Alvén radius (the radius where the inflowing matter couples to the magnetic field lines, i.e. where the ram pressure of the flowing matter equals the magnetic pressure in the dipolar magnetic field of the neutron star -see the derivation in Bhattacharya and van den Heuvel[18]) co-rotates with the neutron star with just the Keplerian velocity around the neutron star at that distance (fig. 31).

Were the neutron star to rotate faster than with P_{eq} the matter would be centrifugally swung out of the magnetosphere and accretion would be impossible. On the other hand, for $P > P_{eq}$ matter can enter the magnetosphere and accretion is possible. Thus, for a given accretion rate $\dot{M}$, P_{eq} is the shortest possible rotation period for the accreting star.

The absolutely shortest rotation period is reached for $\dot{M} = \dot{M}_{Edd}$. So, for a $1.4 M_{\odot}$ neutron star with $R_6 = 1$, $P_{eq,min}$ is given by:

$$P_{eq,min} = (1.89 ms) B_9^{6/7} \tag{64}$$

For $B = 10^{12} G$ ($B_9 = 10^3$) this period is 0.70 sec. Thus the strongly magnetized neutron star in SMC X-1 must be spinning close to its minimum possible spin-period, as it is accreting at a rate close to $\dot{M}_{Edd}$ (its $L_x \sim 2.10^{38}$ erg/s).

K. Peculiar X-ray binaries with a strong radio emission and relativistic gas outflows: SS433 and Cyg X-3.

a. SS433. SS433 is a peculiar 14th magnitude emission-line star at a distance of about $5 kpc$ coinciding with the X-ray source A1909+04. It is centrally located inside the huge radio shell W50, which resembles a supernova remnant and is also observable in X-rays. Two oppositely directed X-ray beams extend from SS433 to the shell, which is elongated in the direction of these beams. In 1978 Margon[163] discovered that a number of the broad emission lines observed in the spectrum of SS433 change drastically in wave length from night to night.

It was found that these changing lines are blue-shifted and red-shifted components of Balmer and HeI lines. In any given spectrum all red shifted components share the same red shift ($\Delta\lambda/\lambda$), and all blue shifted lines share the same blue shift. Figure 32 (after Margon[164]) shows a sample spectrum. The amplitudes

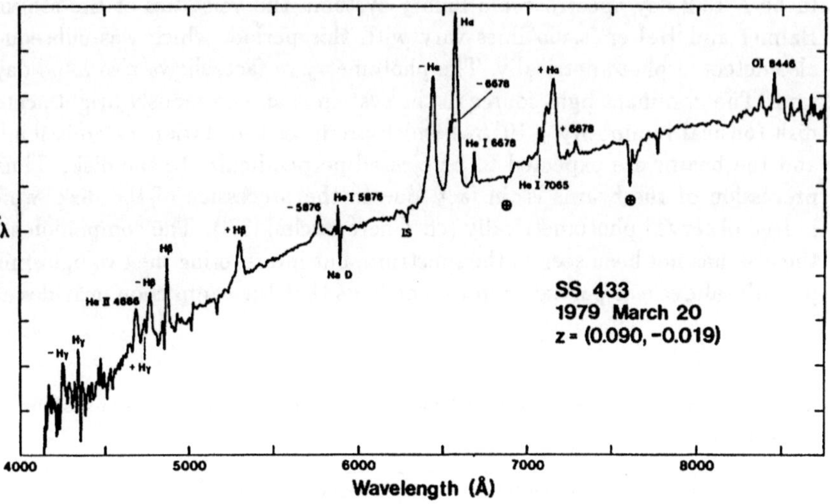

Fig. 32. The spectrum of SS 433 on March 1979, when the pattern of multiple red-and blueshifted emission lines is particularly obvious. Emission lines prefixed with a 'plus' are redshifted features, and the 'minus' prefix denotes blueshifted features. This spectrum was obtained with the Lick 3m Shane reflector. Note the very strong stellar absorptions of Fe II λ 5169 and O I λ 7773, as well as P-Cygni absorption components to Balmer and He I lines in the "stationary" spectral line system. Each division on the ordinate is 0.83 mag (adapted from Margon [164]).

of the Dopplershifts for the red shifted lines range up to 50,000 km/s, those for the blue shifted ones: up to 30,000 km/s. Most surprisingly, these Doppler shifts were found to vary in a cyclical way, with a period of 164 days (see figure 33).

During this cycle, the mean Doppler shift of the red and blue shifted component of each line remains constant $\bar{z} = 0.035$. This was explained by Fabian and Rees[74], who pointed out that narrowly collimated beams ejected in opposite directions from a central object, with constant velocity $v = 0.26c$, could explain the data. The variations in observed Doppler velocities would in this model be due to a change in the inclination angle of the beams with respect to the line of sight. The beam velocity $v = 0.26c$ was derived from the mean red-shift of blue- and red shifted lines, which in this model is simply due to the transverse Doppler effect, and amounts to $\bar{z} = \frac{1}{2}(v/c)^2$ where v is the velocity of the beams (see Milgrom[173]).

Milgrom[173] suggested the 164-day period and explained it in terms of a model in which the beams precess around an axis which is inclined to the line of sight. This model has since been completely confirmed by the observations (see Margon's reviews[164][165]). The beams and their precession have also been observed with radio telescopes, using VLBI techniques.

Table 17 lists the kinematic parameters of the model derived from the stellar observations (Margon 1983); i is the inclination of the precession axis to the line of sight; θ is the opening angle of the precession cone (see Vermeulen[344] for a recent review). An amazing fact is that the outflow velocity in the beams is so constant, $v = 0.26c \approx 80,000 km/s$. Crampton et al.[49]) discovered SS433 to be a 13.2-day spectroscopic binary system: the velocities of the stationary Balmer and HeI emission lines vary with this period, which was subsequently also detected photometrically. The photometry in fact, shows also a 6.5 day period. The dominant light source in the system is an enormously bright accretion disk (optical luminosity $\gtrsim 10^4 L_\odot$) which produces the stationary emission lines, and the beams are expected to be ejected perpendicular to the disk. Thus the precession of the beams is, in fact, due to the precession of the disk, which is indeed observed photometrically (cf. Cherepaschuk[37]). The companion star of the disk has not been seen in the spectrum, but interpreting the two light minima per orbital cycle as partial eclipses, one finds that the companion provides about

Table 17. Kinematic parameters of SS433 (after Margon[164]). Uncertainties quoted are 1-sigma errors.

–	Beam Precession Period	$P = 163^d, 51 \pm 0.13$
–	Outflow velocity in the beams	v/c=0.260±0.0002
–	Inclination angle of the axis of the precession cone with respect to the plane of the sky	$i = 79°.00 \pm 0.15$
–	Angle between the beams and the axis of precession cone	$\theta = 19°.77 \pm 0.15$

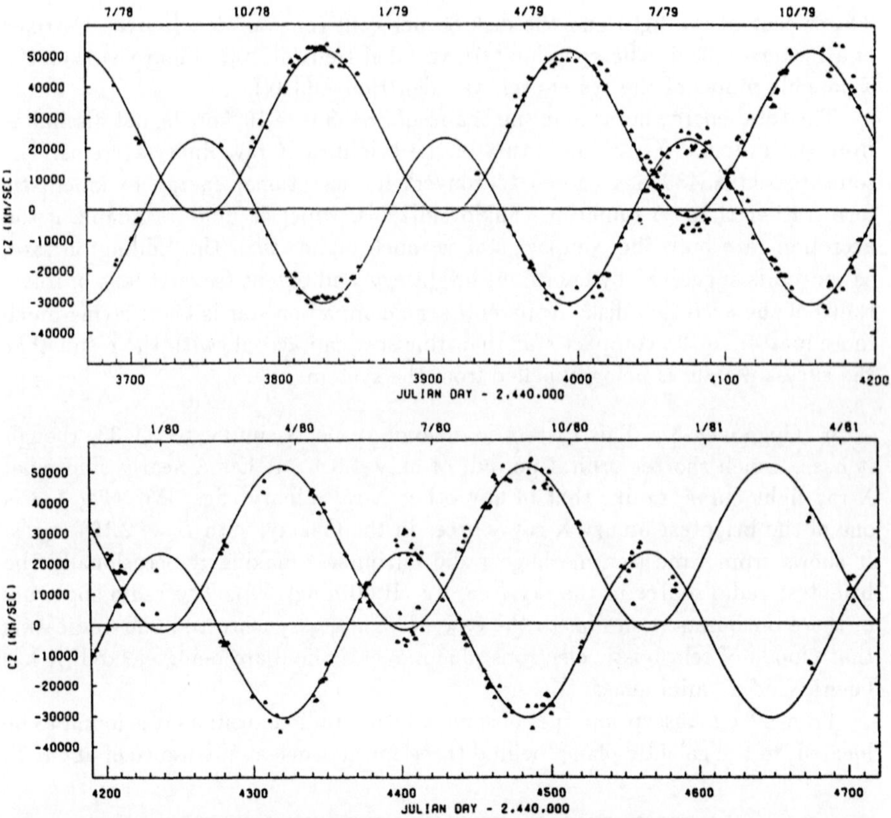

Fig. 33. The values of red- and blueshifts of SS 433 over an almost 3 yr period; displayed are 500 separate Doppler-shift values obtained on 300 separate nights. Most of the data have been obtained by Margon and University of California colleagues, and a few points are from the literature. The solid line is the best-fit to the simple kinematic model, with free parameter values given by table 17 (Margon [165]).

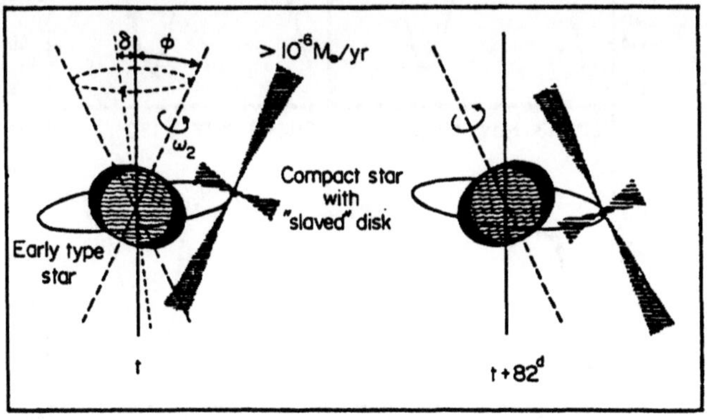

Fig. 34. Outline of slaved-disk model for the 164-day precession of the beams in SS433 (after van den Heuvel et al. [322]).

20 per cent of the light and the disk 80 per cent (cf. van den Heuvel, Ostriker and Petterson[322]; Cherepaschuk[37]; van den Heuvel[310]). Figure 34 sketches a possible model of the system (cf. van den Heuvel[310]).

The total energy emitted in the beams of SS433 is $\sim 10^{39}$ erg/s, not dissimilar from the radiated X-ray luminosity of the brightest X-ray binary systems. For some reason SS433 has chosen to convert its accretional energy to kinematic form rather than to radiation (Shklovski[250]), which is understandable if the accretion rate onto the compact star is much higher than the Eddington rate, as indeed is suggested by enormous brightness and extent (several tens of solar-radii) of the accretion disk. Apparently the companion star is transferring much more matter to the compact star than this star can accept, with the result that the excess matter is being expelled from the system.

b. Cygnus X-3. This system is in some respects similar to SS433, though it has a much shorter orbital period, of only 4.8 h. It has a nearly sinusoidal X-ray light curve, unlike that of any other X-ray binary (fig. 35). Cyg X-3 is one of the brightest binary X-ray sources in the Galaxy, with $L_x \sim 2.10^{38} erg/s$. It shows from time to time huge radio outbursts, making it occasionally the brightest radio source in the sky (see, e.g. Hjellming[109]). The radio spectrum of these outbursts is similar to the case of SS433, a synchrotron one, indicating that clouds of relativistic electrons plus magnetic field are being ejected (it has been called a "mini quasar").

From 21 cm absorption lines observed during radio outbursts it is found to be located, in the galactic plane, behind three spiral arms, at a distance of about 10

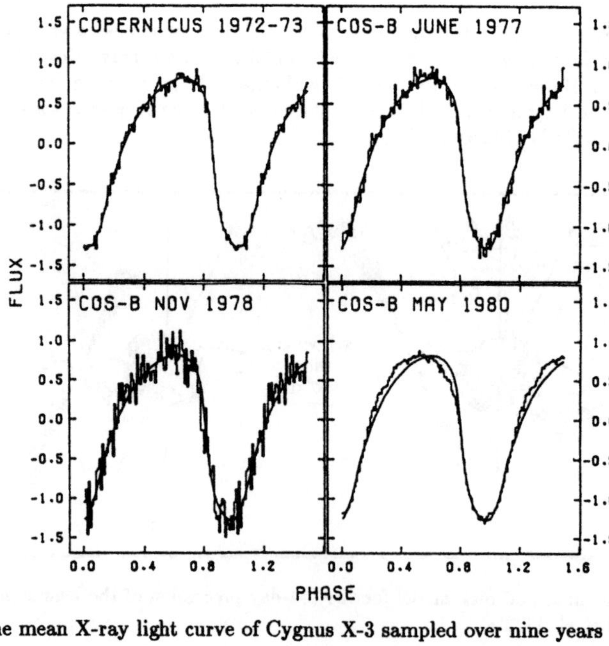

Fig. 35. The mean X-ray light curve of Cygnus X-3 sampled over nine years [27].

kpc. The visual extiction in its direction is at least 20 magnitudes, so no optical counterpart is known. It is, however, a variable infrared source, showing the 4.8^h cyclic orbital variation, in phase with the X-ray light curve. Radio VLBI observations shows that, like SS433, it has small radio jets, presumably due to relativistically outflowing matter.

Recently, high resolution infrared spectroscopy has shown that the infrared spectrum shows only helium and nitrogen lines, no hydrogen, and is very similar to that of a Wolf-Rayet star of type WN7 (see figure 10 and van Kerkwijk et al.[328]). Its absolute infrared luminosity is indeed consistent with it being a population I type Wolf-Rayet star, i.e.: a helium star with a mass of order 5 to 10 solar masses.

The large wind mass-loss rate as observed in Wolf-Rayet stars is consistent with the large rate of increase of its orbital period of $1.6 \times 10^{-6} yr^{-1}$ (see section 2, and van Kerkwijk et al.[328]).

3.6 Binary radio pulsars and related objects.

A. Introduction. Radio pulsars (Hewish et al.[106]) are rapidly rotating strongly magnetized neutron stars, which are formed in a supernova event (Gold[92], [93]; see the review by Srinivasan[257]). The pulsed emission is believed to arise because they emit a highly directional beam of radiation above their magnetic poles, which sweeps around the sky and illuminates a favourably located observer once per rotation period. The energy emission of radio pulsars is rotationally driven.

The rate of increase of the 0.033-second pulse period of the Crab pulsar $\dot{P} = 38 nsec/day$ was measured within a day after its discovery and showed that it loses its rotational energy a rate:

$$\frac{dE_{rot}}{dt} = I\Omega\dot{\Omega} \simeq 10^{38} ergs/s (\sim 2.10^4 L_\odot) \tag{65}$$

where I is the moment of inertia of the neutron star ($\sim 10^{45} g.cm^2$ for a standard neutron star with a mass of $\sim 1.4\ M_\odot$ and a radius of 10 km) and Ω its angular velocity of rotation. With this $\dot{P}$ the pulse period of the Crab pulsar doubles on a timescale of $\sim 10^3$ yrs.

The braking torque is an electromagnetic one: a rotating magnet is expected to emit magnetic dipole radiation (Pacini[191]).

The rate of this emission for the case where the magnetic axis is perpendicular to the rotation axis, and the neutron star is in vacuum is:

$$\frac{dE_{em}}{dt} = \left(\frac{2R^6}{3c^3}\right) B_s^2 \Omega^4 \tag{66}$$

where R is the radius of the neutron star and B_s the strength of the dipole component of the magnetic field at its surface.

Combination of equations (65) and (66), (using $\Omega = 2\pi/P$) yields

$$B_s = \left(\frac{3c^3 I}{8\pi^2 R^6} P\dot{P}\right)^{1/2} = 3.2 \times 10^{19} (P\dot{P})^{1/2} [G] \tag{67}$$

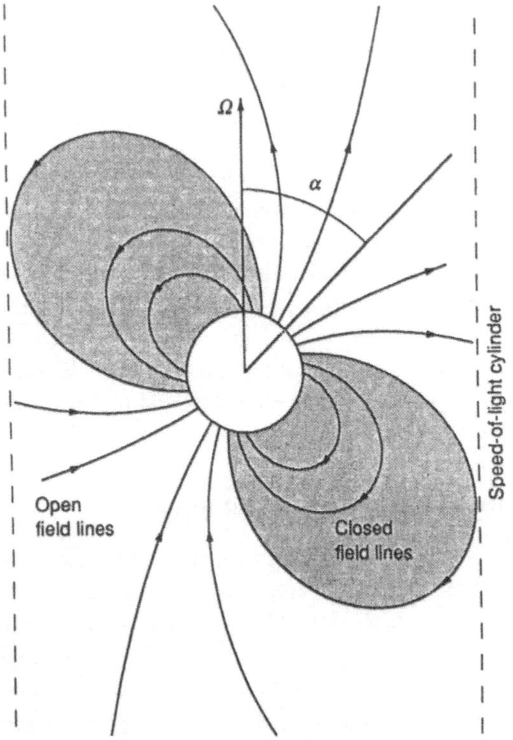

Fig. 36. a.) The basic geometry of a radio pulsar -an inclined dipole rotator. α is the angle between the spin axis and the axis of the magnetic pole. The speed-of-light cylinder corresponds to a distance at which a particle co-rotating with the neutron star will move at the speed of light. Magnetic field lines that are closed within the speed of light cylinder are known as the "closed field lines" (the shaded region) and may contain plasma co-rotating with the star. Ultra-relativistic plasma flows out of the "open" field lines and radiates, generating beamed emission confined to cones around the magnetic poles.

where the numerical coefficient results from using the standard values for the neutron star radius $R \simeq 10^6 cm$ and moment of inertia $I \simeq 10^{45} gcm^2$ with P expressed in seconds ($\dot{P}$ is dimensionless).

Equation (67) shows that the surface magnetic field strength of a neutron star can be estimated from the two observed quantities P and $\dot{P}$ (Ostriker and Gunn, 1969; Gunn and Ostriker, 1970). In deriving eq. (67) it was assumed that the magnetic dipole axis is perpendicular to the rotation axis. The spindown torque is in fact proportional to $B_s^2 \sin^2 \alpha$ where α, called "obliqueness", is the angle between the spin and the dipole axis (figure 36a). This may suggest that eq. (67) underestimates the field strength when α is small, but this appears not to be the case. The reason for this almost certainly lies in the fact that pulsars also lose a substantial amount of energy in a steady wind of relativistic particles. The magnetospheric current associated with this wind gives rise to a spin-down torque that is non-zero even for an aligned rotator (Goldreich and Julian[94]; Beskin et

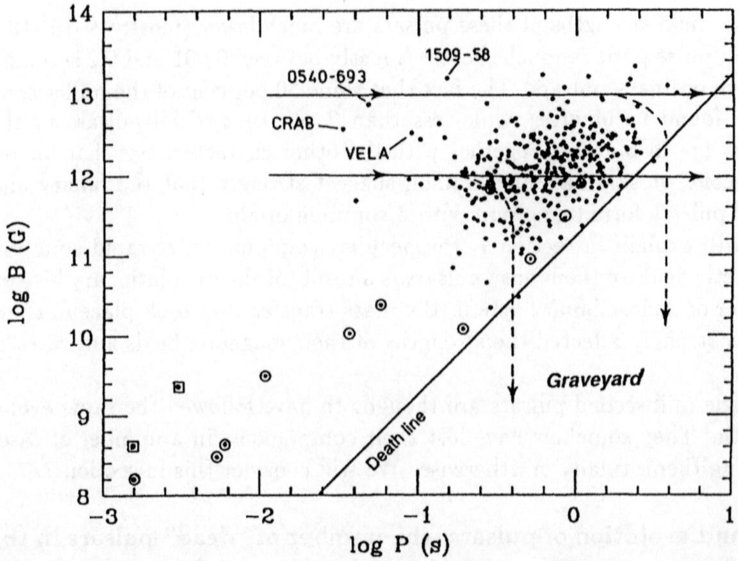

Fig. 36. **b.) The derived magnetic fields ($\alpha\ \sqrt{P\dot{P}}$) of 433 radio pulsars plotted against their spin periods. Pulsars in binaries are shown encircled and millisecond pulsars are shown as squares. The "deathline" corresponds to a polar cap voltage below which pulsar activity is likely to switch off. Four young pulsars in supernova remnants are indicated. Horizontal lines indicate evolutionary tracks of pulsars without magnetic decay. The dashed curves show evolutionary tracks with exponential field decay on 5 Myr timescale.**

al.[17]). The relative importance of the torques due to magnetospheric currents and vacuum dipole radiation is very difficult to predict theoretically because it has not yet been possible to obtain a completely self-consistent and stable model for a pulsar magnetosphere. Nevertheless it was shown by Goldreich and Julian[94] that the spin-down torque on an aligned rotator due to currents can be nearly equal to that on an orthogonal rotator due to vacuum dipole radiation, and this seems to be consistent with recent observations for cases where α is known (for a review and discussion: see Bhattacharya and van den Heuvel[18]). Therefore equation (67) is expected to yield a good estimate of B_s for any value of the obliqueness α.

The spin-down rates of over 400 pulsars are now known. Figure 36b shows the distribution of their spin periods and derived magnetic fields. This is one of the most often used diagrams by means of which the evolution of pulsars can be described.

Pulsars in binaries are marked in figure 36b with open circles and single millisecond pulsars with squares. The figure shows that the majority of pulsars are situated in an "island" with spin periods between 0.2 seconds and several seconds and magnetic fields $\sim 10^{11.5}$ to $10^{13.5}$ G.

It is immediately clear from this diagram that the binary and millisecond pulsars behave differently from the majority of the pulsars: most of the binaries and all of the millisecond pulsars fall outside of the "island" of ordinary pulsars:

the magnetic field strengths of these pulsars are much lower (mostly $\sim 10^8$-10^{11} G) and their pulse periods much shorter (mostly between 0.001 and 0.2 seconds) than those of ordinary pulsars. The fact that some 50 per cent of the millisecond pulsars are found in binaries while less than 2 per cent of all pulsars in the galactic disk are in binaries, together with the other characteristics that binary and millisecond pulsars have in common suggest strongly that the binary and millisecond pulsars form one group with a common origin.

As we will explain in section 7, the peculiar combination of rapid spin and weak magnetic fields of the binary pulsars is a result of their evolutionary history as a member of a close binary sytem: the mass transfer that took place in these systems has strongly affected the strengths of their magnetic fields and rates of their spins.

The single millisecond pulsars are thought to have followed the same evolutionary path. They somehow have lost their companions, in a number of cases by disrupting them, tidally or otherwise. We will consider this in section 7.

B. Birth and evolution of pulsars -the number of "dead" pulsars in the Galaxy. Neutron stars are expected to be the remnants of stars more massive than about 8 $M_\odot$, and are formed in a supernova event, the explosive ejection of a star's outer layers that accompanies the collapse of its nuclearly burned-out core to a neutron star. The youngest known pulsars are still surrounded by their expanding supernova shells (as indicated in fig. 36b).

The tracks in Figure 36b depict schematically the evolution of a newborn pulsar the cases that the magnetic field of the neutron star does not decay (full lines), and that it decays on a relatively short time scale, of order 5.10^6 yrs (dashed lines).

In the first case (B_s=const) the pulsar moves to the right along a horizontal track, as indicated. In this case the pulse period varies with time as:

$$P = 3.1 \times 10^{-20} B_s \sqrt{2t}, \tag{68}$$

as follows from eq. (67), and the characteristic spindown-age of the pulsar is simply given by:

$$t = (3.2 \times 10^{19})^2 P^2 / 2B_s^2 = P/2\dot{P} \tag{69}$$

In the case that the field decays exponentially on a timescale of $\sim 5.10^6$ yrs the track begins to bend downward after a few million years, as indicated by the dashed curves. As the braking torque decays rapidly with time in this case, the rotation period finally becomes constant, i.e. the track becomes vertical. To the right of the diagonal line labelled "deathline" in the $B_s - P$ diagram of figure 36b no pulsars are found. Apparently, after crossing this line, a pulsar stops pulsing. This is understandable because for a certain combination of B and P the electric field generated by the rotating magnetic dipole, becomes too weak to draw charges (electrons) out of the crust of the neutron star.

As the acceleration of these charges along the magnetic field lines is thought to be the source of the beamed radio emission from the pulsar, one expects the pulsar to stop emitting pulsed radiation after it has crossed the deathline. As lines of constant electric field are lines of $B/P^2 = const$, the equation for this line:

$$logB_s = 2logP + const. \tag{70}$$

The region of the $B_s - P$ diagram to the right of the deathline is called the "graveyard".

A typical ordinary pulsar such as the Crab-pulsar, with $B_s \sim 3.10^{12}$ G joins the "island" of pulsars within 10^5 years after its birth and reaches the deathline within $\sim 10^7$ years after its birth.

Most of the ordinary pulsars in the island of normal single pulsars in figure 36b therefore were born between 10^5 and 10^7 years ago as is also indicated by their characteristic spindown ages given by eq. (69).

As the type II supernova rate in a late type spiral galaxy such as our own is about one in 30-50 yrs, one expects the total number of "living" ordinary pulsars in our galaxy to be about 2-3 × 10^5.

This number agrees well with estimates derived from the observed pulsar population within a few kiloparsecs from the sun (cf. Taylor and Stinebring[277]) if a "beaming factor" of order 3 to 5 is assumed. (The latter means that, due to the limited opening angle of the radiation beams of the pulsar, we can observe on Earth only about 20 to 30 per cent of all pulsars; for the others the radiation beams miss the Earth).

It will be clear from the above that the vast majority of the neutron stars that were formed since the birth of our Galaxy, some 2.10^{10}yrs ago, have long since disappeared into the graveyard and thus have become unobservable. Assuming a SN rate of one per 30 to 50 yr throughout the lifetime of the galaxy the total number of defunct pulsars -in the graveyard- will be ~ (4-7) ×10^8. This number should, however, be corrected upwards to take into account the rapid heavy-element enrichment which the Galaxy underwent in the first 2-5 billion years of its life (cf. Sandage[240]). In that time the supernova rate must have been very much larger than than at present. Judging from the heavy-element enrichment in globular clusters in that period a correction factor of 2 or more is needed which brings the total number of "dead pulsars" (old neutron stars) in the graveyard to at least ~ 10^9 .

For comparison, the number of white dwarfs in the galaxy is probably of order 10^{10}, while the total mass in the form of "normal" stars is ~ $(1-2) \times 10^{11}$ $M_\odot$. Hence, neutron stars make up of order one per cent of the mass of the Galaxy -and (very) old neutron stars outnumber active radio pulsars by a factor of order $10^{3.5}$.

C. Neutron Star magnetic fields and their evolution. **Magnetic fields of order $10^{11.5}$ to $10^{13.5}$ G must have been generated in all neutron stars within the first few hundred years following their birth. This follows from the fact that if newborn neutron stars also were to have fields weaker than about $10^{11.5}$ G, one would expect to find many ordinary pulsars in the region below $10^{11.5}$ G in the B_s *vs* P diagram of figure 36b, because pulsars with such fields move much**

slower towards the right through the diagram than pulsars with $B_s \sim 10^{12} - 10^{13.5}$ G (see equation 69). Hence, they remain observable tens to hundreds of times longer than the strong-field pulsars. Consequently, even if only a few per cent of pulsars were born with such weak fields, they would still in the B_s *vs.* P diagram outnumber the much shorter lived pulsars with $B_s \simeq 10^{12} - 10^{13.5}$ G.

Since this is clearly not the case the vast majority (99 per cent or more) of the neutron stars must be born with $B_s \geq 10^{11.5}$ G.

With "born" we mean here that fields of this order are present in them within less than 10^3 yrs after their birth -since all three pulsars known in supernova remnants of that age (the Crab-pulsar, PSR 1509-58 and PSR 0540-693, see figure 36b) have fields in the range $10^{12.5}$ to $10^{13.5}$.

How these fields were generated is not known with certainty although some plausible models have been suggested, such as magnetic flux conservation during core collapse or magneto-hydrodynamic field amplification by dynamo action in the differentially rotating hot liquid neutron star during the first few seconds to minutes after its birth, before its crust solidified (Flowers and Ruderman[80]).

After the crust solidified the electric currents associated with the fields will have anchored the fields in the crust; also a quantized very strong field may exist in the super-conducting core (cf. Pines[208]). Ohmic dissipation of the crustal currents may cause a gradual decay of the observable surface field but although the timescale for this decay was originally thought to be short ($\sim 5.10^6$ yrs, cf. Gunn & Ostriker[97]), modern work shows that the time for decay below 10^{12} G is probably very long, $\geq 10^8$- 10^9 years (Chanmugam[36]).

The recent finding of fields as strong as 10^{10} to 10^{12} G in neutron stars as old as 10^9 yrs (PSR 0655+64, Kulkarni[146]; Hercules X-1, cf. Verbunt et al.[343]) seems to rule out spontaneous magnetic field decay of isolated neutron stars, on timescales much shorter than 10^9 yrs. Also recent statistical analyses of the space distribution and magnetic field distribution of neutron stars -using an improved model for determining pulsar distances- no longer require field decay on timescales shorter than 10^8 years and, in fact, do well without any field decay (Bhattacharya et al.[19]; Wakatsuki et al.[346]). On the other hand, neutron stars in binaries clearly often show evidence for considerably weakened fields. This, however, can be interpreted in terms of external causes such as accretion of matter from a companion star or a special rotation history (cf. Taam & van den Heuvel[268]; Shibazaki et al.[249]; Romani[232]; Srinivasan et al.[258]). See Bhattacharya and van den Heuvel[18] for a recent review.

D. The Binary and Millisecond radio pulsars. Table 18 lists a representative sample of the binary and millisecond pulsars known to date. The -on average- much faster spin and much weaker magnetic fields of the binary pulsars as compared to the "garden variety" single pulsars were already mentioned in section A. The number of binary pulsars known to date in the galactic disk is about a dozen. The binary pulsars appear to fall into two groups that differ in a number of characteristics (van den Heuvel and Taam[325]; van den Heuvel[313] as depicted in figure 37:

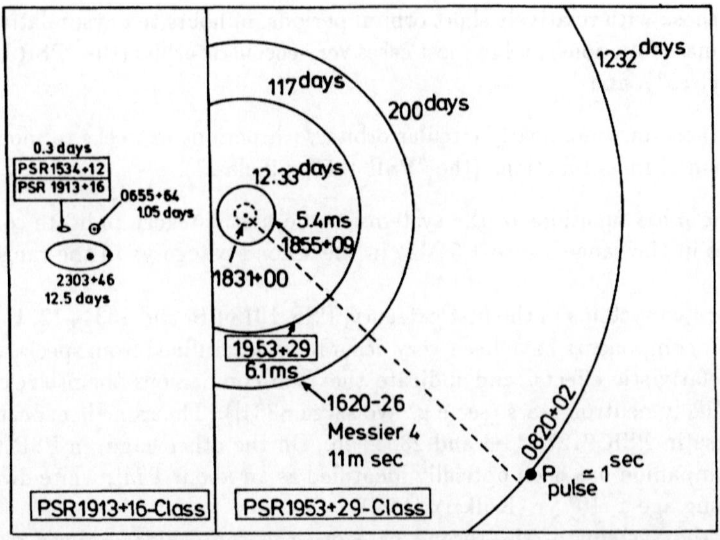

Fig. 37. The two classes of binary radio pulsars. Left: The PSR 1913+16-class of systems tend to have narrow and very eccentric orbits; the companion of the pulsar is itself a neutron star or a massive white dwarf. Right: The PSR 1953+29-class systems tend to have wide and circular orbits; here the companion stars have a low mass, in the range 0.2-0.4 $M_\odot$, or even smaller -and most most probably are helium white dwarfs (see text).

Table 18. Binary Radio Pulsars (see Verbunt and van den Heuvel[342]).

Position	P (ms)	$log\tau_c$ (yr)	$logB$ (G)	P_b (d)	e	M_e ($M_\odot$)
high-mass binary radio pulsar						
1534+12	37.9	8.4	10.0	0.42	0.2737	1.36
1913+16	59.0	8.0	10.4	0.32	0.6171	1.39
0655+64	195.6	9.7	10.1	1.03	<0.00005	>0.7
2303+46	1066.4	7.5	11.9	12.34	0.6584	1.5
low-mas binary radio pulsars						
1957+20	1.6	9.2	8.2	0.38	< 0.001	0.02
1831−00	521.0	8.8	10.9	1.81	<0.005	0.07*
1855+09	5.4	9.7	8.5	12.33	0.000021	0.23
1953+29	6.1	9.5	8.6	117.35	0.00033	0.22*
0820+20	864.9	8.1	11.5	1232.40	0.0119	0.23*
antediluvian radio pulsars (=pulsar before recycling)						
1259−63	47.8			11.33?	0.98	Be
1820-11[a]	297.8	6.5	11.8	357.8	0.795	0.8*
single recycled radio pulsar						
1937+21	1.6	8.4	8.6			
1257+12 [b]	6.2					

[a] this pulsar is tentatively listed as antediluvian; it may be a high-mass binary radio pulsar

[b] this pulsar has two planets

a. those with relatively short orbital periods, of hours to days; relatively large mass functions, and in most cases very eccentric orbits (the "PSR 1913+16-class"), and

b. those in wide, nearly circular orbits with periods of weeks to months, and small mass functions (the "PSR 1953+29-class").

The mass functions of the systems in the first category indicate companion masses in the range 1.0 to 1.5 $M_\odot$, in the second category: in the range 0.2-0.4 $M_\odot$.

For two systems in the first category, PSR 1913+16 and 1534+12, the masses of both components have been very accurately determined from special and general relativistic effects, and indicate that the companions themselves are also most likely neutron stars (see e.g. Wolszczan[364]). The same is expected to be the case in PSR 2127+11C and 2303+46. On the other hand, in PSR 0655+64 the companion has been optically identified as an about 1 $M_\odot$ white dwarf, with a cooling age $> 10^9$ yr (Kulkarni[146]).

In the systems of the second category the companions appear to be low-mass white dwarfs, as follows from the optical identification of PSR 0820+02 (Kulkarni[146]) as well as as evolutionary considerations (see below). Most of the binary pulsars in the second category are millisecond pulsars and their magnetic fields are abnormally weak: $B_s < 10^9$ G.

Figure 36b shows that the positions of these pulsars in the B_s *vs.* P diagram are very similar to those of the isolated millisecond pulsars PSR 1937+21 (which has the extraordinary short period of 1.55 ms) and PSR 1821-24, in the globular cluster M28, which has a period of 3 msec. The extreme stability of the pulse period of PSR 1937+21, which has a spin-down timescale of $\sim 5.10^8$ yrs, has allowed a variety of tests of general and special relativity in the solar system that were previously impossible (Taylor[275] [274]; Backer and Hellings[8]).

About half of all known millisecond pulsars (75% of the ones in the field and 45% of those in globular clusters) are in binaries. his is an extraordinarly large percentage since only 2 per cent of the pulsars in the galactic disk are found in binaries (we exclude here the pulsars in the globular clusters since these are thought to all be very old, and underwent a history different from that of the ordinary pulsars (cf. Bhattacharya and van den Heuvel[18]).

It thus seems that the millisecond pulsars and the binary pulsars are closely related to one another and have a similar origin.

E. Mass determination of binary radio pulsars using general relativity. Pulsars are such accurate clocks, and the double neutron star systems with eccentric orbits PSR 1913+16, PSR 1534+12 and PSR 2127+11 are so close (orbital periods $\sim$ 8 to 10 hours) that four general relativistic effects are observable in them. Three of these four can be measured with high accuracy. These effects turn out to be functions only of the masses M_1 and M_2 of the two stars and of the orbital parameters P_b and e, which can be measured very accurately by Doppler tracking of the pulses.

As a result, the measurement of the four relativistic parameters by careful monitoring of the pulse arrival times on Earth, yields four different equations for the two unknown masses M_1 and M_2. Hence, M_1 and M_2 are in fact over-determined. For the two binary pulsars for which this was done (1913+16 and 1534+12), it has been found that the solutions of these four equations are fully consistent with one another (cf. Taylor and Weisberg[278]; Wolszczan [363] [364]). One of the relativistic effects is the rate of decrease $\dot{P}_b$ of the orbital period due to the emission of gravitational waves. In fact, since the existence of gravitational waves has not yet been demonstrated experimentally (i.e. by using detectors on Earth), the equation for this relativistic effect should not yet be used to determine the masses of the components. Instead, one can determine the masses from the three well-established general relativistic effects:

1. the rate of advance of periastron $\dot{\omega}_{GR}$,

2. the graviatational redshift parameter γ (in fact a combination of redshift and transverse doppler effect),

3. the Shapiro time delay -which yields the orbital inclination i.

Only the first two of these can be measured with very high accuracy (uncertainty of order < 1 per cent). These are therefore used to accurately determine the masses M_1 and M_2. Inserting the thus obtained values into the general relativistic formula for $\dot{P}_b$ one then obtains a theoretical prediction for the value of this parameter. Comparison with the observed value of $\dot{P}_b$ has shown the latter to be -within the observational errors (< 1 per cent) identical with the predicted one. Thus 16 years of Doppler tracking of the pulsar PSR 1913+16 has shown with certainty the existence of orbital shrinking of close binaries by the emission of gravitational radiation (Taylor and Weisberg[278]). For a detailed treatment of the relativistic effects in binary pulsars we refer to the book of Shapiro and Teukolsky[248] and to the review by Backer and Hellings[8]. The most recent reference on the mass determination is Wolszszan[364]. We briefly treat here, as an example, the mass determination from the parameters $\dot{\omega}_{GR}$ and γ.

Figure 38 shows how the longitude of periastron ω_o is determined from the *shape* of the observed radial velocity curve of the pulsar. For $\omega_o = 0°, 180°$, the curve is symmetrical in shape with respect to periastron passage, for $\omega_o = 90°, 270°$ it has a saw-tooth shape.

The observed Doppler delay curve yields the five orbital elements: $K_1 = v_1 \sin i$ (radial velocity amplitude), P_b, e, ω_o and T_o, where T_o is the time of periastron passage. e and ω_o follow from the shape of the curve.

Integration of the radial velocity over the orbital period yields $a_1 sini$ (the projected semi-major axis of the observed pulsar). Also, the mass function $f(m)$ can be determined (see table 19).

The equation for the General Relativistic periastron motion is (cf. Shapiro and Teukolsky[248]):

$$\dot{\omega}_{GR} = \frac{6\pi G M_2}{a_1(1-e^2)P_b c^2} \tag{71}$$

Table 19. Parameters of the two close and eccentric binary radio pulsar systems PSR 1913+16 and PSR 1534+12 (after Taylor and Weisberg[278] and Wolszczan[364]). Figures in parantheses are uncertainties in last digits quoted. æ æ

	PSR 1913+16	PSR 1534+12
-Pulsar Period, P (s)	0.059029997929883 (7)	0.037904403665 (4)
-Period derivative $\dot{P}$ (ss^{-1})	8.62629 (±8) × 10^{-18}	2.43 (8) ×10^{-20}
-Right Ascencion (1950.0)	$19^h13^{min}12^s.46549$ (15)	$15^h34^{min}37s.686$ (3)
-Declination	$16^{circ}01'08'$ (3)	$12^{circ}05'45''.23$ (3)
-Epoch (JED)	2,445,888.745517962	2,448,200.0
-Projected Semi-major axis, $a_1 \sin i$ (lights)	2.341749 (19) = 1.0094 $R_\odot$	3.72943 (3)
-Eccentricity e	0.617127	0.2736761 (7)
-Epoch of periastron τ_0 (JED)	2331.446132 (3)	2448199.8121643 (3)
-Orbital period, P_b (s)	27906.98163 (2) (7.75h)	36351.7027 (3) (=10.1h
-Longitude of periastron, ω	178°.983 (4)	264°.672 (6)
-Mass function $f(m_1, m_2)$	0.132 $M_\odot$	0.3146 (2) $M_\odot$

Measured Relatitivistic parameters and derived masses and other parameters.

-Periastron advance, $\dot{\omega}$	4 °.22660 (4)	1.°7562(15)/yr
-Time dilation, Gravitational Redshift, γ	4.288 (10) ms	1.7 (4) ms
-Rate of decrease of orbital period wave emission, $\dot{P}_b$ (ss^{-1})	2.427 (30) ×10^{-12}	?
-Total mass $M(M_\odot)$	2.82827 (4)	2.679 (3)
-Companion mass, m_2 $(M_\odot)$	1.442 (3)	1.36 (3)
-Pulsar mass, $m_1 = M - m_2$	1.386 (3)	1.32 (3)
-Orbital inclination, m_i	0.73 (4)	~ 74°
-Pulsar surface magnetic field, B_s (G)	2 ×10^{10}	1 ×10^{10}
-Pulsar characteristics age, τ_c (yr)	10^8	2.4 ×10^8

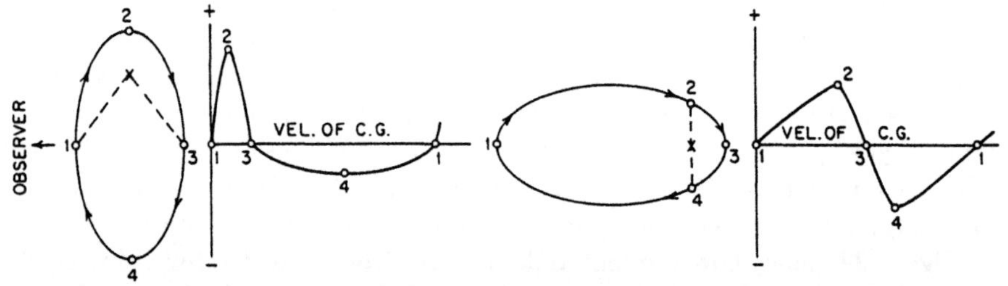

Fig. 38. The shape of the radial velocity curve of an eccentric binary depends on the orientation of the major axis of the orbit with respect to the direction to the Earth. Above, on the left, a star travels in an elliptical orbit whose minor axis points toward the observer. The speed in the orbit varies in accordance with the law of areas, and the time taken to travel from 1 to 3 is much smaller than from 3 to 1. On the right, the major axis of the orbit points toward the observer. The time from 1 to 3 is now equal to the time from 3 to 1, but the time from 1 to 2 and 4 to 1 is longer than from 2 to 3 and 3 to 4: the velocity curve has a saw-tooth shape.

With Kepler's laws: $\frac{M_2}{a_1} = \frac{(M_1+M_2)}{a_1+a_2}$ and $(a_1 + a_2)^3 = 4\pi^2 G(M_1 + M_2)P^2$, such that eq. (71) transform into:

$$\dot{\omega}_{GR} = \frac{(M_1 + M_2)^{2/3} 6\pi G^{2/3}}{(1 - e^2)P_b} \left(\frac{2\pi}{P_b}\right)^{2/3} \tag{72}$$

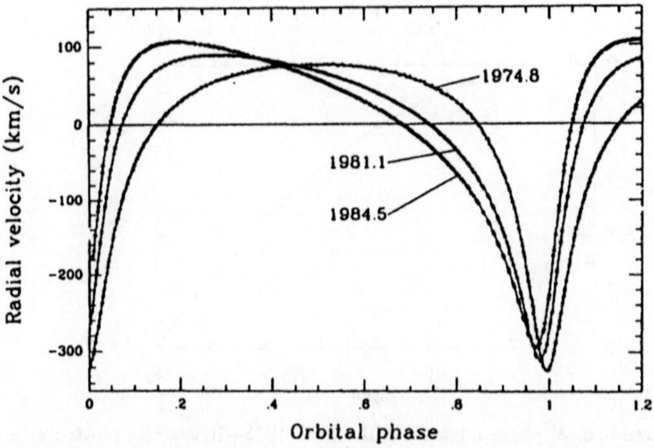

Fig. 39. Velocity curves observed for PSR 1913+16 at three different epochs over 10 years (Taylor [275]).

In the right hand member all quantities except $(M_1 + M_2)$ are known; since $\dot{\omega}_{GR}$ is known from the observations, $(M_1 + M_2)$ can be determined very accurately. For PSR 1913+16, $\dot{\omega}_{GR}$ turns out to be 4.22660 degrees per year, yielding $M_1 + M_2 = 2.82827(\pm 4) M_\odot$.

Figure 39 shows how the shape of the radial velocity curve changed between 1974 and 1984 due to the change in ω_o; from the comparison of these graphs $\dot{\omega}_{GR}$ was determined.

Similarly, γ is given by the equation:

$$\gamma = \frac{GM_2(M_1 + 2M_2)eP_b}{2\pi c^2(M_1 + M_2)^2 a_1} = \frac{G^{2/3}(M_1 + 2M_2)eP_b^{1/3}}{(2\pi)^{5/3}c^2(M_1 + M_2)^{1/3}} \tag{73}$$

and can be measured accurately. Since all quantities in the right hand member except $M_1 + 2M_2$ are known, $M_1 + 2M_2$ can be solved by measuring γ, and M_1 and M_2 can now be solved by combining equations (72) and (73). This yielded the values in table 19 for the systems PSR 1913+16 and PSR 1534+12. These are the most accurately known stellar masses in all of astronomy up to now.

F. Confirmation of the existence of gravitational radiation. The rate of decrease of the orbital period due to gravitational radiation losses according to Einstein's General Relativity is given by (Shapiro & Teukolsky 1983):

$$\dot{P}_b = -\frac{192}{5}\frac{\pi G^{5/3}}{c^5(M_1 + M_2)^{1/3}}\frac{M_1 M_2 f(e)}{} \left(\frac{2\pi}{P_b}\right)^{5/3} \tag{74}$$

with:

$$f(e) = (1 + \frac{73}{24}e^2 + \frac{37}{96}e^4)(1 - e^2)^{-7/2} \tag{75}$$

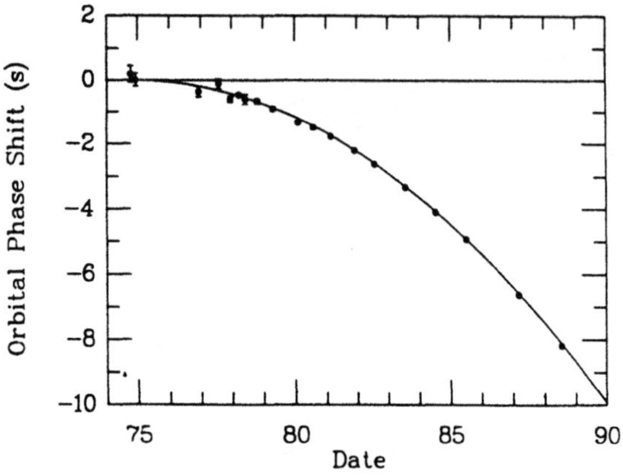

Fig. 40. Measured orbital phase advance in the PSR 1913+16 system, plotted as a function of the time elapsed in years since 1974.8. The parabola gives the expected rate of orbital decay from gravitational radiation (Taylor & Weisberg [278]).

$\dot{P}_b$ can be accurately measured by monitoring the times of periastron passage. Figure 40 (the dots) shows the measured times of periastron passage, as compared to those for a constant orbital period.

The thus measured value is: $\dot{P}_b = (-2.427 \pm 0.026) \times 10^{-12} ss^{-1}$.

The theoretically predicted value, obtained by inserting the masses obtained from $\dot{\omega}_{GR}$ and γ for PSR 1913+16 into eq. (74) yields $\dot{P}_b = -2.40216 \times 10^{-12}\ ss^{1}$. The observed and theoretical values are in excellent agreement (within the observational error of $\pm$ 1 %), confirming the existence of gravitational radiation (Taylor & Weisberg[278]).

4. Overview of the evolution of single stars.

4.1 Why stars shine and evolve: the virial theorem.

A non-rotating globe of ideal monatomic gas without energy sources and in hydrostatic equilibrium obeys the simplest form of the virial theorem

$$2E_{th} + E_{pot} = 0 \tag{76}$$

where E_{th} is the thermal energy content of the globe and E_{pot} its gravitational potential energy ("binding energy"). Equation (76) can be derived from the equation of hydrostatic equilibrium, as can be found in any textbook on stellar interiors (e.g.: Chiu[41]; Cox and Giuli[48]; Clayton[43]; Kippenhahn and Weigert[133]). E_{th} is given by

$$E_{th} = \frac{3}{2} Nk\overline{T} = \frac{3}{2} M(\Re/\mu)\overline{T} \tag{77}$$

where N is the number of particles in the star, k is Boltzmann's constant, $\overline{T}$ is the mean temperature of the gas, M is the mass of the globe, $\mathcal{R} = k/m_H$ the ideal gas constant, and μ is the mean particle mass, in units of the mass m_H of the hydrogen atom.

E_{pot} is given by

$$E_{pot} = -\alpha \frac{GM^2}{R} \tag{78}$$

where R is the stellar radius, G is the constant gravity and α is a constant of proportionality of order unity, which depends on the precise density distribution in the star (see e.g. Cox and Giuli[48]).

Substitution of equations (77) and (78) into eq. (76) yields

$$\overline{T} = \alpha \frac{GM}{R} \left(\frac{\mu}{3\Re}\right) \tag{79}$$

which shows that for a given stellar mass, the mean interior temperature depends only on the radius, and increases when the star contracts. Equations (76) and (79) imply that a gas globe in hydrostatic equilibrium and without energy sources must contract indefinitely -or until it ceases to obey the ideal gas law- for the following reasons.

Its temperature is above 0 K, which implies that it will radiate, and thus will lose energy. This energy loss goes at the expense of its total energy content which, by using eqs. (76) and (78), is given by:

$$E_{tot} = E_{th} + E_{pot} = \frac{1}{2} E_{pot} = -\frac{\alpha GM^2}{2R} \tag{80}$$

Since E_{tot} is < 0, a decrease in E_{tot} implies according to eq. (80) that R must decrease. The heat losses from the surface therefore force the star to contract. But, according to eq. (79), this contraction implies that the mean interior temperature $\overline{T}$ of the star will increase. Thus, while the star tries to cool itself by radiating away energy from its surface, it gets hotter instead of cooler! The star has a "negative heat capacity". Clearly, this cannot be a stable situation: the gas globe, starting out with a very large radius -as part of interstellar cloud- is forced by radiation losses into a vicious circle of contraction and heating. The more it radiates to cool itself, the more it will contract, the hotter it gets and the more it is forced to go on radiating.

It is this "vicious virial circle" (depicted in figure 41) which in a nutshell explains how and why an interstellar gas globe, once it has started contracting, must finally end its life as a compact object, either a white dwarf, black dwarf, neutron star or black hole.

In the meantime the globe may spend a considerable amount of time in intermediate stages, which are called "the main-sequence", "the giant branch" etc. These stages are, however, only temporary "stations" in the contraction history of the globe. It is important to realize that stars do not shine "because they are burning nuclear fuel". They shine because they are hot, and they obtained their high interior temperature due to their history of gravitational contraction.

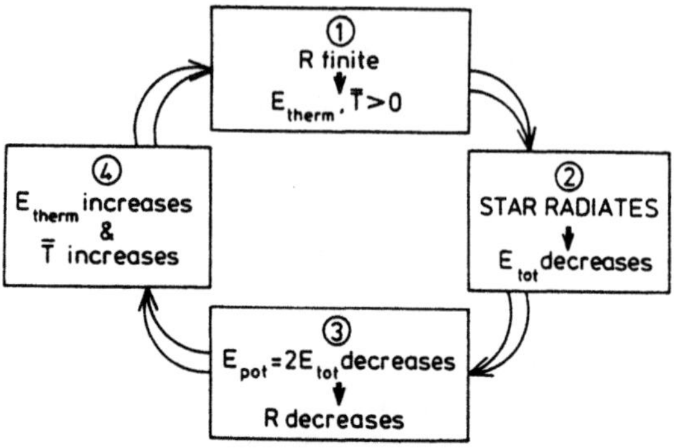

Fig. 41. The virial theorem (VT) makes that a globe of ideal gas ($\gamma = 5/3$) in hydrostatic equilibrium once it has a finite radius is forced to contract indefinitely. (1) Because of its finite radius it has, according to the VT, a positive temperature $\overline{T}$, and therefore it radiates, causing its total energy to decrease (2) which, according to the VT causes its potential energy and radius to decrease (3) which, according to the VT causes its mean temperature $\overline{T}$ to increase (4) which causes it to radiate even stronger (1), causing it to contract further, etc.

Equations (76) and (80) show that for an ideal monatomic gas globe half of the gravitational potential energy lost during the contraction of the globe is converted into thermal energy, and the other half -the loss in total energy- is radiated away.

In the above we have considered only the case of an ideal monatomic gas, i.e.: $\gamma = C_p/C_v = 5/3$. This is an excellent approximation for globes of non-degenerate ionized hydrogen and helium, which all normal stars are. It is useful, however, also to have the above equations for other values of γ. The generalized forms of equations (76), (77) and (80) are respectively, as one can easily derive, see for example Cox and Giuli[48]:

$$3(\gamma - 1)E_{th} + E_{pot} = 0 \tag{76a}$$

$$E_{th} = \frac{1}{(\gamma-1)} NkT = \frac{1}{\gamma-1} M(\Re/\mu)\overline{T} \tag{77a}$$

and $$E_{tot} = E_{th} + E_{pot} = \frac{(3\gamma-4)}{3(\gamma-1)} E_{pot} \tag{80a}$$

To have a gravitationally bound star, E_{tot} should be negative which, since $E_{pot} < 0$ implies that $\gamma > 4/3$. For $\gamma \leq 4/3$ the star cannot be in hydrostatic equilibrium, i.e.: it collapses or explodes. This is important for example when the star becomes fully convective, or if it becomes relativistically degenerate: in both cases γ=4/3, and hydrostatic equilibrium becomes impossible.

4.2 The onset of nuclear burning: the hydrogen-burning station

The vicious circle of contracting and getting hotter of a globe of ideal gas can only be broken if in the interior an energy source appears that produces precisely as much radiation energy as the globe loses from its surface. This occurs when the central temperature of the globe reaches a value of $\sim 10^7$ K, and H-fusion begins. In view of the steep temperature dependence of the nuclear energy generation rate:

$$L_{nuc}(:)T^n \tag{81}$$

where n has a value ranging from 4 (p-p reactions) to 12 (CNO cycle), a slight further contraction and temperature increase causes an enormous increase in L_{nuc}, such that the star will be able to reach an equilibrium situation at which

$$L_{nuc,ph} = L_{out} \tag{82}$$

where L_{out} is the heat loss at the outer surface of the star and $L_{nuc,ph}$ is the photon part of the nuclear luminosity (fig. 43). A part $L_{nuc,\nu}$ of the luminosity is lost in the form of neutrinos, which freely escape into space.

When eq.(82) is fulfilled, the contraction stops. The nuclear burning equilibrium that now is reached is stable, since the virial theorem works as a safety valve. If too much heat were to be produced by nuclear buring, the interior temperature would rise, causing an increase in gas pressure which will lead to expansion and cooling of the star.

4.3 The three basic stellar timescales.

Three timescales are of fundamental importance in stellar evolution (cf. Cox and Giuli[48]). When the hydrostatic equilibrium of a star is disturbed (e.g. because of sudden mass loss), the star will restore this equilibrium on a so-called dynamical timescale (also called pulsational timescale), which is of the order of the time it takes a sound wave to cross one stellar radius. Numerically the dynamical timescale is given by

$$\tau_d = 50(\bar{\rho}_\odot/\bar{\rho})^{1/2}\ min \tag{83}$$

where $\bar{\rho}$ and $\bar{\rho}_\odot$ denote the mean density of the star and the sun, respectively ($\bar{\rho}_\odot = 1.4$ g/cm^3). When the thermal equilibrium of the star is disturbed, it will restore this equilibrium on a thermal timescale τ_{th} (also called the Kelvin–Helmholtz timescale), which is the time it takes to emit its thermal energy content (GM^2/R) at its present luminosity L.

Thus

$$\tau_{th} = GM^2/RL. \tag{84}$$

Stars with nuclear burning obey the so-called mass–luminosity relation, which for $M > M_\odot$ is well approximated by $L \propto M^{3.5}$, and for $M < M_\odot$ by $L \propto M^3$. Main-sequence stars, furthermore, obey a mass-radius relation, which for $M > M_\odot$ follows $R \propto M^{0.5}$, and for $M \leq M_\odot$, $R \propto M$. Substituting these expressions in eq. (84) one finds that τ_{th} can be approximated numerically as

$$\tau_{th} = 3 \times 10^7 (M/M_\odot)^{-2} yr. \tag{85}$$

The third stellar timescale is the nuclear one, which is the time needed for the star to exhaust its nuclear fuel reserve (which is proportional to M), at its present fuel consumption rate (which is proportional to L), so

$$\tau_{nuc} = 10^{10}(M/L)(L_{\odot}/M_{\odot})yr \tag{86}$$

With the above expressions for the mass-luminosity relation one finds that numerically τ_{nuc} can be approximated as

$$\tau_{nuc} = 10^{10}(M/M_{\odot})^{-2.5}yr \tag{87}$$

for $M > M_{\odot}$, and

$$= 10^{10}(M/M_{\odot})^{-2}yr \tag{88}$$

for $M \leq M_{\odot}$.

4.4 Evolution beyond hydrogen burning.

The further evolution of the star after the exhaustion of its hydrogen fuel can be described again by applying the virial theorem, as follows.

When hydrogen is exhausted L_{nuc} drops to zero. However, the star keeps radiating because it is hot. So L_{out} does not decrease. This implies that the star must start contracting again, which will cause its interior temperature to rise.

The contraction of the central parts will go on until a central temperature is reached at which the next nuclear fuel is ignited. This is helium, which burns at a temperature of $1 - 2.10^8$ K. Information about the sequence of burning stages through which the stellar interior subsequently passes is given in table 20 (after Arnett[6]; Weaver and Woosley[348]; Trimble[285]). It should be noticed that for the radius R in eq. (79) one should in fact read the radius of the dense burning core. The outer layers of the star, which in most cases comprise some 70 to 80 % of the stellar mass, do not take part in the nuclear burning. In the stages beyond hydrogen burning they have a tendency to expand while the core further contracts. The presence of the outer layers hardly affects the value of E_{pot} in eq. (78) as the largest contribution to the binding energy is made by the dense core, due to the small radius of this core. See for example Kippenhahn and Weigert[133].

The table shows that during carbon burning and beyond, most of the nuclear energy generated in the interior is liberated in the form of neutrinos. These represent a "leak" of energy from the stellar interior as they freely escape without interaction with the stellar gas. This causes an acceleration of the nuclear burning, because only the photon fraction of the liberated energy is of interest for maintaining the hydrostatic equilibrium, and for preventing the star from contracting. Beyond carbon burning the fraction of the energy lost in the form of neutrinos becomes so large that the duration of these stages becomes negligible (of the order of 1 year or less). As an example, the last column of table 20 lists

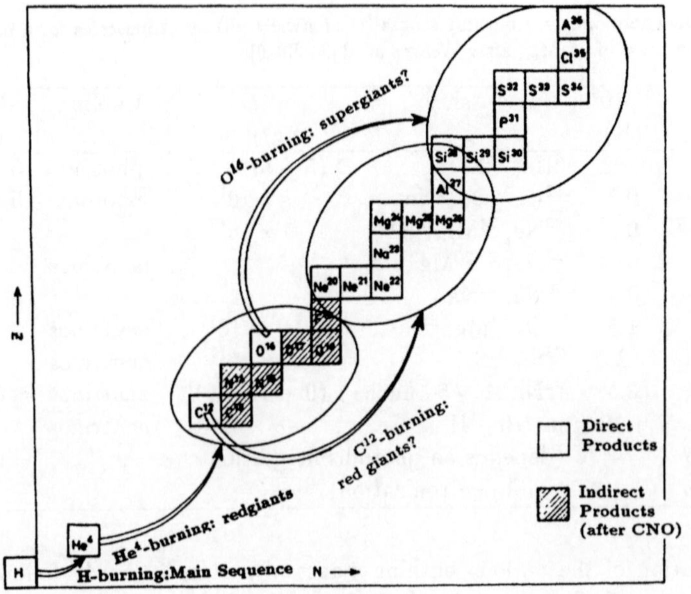

Fig. 42. The products of nuclear burning produced by the subsequent evolutionary phases of a star. N and Z denote the numbers of neutrons and protons in the nucleus (from Reeves [225]).

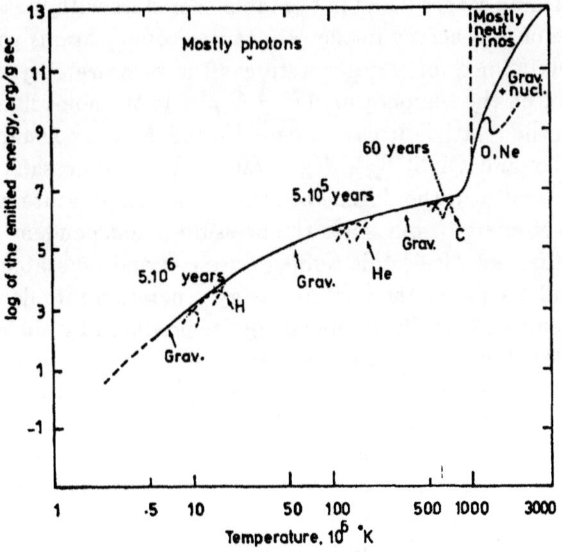

Fig. 43. Schematic representation of the evolution of the stellar luminosity as a function of central temperature. The total luminosity on the "photon" branch is equal to the outer luminosity (heat radiation) of the star. It is provided by alternating stages of gravitational contraction and nuclear burning, up until carbon burning. During carbon burning and beyond, most of the energy generated in the interior escapes in the form of neutrinos, causing the duration of these stages to become very short, and the neutrino luminosity to become very large (after Reeves [225]). The time scales are for a star of 25 $M_\odot$, cf. Table 20.

Table 20. Thermonuclear burning stages (after Arnett [6]) and timescales for a population I star with a mass of 25 $M_\odot$, after Weaver at al.[348][349]

Fuel	$T/10^9$ (K)	Ashes	E (erg/g fuel)	Cooling	Time (yr)
^{1}H	0.02	^{4}He, ^{14}N	$(5-8)\times10^{18}$	photons	5×10^6
^{4}He	0.2	^{12}C, ^{16}O, ^{22}Ne	7×10^{17}	photons	5×10^5
^{12}C	0.8	^{20}Ne, ^{24}Mg, ^{16}O	5×10^{17}		
		^{23}Na, ^{25,26}Mg		neutrinos	60
	0.4	^{20}Ne, ^{23}Na	–		
^{20}Ne	1.5	^{16}O, ^{24}Mg, ^{28}Si	1.1×10^{17}	neutrinos	1
^{16}O	2	^{28}Si, ^{32}S	5×10^{17}	neutrinos	0.5
^{28}Si	3.5	^{56}Ni, $A\sim56$ nuclei	$(0-3)\times10^{17}$	neutrinos	0.01
^{56}Ni	6–10	n, ^{4}He, ^{1}H	-8×10^{18}	neutrinos	
$A\sim56$ nuclei		(depends on photodisintegration and neutronization)			10^{-6}

the duration of the various burning stages of a star of 25 $M_\odot$ following from detailed numerical computations[349][348]. We will not consider here the stages beyond silicon burning as these last less than a few days. For recent reviews of these stages we refer to Hillebrandt[108] and Woosley and Weaver[366].

Only massive stars ($M > 12M_\odot$) pass without complications through the sequence of burning stages listed in the table. For stars with masses below 7 $M_\odot$ the occurrence of degeneracy in the core causes complications with the ignition of nuclear fuels. Since for a degenerate gas the pressure only depends on the density and not on the temperature ($P = K_1\rho^{5/3}$ in the non-relativistic case and $P = K_2\rho^{4/3}$ in the relativistic case, where K_1 and K_2 are constants and ρ is the density (cf. Cox and Giuli[48])), degenerate cores have no safety valve. When a nuclear fuel ignites in the degenerate state, the temperature will rise (due to the liberation of energy), but since the pressure is independent of temperature the pressure does not "feel" this temperature rise and no stabilizing expansion and cooling will occur. As the nuclear energy generation rate depends on a high power of the temperature, the temperature rise produced by nuclear burning will cause the nuclear luminosity to rise enormously.

As a result, the ignition of a degenerate fuel causes a runaway nuclear energy generation, and a runaway rise of the interior temperature, producing a so-called "flash". Only when the temperature becomes so large that the degeneracy boundary is crossed (the double-dashed line in fig. 44) will the gas be able to expand and cool. In stars with $M < 2.3M_\odot$ the helium core becomes degenerate during hydrogen shell burning and, when M_{He} reaches $0.45M_\odot$, helium ignites with a flash; the temperature rises up to $\sim 10^9$ K, at which the degeneracy is removed. The helium flash is not violent enough to cause a disruption of the star (cf. Cox and Giuli[48]). Stars with masses in the range 2.3 to about $7-8M_\odot$ ignite carbon with a flash.

Figure 44 depicts the evolutionary track of the central density and temperature of stars of various masses (cf. refs. Trimble[285], Maeder & Meynet[160]).

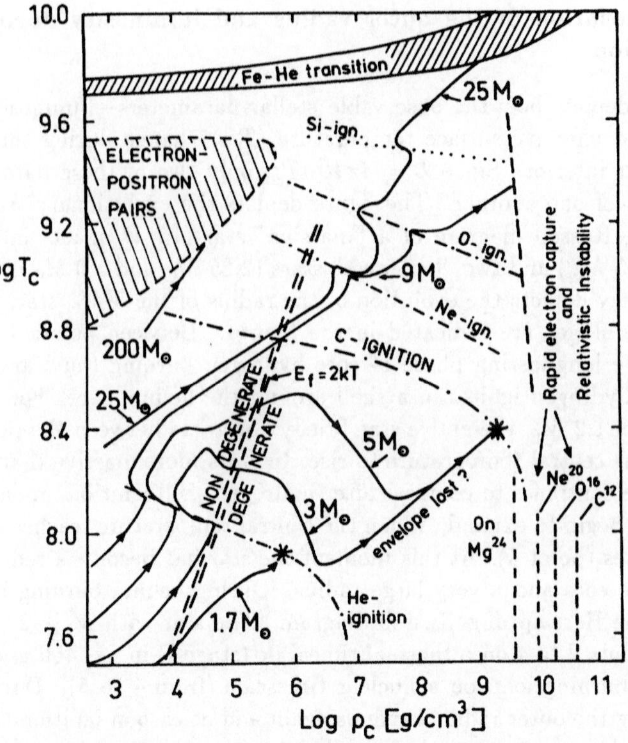

Fig. 44. Schematic evolutionary tracks of stars of different masses in the central temperature versus central density diagram. Dash–dotted lines indicate loci of ignition of nuclear fuels. Dashed lines indicate regions of instability due to pair production and photodisintegration of iron (upper left) and electron capture and relativistic instabilities (far right). Solid lines are (slightly simplified) evolutionary tracks of stars of 1.7, 3, 5 and 9 $M_\odot$ after Maeder and Meynet [160], or 25 $M_\odot$ after Weaver et al. [349] and Weaver and Woosley [348], and 200 $M_\odot$ according to Mazurek and Wheeler [167]. Collapse triggered by electron–positron production gives way to collapse triggered by photodisintegration at about 100 $M_\odot$; photodisintegration of iron gives way to electron capture on lighter elements at 12 $M_\odot$; stars below 8 $M_\odot$ either completely disrupt upon C ignition, or lose their envelopes before this happens. Diagram adapted from Trimble [285].

Only above $\sim 8M_\odot$ is carbon ignited sufficiently close to the boundary of the non-degenerate region that the star can avoid the carbon flash. All evidence suggests that in stars of mass in the range $3 - 8M_\odot$ the carbon flash generates sufficient energy to disrupt the entire star in a so-called carbon-deflagration supernova. The stellar material is converted into ^{56}Ni, which decays in 120^d to ^{56}Fe. It has been suggested that a carbon-deflagration supernova may be identified with a type I supernova as both the liberated energy ($\sim 10^{51}$ erg) and the expected shape of the light curve fit the observations (cf. refs. Chevalier[38], Nomoto et al.[183], see especially the reviews Woosley and Weaver[366]; Branch[29]). However, it is not certain that stars ever will reach this flash before they have lost their envelopes (see below).

4.5 The variation of the outer radius and luminosity during stellar evolution.

Figure 45a depicts how the observable stellar parameters—luminosity (L), radius (R) and effective surface temperature (T_e)—change during the evolution of the stellar interior. Since $L = 4\pi R^2 \sigma T_e^4$, only two of these parameters are independent of one another. The figure depicts the evolutionary tracks in the Hertzsprung–Russell diagram of a "massive" star (15 $M_\odot$), an "intermediate-mass" star (5 $M_\odot$) and two "low-mass" stars (2.25 $M_\odot$ and 1.0 $M_\odot$). Figure 45b also separately depicts the evolution of the radius of the 5 $M_\odot$ star. Important evolutionary stages are indicated in the figures. Between points 1 and 2 the star is in the long-lasting phase of core hydrogen burning (nuclear timescale). At point 3 hydrogen ignites in a shell around the helium core. For stars more massive than 1.2 $M_\odot$ the entire star briefly contracts between the points 2 and 3, causing its central temperature to rise. In stars more massive than about 2.3 $M_\odot$ the core continues to contract after hydrogen-shell ignition, upon which the outer layers begin to expand. When the central temperature reaches $T \sim 10^8$ K, helium ignites (point 4). At this moment the star has become a red giant, with a very dense core and a very large radius. During helium burning it describes a loop in the Hertzsprung–Russell diagram. The star with $M \geq 2.3 M_\odot$ moves from the points 2 to 4 on a thermal timescale (stage II in fig. 45b and describes the helium-burning loop on a nuclear timescale (from 4 to 5). During helium shell burning the outer radius expands again and at carbon ignition the star has become a red supergiant, on the so called Asymptotic Giant Branch (AGB).

The evolution of *low-mass* stars ($M \lesssim 2.3 M_\odot$) takes a somewhat different course. After hydrogen-shell ignition the helium core becomes degenerate and the hydrogen burning shell generates the entire stellar luminosity. While its core mass grows, the star gradually climbs upwards along the giant branch until it reaches helium ignition with a flash (point 4). For all stars less massive than about 2.3 $M_\odot$ the helium core at helium-flash ignition has a mass of about 0.45 $M_\odot$.

The evolution described above holds for an initial chemical composition of 70% hydrogen, 28% helium and neglecting the effects of convective overshooting (for details see ref. Maeder & Meynet[160]).

4.6 Final evolution of stars in the mass range 3–8 $M_\odot$: competition between carbon deflagration and mass loss; evidence from white dwarfs in clusters.

When stars in this mass range are approaching C ignition, they are on the Asymptotic Giant Branch (AGB) appearing as bright red giants (see fig. 45a). Such stars lose much mass in the form of stellar winds. White dwarfs have been found in galactic clusters that still contain stars as massive as 5–6 $M_\odot$ (Romanishin and Angel[235]; Reimers and Koester[227]) This indicates that stars as massive as at least 6 $M_\odot$ still terminate life as a white dwarf, i.e., are apparently able to shed their envelopes in the AGB phase before carbon violently ignites. This

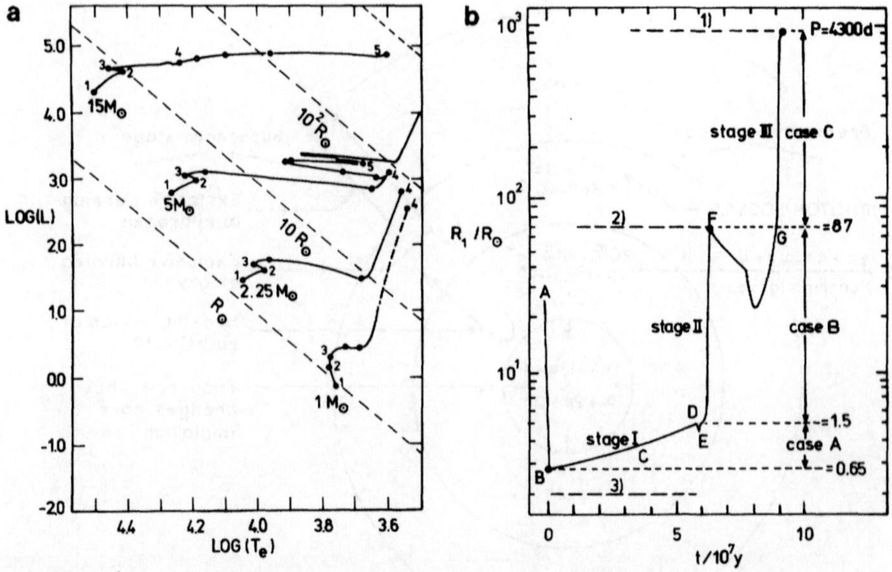

Fig. 45. **a) Evolutionary tracks in the Hertzsprung–Russell diagram of stars of 1 $M_\odot$, 2.25 $M_\odot$, 5 $M_\odot$ and 15 $M_\odot$, after Iben [118]. b). Secular change of the radius of a star of 5$M_\odot$, from pre-main-sequence contraction (A) to carbon ignition. Also indicated are the orbital periods (in days) of close binaries in which the star just fills its Roche lobe at various evolutionary stages, for binaries with a mass ratio $q = M_2/M_1 = 0.5$. The ranges of binary periods for evolution according to the cases A, B and C (see section 5) are indicated. 1) carbon ignition, 2) helium ignition, 3) main-sequence model (after Paczyński [200]).**

confines the possible mass range for carbon deflagration "supernovae" of single stars to at most 6–8 $M_\odot$. (It is possible, however, that even up to 8 $M_\odot$ single stars never reach this phase, and that carbon-deflagration supernovae in single stars do not occur in nature.)

4.7 Final evolution of stars more massive than about 8 $M_\odot$.

In the mass range between 8 and 10–12 $M_\odot$, C ignition is non-violent, but the O–Ne–Mg core which is formed becomes degenerate. Here, however, after further fuel ignition the threshold for electron capture is expected to be crossed (see fig. 44). This causes the onset of core collapse, as the removal of electrons causes the pressure of the electron-degenerate gas to drop. At the time of the collapse the iron-element core is relatively small ($\sim 0.6 M_\odot$, cf. Hillebrandt[107]). For stars more massive than $\sim 10 - 12 M_\odot$ an iron core of mass $> 1.2 M_\odot$ is formed, which collapses as a consequence of the iron–helium transition (see fig. 44 and Hillebrandt[107]). Thus for M $\gtrsim 8 M_\odot$, stars are expected to leave neutron stars as remnants. Figure 46 shows a cross-section through a star of 25 $M_\odot$ at the moment of the collapse of its iron core. At masses $\gtrsim 100 M_\odot$ stars are expected to undergo an $e^\pm$ pair-production collapse, as has been suggested to possibly cause complete disruption of the star. According to present insights the upper

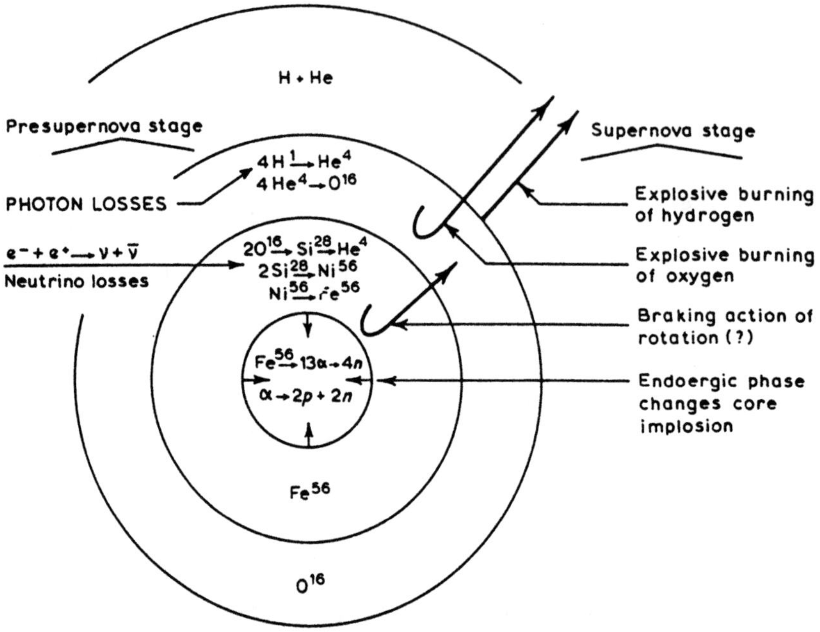

Fig. 46. The supernova phenomenon (after Fowler and Hoyle) [82]).

mass limit for leaving a neutron star remnant (which is the lower mass limit for leaving a black hole remnant, is between 40 and 60 $M_{\odot}$ (van den Heuvel and Habets[320]; De Kool et al.[59]). As it seems most unlikely that the 3 recently discovered Low Mass X-ray binaries that contain black holes originate from binaries with a primary mass $\gtrsim 60 M_{\odot}$ (van den Heuvel[317]) this limit is $50 \pm 10 M_{\odot}$. So, stars in the main-sequence mass range 50($\pm$ 10) to at least 100 $M_{\odot}$ are expected to leave a black hole.

Supernova 1987A, the $^{12}C(\alpha, \gamma)^{16}O$ reaction rate and the metal abundance Z. During later phases of core-helium burning, some of the produced ^{12}C is converted into ^{16}O reaction. Recently, it was found that this reaction rate is much larger than previously thought. With this revised rate, Hillebrandt[107] found that in stars more massive than 20 $M_{\odot}$, helium burning produces mainly ^{16}O. This implies that in such stars, a stage of carbon-burning will be absent. After helium burning they go immediately into oxygen burning. As the duration of oxygen burning and later stages is negligible, such stars explode as supernovae immediately following helium burning. They might do this either as a red or as a blue supergiant, depending on the heavy element abundance Z and the convection criterion used (see figure 47 after Arnett[7]).

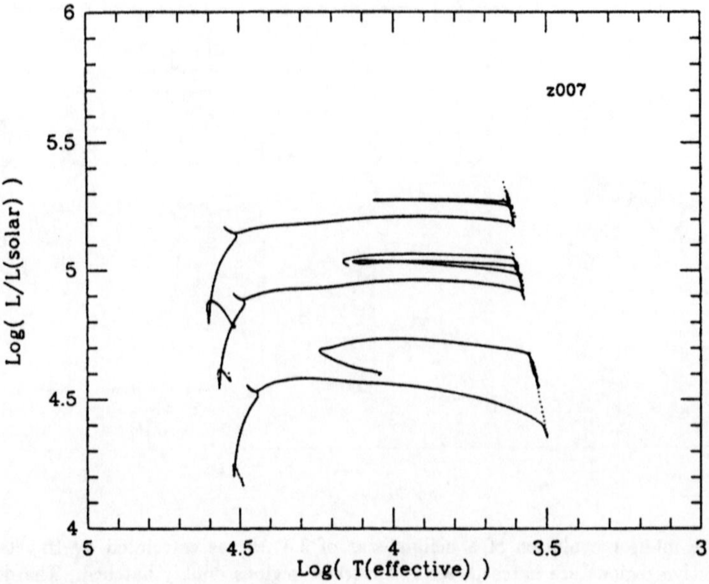

Fig. 47. The evolution in the HR-diagram for stars of $M/M_{\odot} = 15$, 20 and 25, for abundances of the large Magellanic Cloud (Arnett [7]). Only the $M/M_{\odot}$=25 model dies as a red supergiant. Both the $M/M_{\odot} = 15$ and 20 die as progenitors similar to Sanduleak -60°202 (SN 1987A) after spending time as a red supergiant.

4.8 The evolution of helium stars.

As the evolution of the helium core in post-main-sequence stars is practically independent of the presence of the extended hydrogen-rich envelope (cf. ref. Arnett[6]), it is, for the study of the final evolution, practical to study only the evolution of the helium core. Pioneering studies in this field are those of Paczyński[201] and Arnett[6]. The results of recent studies of helium stars by Habets[99][100] are shown in figs. 48, 49 and 50.

Figure 48 shows the interior evolution of a helium star of 3.5 $M_{\odot}$, which corresponds to a hydrogen star of about 13 $M_{\odot}$. The figure shows that the heavy-element core formed in the star has a mass $\geq 2.2 M_{\odot}$ and will therefore certainly collapse to a neutron star.

Figure 49 and 50 show the evolutionary tracks for helium stars with masses 2.0–4.0 $M_{\odot}$ in the Hertzsprung Russell Diagram and the radius versus core-mass diagram. C denotes C ignition. Figure 50 shows that in all helium stars with masses $> 2.2 M_{\odot}$ the heavy element core becomes larger than the Chandrasekhar mass, thus these stars are expected to leave neutron star remnants. Table 21 summarizes the above expectations for the types of remnants from main-sequence stars of various masses.

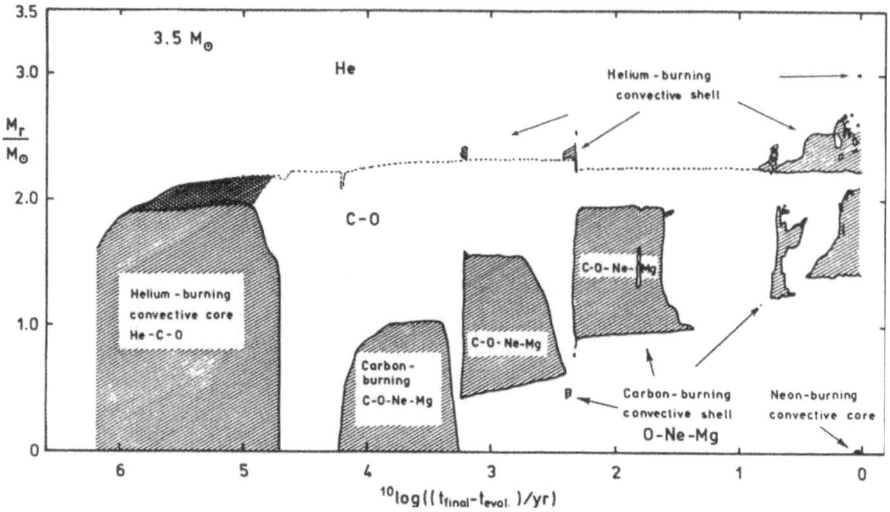

Fig. 48. The interior evolution of a helium star of 3.5 $M_\odot$ as calculated by Habets [99] [100]. Convective regions are hatched, semi-convective regions doubly hatched. The dotted line indicates the region of maximum energy generation in the helium-burning shell. This dotted line is expected to be the boundary of the C–O core. This core mass becomes so large (>2.2 $M_\odot$) that this core will certainly collapse to form a neutron star.

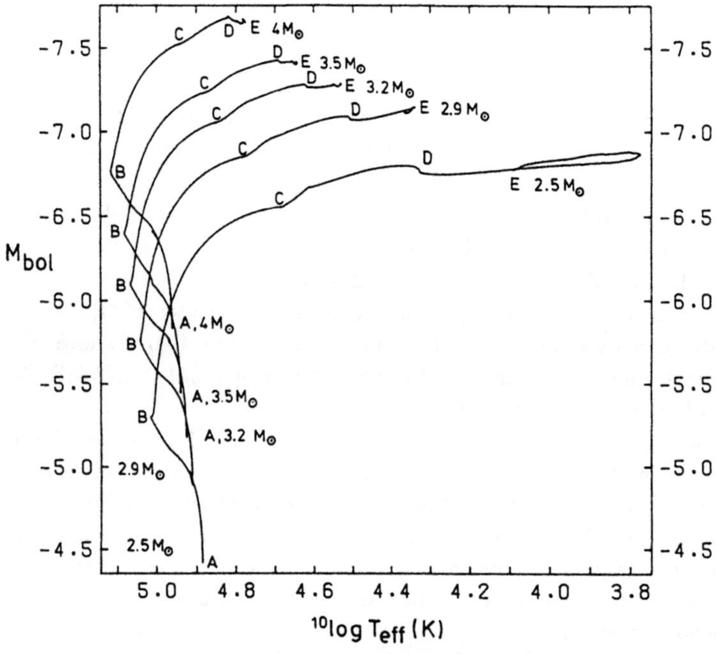

Fig. 49. The evolutionary tracks of helium stars of 2.5-4.0 $M_\odot$ in the Hertzsprung Russell diagram. The chemical composition is $Z = 0.03$, $Y = 0.97$. For the meaning of the letters: see figure 50 (after Habets [99] [100]).

Table 21. Types of final evolutionary products expected as a function of stellar mass.

He core mass	Main-sequence mass	Final product	Final product in close binary
1.4–1.9 $M_\odot$	3–8 $M_\odot$	C flash (disruption?) or CO white dwarf	CO white dwarf
1.9–2.2 (3.0)$M_\odot$	8–10 (12) $M_\odot$	Degenerate O–Ne–Mg core: Ne ignition/collapse	O-Ne-Mg white dwarf
> 2.2(3.0)$M_\odot$	> 10(12)$M_\odot$	Collapse of iron core	neutron star or black hole > $60 M_\odot$
> $100 M_\odot$	Pair-creation instability (no remnant?)		

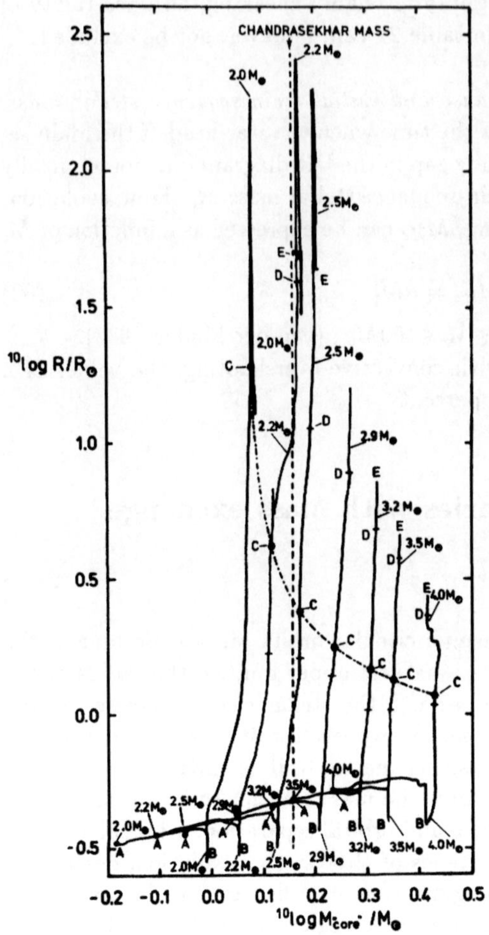

Fig. 50. Helium core mass versus radius relation of helium stars in the mass range 2.0 to 4.0 $M_\odot$ undergoing C ignition and neon ignition (the 2.9 $M_\odot$ helium star was followed until very very close to neon ignition; the 2.0 $M_\odot$ helium star was only followed up to the time at which the convective carbon-burning shell reaches the center). During convective core helium burning from letter A to B the core mass is defined as the mass of the convective helium-burning core including adjacent semi-convective regions. After convective core-helium burning the core mass is defined as the mass of the core (consisting of ^{12}C, ^{16}O, and possibly ^{20}Ne and ^{24}Mg and some ^{4}He at the outer boundary of the core) which is limited by the helium-burning shell with maximum net energy generation. The open circles mark the onset of either central or off-center convective core-carbon burning (at C). The dash-dotted curve has been drawn through these circles. The horizontal marking between B and C indicates the point where carbon burning begins in the radiative core (after Habets[99] [100]). D is neon ignition.

4.9 Discussion: the effect of convective overshooting.

The mass ranges given in table 21 for the various types of final evolution depend quite strongly on the treatment of convection in the evolutionary computations. For example, Bertelli et al.[16], Castellani et al.[35] and Maeder and Meynet[160] have pointed out that if convective overshooting is properly taken into account, the helium cores in moderately massive stars become some 20% larger than found in earlier computations. This would reduce the boundary values of the main-sequence mass ranges given in table 21 by about the same fraction. For example, the lower mass limit for terminating with a neutron star would move down to about 6.5 $M_\odot$. Also the treatment of semi-convection, important in helium stars, may affect the values of the mass boundaries. Since Habets[99] [100], however, took the latter effects carefully into account, we do not expect this to be a major source of uncertainty. Roughly speaking, an uncertainty of some 20% in the mass boundaries in table 21 can, however, not be excluded.

The relation between helium core mass and initial main-sequence stellar mass. The mass of the helium core M_{He} at the time when the star has left the main sequence and is crossing the Hertzsprung gap in the HR-diagram is a monotonically increasing function of the initial main-sequence stellar mass M_i. From evolutionary calculations without overshooting, M_{He} can be expressed as a function of M_i (for X=0.70, Z=0.03) as:

$$M_{He} = aM_1^b \tag{89}$$

where a=0.10, b=1.40 for $3M_\odot \lesssim M_i \lesssim 100M_\odot$, (van der Linden 1982).

For evolutionary calculations with convective overshooting, the values of a should be in increased by about 25 percent.

5. Evolution of close binaries with mass exchange.

5.1 Introduction.

Throughout this chapter we will assume circular orbits. In a close binary, the presence of the Roche lobes presents a natural upper limit to the radius of the compact stars. As shown by figures 45 a, b the stellar radius increases during evolution. As the more massive component evolves fastest, it will be the first to fill the Roche lobe, and to start to transfer matter to its companion.

The type of remnant that is left after the mass transfer turns out to depend on the initial primary mass, mass ratio and orbital separation. Kippenhahn and Weigert[132] defined the three basic types of close binary evolution, designated as cases A, B, and C –depicted in figure 51 and in figure 45b. In case A the system is so close that the primary star already begins to overflow its Roche lobe when it is still on the main sequence, in case B the system is wider, and the primary begins to overflow its Roche lobe after it has terminated core-hydrogen burning, but before the onset of helium burning. In case C the system is so wide

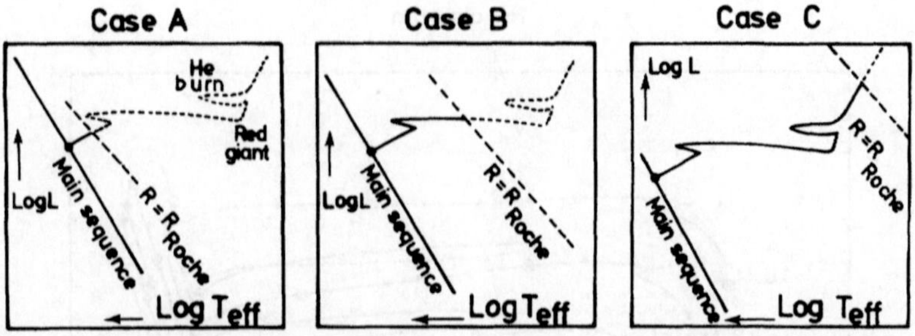

Fig. 51. The three basic cases A, B and C of mass exchange in a close-binary system, as defined by Kippenhahn and Weigert [132], illustrated on the evolutionary track of the primary star in the Hertzsprung-Russell diagram.

that the primary star only begins to fill its Roche lobe after it has terminated core-helium burning (i.e.: during helium-shell burning or beyond).

From the distribution of orbital periods and separations of binaries (dealt with in section 9) it appears that case A is relatively rare (of order ten per cent or less of all mass exchange binaries) and case B is very common: more than half of all observed unevolved close binaries will evolve according to this case. Because of strong observational selection effects against the detection of long-period spectroscopic binaries, however, the precise incidence of case C is still not well known, but is expected to be high as well (see section 9).

Figure 45b shows that for a system with primary mass $5M_\odot$ and mass ratio 0.5, case A occurs only for orbital periods $< 1.5^d$, case B for orbital periods between 1.5 and 87 days, and case C for orbital periods $> 87^d$.

Figure 52 illustrates how the orbital period ranges for cases A, B and C depend on the initial primary mass (for an assumed mass ratio $Q = M_1/M_2 = 2$). The various symbols indicate the orbital periods at which the primary star fills its Roche-lobe, for stars with masses ranging from $1M_\odot$ to 16 $M_\odot$ (after Pols[213]). The figure shows that for $M_1 = 12M_\odot$, case B occurs for all systems with orbital periods between 2.5^d and about 600^d.

The evolutionary tracks in the HR-diagram of figure 52 can also be transformed into a radius versus primary mass diagram as depicted in figure 53 (after Webbink[352]). In this diagram are indicated the radii where hydrogen burning starts (ZAMS) and where core-hydrogen exhaustion occurs, where helium and carbon are ignited, and where the core collapses. Also indicated are the boundary where a deep convective envelope develops in stars with helium cores (labelled "convective boundary") and in stars with CO-cores (labelled "Tip of Giant Branch").

In this diagram lines of constant orbital period are drawn, corresponding to systems with mass ratio $q = M_2/M_1 = 0.5$ in which the primary star just fills its Roche lobe.

Figure 53 enables one to easily see, for a given primary mass, for what orbital periods the cases A, B, and C do occur. This is also indicated in the figure.

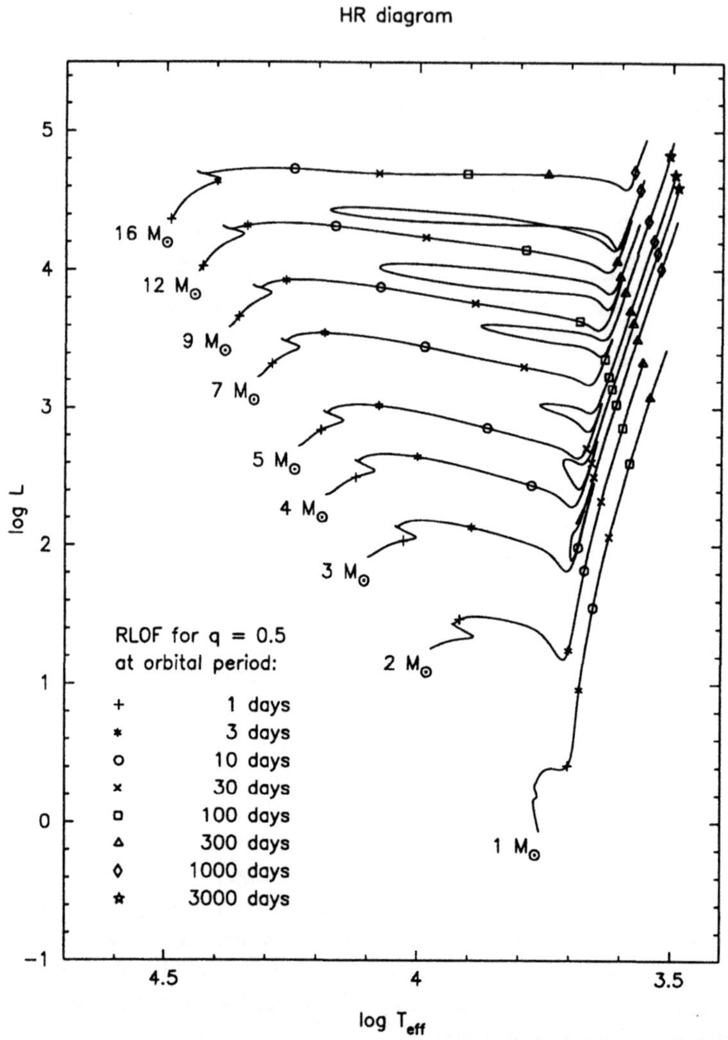

Fig. 52. Evolutionary tracks in the HR diagram of star of 1 $M_\odot$ to 16 $M_\odot$ calculated by O. Pols using the evolutionary code of P.P. Eggleton. On the tracks are indicated the radii at which the stars fill their Roche lobes in binaries with the indicated orbital periods, for $q = M_2/M_1 = 0.5$.

It should be noticed here that for $M_1 > 15 M_\odot$, due to uncertainties in convection theory and opacities, and due to the effects of the large stellar wind mass loss occurring in these stars, the location of the helium ignition line in figure 53 is rather uncertain.

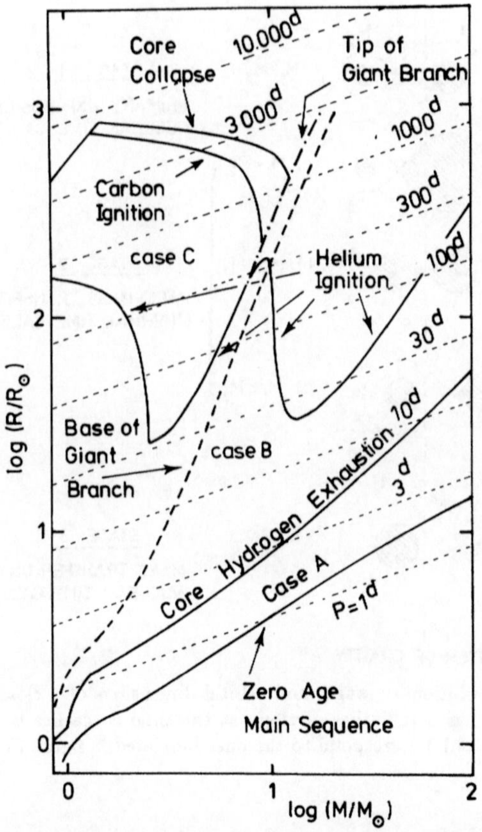

Fig. 53. Stellar radii as a function of stellar mass, at various evolutionary stages. The regions for close binary evolution cases A, B and C are indicated, as well as the binary period at which the stars fill their Roche lobes (for $q = 0.5$): these are the dashed lines. Above the convective boundary mass transfer from primary to secondary is dynamically unstable (after Webbink [352]).

5.2 Changes in orbital period and separation during "conservative" mass transfer.

If one assumes that during the mass transfer from star 1 to star 2 the total mass and orbital angular momentum of the system are conserved, the orbital separation a and period P will change as given by equations (21) through (26), i.e.

$$\frac{da}{a} = -2(\frac{1}{M_1} - \frac{1}{M_2})dM_1 \qquad (90)$$

which implies that the orbit shrinks when mass is transferred from the more massive to the less massive component and expands in the opposite case (since $dM_1 < 0$). Figure 53 depicts how the separation a changes as a function of the decreasing primary mass M_1. Figure 54 depicts the actual size of the Roche-lobes of the components during the exchange. The minimum separation is, of course,

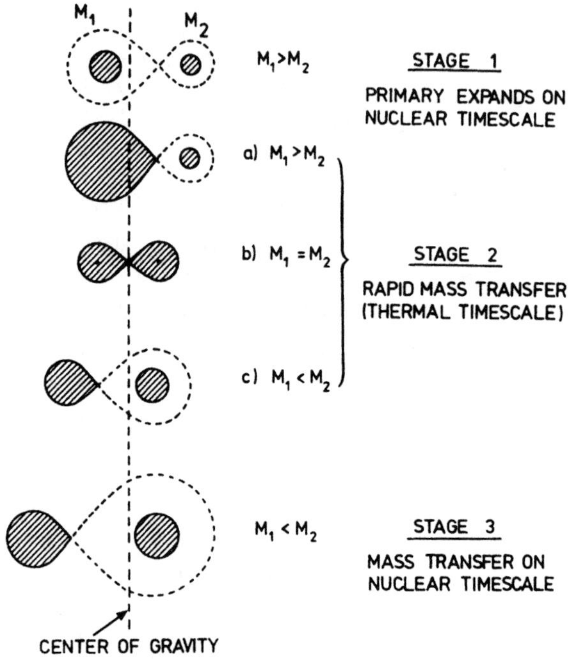

Fig. 54. Subsequent evolutionary stages and orbital dimensions of a close binary system in which the primary star has a radiative envelope at the time it reaches ist Roche lobe. The evolutionary stages 1, 2 and 3 correspond to the ones indicated in figure 57.

reached for $M_1 = M_2$. The minimum Roche-lobe radius of the primary star is reached for $M_1 = 0.4545M$, where M is the total mass of the system. This can be derived as follows.

From equation (90) it follows that

$$a = a_o(M_1^\circ(M - M_1^\circ)/M_1(M - M_1))^2 = const.(1+q)^4/q^2 \tag{91}$$

where $q = M_2/M_1$. Combining this equation with the equation for the Roche-lobe radius of star 1 (see Livio, this volume section 2):

$$R_L = 0.462a(1/(1+q))^{1/3} \tag{92}$$

one finds by setting the derivative of R_L equal to zero, that R_L reaches its minimum value for $q = 6/5$, which corresponds to $M_1 = 0.4545M$.

5.3 Response of the stellar radius to mass loss.

Before the primary star reached its Roche lobe and began to transfer matter, it was in state of hydrostatic and thermal equilibrium. When it loses an amount of mass ΔM from its outer layers, both of these equilibria are disturbed.

A. Restoration of hydrostatic equilibrium. The star will restore its hydrostatic equilibrium very quickly, on a dynamical (=pulsational) timescale which, according to eq. (83) is only of the order of hours to days. After this adiabatic response to the change in its pressure equilibrium, the radius of a star with a radiative envelope is always smaller than the one before the mass loss. For stars with a convective envelope, however, the new radius may be *larger* than the original one. Figures 55ab (after Hjellming[110]) show the adiabatic response of the stellar radius as a function of the mass taken away, for a star of $1.5M_\odot$ for various evolutionary stages at the onset of the mass loss. The curves labelled ZAMS, TAMS, BGB and TGB designate the response for a star on the Zero Age Main Sequence, at the turnoff of the Main Sequence, at the Base of the Giant Branch and the Tip of the Giant Branch, respectively. Also the response for mass loss starting at some phases between the BGB and the TGB is shown. The figures show that for mass loss starting before reaching the BGB (and also still somewhat beyond the BGB) the hydrostatic equilibrium radius decreases when mass is lost, whereas on the TGB the radius rapidly *increases* when mass is lost. The latter is due to the presence of a deep convective envelope in the star.

The reason for this violent response of an almost completely convective star to mass loss can be understood as follows. The matter in a state of full adiabatic convection obeys a simple polytropic equation of state:

$$P = const.\rho^\gamma \tag{93}$$

where $\gamma = 5/3$ (we assume the stellar gas to be fully ionized). This is exactly the same equation of state as that of a non-relativistically degenerate star (see section 4), and such stars are known to have an inverted mass radius relation:

$$R \propto M^{-1/3} \tag{94}$$

(see e.g. Cox and Giuli[48], p. 874 ff.) In the fully convective star, the reason why equation (93) holds is that the convection proceeds rapidly, and therefore adiabatic: there is not enough time available for the star, on a dynamical timescale, to lose a non-negligible fraction of its internal heat content. Therefore, its total heat content is conserved, and the convection is in good approximation adiabatic. (See for example, Cox and Giuli[48], p. 743 ff).

Equation (94) implies that if mass is taken away on a timescale shorter than a thermal one, the radius of a convective star expands. The $1.5M_\odot$ star was taken here as an example because, as figures 55ab nicely demonstrate, it shows that as long as most of the envelope is radiative, the hydrostatic equilibrium radius decreases upon mass loss, whereas when most of the envelope is convective, it expands.

In figures 55ab also the radius of the Roche lobe as a function of the (decreasing) primary star mass M_1 is shown, for q-values ranging from 0.5 to 2.0. Figure 55a shows that for primary stars (with these q-values) that start mass loss on the ZAMS and TAMS, the hydrostatic equilibrium radius always remains smaller than the radius of the Roche lobe; this is still true for stars on the BGB.

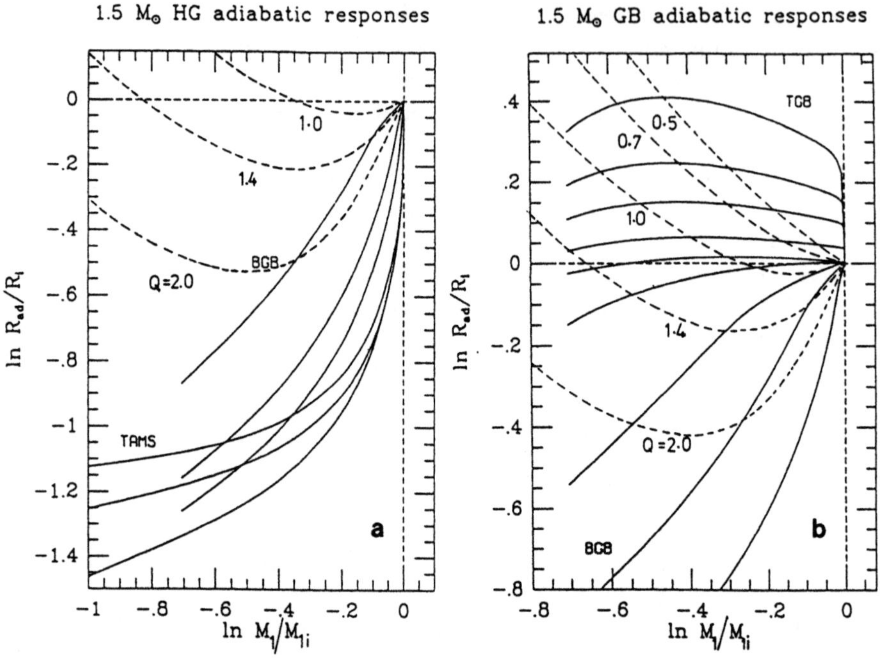

Fig. 55. a.) Radius after restoration of the *hydrostatic* equilibrium in a star with initial mass $M_{1,i}$=1.5 $M_\odot$, as a function of the amount of mass lost, for mass loss starting at various evolutionary stages between Zero Age Main Sequence (ZAMS) and the Base of the Giant Brach (BGB). TAMS indicates the turn-off point of the main sequence. M_1 denotes the new mass and R_{ad}/R_i the new hydrostatic equilibrium radius divided by the original radius. The dashed curves show how, when the star is the primary of a binary system, its Roche-lobe radius changes during the mass loss, for various values of the initial mass ratio $Q = M_1/M2$ (after Hjellming[110]).

Fig. 55. b.) The same as figure 55a, for mass loss starting between the Base of the Giant Branch (BGB) and the Tip of the Giant Branch (TGB).

However, figure 55b shows that a primary star on the TGB is unable to stay inside its Roche lobe upon mass loss, not even if $q > 2$, i.e. when it is more than twice as small in mass as its companion, and thus its Roche lobe is rapidly expanding upon mass loss (note: $q = M_2/M_1 = 1/Q$).

Clearly, stars with deep convective envelopes run into a violent mass loss instability when they begin to overflow their Roche lobes, even when they are the less massive component of the system!

This implies that the primary stars in all systems that are located above the convective boundary in figure 53 run into catastrophic mass loss on a dynamical time scale as soon as they begin to overflow their Roche lobes.

On the other hand, for stars with a radiative, or not too deep convective, envelope there is no problem to stay inside the Roche lobe during the mass tranfer, as their hydrostatic equilibrium radii decrease faster upon mass loss than their Roche lobe radii.

B. Restoration of thermal equilibrium. We consider stars which after transferring a small amount of mass ΔM, and the subsequent restoration of their hydrostatic equilibrium (on a dynamical timescale) have a radius inside that of their (new) Roche lobe. Due to the mass loss their thermal equilibrium will be disturbed and the first thing the star will now start to do is restore its thermal equilibrium. The time scale required for restoring the thermal equilibrium will be of order of the thermal time scale of the stellar envelope, which is much longer than the dynamical time scale (see section 4).

Whether the new thermal equilibrium radius will be smaller or larger than the original one depends on the evolutionary state of the star at the onset of the mass transfer: for main-sequence stars in not too advanced stages of core-hydrogen burning, the thermal equilibrium radius decreases upon mass loss. For stars in advanced phases of core-hydrogen burning ($X_c < 0.2$) and beyond, the thermal equilibrium radius increases when the star loses mass. This is illustrated by figure 56 (after Hjellming[110]) which shows the response to mass loss of the thermal equilibrium radius of a star of $3M_\odot$, for mass loss starting at various evolutionary stages, ranging from Zero Age Main Sequence (ZAMS, $X_c = 0.70$) to the Tip of the Giant Branch. The response for the ZAMS, TAMS and TGB is indicated. Like in figures 55, also the variation of the Roche lobe radius during the mass loss is drawn in figure 56, for Q ranging from 0.5 to 2.0.

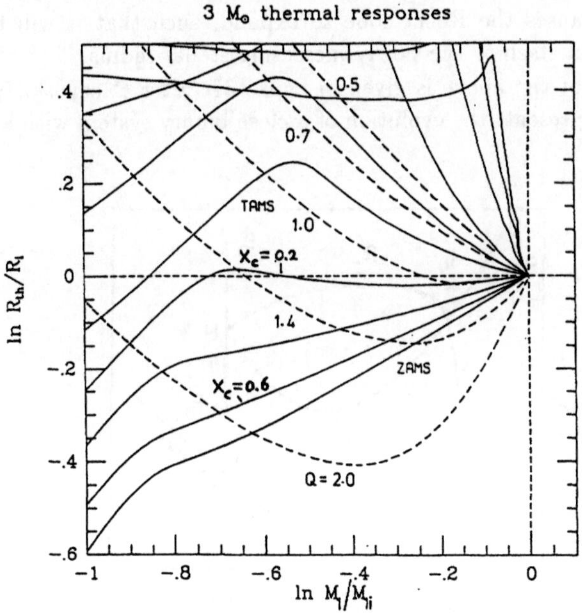

Fig. 56. Radius after restoration of *thermal* equilibrium in a star of initial mass 3 $M_\odot$, as a function of the amount lost, for mass loss starting at various evolutionary stages. X_c is the central hydrogen content. Curves above the TAMS have helium cores and convective envelopes of increasing depth. At TGB the entire envelope is convective. Roche-lobe radii for various Q-values are indicated (after Hjellming [110]).

5.4 Reasons for the large scale mass transfer during the evolution of a close binary.

Figure 56 contains all information required to understand why large scale mass transfer will ensue and the role of primary and secondary will be reversed once the primary star in a close binary begins to overflow its Roche lobe. For the sake of argument we consider a system with $Q = 2.0$. The figure shows that even when the star is on the ZAMS when the mass transfer begins, it will have to lose a large fraction of its mass before its thermal equilibrium radius becomes smaller again than that of its Roche lobe: this occurs for $ln\ (M_1/M_{1,i}) < --.625$, i.e. $M_1 = 0.535\, M_{1,i} = 1.60 M_\odot$. Thus, the $3 M_\odot$ primary star has to lose 1.40 $M_\odot$ to its $1.5 M\odot$ companion, which then becomes a star of 2.90 $M_\odot$, i.e.: the roles of the primary and secondary star will be reversed after the mass transfer.

Similarly, for $X_c = 0.40$ at the onset of the mass transfer, one finds $ln(M_1/M_{1,i})=$ -0.83 at the end of the transfer, i.e. M_1=1.31 $M\odot$, $M_2 = 3.19\ M_\odot$. For $X_c = 0.00$, $ln(M_1/M_{1,i}) = -0.96$ at the end of the transfer, and $M_1 = 0.383\, M_{1,i} = 1.15 M_\odot$, $M_2 = 3.35 M_\odot$. The timescale of the mass transfer will be of the order of the thermal timescale of the primary star, given by eq. (85), which for a star of 3 $M_\odot$ is about 3.10^6 yr.

After the end of the rapid (thermal timescale) mass transfer, the original primary star will evolve further on a nuclear timescale. During this evolution its radius will gradually expand, causing further *slow* mass transfer. This transfer is *stable* as it causes the Roche lobe to expand, such that it will be able to continously adapt itself to the slowly increasing stellar radius.

A summary of the above is given in figure 57 (after Paczynski[199]) which schematically represents the evolution of a close binary system with a given pri-

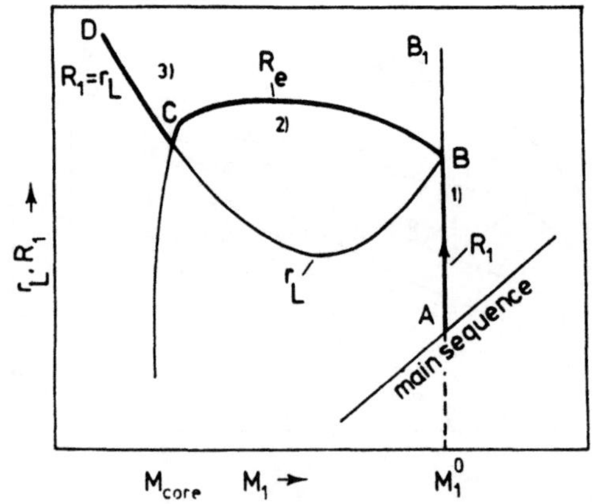

Fig. 57. Variation of the thermal-equilibrium radius of the primary component R_e and the radius of its Roche lobe r_L ($=R_L$) during the mass transfer from the primary to the secondary. Only after the primary has become less massive than the secondary will r_L begin to increase. Consequently, equilibrium is not possible before the primary has lost so much matter that it has become the less massive component (after Paczynski [199]).

mary mass, mass ratio and initial orbital separation (for the case that the primary star at the onset of the exchange does not have a deep convective envelope). In stage 1 the primary star of mass M_1° is smaller than its Roche lobe. During this stage its radius gradually increases on a *nuclear timescale*. In point B the star begins to overflow its Roche lobe and mass transfer begins: as during this stage the thermal equilibrium radius of star 1 is always larger than its Roche lobe radius, the star loses mass on a *thermal timescale* up to point C. In C the Roche lobe radius becomes larger than the thermal equilibrium radius, and the star is able to stabilize again. In the subsequent stage (*stage 3*) the initial primary evolves on a nuclear timescale again, gradually transferring mass on this timescale.

5.5 Some examples of the results of detailed calculations of close binary evolution with mass transfer

Case A. Figure 58 represents the evolution of a close binary system with initial component masses of 9+5 $M_\odot$ and initial orbital period of 1.20 days as calculated by Kippenhahn and Weigert[132], for initial chemical composition X=0.60, Z=0.044.

The HR diagram of figure 58 shows the evolutionary track which the 9 $M_\odot$ would have followed if it had been single. In point *a*, 1.25×10^7 yr after the birth of the system, the primary star begins to overflow its Roche lobe. At that time its central hydrogen content has decreased to X_c= 0.35. It now begins a phase of rapid thermal-timescale mass transfer and in 10^4 yrs transfers 5.27 $M_\odot$ to its companion, which now becomes a star of 10.3 $M_\odot$ (indicated on the main-sequence in figure 58). The rapid mass transfer ends and the star stabilizes in point *b*: the mass of the primary star is now only 3.73 $M_\odot$. The star is still in the phase of core-hydrogen burning and still fills its Roche lobe. It is a subgiant, i.e. has a radius and luminosity larger than that of a main-sequence star of the same mass. From here on it now continues to evolve on a nuclear timescale, and its radius gradually increases, causing the star to slowly lose mass and the system to slowly expand on this timescale. The calculations were terminated at $t = 3 \times 10^7\,yr$ (point *c*) when the primary mass had decreased to 3.02 $M_\odot$, and the primary was still in the hydrogen-burning phase.

Between the points *b* and *c* (and also still beyond *c*) the system closely resembles the massive semi-detached Algol-type close binaries listed in table 3, such as U Her, V356 Sgr and Z Vul. In all of these systems the Roche-lobe filling component is the less massive of the two and is a subgiant which is overluminous for its mass. As figure 58 shows the same is true in the calculated post mass transfer system: the subgiant with its mass decreasing from 3.73 to 3.02 $M\odot$ has a luminosity similar to that of a 5 $M_\odot$ main-sequence star.

Case B. For systems evolving according to case B there appear to be two possible kinds of outcome depending on the mass M_1° of the initial primary star: the so-called Low Mass Case B ($M_1^\circ \lesssim 2.8 M_\odot$) and the Higher Mass Case

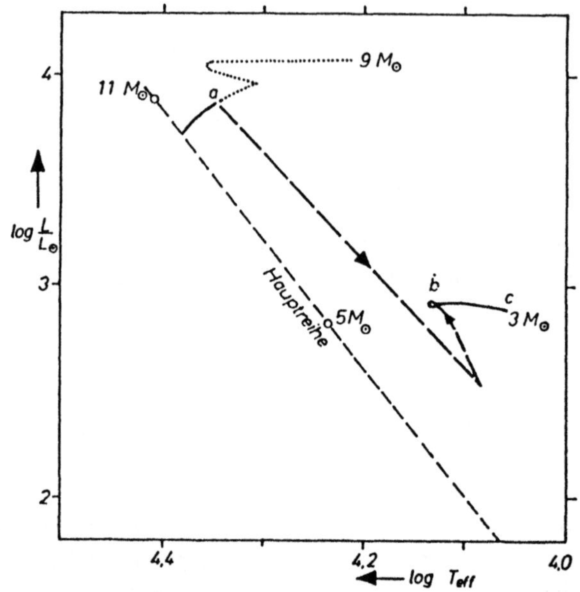

Fig. 58. The evolution of a binary in case A, calculated by Kippenhahn and Weigert[132]. The primary star begins with $M_1 = 9M_\odot$ on the Zero Age Main Sequence ($X_c = 0.60$) and evolves to the point a where it begins to overflow its Roche lobe. (If it had been single it would have followed the dotted evolutionary track). It now begins to rapidly transfer matter and moves down in the HR diagram along the dashed line until in b the rapid mass transfer ends and a phase of slow mass transfer follows (on a nuclear timescale). The secondary star, which started out with $M = 5M_\odot$ has grown to about 11 $M_\odot$ by the end of the mass transfer and has moved upwards along the main sequence as indicated. The parameters of the system in the points a, b and c are as follows:

age (yr)	$t_a = 1.25 \times 10^7$ yr	$t_a + 1.10^4 yr$	3.10^7 yr
M_1	9	3.73	3.02
M_2	5	10.27	10.98
log a (cm)	11.965	12.11	12.34
log R_L1	11.60	11.57	11.67

B ($M_1^\circ > 2.8M_\odot$). This dichotomy is a direct result of the fact that in stars with low mass ($< 2.3M_\odot$) the post main-sequence evolution proceeds differently from that of stars with higher mass, as is explained in the previous chapter and depicted in figures 44 and 45a. (The limiting mass in the case of binaries is higher than in single stars, due to the loss of the envelope; the precise value depends somewhat on the input physics. For example, if one assumes convective overshooting to be important, this limit may be lower, cf. Chiosi et al.[39].

In post main-sequence stars with $M < 2.8M_\odot$ the contracting helium core becomes degenerate and does not reach helium ignition. The energy generation in the hydrogen-burning shell around the core is sufficient to supply the star's increasing luminosity when it gradually climbs up along the giant branch. Due to the hydrogen-shell burning the core mass gradually increases and when the degenerate core reaches a mass of $\sim 0.45 M_\odot$ (the star is then at the TGB), helium ignites with a flash. The star thus spends a very long time as a giant with a degenerate helium core.

On the other hand, post-main-sequence stars with $M \gtrsim 2.8 M_\odot$ do not develop a degenerate helium core and go immediately to core-helium burning, as depicted in figures 44 and 45a. The higher-mass Case B variant itself appears to have two variants still:

- the intermediate mass Case B ($2.8 \lesssim M_1 \lesssim 15 M_\odot$)
- massive Case B ($M_1 \gtrsim 15 M_\odot$)

The difference between these two cases is that in the intermediate mass case, there is a second phase of mass transfer (so called Case BB or BC transfer) before the primary star terminates its evolution, whereas in the massive Case B ($M_1^\circ > 15 M_\odot$) such a second mass transfer phase is absent. Table 22 lists these 3 cases.

Table 22. The three types of Case B mass transfer.

M_1^0	Type	Number of mass transfer phases
$< 2.8 M_\odot$	Low-Mass Case B	one
$2.8 \lesssim M_1^0 \lesssim 15$	Intermediate Mass Case B	two
$M_1^0 > 15$	Massive	one

Low Mass Case B. Figure 59ab and table 23 show as an example the first calculated evolution of a low-mass Case B system, by Kippenhahn, Kohl and Weigert (1967). The system initially consisted of stars of 2+1 $M_\odot$ with an orbital period of 1.14 days.

After $\sim 5.75 \times 10^8\, yr$ the primary star has left the main-sequence and begins to overflow its Roche lobe. Now a phase of fast (thermal timescale) mass transfer ensues, lasting 6.10^6 yrs after which the mass of the primary has been reduced to $0.96 M_\odot$ and that of the secondary increased to 2.04 $M_\odot$. After this a long phase ($\sim 10^8$ yr) of nuclear timescale mass transfer sets in during which the degenerate helium core-mass gradually grows and the Roche-lobe filling subgiant gradually expands and transfers mass, roughly at a rate $\sim 10^{-8} M_\odot/yr$. Energy is generated in the subgiant by hydrogen-shell burning. 6.68 $\times 10^8$ yr after the beginning the primary mass has been reduced to 0.28 $M_\odot$. The star now consists of a 0.25 $M_\odot$ degenerate helium core and a 0.03 $M_\odot$ hydrogen rich envelope which still fills the Roche lobe. Hydrogen-shell burning continues to provide the luminosity and drive the mass transfer. Another 0.22 $\times 10^8$ yrs later (6.9 $\times 10^8 yr$ after the beginning) the hydrogen-rich envelope is exhausted and the remant of the envelope ($< 0.001 M_\odot$) contracts to form a white-dwarf envelope around the remaining degenerate core which has a mass 0.26 $M_\odot$.

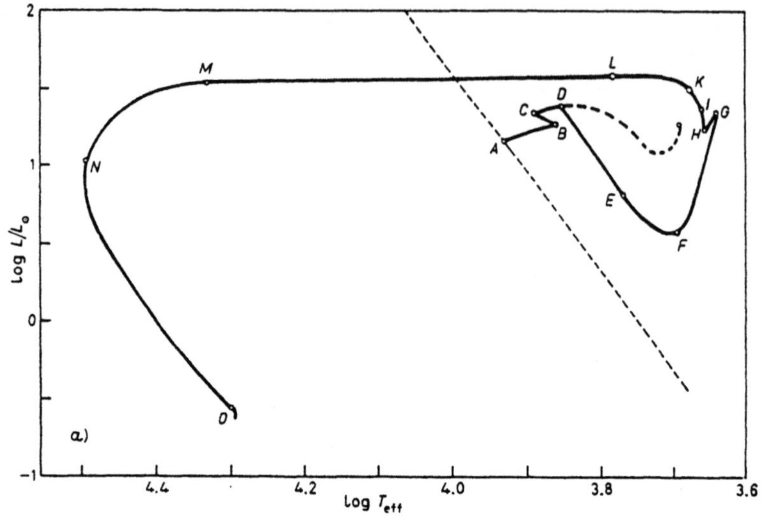

Fig. 59. a.) Evolution in the HR diagram of the primary of a close binary with $M_1 = 2.0M_\odot$, $M_2 = 1.0M_\odot$, $P = 1.15d$ (after [131]). At point D, case B evolution begins. At K, the star becomes a 'subgiant', and at O a white dwarf. The period at K and O is 25.41 d. —main sequence.

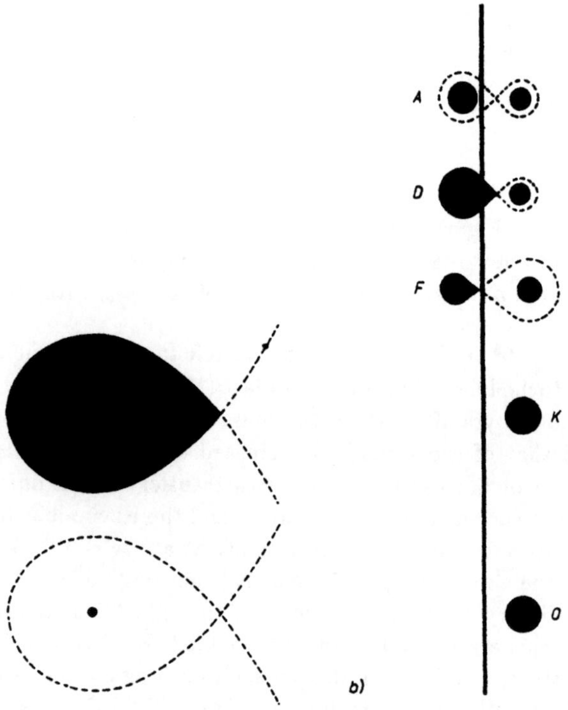

Fig. 59. b.) Dimensions of the system of fig. 59a, at various evolutionary stages.

Table 23. The low-mass Case B binary system of 2+1 $M_\odot$ from figure 59 a,b at various evolutionary stages, as calculated by Kippenhahn, Kohl and Weigert[131]. The initial abundances were $X = 0.602$, $Y = 0.354$, $Z = 0.044$. The letters refer to the evolutionary stages, indicated in figure 59a.

Stage	Age (in 10^8 yr)	M_1 ($M_\odot$)	M_2 ($M_\odot$)	P_{orb} (d)	Status	$log\ (L/L_\odot)$	$log\ T_e$
A	0.00	2.00	1.00	1.14	D	1.307	3.972
B	4.8179	2.00	1.00	1.14	D	1.405	3.909
C	5.0591	2.00	1.00	1.14	D	1.500	3.947
D	5.6968	2.00	1.00	1.14	SD	1.567	3.899
E	5.6989	1.55	1.45	0.80	SD	0.715	3.781
F	5.7509	0.96	2.04	1.22	SD	0.518	3.710
G	6.3538	0.28	2.72	21.40	SD	1.444	3.634
H	6.4940	0.28	2.72	21.40	SD	1.280	3.653
I	6.6808	0.28	2.72	21.40	SD	1.489	3.642
K	6.8176	0.26	2.74	23.74	D	1.634	3.667
L	6.8710	0.26	2.74	23.74	D	1.698	3.795
M	6.9044	0.26	2.74	23.74	D	1.588	4.329
N	6.9328	0.26	2.74	23.74	D	1.024	4.492
O	6.9651	0.26	2.74	23.74	D	-0.572	4.297

Two things are clear from this evolution.

1. Low-mass case B evolution produces long-lived classical Algol-type binaries like Algol itself and UCep. The subgiants in these systems have a degenerate helium core and their luminosity is generated by hydrogen-shell burning. Probably all of the classical Algol systems listed in table 4 are of this type (cf. De Greve[52]).

2. This calculation is the first one in which stellar evolution calculations straightforwardly produced a white dwarf.

Intermediate-mass and massive case B evolution. For $M_1^\circ > 2.8M_\odot$, the helium core in the post-main-sequence star is sufficiently massive ($> 0.5M_\odot$) to avoid degeneracy and to contract immediately to helium ignition (see figure 44). It appears that practically independently of the mass of the envelope the thermal equilibrium radius of a star with a helium burning core and a hydrogen-rich envelope is *always* very large, as can be seen from figures 56 and 60. The latter figure (by Giannone et al.[87]) shows the radii of stars with a helium-burning core mass of 0.8 $M_\odot$ and a total mass ranging from 0.8 to 5 $M_\odot$. (More such models of stars with helium-burning cores or with evolved hydrogen-burning cores are given in figures 85 and 86). The figure shows that for total masses $> 0.9M_\odot$ (i.e. hydrogen-rich envelope masses $> 0.1M_\odot$) the stars all have very large giant-like radii. From this it follows that in a mass-transfer binary such a star can only stay inside its Roche lobe after it has lost practically its entire hydrogen-rich envelope. One thus expects, that in case B systems with $M_1^\circ > 2.8M_\odot$ practically only the helium core of the mass-losing star remains after the mass transfer. This is indeed shown by detailed calculations for this case initiated by

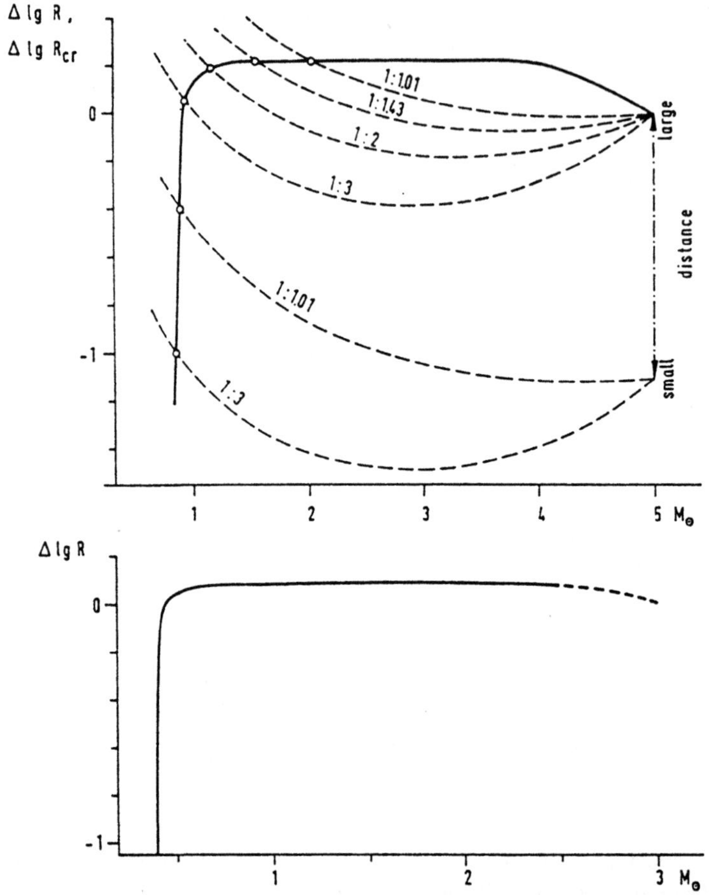

Fig. 60. End of mass loss in case *B* for two initial masses (5 and 3 $M_\odot$) of the originally more massive component, for different mass ratios of secondary and primary (from 1:1.01 to 1:3) and for different initial distances of the components. Solid curves (∇lgR): Plotted over the mass *M* of the originally more massive component are the radii *R* of possible thermally adjusted remnants of that star. Dashed curves: (∇lgR_{cr}): Over the same scale are plotted the radii R_{cr} of the critical Roche volume of this star. Conservation of total mass and total angular momentum of the system is assumed. For different initial distances of the components, the R_{cr} curves have to be shifted parallel up or down (as indicated by arrows in the upper figure). For a given case, the mass loss stops at the intersection of the corresponding dashed curve and the solid curve (open circles). For the lower part of the diagram (initially 3 $M_\odot$), the dashed R_{cr} curves can be shifted parallel downwards from the upper part. Figure 86 gives similar plots for systems transferring mass according to case A. Figure 85 gives additional information about remnants after case B mass transfer (from Giannone et al. [87]).

Paczynski[193] and depicted here in fig. 61a, b, c after van der Linden[331] [332]. These figures show the evolution of five case B systems, with an initial mass ratio $Q_o = M_1^o/M_2^o = 1.5$ and initial primary star masses of 3 $M_\odot, 4M_\odot, 6M_\odot, 9M_\odot$ and $12M_\odot$, respectively and initial orbital periods of 2.00^d, 1.78^d, 3.00^d, 4.00^d and 5.00^d, respectively. Figure 61a shows the evolutionary tracks of the five

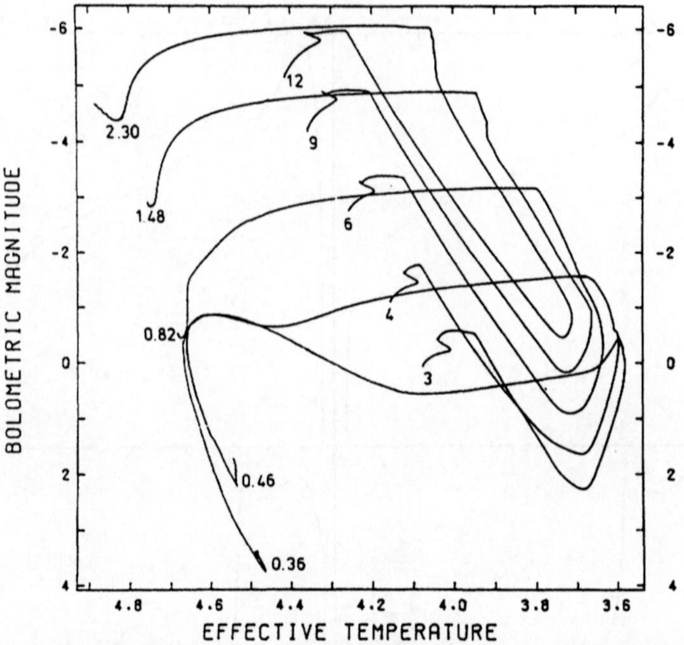

Fig. 61. a.) Theoretical tracks of the primary component of five case B binary systems in the HR-diagram. Indicated at the beginning of each track are primary masses. Remnant masses are indicated at the end end of each track (van der Linden [331], [332]).

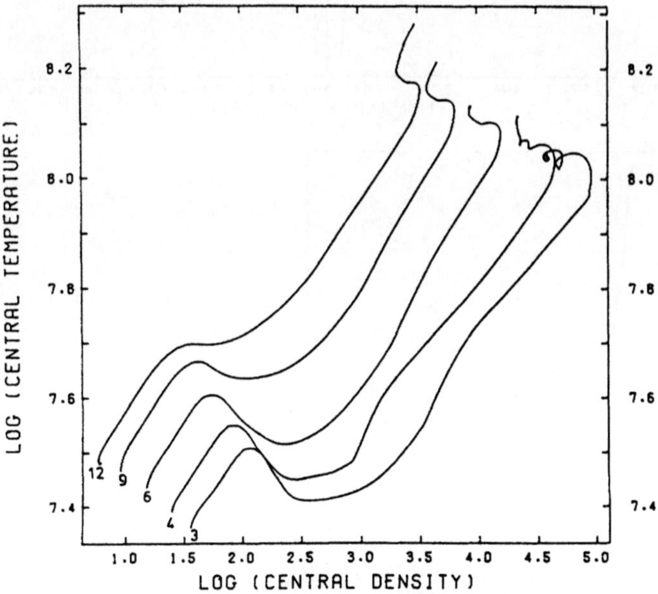

Fig. 61. b.) Temperature-density history of the center of the mass-losing star. Notice the loops in the tracks for the 3 $M_\odot$ primary, which occurs at helium ignition (van der Linden [331], [332]).

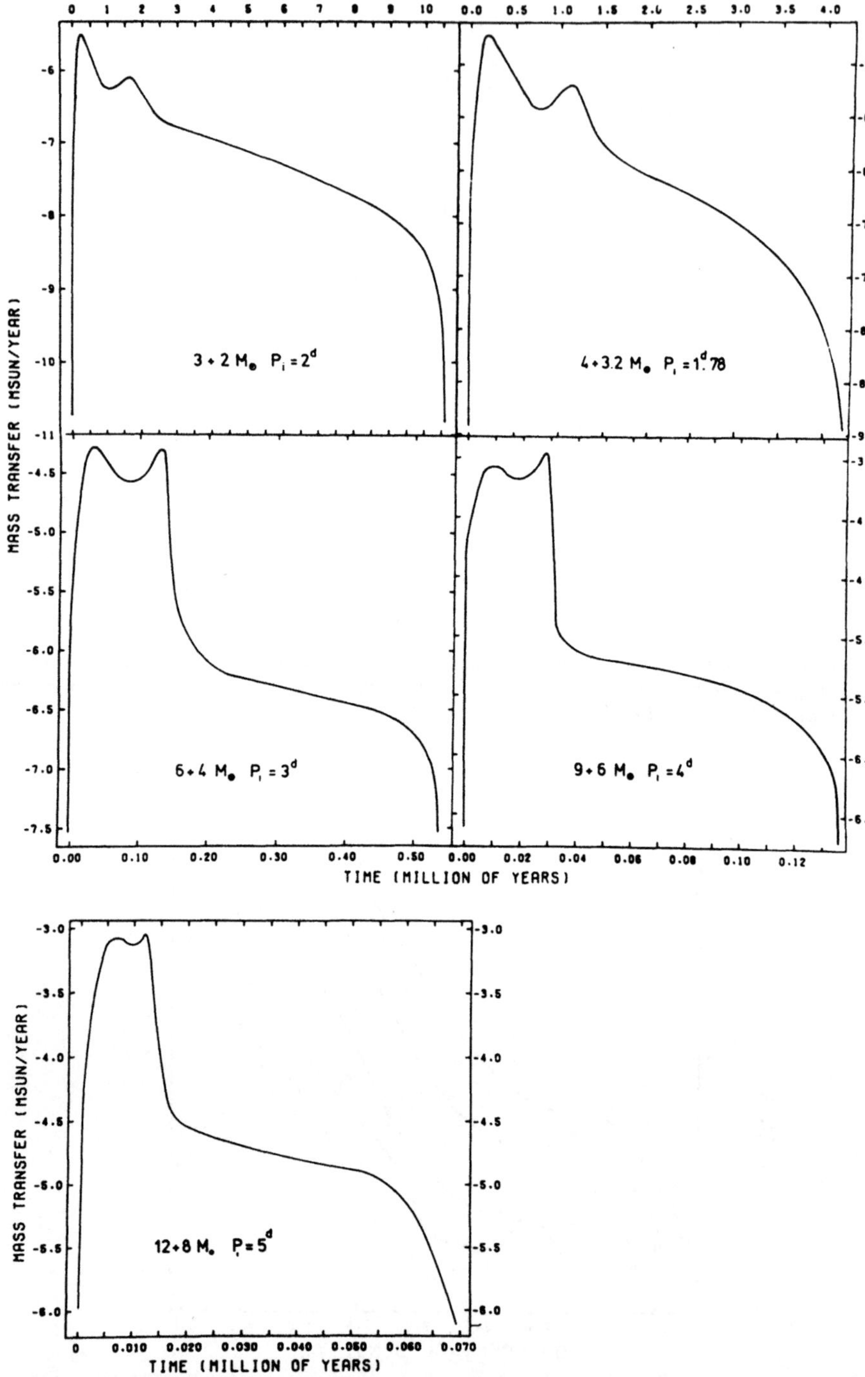

Fig. 61. c.) **Mass transfer rate as a function of evolutionary time since the beginning of mass transfer. Indicated are the masses and initial period of thye systems (van der Linden [331], [332]).**

Table 24. Durations of the fast mass transfer phase and the amounts of mass transferred in the five systems of figures 61a, b, calculated by van der Linden[331],[332].

	Duration of fast mass transfer ($\times 10^6$ yr)	τ_{th} ($\times 10^6$) yr	$\dot{M}$ during fast phase ($M_\odot/yr$)	Mass transferred during fast phase $M_\odot$
$M_1^0 = 3$ M $_\odot$	1.6	1.0	$10^{-6.0}$	1.6
$M_1^0 = 4$	1.0	0.7	$10^{-5.7}$	2
$M_1^0 = 6$	0.12	0.26	$10^{-4.7}$	2.5
$M_1^0 = 8$	0.026	0.047	$10^{-3.6}$	6
$M_1^0 = 12$	0.010	0.030	$10^{-3.1}$	8

primary stars and figure 61c shows the mass-transfer rate $\dot{M}$ as a function of time for the systems. All systems show two maxima of $\dot{M}$ during the transfer. The first one is simply due to the fact that before the reversal of the mass ratio, the mass transfer is very unstable since the star wishes to expand while its Roche lobe is contracting due to the mass transfer. During the rapid mass transfer the luminosity goes down, as can be seen in the HR diagram, as most of the nuclear energy generated in the interior is absorbed in the expanding envelope (80-90 % of the energy) to drive the mass transfer. The second rise of $\dot{M}$ to its maximum occurs at minimum luminosity and is due to the formation of a small convection zone in the envelope of the primary which -as outlined above in section 5.3- causes enhanced mass transfer even after the mass ratio has been reversed. The extent of this zone is regulated by the luminosity of the star. Therefore, when the star tries to return to thermal equilibrium and its luminosity and radius increase, also the extent of the convection zone increases. The growth of this zone makes the mass transfer more unstable and therefore boosts $\dot{M}$ to its second maximum. Hereafter follows a phase when the thermal equilibrium is restored and $\dot{M}$ drops below $10^{-5} M_\odot/yr$.

The durations of the fast mass-transfer phase (from the first peak till after the second one) for the five systems, together with the average mass transfer rates $\dot{M}$ and the total amounts of mass transferred during this fast phase, are listed in table 24. The table also lists the total duration of the mass transfer phase and the thermal timescales of the stars, at the time of minimum distance ($M_1 = M_2$), which is roughly in the middle of the mass transfer phase. For the luminosity we took the nuclear luminosity of the star, which is equal to the luminosity at the onset of the mass transfer. The table shows that indeed the timescales of the fast mass transfer phase are, within about a factor of two, equal to the thermal timescales.

As figure 61a shows, the remnants of the primary stars in all cases are the helium cores of the stars (with only a very thin hydrogen-rich envelope), which in the HR-diagram are situated close to the pure-helium-star main sequence.

The remnant masses range from 0.36 $M_\odot$ to 2.3 $M_\odot$. The 0.36 $M_\odot$ helium star still reaches core helium burning. Since at helium ignition its core is partly degenerate, this ignition leads to the occurrence of several thermal pulses during which the convective core mass in this star varies between 10 and 50 per cent of the stellar mass (van der Linden[331] [332]). Only helium cores below 0.33 $M_\odot$

We will consider the further evolution of helium stars below. Figure 61b shows the evolution of the central temperature and density of the primary stars in the five systems, up till helium ignition.

5.6 Comparison with the observations.

Intermediate-mass case B evolution and Bêta Lyrae, the formation of Be stars in binaries and helium star binaries. The system of Bêta Lyrae consists of a 2 $M_\odot$ B8II bright giant primary component which is overflowing its Roche lobe and a 12 $M_\odot$ companion which is dimmer than the B8II star and has the appearance of a rapidly rotating disk. The latter follows from the shape of the light curve, the underluminosity of this star, and from the enormous rotational broadening of its spectral lines. As shown in section 2.3 the mass transfer rate in light curve, the underluminosity of this star, and from the enormous rotational broadening of its spectral lines. As shown in section 2.3 the mass transfer rate in the system is close to $10^{-4} M_\odot/yr$ which implies that the B8II star is still losing mass to its more massive companion at a very high rate.

Most remarkable is that in the near-infrared spectrum of Bêta Lyrae the absorption lines of CI, which normally are strong in a B8II star, are totally absent (see figure 62), which indicates that carbon is underabundant by a factor 25 or more.

From the strength of its helium lines one finds that He/H= 1.5, i.e. five times larger than normal. These two facts together imply that the atmospheric material of the primary star has been nuclearly processed by the CNO cycle: it was part of the hydrogen-burning core of the original primary star. It is well-known that during the main-sequence evolution the size of this convective core, in which hydrogen-burning proceeds by the CNO-cycle, gradually shrinks, leaving behind a region enhanced in helium and depleted in C. Figure 63 shows this for models calculated by van der Linden[331]. From this information, Tomkin and Lambert[282] conclude that the B8II primary star consists of a 1.6 $M_\odot$ helium core surrounded by a 0.4 $M_\odot$ envelope with He/H=1.5. When the latter envelope will have been transferred, some 4,000 years from now, the system will consist of a 1.6 $M_\odot$ helium star and a $\sim 12.4 M_\odot$ companion with an orbital period of about 20 days.

After the disk matter surrounding the companion has settled, this star will become a very rapidly rotating early B-type main-sequence star (spectral type about B2V). Since very rapidly rotating main-sequence B stars in general have

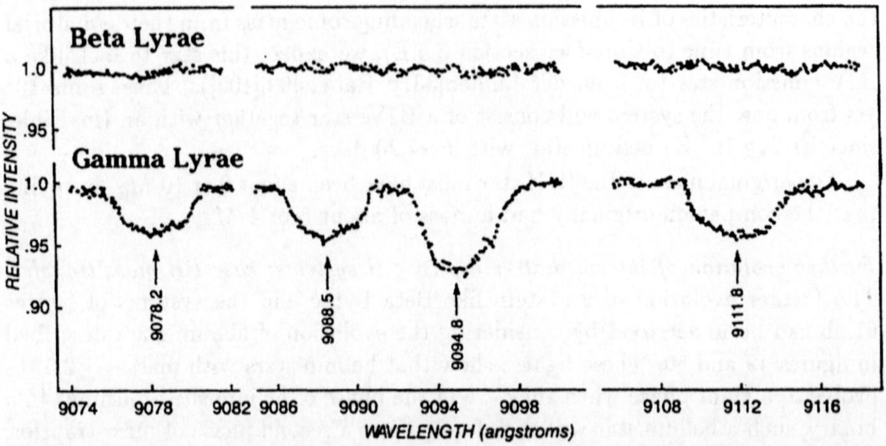

Fig. 62. The 9100-angstrom absorption lines of neutral carbon in Beta (top) and Gamma Lyrae. Gamma is of normal chemical composition, but otherwise similar to Beta. The complete absence of the lines in the latter indicates an unusually low abundance of carbon in the B8 primary, just as expected if it is the evolved core of a once more massive star. Diagram from Tomkin and Lambert [282]).

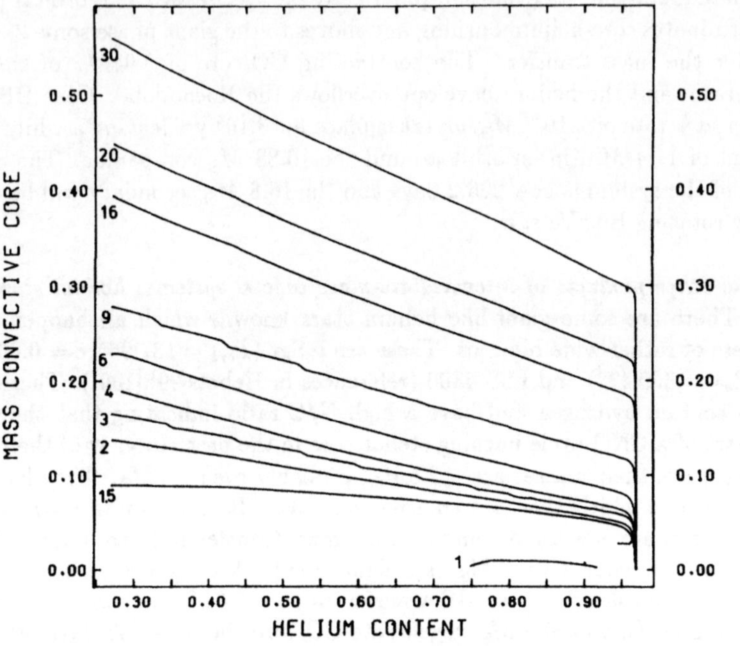

Fig. 63. Relative mass of the convective core during the main-sequence evolution, plotted as function of the helium content in the core. Indicated are the stellar masses. The stars have $X = 0.70$, $Z = 0.03$; van der Linden [332].

the characteristics of B-emission stars -shedding some mass from their equatorial regions from time to time (see section 3.5.E), we expect this star to look like a B2V-emission star (cf. van der Linden[331]; Habets[99][100]). Thus, some 10^4 yrs from now the system will consist of a B2Ve star together with an (invisible, since its $T_e \gtrsim 10^5$ K) helium star, with $P \sim 20$ days.

The original mass of the B8II star must have been about 9 or 10 $M_\odot$, implying that the companion originally had a mass of about 5 or 4 $M_\odot$.

Further evolution of intermediate-mass case B systems: case BB mass transfer. The further evolution of a system like Bêta Lyrae and the systems of figures 61.ab can be understood by considering the evolution of helium stars described in figures 49 and 50. These figures show that helium stars with masses $\lesssim 2.5 M_\odot$ evolve to a giant phase when they are in the phase of helium shell-burning. In a binary, such a helium star will therefore go into a second phase of mass transfer, called case BB transfer by Delgado and Thomas[61]. The first persons to actually calculate the evolution of such a system up till the end were De Loore and De Greve[64] who considered the evolution of a system with initial masses 10+8 $M_\odot$ and initial orbital period 3.15 days. This system is not much different from the alleged progenitor of the Bêta Lyrae system.

After the first transfer the secondary, now a star of 16.34 $M_\odot$, is practically rejuvinated, starting its evolution practically on the ZAMS. The orbital period star terminates core-helium burning and moves to the giant phase some 2.2×10^6 yrs after the mass transfer. The contracting CO-core of $0.94 M_\odot$ of this star degenerates and the helium envelope overflows the Roche lobe. Case BB mass transfer at a rate of $2.10^{-5} M_\odot/yr$ takes place for 3.10^4 yrs leaving a white dwarf remnant of 1.14 $M_\odot$, in an orbit around the 16.88 $M_\odot$ companion. The orbital period of the system is now 226.2 days and the 16.8 $M_\odot$ secondary will be a very rapidly rotating B0.5Ve star.

Possible other products of intermediate-mass case B systems: helium star binaries. **There are some giant-like helium stars known, which all happen to be members of rather wide binaries. These are ν Sgr ($P_{orb} = 137.96^d, e = 0.06$), KS Per ($P_{orb} = 360.47^d$) and LSS 4300 (references in Habets[99],[100]). These stars do not contain hydrogen and have a high N/C ratio indicating that they have been part of a CNO-cycle burning stellar core in the past. In view of their giant like structure, their masses are $\lesssim 2.5 M_\odot$, probably even $\lesssim 1 M_\odot$. They have hot companions to which they are transferring mass. It seems quite possible that these systems are now undergoing case BB mass transfer and had initial primary star masses of order 4 to 6 $M_\odot$ (see figure 61a). A $6 M_\odot + 4 M_\odot$ system with $P = 7^d$ will finish after case BB transfer as a $\sim 0.6 M_\odot$ helium shell-burning giant remnant (with a 0.5 $M_\odot$ degenerate CO core) in an $\sim$ 137 days orbit like ν Sgr.**

Massive Case B systems and Wolf-rayet binaries. For primary stars more massive than $15 M_\odot$ the evolution through case B mass transfer proceeds in a

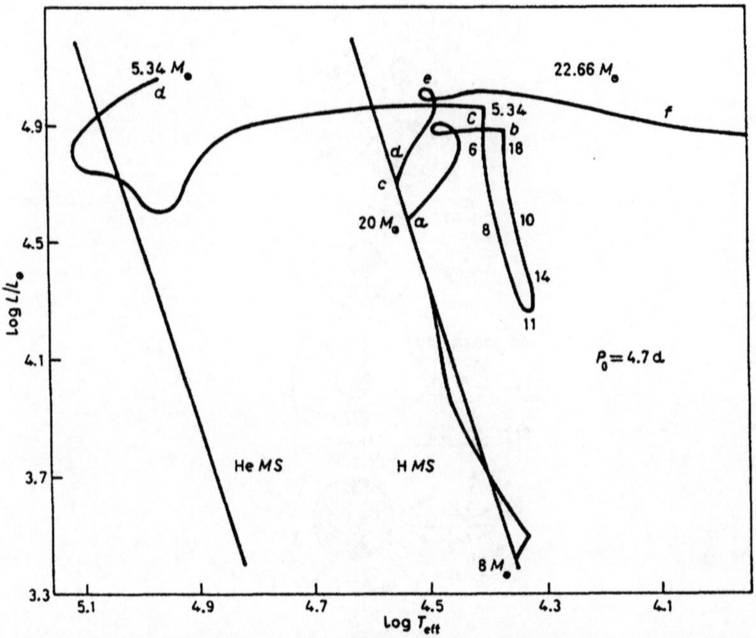

Fig. 64. a). Evolutionary tracks in the Hertzsprung-Russell diagram of the components of a close binary with initial masses of 20 $M_\odot$ and 8 $M_\odot$ and an initial period of 4.70^d computed by De Loore et al. [63] with the assumptions of conservation of total mass and angular momentum. Letters correspond to the evolutionary phases depicted in fig. 64b. The first stage of mass transfer begins in *b* and terminates in *c*. In *d* the primary (a 5.34 $M_\odot$ helium star) explodes as a supernova. In the *f* the secondary reaches its Roche lobe and the second phase of mass exchange begins. Some values of the mass of the primary are indicated along its mass-loss track.

way similar to that of the intermediate-mass case B systems, producing more and more massive helium stars.

Pioneering calculations for massive case B systems were made by Paczynski[198], who was the first to point out that the systems resulting from such evolution closely resemble Wolf-Rayet binaries. Later on such systems are expected to evolve into high-mass X-ray binaries as was first pointed out by van den Heuvel and Heise[321] and van den Heuvel[303]. As an example, figures 64a,b show the evolution of a binary with initial masses of 20 $M_\odot$ and 8 $M_\odot$ and an initial orbital period of 4.70 days (after De Loore et al.[63]). At $t = 6.17 \times 10^6$ yrs (phase b) the primary begins to overflow its Roche lobe and after 3.10^4 yrs a close binary is left consisting of the 5.34 $M_\odot$ helium core of the initial primary star and a 22.66 $M_\odot$ companion in an 10.86 day orbit (phase c).

Figure 64a shows the evolution of this system in the HR diagram. The figure show that the (bolometric) luminosity of the $5.34 M_\odot$ helium core (with its hydrogen envelope of negligible mass) is similar to that of its $22.66 M_\odot$ main-sequence companion.

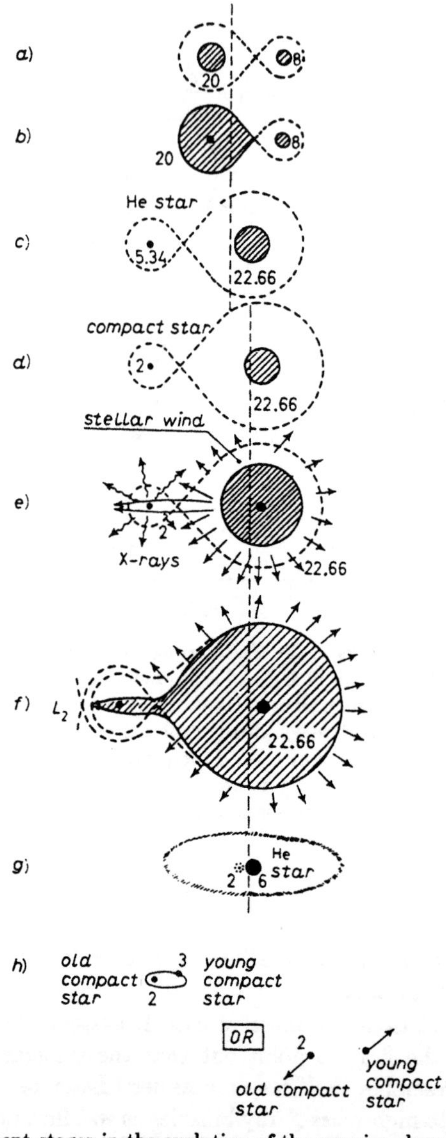

Fig. 64. b.) Subsequent stages in the evolution of the massive close binary of fig. 64a. It is assumed that the supernova explosion of the primary leaves a 2 $M_\odot$ compact star(neutron star or black hole). a) $t = 0$, $P = 4.70\,d$; b) $t = 6.17 \times 10^6\,y$, $P = 4.70\,d$, onset of the first stage of mass exchange; c) $t = 6.20 \times 10^6\,y$, $P = 10.86\,d$, end of first stage of mass exchange(beginning Wolf-Rayet stage); d) $t = 6.76 \times 10^6\,y$, $P = 12.63\,d$, He star (=Wolf-Rayet star) has exploded as supernova; e) $t = 10.41 \times 10^6\,y$, $P = 12.63\,d$, the normal star becomes a supergiant; its strong stellar winds turns the compact star into powerful X-ray source; f) $t = 10.45 \times 10^6\,y$, $P = 12.63\,d$, onset of second stage of mass exchange, the X-ray source is extinguished and large mass loss from the system begins; g) $t \sim 10.47 \times 10^6\,y$, $P \sim 4h$, onset of second Wolf-Rayet stage; h) $t \sim 11 \times 10^6\,y$, the second He star has exploded as a supernova, survival or disruption of the system depends on the mass of the remnant.

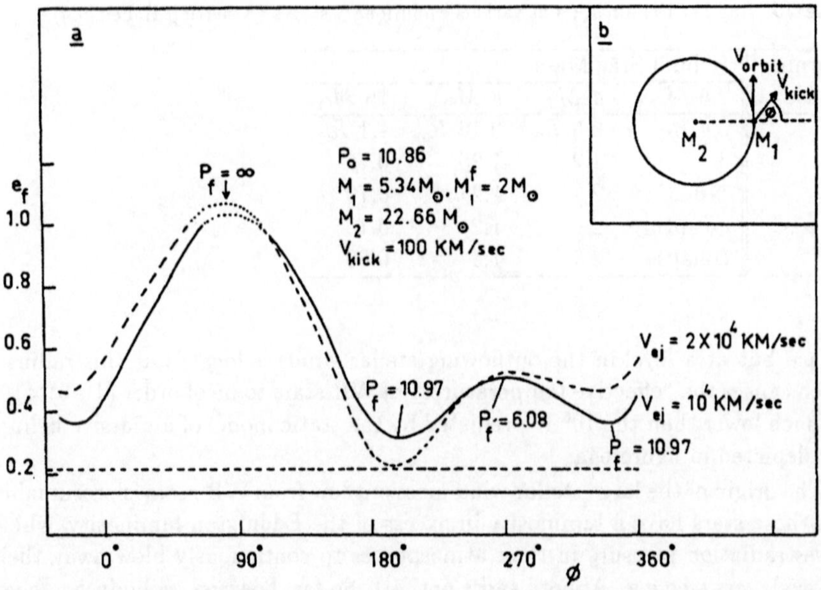

Fig. 65. The effects of asymmetric mass ejection on the post-supernova orbital eccentricity of the system of figure 64ab, for two values of the ejection velocity of the supernova shell. It is assumed that the compact star receives at birth a kick in the orbital plane of 100 km^{-1}, directed under an angle Φ, as defined in figure b. Effects of impact were taken into account. The straight lines represent the eccentricities resulting from symmetric mass ejection.

Such a system consisting of two components differing by a factor three or four in mass, though having similar bolometric luminosity, closely resembles the Wolf-Rayet binaries listed in table 5 (although most of the systems in the table are somewhat more massive than the system of figures 64a, b). Indeed, just like massive helium-burning helium stars Wolf-Rayet stars have in general little or no hydrogen in their atmospheres, and are overluminous for their masses by a factor of order 30 (in comparison to the luminosity expected for the standard mass-luminosity relation for hydrogen-rich stars). It should be pointed out that although these evolutionary calculations do correctly predict the masses and luminosities of Wolf-Rayet stars, they do not predict the very large mass loss rates by high velocity stellar winds observed in these stars, $\dot{M} \gtrsim 10^{-5} M_\odot/yr$ at v $\simeq$ 2,000-4,000km/s, which are the distinguishing characteristic of a Wolf-Rayet star. It is just this dense high-velocity wind in combination with its large ultraviolet luminosity that produces the characteristic Wolf-Rayet emission spectrum (see e.g. Underhill[292]; De Jager[54], van der Hucht[327]).

The evolutionary calculations predict for helium stars a static stellar model with a a small radius, of order $0.5 - 1 R_\odot$ (see table 25). However, due to the high density of the wind, this wind is opaque to optical radiation for distances closer than 5 to 10 $R_\odot$ to the stellar center (the surface of opical depth $\tau = 2/3$ in the wind is located at approximately these radii). Therefore the pseudo-photoshere of Wolf-Rayet star ("pseudo" because we are, in fact, not looking at a stellar

Table 25. Radii of Helium stars at various evolutionary stages (Arnett[6]; Habets[99]).

Burning	Helium Star Mass			
Phase	2.5 $M_\odot$	4 $M_\odot$	8 $M_\odot$	16 $M_\odot$
He	0.4 $R_\odot$	0.5 $R_\odot$	0.76 $R_\odot$	1.1 $R_\odot$
C	3	1.2	1.05	0.75
Ne	110	2	1.2	0.70
O	e-capture	2	1.2	0.75
Si	collapse	2	1.3	0.80

surface but at a level in the outflowing stellar wind) is located at this radius - which causes the "effective temperatures" of WR-stars to be of order $(4-5) \times 10^4$ K, much lower than the 10^5 K predicted by the static model of a massive helium star depicted in figure 64a.

The origin of the large stellar wind mass outflow from WR-stars is presumably that these stars have a luminosity in excess of the Eddington luminosity, which causes radiation pressure in their atmospheres to continuously blow away their surface layers (see e.g. Abbott and Conti[4]). So far, however, nobody has been able to self-consistently predict the high mass-loss rates from these stars by model calculations.

The final evolution of case B systems with primary masses $\gtrsim 12$ ($\pm$ 1) $M_\odot$: effects of supernova explosions on the orbits. The helium cores of these stars left after case B mass transfer are expected to be more massive than about $2.2 M_\odot$ and will, according to figure 50, develop heavy-element cores larger than the Chandrasekhar mass, which will certainly collapse to a neutron star. For helium core masses $\gtrsim 3.5 M_\odot$ the helium star never reaches a radius larger larger than 4 $R_\odot$ before its core collapses. Hence, in a binary -unless it is very close- it will not go through a second (case BB) phase of mass transfer.

On the other hand, helium stars in the mass range 2.2 $M_\odot$ to 3.5 $M_\odot$ - corresponding to initial primary star masses between 12 $(\pm 1) M_\odot$ and about 14 $(\pm 1) M_\odot$, will go through case BB mass transfer before their cores collapse. Figure 66 shows the evolution of such a system with initial component masses of 13 + 6.5 $M_\odot$ and an initial orbital period of 2.58 days, calculated by Habets[99],[100]). After the first case B mass-transfer phase starting at $t = 1.6 \times 10^7$ yr the component masses are 2.5 + 17 $M_\odot$ and the orbital period is 20.29 days. Two million years later case BB mass transfer takes place, which takes 2.10^3 yrs and leaves component masses 2.2 $M_\odot$ and 17.3 $M_\odot$ with $P = 25.1$ days. 4.10^3 yrs later the core of the 2.2 $M_\odot$ remnant collapses and 0.8 $M_\odot$ are expected to be explosively ejected in a supernova.

The explosive mass ejection increases the orbital period to 30.6 days and imparts a runaway velocity of 10 km/s to the center of gravity of the system (see section 2).

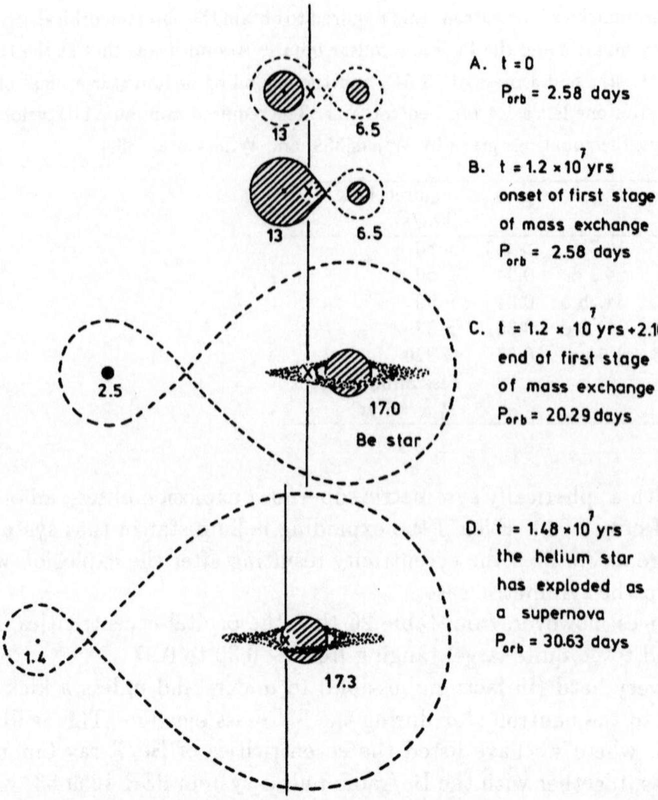

Fig. 66. Conservative scenario for the formation of a Be/X-ray binary out of a close pair of early B stars with masses of 13.0 $M_{\odot}$ and 6.5 $M_{\odot}$. The numbers indicate mass (in units of $M_{\odot}$). After the end of the mass transfer the Be star presumably has a circumstellar disk or shell associated with its rapid rotation (induced by the previous accretion of matter with high angular momentum; from Habets [99]).

The final system consisting of a neutron star and a rapidly rotating early B-type star closely resembles the B-emission X-ray binaries (see table 15 and figure 25).

The impact of the SN-shell onto the massive companion is not expected to have much effect, neither on the companion, nor on the orbit, as was shown by the hydrodynamic calculations by Fryxell and Arnett[85]. A small amount of matter will be ablated from the massive star, but over-all the pure mass-loss effects on the orbit, dealt with in section 2, are dominant. (The same is not necessarily true if the companion is a low-mass star, cf. Taam and Fryxell[267], although also here in most cases these effects are smaller than one might have expected). In view of this, it is justified to neglect the impact effects in a wide massive system like the one in figure 66.

The orbital eccentricities in Be/X-binaries expected with and without asymmetries in the SN mass ejection: Evidence for kicks. The system depicted in figure

Table 26. Minimum kicks to neutron stars required to obtain the observed orbital eccentricities of the Be/X-ray binaries and the Be/radio pulsar binary. Assumed was that at the time of the explosion the Be-star had a mass of 15 $M_\odot$ and the exploding helium star a mass of 2.5 $M_\odot$, and that the latter one left a 1.4 $M_\odot$ neutron star. The required minimum kick velocities were calculated using the equations given by Wijers[359] and Wijers et al.[360].

Be/X-ray binaries	P_{orb}	e	required kick (km/s)
A0538-66	16.7 d	> 0.4	> 80
A0115+63	24.3 d	0.34	> 60
A0332+53	34.25 d	0.31	> 50
A0535+26	111.0 d	0.30	> 32
PSR 1259-63	3-6 yr	0.97	> 220 (assuming present P_{orb} =6 yr)

66 would with a spherically symmetric supernova explosion obtain an orbital eccentricity of only 0.043. Even if the exploding helium star in that system had a mass as large as 3.0 $M_\odot$, the eccentricity resulting after the explosion would be only 0.086 in the symmetric case.

One notices, however, from table 26 that the orbital eccentricities in Be/X-binaries tend to be quite large, ranging from $\sim$ 0.30 to 0.97.

This is very hard (in fact: impossible) to understand unless a kick velocity is imparted to the neutron star during the SN mass ejection. This is illustrated by table 26, where we have listed the eccentricities of Be/X-ray binaries with known orbits, together with the Be/radio pulsar system PSR 1259-63, which has $e = 0.97$ (Johnston et al.[125].) Since the Be stars in Be/X-ray binaries have spectral types B0-2V-III, their masses are expected to be between 10 and 20 $M_\odot$, i.e. on average $\sim 15 M_\odot$, which very well fits the scenario depicted in figure 66. From eq. (16) it follows that for a symmetric explosion $e_f = (1 - \mu_f)/\mu_f$ with $\mu_f = (M_1^f + M_2^f)/(M_1^0 + M_2^0)$. Assuming $M_1^f = 1.4 M_\odot$, $M_2^f = M_2^0 = 15 M_\odot$, the observed values of e_f allow one to calculate M_1^0, the mass of the helium star before the assumed symmetric explosion. The thus obtained values of M_1^0 range from 6.3 $M_\odot$ (for $e = 0.30$) to 17.3 $M_\odot$ (for $e = 0.97$). These values are, obviously, much too high: they would imply that the helium stars in the progenitors of all Be/X-ray binaries were massive Wolf-Rayet stars. Since the masses listed in the table are the ones at the moment of the explosion, and WR stars have very high mass loss rates ($> 10^{-5} M_\odot/yr$) for several times 10^5 yrs, these helium cores would have had initial masses in the range 10 to 25 $M_\odot$, implying main-sequence progenitor masses between 30 and 60 $M_\odot$. Obviously this is most unlikely since, first of all, the Be star (the initial secondary) would in that case have captured much more of the envelope matter of this massive companion- and thus would have had a mass much larger than the observed values $\sim 10 - 20 M_\odot$, and secondly, in that case -since Be/X-ray binaries are very numerous- one would expect to observe many WR stars with early B-type companions, which is not the case (see table 5): in practically all WR binaries the companion is an O-star.

The sheer abundance of Be/X-ray binaries (with an estimated total galactic number $\sim 10^4$, see section 9) implies that they must originate from binaries with primary stars $\sim 10 - 20 M_\odot$ (see e.g. Meurs and van den Heuvel[171]) which implies helium-star progenitors of their neutron stars, in the mass range 2.2 $M_\odot - 5 M_\odot$. With such helium stars the observed eccentricities of Be/X-binaries cannot be reached with a symmetric SN explosion.

The high observed eccentricities of the Be/X-ray binaries therefore provide unequivocal and compelling evidence for asymmetric mass ejection ("kicks to neutron stars") in supernova explosions (see also Dewey & Cordes[67] and Bailes[9] for independent evidence).

Assuming helium star progenitors in Be/X-ray binaries to have M $\sim 2.5 M_\odot$ at the time of the explosion (implying an initial helium-star mass ~ 3.0 to $3.3 M_\odot$ before case BB mass transfer), one needs kick velocities of the order of those listed in table 26 to achieve the presently observed orbital eccentricities of the Be/X-ray binaries. These kicks were calculated for the most favourable case, i.e.: a kick exactly in the direction of the orbital motion of the exploding star. Since, however, kicks are expected to be randomly directed in space, the really required kick velocities will be 1.5 to 2 times larger. One thus obeserves that the required kick velocities are typically of order 50 to 150 km/s, i.e. of the same order as observed pulsar velocities (Harrison et al.[103] and references therein). Only for the Be/radio pulsar system PSR 1259-63 a very high kick velocity of order > 220 km/s is required.

The fate of the most Massive Case B systems ($M_1^0 \geq 15 - 20 M_\odot$): Formation of "standard" high mass X-ray binaries. Since helium stars with $\gtrsim 4 M_\odot$ do not undergo much radius expansion before their final core collapses to a neutron star (or a black hole), one expects no case BB mass-transfer in these systems.

Figure 64b shows the further evolution of such a system in which the 5.34 $M_\odot$ helium star that resulted from 20 $M_\odot$ primary exploded as a supernova. The resulting system has a runaway velocity of $\sim 30 - 50 km/s$ (depending on whether or not the neutron star received a kick at birth and on the direction of the kick, as explained in the caption of figure 65).

The resulting system which has an orbital period of 12.63^d (or somewhat shorter or longer if a kick to the neutron star is included) is an excellent progenitor for an HMXB such as Vela X-1 (4U 0900-40).

For the system of figure 64b it will take another 3.6 million years after the SN-explosion of the helium star, before the 22.66 $M_\odot$ companion has exhausted the hydrogen in its core and evolves into a blue supergiant with a strong wind (see also figure 64a). When this happens, the system will become a strongly X-ray emitting "standard" HMXB system like Vela X-1 for a short period of time. This phase lasts until the blue super giant begins to overflow its Roche lobe and a very high -thermal time scale- mass transfer rate of order $10^{-3} M_\odot/yr$, will ensue. This will completely extinguish the X-ray source. The further evolution of this system, after this stage, will be dealt with in chapter 6.

Wolf-Rayet binaries, standard HMXBs and the origin of runaway OB stars.

Wolf-Rayet stars and HMXBs. The evolution of the system of figure 64b into a "standard" HMXB is straight forward. However, there are in massive systems some complications due to stellar wind mass loss which we so far ignored. This will not qualitatively alter the evolutionary pattern depicted in figure 64b, but only somewhat change the numerical values of the masses of the components (and also the orbital period) at various evolutionary stages. But the main line of the evolution will remain the same.

Wolf-Rayet stars are helium stars with masses $\gtrsim 5 - 6 M_\odot$ and have strong stellar wind mass loss at rates of order 10^{-5} to $10^{-4} M_\odot/yr$. As they live for several times 10^5 yr they will lose amounts of mass ranging from a few to several tens of solar masses (in the most massive systems).

We have taken this strong wind mass loss into account for the observed WR+O binaries listed in table 5. We have assumed that WR stars with masses $> 10 M_\odot$ will at the time of their SN explosion have a mass of $10 M_\odot$, and that WR stars with masses $< 10 M_\odot$ will explode with their present masses. We assumed them to leave a 1.4 $M_\odot$ neutron star, and assumed the SN mass-ejection to be symmetric and instantaneous and calculated the resulting orbital periods P_f, eccentricities e and runaway velocities imparted to the center of mass of the systems, v_g. These are listed in table 27.

Table 27. Orbital periods (P^f), eccentricities (e) and system runaway velocities (v_g) that known Wolf-Rayet Binaries (cf. Table 5) will obtain after the supernova explosion of their WR star, assuming a 1.4 $M_\odot$ neutron star to be left. It is assumed that the mass of the WR star at the time of the explosion is 10 $M_\odot$ (or present WR mass, if this is less than 10 $M_\odot$), and that the mass ejection is instantaneous and symmetric. For short-period systems also the period after tidal circularization of the orbit is given, in parentheses. The last column lists the radial velocity amplitude of the O-star induced by the presence of the neutron star remnant of the WR star.

Name	$P(d)$	$M_{wr} + M_{OB}$ $M_\odot$	$P_f(d)$	e	$V_g(km/s)$	$K(km/s)$
AB 6	6.861	6.4 +37	8.99	0.13	43.4	12
AB 8	16.644	13.5 + 50	23.66	0.17	45.2	8
γ^2 Vel	78.5	25 + 35	130.8	0.24	33	5.4
HD94305	18.82	16 + 35	31.33	0.24	52	9
HD97152	7.886	20 + 35	13.1	0.24	70	11.6
HDE311884	6.34	50 + 60	8.5 (8.2)	0.14	57	9.6
HD152270	8.893	20 + 60	11.9	0.14	51	8.4
HDE320102	8.83	11 + 35	14.7	0.24	67	11.2
HD168206	29.707	13 + 25	61.8	0.33	53	8.6
HD186943	9.555	13 + 25	19.9	0.33	77	12.4
HD190918	112.8	9 + 35	176.2	0.21	26	4.9
V444Cyg	4.212	11 + 30	6.2 (5.9)	0.19	77	13.1
GPCep	6.69	10 + 45.5	9.9 (9.6)	0.18	65	10.9
CQCep	1.64	23 + 19.3	4.3 (3.2)	0.41	155	27
CXCep	2.217	11 + 25.6	4.3 (3.7)	0.32	126	21.6

The table shows that the resulting orbital periods range from 4^d to 176^d, with the majority in the range 4-15^d, as observed in most of the standard HMXBs. The eccentricities range from 0.14 to 0.56 and the imparted runaway velocities range from 26 to 172 km/s, with most systems receiving velocities in the range 50-120 km/s.

After tidal circularization of the orbits, most orbital periods will be between 10 and 25 per cent shorter (as indicated in parentheses for a few systems; for calculation of these post-circularization periods, see e.g. De Greve et al.[53], cf. Tassoul and Tassoul[272]).

Runaway Velocities and Runaway OB-Stars. The runaway velocities of the O-type spectroscopic binaries with a compact companion listed in table 27 are typically in the range observed for runaway O-stars (see, for example, Gies[88]; Gies and Bolton[89]; Stone[263]). The spectroscopic binary character of these O-stars will be very hard to recognize as the velocity amplitudes induced by the presence of the neutron star are mostly below the detection limit of 15-20 km/s in these broad-lined stars (see e.g. van den Heuvel 1985[314] for a discussion).

Already over 30 years ago Blaauw[22] suggested that runaway OB stars resulted from close binaries in which one of the components exploded as a supernova. In this original paper the occurrence of mass transfer prior to the SN-explosion was not yet taken into account, as this was not yet known at the time. It was thus assumed that it was the more massive component which exploded, and that the systems were all disrupted. However, as table 27 shows, also if the mass transfer is properly taken into account, runaway velocities mostly in the range 50-120 km/s result, while the systems are not disrupted.

In a recent important paper Blaauw[23] showed that the seven best studied runaway OB stars -for all of whom the parent association from which they were ejected is known- indeed show clear traces of prior close binary mass transfer, namely:

- a higher than normal helium abundance;
- very rapid rotation in 6 out of the 7 cases;
- in the HR diagrams of their parent associations these stars are "blue-stragglers", just as one would expect from close binary mass transfer, as this mass transfer will have rejuvinated the mass-receiving component (see chapter 9, figure 89).

Most of these runaway stars are O-stars and it thus seems most logical to associate them with WR-binary progenitors like the systems in table 27.

It is important to notice that also the HMXBs as a group share the characteristics of the runaway OB stars, as was shown by van Oyen[333]; see also van den Heuvel[314]), namely:

- they are practically never found in OB-associations;
- they have a significantly wider galactic $|z|$-distribution than the ordinary OB stars;

- several of them have the characteristics of extreme runaway OB stars by either being at distances of 200 to 300 pc from the galactic plane, or by having a very high system radial velocity (e.g. QV Nor (4U 1538-52) which has an excess radial velocity of $> 100 km/s$ with respect to its local center of rest).

In addition recent ROSAT observations of M31 have nicely confirmed that the HMXBs in that galaxy are never found in OB associations. There is no positional correlation between the X-ray sources in the regions of the spiral arms -presumably these are the HMXBs like in our galaxy- and the OB associations (Lewin and van Paradijs[153]). This indicates clearly that the HMXBs in M31 must be runaway objects.

It thus seems that the the runaway OB stars and the HMXBs form one and the same group of objects.

Since kicks to neutron stars in SN explosions are important, as we saw with the Be/X-ray binaries, one might think that a number of the systems may have been disrupted in the SN explosions. However, since the kicks are expected to be randomly directed in space, certainly not more than 50 per cent of the systems can have been disrupted in this way. As a matter of fact, the fraction of systems that is disrupted is much smaller than 50%, as is shown by calculating the "kick"-velocities directed in the same way as the orbital motion (so-called prograde kicks) that are required to disrupt the WR systems listed in table 27. One can easily verify that only in three out of the 14 systems the purely prograde kick velocities required to disrupt the systems are $< 200 km/s$, and in all other cases they range from 200 to 400 km/s. Since purely prograde kicks represent the most favourable conditions for disruption, these kick velocities represent the required rock-bottom minimum values. Since the bulk of the pulsars probably have runaway velocities not larger than 200 km/s and since half of all kicks will be retrograde, we expect less than 20 per cent of the systems in table 27 to be disrupted in the SN explosions of their WR stars even if kicks are taken into account.

We thus expect many O-type runaway stars to be accompanied by a quiet compact object (quiet, because as long as the O-star is on the main sequence, wind accretion will be negligible).

We conclude from the above that the number of X-ray quiet systems is very much larger than the number of standard HMXBs, since the quiet phase lasts of order 2 to 5 million years (see figure 64b) whereas the X-ray phase lasts probably not longer than $\sim 0.5 \times 10^5$ yrs, i.e.: some hundred times shorter. So, with some ~ 50 standard HMXBs in the Galaxy, the number of O-type systems with a quiet compact companion will be some 40 to 100 times larger, i.e. some $(2 - 5) \times 10^3$ (van den Heuvel[304][306][308]).

From the above-mentioned new findings by Blaauw[23] and the new ROSAT findings on M31 mentioned above (Lewin and van Paradijs[153]) it seems now most plausible to identify these quiet systems with a large fraction of the classical runaway OB stars.

Reasons for the systematically much longer orbital periods of the Be/X-ray binaries relative to those of the standard HMXBs. Equation (89) shows the mass of the helium core as a function of the initial main-sequence stellar mass for stars with X=0.70, Z=0.03 without convective overshooting. The value of the mass core M_c is taken at the time when the star leaves the main sequence. According to equation (89) this mass is given by

$$M_c = 0.10 M^{1.40} \tag{95}$$

The mass of the convective burning core M_{cc} on the Zero Age Main Sequence is much larger (see figure 63), and is roughly given by:

$$M_{cc} = 0.17 M^{1.40} \tag{96}$$

Equation (95) implies that after case B mass transfer, a star of lower mass leaves a much smaller fraction of its mass as a helium star, than a star of larger mass.

Because of equation (91) the orbital change in the case of conservative case B mass transfer will therefore be much larger in systems of lower mass than in high-mass systems, for systems that started out with the same mass ratio Q_0. Using eq. (91) the change in orbital period is given by

$$\mathcal{P}/\mathcal{P}_0 = (M_1^0.M_2^0/M_1.M_2)^3 \tag{97}$$

where indices zero indicate the initial situation. Since $M_1^0 = Q_0 M_2^0$ and $M_1 = M_c = pM_1^0$, one has:

$$\mathcal{P}/\mathcal{P}_0 = [p(1+(1-p)Q_0]^{-3} \tag{98}$$

For the same Q_0, the mass fraction p in the core therefore determines the change in orbital period very sensitively: it goes roughly with p^{-3}. Table 28 illustrates this: for $Q_0 = Q = 1/q_0 = 1/0.7$ one observes that for $M_1^0 = 8M_\odot$ the orbital period increases by a factor 9.4, whereas for $M_1^0 = 25M_\odot$ the increase is only by factor 3.1.

Table 28. Increase in orbital period due to conservative mass transfer in systems with initial primary star masses 8 $M_\odot$ and 25 $M_\odot$, respectively, for four values of the initial mass ratio $q_0 = M_2^0/M_1^0$. Case B mass transfer is assumed, such that the helium core, of mass M_{core}, is left after the exchange.

	P^f/P_o			
	q_0 =0.4	0.5	0.6	0.7
$M_1^0 = 8\ M_\odot$ M_{core} =1.8 $M_\odot$	3.5	5.3	7.3	9.4
M_1^0 =25 $M_\odot$ M_{core} =9 $M_\odot$	1.2	1.8	2.4	3.1

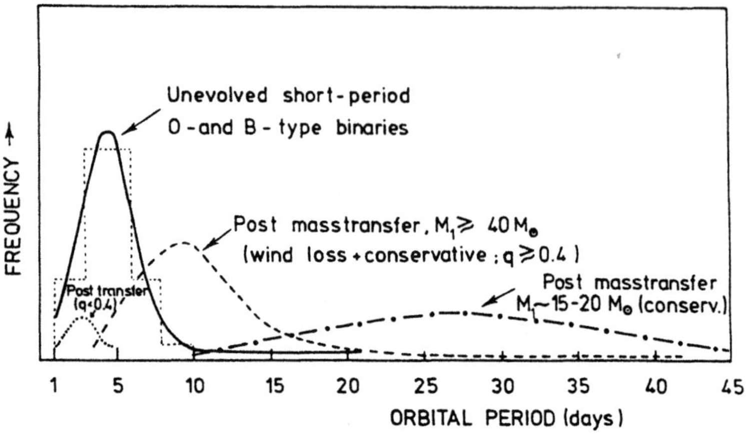

Fig. 67. Solid line and histogram: distribution of orbital periods of unevolved O- and B-type close binaries ($P \leq 30\,d$) after Garmany et al.[86]. Dashed and dash-dotted lines: resulting expected post-mass-transfer distributions for very massive and moderately-massive systems, as explained in the text. After the supernova explosion of the primary star the post-mass-transfer periods are increased by $\sim$ 60 per cent for the very massive and $\sim$ 40 per cent for the moderately massive systems.

Since the distribution of the orbital periods of unevolved early-type spectroscopic binaries peaks around 3-6 days, one expects the progenitors of Be/X-ray binaries to have, after the first phase of mass transfer, orbital periods between $\sim$ 30 and 60 days, whereas the massive post-mass transfer systems (WR+O binaries) are expected to have periods roughly peaking between about 6 and 12 days, in good agreement with the observations. After the SN explosion of the helium star this difference in orbital period range will persist.

Figure 67 schematically illustrates the resulting difference between the periods Be/X-ray binaries and standard HMXBs.

Overview of remnants of primaries after close binary evolution with mass transfer, as a function of initial primary mass and orbital period. The type of final remnants that can be expected after case B and case BB mass transfer as a function of helium core mass are as follows (see also sections 4.8 and 4.9 and table 21):

1. Helium stars less massive than 1.9 $M_\odot$ leave CO white dwarfs.

2. In the mass range $M_{He} = 1.9 M_\odot - 2.2 M_\odot$ they develop a degenerate O-Ne-Mg core which, if the star loses no further mass, will collapse to a neutron star, due to *e*-capture.

   In a binary, however, case BB mass transfer will remove the envelope of this core before it reaches the critical mass for collapse. In that case an O-Ne-Mg white dwarf is left. *This type of white dwarf, with a mass 1.0-1.2 $M_\odot$ can therefore form only in a close binary!*

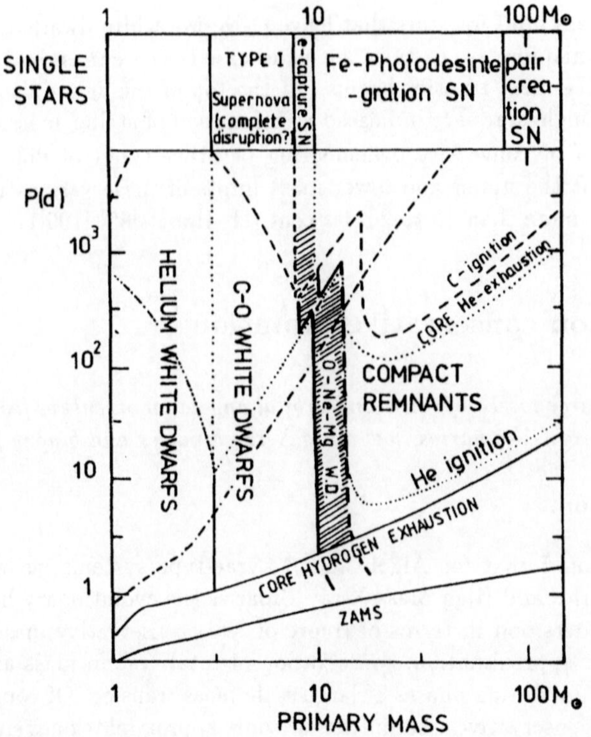

Fig. 68. Classification of expected final evolutionary states of primary stars of close binaries as a function of primary mass and initial orbital period (for mass ratio 0.5 and hydrogen fraction X=0.70). At the top of the figure the expected final evolution of single stars is indicated. In a subsequent phase of reversed mass transfer CO white dwarfs may be triggered to explode as type I supernova (complete disruption): some CO white dwarfs with a mass close to the Chandrasekhar limit, and O-Ne-Mg white dwarfs may be triggered by accretion to implode to a neutron star (see text).

3. Helium stars in the range $M_{He} = 2.2 - 2.5 M_\odot$ may also undergo case BB transfer. The close systems in this mass range may still leave an O-Ne-Mg white dwarf, whereas in the wider systems the core will grow sufficiently before the mass transfer begins, to allow it to collapse to a neutron star (see figure 50 for the growth of the core mass as a function of stellar mass and radius).

4. In the mass range $M_{He} = 2.5 - 3.5 M_\odot$ the stars will still undergo case BB transfer but their cores will collapse due to Fe-photodisintegration.

5. For $M_{He} \gtrsim 3.5 - 4 M_\odot$ the same occurs without case BB mass transfer.

Figure 68 indicates very schematically the types of remnants of primary stars expected as a function of initial mass and orbital period.

The lower mass limit for stars that leave O-Ne-Mg white dwarfs is larger for case B systems than for case C, as in the latter case the core mass had more time to grow before the mass transfer begins. At the top of the figure also the types of remnants of single stars are indicated as a function of stellar mass.

The inclusion of convective overshooting or other types of different input physics may shift the upper and lower mass limits of these various ranges but probably by not more than 10 to 20 per cent (cf. Habets[99] [100]).

6. Highly non conservative evolution

Evolution with large losses of mass and orbital angular momentum from the systems; formation of CV-binaries, low-mass X-ray binaries and binary pulsars.

6.1 Introduction.

We saw in section 5 that for Algol- and β-Lyrae-type systems, as well as for Wolf-Rayet binaries and High Mass X-ray Binaries the evolutionary history can be quite well understood in terms of (more or less) conservative mass transfer, i.e. with, in first approximation, conservation of total system mass and orbital angular momentum during phases of large-scale mass transfer. Of course, we do know that these conservative assumptions are only approximate ones since during the evolution also mass loss by stellar winds occurs (which may be substantial in the case of WR binaries) and since also some mass is lost in the SN-explosios of the helium stars. Still the conservative approach turns out to give a good over-all description of the formation of these systems.

On the other hand, there are also many types of close binaries, discussed in section 3, for which the evidence is clear that they have lost a very large fraction of their original mass and orbital angular momentum during the evolution. In such systems the "conservative" approach certainly does not lead to an understanding of how these systems were formed. Examples of such systems are:

1. *The close binary radio pulsars* such as PSR 1913+16 ($P = 7^h45^m, e = 0.61$; PSR 1534+12, $P = 10.1^h, e = 0.27$). To form a pulsar in a binary one has to start out with a star of mass $\gtrsim 8 - 10 M_\odot$ and to form two pulsars in a binary the original system mass must be at least about $15 M_\odot$, whereas these systems now have a total mass of only $\sim 2.8 M_\odot$. Also these binaries cannot have had orbital periods shorter than 1.5 to 2 days as one can see from figure 68. So they lost over 80 per cent of their original mass and over 90 per cent of their original orbital angular momentum.

2. *The Low Mass X-ray binaries.* These systems now have total masses mostly $\lesssim 2 M_\odot$, while the progenitors of their neutron stars must have been more massive than $\sim 8\ M_\odot$. Their very short orbital periods (the shortest ones being 11 and 41 minutes; many of them having periods between 3^h and 12^h)

indicate that also here the bulk of the original orbital angular momentum has been lost.

3. *The Cataclysmic Variable binaries, pre-CV binaries and double nuclei of planetary nebulae.* In most cases the white dwarfs are quite massive, in the range 0.5 to 1.2 $M_\odot$. To form such a CO-(or in some cases: O-Ne-Mg) white dwarf, the progenitor must have been able to evolve to the Asymptotic Giant Branch (AGB) and must have had a mass $\gtrsim 2 - 3M_\odot$. This implies initial orbital radii of one AU or more, whereas the present separations are only a few solar radii.

   These systems must therefore have lost an enormous amount of orbital angular momentum and a large fraction of their total mass, as was first pointed out by Paczynski[202] and Ostriker[188]. Very clear examples of this large loss in mass and orbital angular momentum are the three pre-CV systems in the Hyades (see table 8). Their orbital periods are only $\sim$ 0.5 day and their total system masses range from 0.76 $M_\odot$ to 1.6 $M_\odot$, the white dwarfs having masses of $\sim$ 0.6 to 0.8 $M_\odot$.

   As the main-sequence turn-off point of the Hyades is at $\sim 2M_\odot$ these white dwarfs originate from stars more massive than $2M_\odot$ that have evolved into giants. This again indicates large losses of mass and orbital angular momentum from the systems.

   In these systems, as well as in the double nuclei of planetary nebulae the masses of the normal companion stars, mostly M- or K-dwarf main-sequence stars, are so low (0.1 to 0.8 $M_\odot$) that these stars can hardly have captured any mass from their (initially much more massive) companions.

6.2 Reasons for the occurrence of highly non-conservative evolution and of a common-envelope phase.

The conservative assumptions are good approximations only if:

a. the two stars do not differ too much in mass, i.e. by not more than a factor 2 to 3, and

b. the envelope of the mass-losing star is (largely) in radiative equilibrium.

The condition (a) arises from the fact that if the thermal time scales of the two stars are very different, the mass-receiving secondary star will not be able to quickly accept the large amount of mass transferred to it on the (short) thermal time scale of its companion. The secondary will require its own thermal time scale to readjust its thermal equilibrium to the large amounts of mass which it is receiving. If the time scale on which it receives mass is much shorter than its own thermal time scale, it will rapidly swell up and begins to resemble a pre-main-sequence star on the Hayashi-track. This was shown by numerical calculations by Benson[15], Ulrich and Burgher [291] and Flannery and Ulrich[78].

For $Q_0 \gtrsim 2.5$ this will always occur. Both stars will then overflow their Roche lobes and further mass loss from the primary (which continues to expand towards the giant phase) leads to the formation of a Common Envelope (CE) around the system, from which -presumably- mass and angular momentum loss will start through the outer Langrarian points L_2 and L_3.

Condition (b) is obvious and has been dealt with already extensively in section 5.3.

As figure 55b shows, even if $Q < 1$ mass transfer from a convective envelope leads to a violently unstable situation in which most of the envelope is lost to the secondary on a dynamical time scale (i.e.: hours to days).

Obviously, a companion will never be able to accept this large amount of mass on such a short time scale, and again a Common Envelope (CE) will form, now around the radiative (and in many cases: degenerate) core of the mass-losing star and the companion star.

6.3 Tidal Instability.

An additional reason for entering a CE-phase in systems with a large difference in mass between the components may be the occurrence of a *tidal instability*, first discovered by Darwin[51] and rediscovered by Counselman[46].

When a star evolves, its radius and moment of inertia will increase, which will cause its rotation to slow down, due to conservation of (rotational) angular momentum.

If the star is in a close binary, the strong tidal forces occurring in such a system will attempt to make the rotation of the components synchronous with the orbital revolution. Indeed, in close binaries with $P \lesssim 10$ days in most cases the components are observed to rotate synchronously (see, van den Heuvel[299]; Batten[12]; Tassoul[271]).

So, if a component's rotation slows down due to its evolutionary increase in radius, tidal forces will speed it up again, at the expense of the orbital angular momentum: orbital angular momentum is transformed into a rotational angular momentum, which causes the orbital period to decrease and the orbital angular velocity to increase. This, in turn, will cause a further speed up of the rotation of the star by tidal forces, causing further shortening of the orbital period, etc.

In general this process converges, and the system is able to maintain a state of (near-) synchronous rotation.

However, Darwin[51] showed that if the mass of the companion star is too small, the companion's orbit spirals further and further inwards and the two stars merge in a "tidal catastrophe".

Darwin and Counselman showed that the condition for this catastrophe to occur is that

$$J_{orb} \leq 3J_{rot} \tag{99}$$

if the rotation is assumed to be synchronous with the orbital motion, where J_{orb} is the orbital angular momentum in this synchronous orbit and J_{rot} is the

Table 29. Lower limits to k^2 of the optical component required for tidal instability in five short-period massive X-ray binaries.

System	P(d)	$a/R_\odot$	R_{opt}	$M_{opt}/M_\odot$	Unstable if $k^2 \geq$
Cen X-3	2.087	18.3	12.2	19	≥ 0.050
LMC X-4	1.402	16.9	10.5	32	≥ 0.031
SMC X-1	3.89	27.4	16.5	17	≥ 0.057
4U1538-52	3.73	26	16.0	16	≥ 0.069
4U1700-37	3.41	19.9	24	30	≥ 0.023

corresponding rotational angular momentum of the primary star in synchronous rotation. If condition (99) holds, any slight disturbance from the synchronous situation will cause the system to become tidally unstable and the companion will spiral-in catastrophically: coalescence of the two stars will be unavoidable.

In several HMXBs this condition may be fulfilled and these systems may therefore evolve into a coalescence. Writing

$$J_{rot} = k^2 M_1 \omega R_1^2 \tag{100}$$

and

$$J_{orb} = \omega M_1 M_2 a^2/(M_1 + M_2) \tag{101}$$

where a is the orbital radius, ω the orbital angular velocity and k^2 the so-called radius of gyration of the star, condition (99) becomes:

$$k^2 > M_2 a^2/3R_1^2(M_1 + M_2). \tag{102}$$

For five HMXBs the values of M_1, R_1 and a are listed in table 29 together with the resulting lower limits for k^2, for an adopted M_2-value of $1.4 M_\odot$. For stars with M_1 between 15 and 30 $M_\odot$ the theoretically calculated values of k^2 decrease from 0.075 to 0.030 between zero-age and core-hydrogen exhaustion (see de Greve et al.[53]).

Since the optical stars in LMC X-4, Cen X-3 and 4U 1700-37 are still in the core-hydrogen burning phase these systems may well be tidally unstable and may spiral in to coalesence.

6.4 Spiral-in of HMXBs: Common Envelope formation and its possible alternatives: SS 433 and Cygnus X-3.

Tidal instability may be one of the reasons for the occurrence of the spiralling-in of the neutron star during the later evolution of an HMXB.

On the other hand, in these systems also the neutron star certainly cannot accept all the matter which the massive star will transfer to it once thermal-timescale Roche-lobe overflow occurs, as the expected mass-transfer rates to the compact star are $\sim 10^{-3} M_\odot/yr$, while due to the Eddington limit this star cannot accept more than $\sim 2.10^{-8} M_\odot/yr$. This has been suggested as a reason why a CE phase cannot be avoided following the X-ray binary phase. A CE phase will, when it occurs, certainly lead to rapid spiral in of the compact star into the envelope of its companion, due to the large frictional drag which the star experiences when it is immersed in the companion's envelope (Taam et al.[266]; Bodenheimer and Taam[25] see Livio this volume, section 9).

Whether indeed in HMXBs a CE phase will occur is, however, not certain since the only HMXB system for which we know that certainly super-Eddington mass-transfer is taking place, the system of SS 433, perhaps behaves in a different way.

Here an enormously bright accretion disk has formed with a radius $\gtrsim 10R_\odot$ and a luminosity $\sim 10^5 L_\odot$. The excess matter that is spiralling in through the disk is ejected from the vicinity of the compact star in two opposite relativistic beams with $v = 0.26c$ (see section 3.5.K).

This type of mass transfer followed by ejection from the compact star resembles the "isotropic re-emission" mode of mass loss dealt with in section 2. According to eq. (40) one has here

$$a/a_0 = \left(\frac{1+Q_0}{1+Q_0+\alpha(Q-Q_0)}\right)(1+(1-\alpha)(Q_0-Q))^{\frac{-2}{1-\alpha}} \tag{40}$$

where α is the fraction of an exchanged mass element that is isotropically ejected from the compact star and (1-α) the mass fraction captured by this star; Q_0 and Q are the initial and present mass ratio.

Taking $Q_0 = 10$, as in a typical HMXB, and $\alpha = 0.99$ as expected in SS433, one finds that after the transfer of 1.4 $M_\odot$ or 2.8 $M_\odot$ (i.e. Q=9 and 8 respectively, the orbital radius goes down to $0.15a_o$ and $0.023a_o$, respectively. Hence, the orbit shrinks rapidly.

It is, of course, also possible that the bulk of the mass loss from the massive star goes in the form of a high-velocity isotropic stellar wind, which will cause a widening of the orbit ("Jeans mode", see section 2). This will partly counteract the shrinking of the orbit produced by the mass loss in the beams. Still one expects for almost any realistic combination of mass loss modes that the orbits of HMXBs, in the rapid mass-transfer phase following the X-ray phase, will shrink dramatically. This was already pointed out by van den Heuvel and De Loore[319] before CE-models were first dicussed. These authors concluded that in any case the orbit of an HMXB like Cen X-3 will shrink very much and that the entire hydrogen-rich envelope of the companion will be lost, resulting in a system that consists of just the He-core of the massive star together with the compact star, in an orbit of a few hours period. This is schematically depicted in figure 64b, as the evolutionary stage following the HMXB-phase. Van den Heuvel and De Loore[319] suggested that the system of Cyg X-3 (P=4.8 hours) may be in this post-spiral-in phase. This has recently been confirmed by infrared spectroscopy (van Kerkwijk et al.[328]), which showed that the companion of Cyg X-3 is a nitrogen-type Wolf-Rayet star (see section 3.5.K). Such stars are helium stars with masses typically in excess of 5-6 $M_\odot$ (see tables 5 and 27).

After these words of caution concerning whether or not CE-evolution may occur, we now consider how the orbits will change when CE-evolution really occurs.

6.5 Common Envelope Evolution.

Introduction. As described in section 6.2 in systems with mass ratios far from unity ($Q_0 > 2.5$) or in which the mass losing star has a convective envelope, a CE phase cannot be avoided. The only possible exception is, as mentioned in section 6.4, the case of the HMXBs, where due to the release of enormous amounts of energy by accretion onto the compact object, relativistic mass outflows may be driven that may prevent the formation of a common envelope. But also in this case large amounts of mass and angular momentum are lost and spiral-in of the orbit is expected on quite general grounds, as argued above.

Results of numerical and semi-empirical studies of CE evolution. These are described by Livio (this volume, section 9) to which we refer here. In these studies it was found that the so-called efficiency factor for CE evolution

$$\alpha_{ce} = \Delta E_b / \Delta E_{orb} \tag{103}$$

is close to unity, where ΔE_b is the binding energy (gravitational minus thermal) of the ejected envelope material and ΔE_{orb} is the drop in the system's orbital energy during spiral-in. Also studies of the orbits of double nuclei of planetary nebulae suggest $\alpha_{ce} \simeq 1$ (de Kool[56];[57]). $\alpha_{ce} \simeq 1$ means that the mechanical energy required to eject the envelope is completely provided by the loss of the orbital energy during spiral-in.

This high efficiency of conversion of orbital potential energy into mechanical energy for ejecting the envelope is, in fact, due to the very short time scale of the spiral-in process following from these numerical studies: of order 10^2 to 10^3 years. On such a short time scale ($\ll \tau_{th}$) there is no time to convert much of the orbital energy into other forms of energy such as radiation.

Also, the amount of the envelope matter which a companion (normal or compact) can capture on such a short time scale is negligible.

There is the realistic possibility that α_{ce} becomes even larger than unity, namely when -in addition to the orbital energy- other forms of energy are available that can assist in driving off the envelope matter. This can be, for example, recombination energy of ions and molecules, or accretion energy as mentioned in section 6.4.

Since planetary nebulae are ejected also in the absence of an in-spiralling companion, here α_{ce} may be quite much larger than unity -perhaps it can be as large as 5 to 10. The same may be true in HMXBs.

In terms of α_{ce} the change in orbital separation during spiral in can be expressed simply as follows (after Webbink[350]; see also de Kool[55]).

We define as M_1 the mass of the small star and M_{2e}, M_{2f} the envelope mass and the core mass of the large star and a_1, a_2 the orbital separation before and after spiral-in, respectively. At the onset of spiral-in the large star fills its critical lobe or tidal lobe, which means that it has a radius $R_2 = a_1 r_c$, where r_c is the critical lobe radius, in units of a_1 (see figure 69). Thus the binding energy of the envelope of the star at the time is

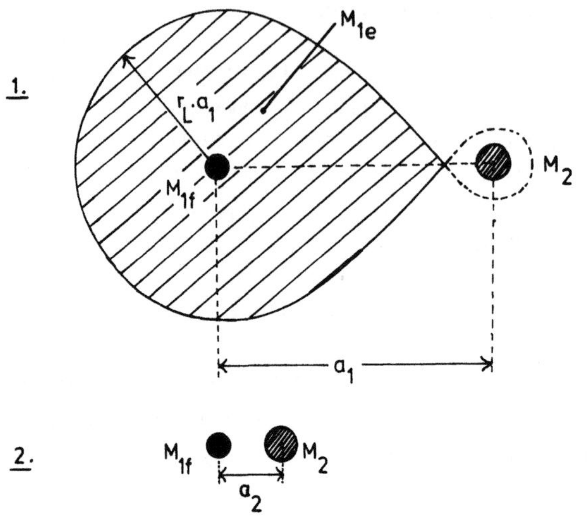

Fig. 69. Outline of the various system parameters at the onset of Common Envelope evolution. After CE evolution the system consists of stars with masses M_2 and M_{1f} with separation a_2 (further description in the text).

$$\Delta E_b = G(M_{2f} + M_{2e})M_{2e}/\lambda r_c a_1 \tag{104}$$

where λ is a weighting factor, of order unity, that depends on the density profile in the envelope; r_c is a function only of M_1/M_2. The change in orbital binding energy during spiral-in is

$$\Delta E_{orb} = \frac{GM_1M_{2f}}{2a_2} - \frac{G(M_{2f} + M_{2e})M_1}{2a_1} \tag{105}$$

Setting, according to eq. (103):

$$\Delta E_b = \alpha_{ce}\Delta E_{orb} \tag{106}$$

one obtains

$$a_2/a_1 = \frac{M_1M_{2f}\alpha_{ce}}{(M_{2f} + M_{2e})}/(M_1\alpha_{ce} + 2M_{2e}/\lambda r_c) \tag{107}$$

We now apply this expression to two examples.

Example 1. A system consisting of an $M_1 = M_\odot$ main-sequence star and an $M_2 = 5M_\odot$ red giant with a degenerate core of mass $M_{2f} = M_\odot$.

The CE evolution of this system was studied by Meyer and Meyer-Hofmeister[172]. Let us assume $r_c = 0.5 = \lambda$ and $\alpha_{ce} = 1$. Then one finds with eq. (107) $a_1/a_2 = 165$, and $P_1/P_2 = 1300$. So, if the system started out with $P_1 = 650^d$ ($a_1 = 2.67$ AU), it finishes with $P_2 = 0.5^d$. The resulting system closely resembles a pre-CV binary like V 471 Tau (BD+16516) in the Hyades (see table 8).

Example 2. A system consisting of an $M_1 = 1.4 M_\odot$ neutron star and a 13.2 $M_\odot$ red giant with a 3.2 $M_\odot$ helium core. This is a later evolutionary stage of a typical Be/X-ray binary. With the same α_{ce}, r_c and λ as in example 1, one obtains: $a_1/a_2 = 250$.

Hence, if the system started out with $a_1 = 2.5$ AU ($P = 1.03$ yr) it finishes with $a_2 = 0.01$ AU=2.16 $R_\odot$ ($P = 4.09$ h).

The radius of the 3.2 $M_\odot$ helium star at that time is, indeed, smaller than the $0.98 R_\odot$ radius of its Roche lobe. (However, later the radius of this helium star will increase and a second spiral-in phase will ensue -see section 4).

This example shows that, with $\alpha_{ce} \simeq 1$, an X-ray binary should start out as a very wide system in order to survive as a binary after spiral in. This result was already found by Taam et al.[266]. All closer HMXBs (i.e. all standard HMXBs) would, in this model, not be able to survive as a binary after spiral-in: the neutron star in such a system is expected to spiral down completely into the center of the massive companion, without the envelope of this star being thrown off.

The resulting object, a massive star with a neutron star in its center, is called a Thorne-Zytkow object (TZO), after the first authors who calculated models for such objects (Thorne and Zytkow[280]; see also Cannon et al.[33]; Biehle[20]). TZOs have the structure of a red supergiant with radii of several AU, a low effective temperature and a luminosity of order $10^5 L_\odot$. Figure 70 graphically illustrates the two possibilities for the CE-evolution of a high mass X-ray binary.

It should be kept in mind here, however, that the limiting orbital period that divides between these two cases depends critically on the value of α_{ce}. If this value is much larger than unity -as will be the case when SS 433-like mass ejection occurs and, in addition, the star has internally driven very strong wind mass loss- this critical period may be much shorter than one year, because in that case the orbits may shrink by a much smaller factor than with $\alpha_{ce} = 1$. This may well be the case if the non-degenerate component is very massive, e.g. $\geq 20 - 25\, M_\odot$.

6.6 The final evolution and fate of HMXBs: formation of binary and runaway radio pulsars.

The helium stars in the post-spiral-in systems resulting from the wider HMXBs, as depicted in figures 70, 64b and 66, will if their masses are M_{He} are $\gtrsim 2.0$ to $2.5 M_\odot$ (depending on the value of the orbital period) finally also explode as supernovae.

If $M_{He} < 3.5 - 4.0 M_\odot$ these systems will probably go through a second (case BB) spiral-in phase when the envelope of the helium star expands during helium-shell burning (see section 4). Assuming here again eq. (107) with $\alpha_{ce} = 1$ to hold, the orbits will now shrink by another substantial factor, easily of order 5 to 10, before the final supernova explosion. After the explosion here a bound system of two neutron stars, in a close eccentric orbit, will result. On the other hand, for $M_{He} \gtrsim 3.5 - 4.0 M_\odot$ not much expansion of the helium star will occur

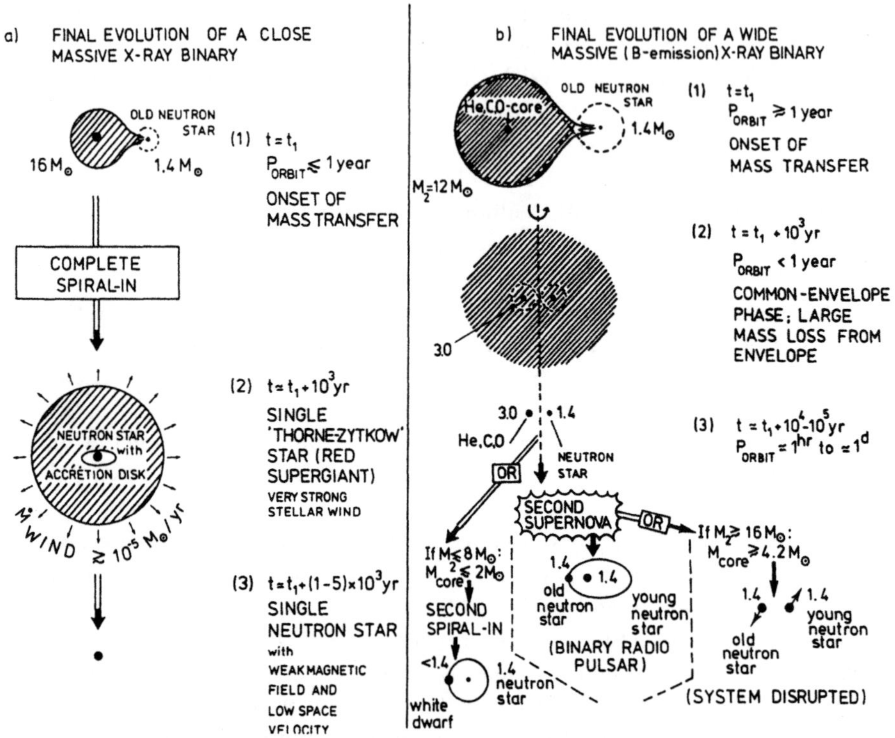

Fig. 70. The various possibilities for the final evolution of a High-Mass X-ray Binary. In all cases the onset of Roche-lobe overflow leads to formation of a common envelope and the occurrence of spiral-in. (a) In systems with orbital periods less than about 1 year there is most probably not enough energy available in the orbit to eject the common envelope, and the neutron star spirals down into the core of its companion. Subsequently, the envelope is ejected by the liberated accretion energy flux, leaving a single recycled radio pulsar. (b) In systems with orbital periods longer than about a year the common envelope is ejected during spiral-in, and a close binary can be left, consisting of the neutron star and the core, consisting of helium and heavier elements, of the companion. Companions initially more massive than 8–12 $M_\odot$ leave cores that will explode as a supernova, leaving an eccentric-orbit binary pulsar or two runaway pulsars. Systems with companions less massive than $\sim 8-12M_\odot$ leave close binaries with a circular orbit and a massive white dwarf companion, similar to PSR0655+64.

before the final explosion and the post-spiral-in period will remain as it was. If $M_{He} > 4.2M_\odot$ (for an assumed neutron star mass of 1.4 $M_\odot$) the systems wil be disrupted in a symmetric explosion. In this case two runaway pulsars will result, one newborn, the other one already several million years old, with a complex history of accretion and spiral in behind it.

If $M_H \lesssim 8-10M_\odot$, M_{He} will be $\lesssim 2M_\odot$ and a CO or O-Ne-Mg white dwarf will be left behind, in a circular orbit, after one or two spiral-in phases.

Figure 70 graphically depicts these three possible kinds of outcome. As mentioned in section 6.5 the limiting orbital period of one year required for these three types of outcome to result is rather uncertain and may easily be considerably shorter.

6.7 The fate of Thorne-Zytkow Objects.

The TZOs resulting from HMXBs with short orbital periods will not have a long life. Red super giants with radii and luminosities like those of TZOs have strong mass loss by stellar winds, at a rate of order 10^{-6} to $10^{-5} M_\odot/yr$. Kudritzski and Reimers[143] give a semi-empirical wind mass loss formula for red stars:

$$\dot{M}_w = -5.5 \times 10^{-13}(R/R_\odot)(L/L_\odot)(M/M_\odot)^{-1} M_\odot/yr \qquad (108)$$

So, the TZO will lose its $\sim 13M_\odot$ envelope in about 4.10^6 yrs or less. What will be left is a single old neutron star with a moderate space velocity -like that of its Be/X-ray binary progenitor, at most a few tens of km/s (see figure 66), which is much lower than the average for pulsars.

6.8 The origin of the Low-Mass X-ray binaries.

Introduction: the need for a very rare formation process. As mentioned in section 3.5.H, since the mass donor in an LMXB is less massive than the compact star, mass transfer in these systems will be stable, and will therefore take place on a long time scale. As we will explain in section 7 the mass-transfer time scale in LMXBs is -like in CV binaries- of order 10^8 yrs or longer. In most of the systems the mass transfer is driven by by orbital shrinking due to gravitational radiation losses or other ways of angular momentum loss.

With such a lifetime the total number of LMXBs in the Galaxy, of $\sim 10^2$ indicates that their formation rate in the Galaxy is of order 10^{-6} per year, some 4.5 orders of magnitude lower than the type II supernova rate. Hence, not more than one out of every $\sim 10^{4.5}$ neutron stars finishes as a member of an LMXB. This is in great contrast with the case of the standard HMXBs, which with a lifetime of $\sim 5.10^4$ yrs (or less) number ~ 50 in the Galaxy. This indicates a Galactic formation rate of order $\sim 10^{-3} yr^{-1}$, only about one order of magnitude smaller than the type II SN rate. In addition also the Be/X-ray binaries have a formation rate of order $(1-2).10^{-3}/yr$ in the Galaxy (see section 9). Hence, between 1 and 3 out of every 10 neutron stars is formed in some type of high-mass binary system. Thus, while the production of an HMXB is, apparently, a normal event in the life of almost every high-mass close binary system, and requires no special initial conditions, the production of an LMXB is clearly a most unusual event which occurs extremely rarely in nature, and therefore must require very special conditions in order to take place.

The following three processes have been suggested:

1. Common-envelope evolution of a fairly massive close binary with an extreme initial mass ratio.
2. Accretion-induced collapse of a massive white dwarf in a CV-like binary.
3. Origin from a close HMXB with a much more distant low-mass main-sequence (G- or K-dwarf) companion (Eggleton and Verbunt[72]).

We briefly discuss each of these processes.

Common-Envelope evolution of a fairly massive close binary with an extreme initial mass ratio. This model was first proposed by Sutantyo[264] to explain the origin of the Hercules X-1 system. This system consists of an X-ray pulsar together with an about 2 $M_\odot$ companion A-star, with an orbital period of 1.70 days.

The system is located 3 kpc outside the galactic plane. Nevertheless, it must be a fairly young object, since a $2M_\odot$ star cannot live longer than $\sim$ 8.10^8 yrs. Stars of that age belong to population I and are found always at $|z| \lesssim 200$ pc. The system must therefore have originated close to the galactic plane, but in order to reach its z-distance of 3 kpc it must have been shot out of this plane with a z-velocity> $120km/s$. This large runaway velocity is undoubtedly due to the sudden mass ejection in the supernova in which the neutron star was formed (see section 2). Verbunt et al.[343] have shown that a system runaway velocity of this magnitude can never be reached if the neutron star was formed by accretion-induced collapse of a white dwarf, even if the neutron star received a kick of several hundred km/s at birth. The point is that the kicks required to give it the necessary z-velocity are so large that they either would have disrupted the system or had shot the neutron star into its companion.

Therefore, for the formation of this system the only model possible is that suggested by Sutantyo[264] [265], as depicted in figure 71.

In this model the initial system consists of a fairly massive star ($\sim 12-15M_\odot$) together with a 2 $M_\odot$ star, in a fairly wide orbit ($P > 1$ yr). When the massive component becomes a red super giant, at age $\sim 10^7$ years, the system enters a CE phase in which the 2 $M_\odot$ star spirals inward and the hydrogen-rich envelope is ejected. The post-CE system consists of a helium star of 3-4 $M_\odot$ and the 2 $M_\odot$ star in an orbit with $P \sim 0.5 - 1.0^d$. Some 1.5×10^6 yr later the helium star explodes as a supernova. In a a symmetric explosion the system is not disrupted, but it receives a large runaway velocity due to the sudden mass ejection (1.6 to 2.6 $M_\odot$), and the orbit becomes very eccentric.

As the system is now 1.1 $\times 10^7$ yr old and the 2 $M_\odot$ star remains on the main sequence for 8.10^8 yrs, the orbit has ample time to circularize by tidal forces, and the system has sufficient time to travel to z=3 kpc, before the 2 $M_\odot$ star overflows its Roche-lobe and the system becomes an X-ray source.

As a matter of fact, the sytem probably already completed several oscillations around the galactic plane during its life time.

The Sutantyo scenario of figure 71 does not work for LMXBs with companion stars of $M \lesssim 1.0M_\odot$ for the following reason.

The minimum post-spiral in orbit of a system consisting of the helium star plus its low-mass companion (M_2) is given by the condition that the latter star should not overflow its Roche lobe. This yields, for a given value of M_2 and of the helium-core mass M_{He}, a minimum orbital period. Also, with eq. (107), one can calculate for these mass-combinations the ratio a_1/a_2 of initial and final orbital radius (as $M_{He} = M_f$ and the envelope mass is uniquely determined by M_{He}).

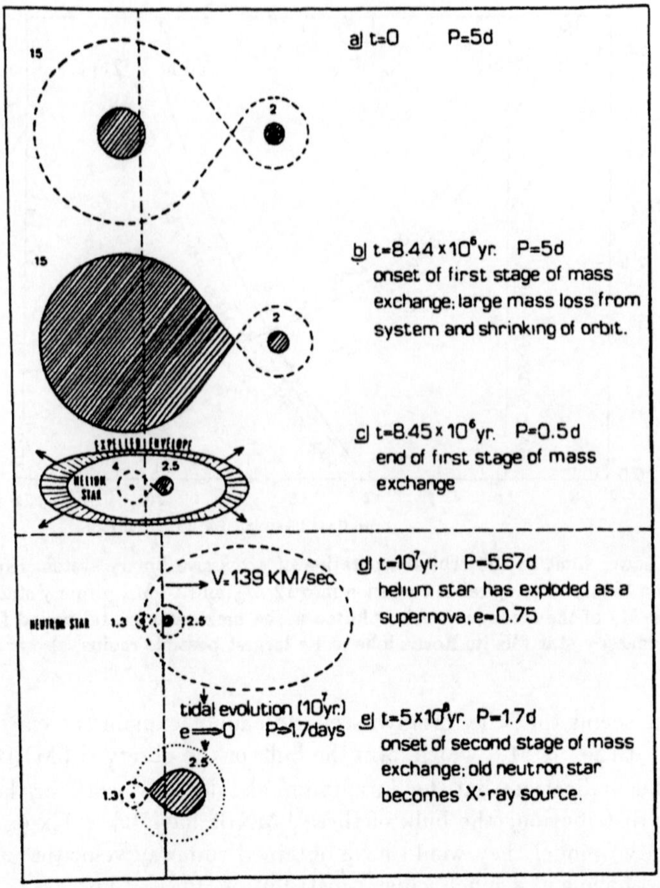

Fig. 71. Scenario for the formation of the Her X-1 system out of an initially massive close binary with an extreme mass ratio. When the 15 $M_\odot$ primary begins to overflow its Roche lobe, the 2 $M_\odot$ secondary spirals in and causes the hydrogen-rich envelope of the primary to be expelled, leaving a 4 $M_\odot$ helium star plus the 2 $M_\odot$ secondary, which is expected to have accreted only 0.5 $M_\odot$ in this process. The supernova mass ejection and impact effects imparted a 139 km s^{-1} velocity to to the center of mass of the system, and induced a large orbital eccentricity. After the supernova explosion of the helium star the system is 'quiet' for about 5 $\times 10^8$ yr and the orbit is circularized by tidal friction. When the star begins to transfer mass to the neutron star the system becomes an X-ray source (after Sutantyo [264, 265]; explanation in the text).

The thus obtained values of the orbital radius a_1 can be then converted into a minimum pre-CE orbital period. These minimum orbital periods are depicted in figure 72 as a function of M_2, for $M_1 = 12M_\odot$ and $15M_\odot$.

Also depicted are here the maximum orbital periods at which the primary star still overflows its Roche lobe before exploding as a supernova. The figure shows that for $M_1 = 12M_\odot$ only systems with $M_2 > 1M_\odot$ can just fulfill the condition that in the post-spiral in system the low-mass star does not overflow its Roche lobe. For $M_1 = 15M_\odot$ this condition becomes $M_2 > 1.2M_\odot$.

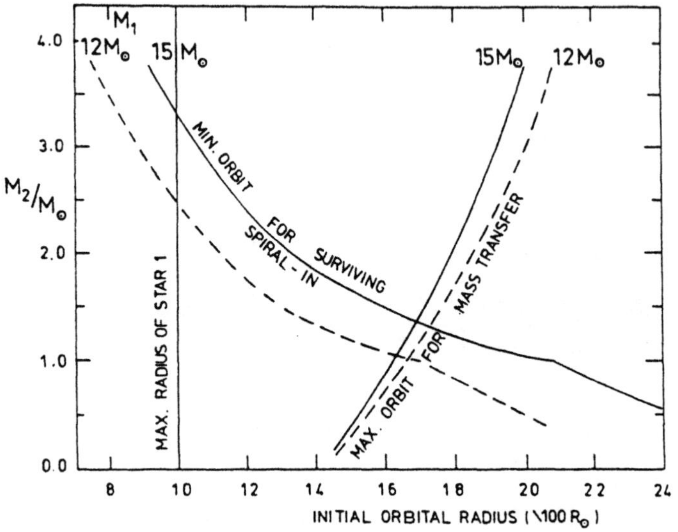

Fig. 72. Lower limit to the orbital separation of a massive binary system required for the survival as a binary after spiral-in evolution into 12 $M_\odot$ and 15 $M_\odot$ primary stars. Parameter is the mass M_2 of the secondary. Also plotted is the maximum orbital period for systems in which the massive star fills its Roche lobe at its largest possible radius, also as a function of M_2.

If thus seems that the Sutanyo scenario cannot explain the existence of the LMXBs with $M_2 < M_\odot$, which form the bulk of the observed LMXBs.

Another indication that the Sutantyo model does not work for these systems is their z-distribution: the bulk of these LMXBs have $|z| < 1$ kpc, whereas in the Sutantyo model they would have obtained runaway velocities > 100 km/s, and would have had a much wider z-distribution, up to 3 kpc.

A word of caution as to the latter argument should, however, be added as it may not be one hundred per cent water-tight. Namely since the orbital eccentricity induced by the SN effects is proportional to the runaway velocity imparted to the center of mass of the system, there is an upper limit to the space velocity that a system can obtain without being disrupted. This upper limit is lower if the mass of the companion of the neutron star is lower. Thus bound systems with a low-mass companion will always have space velocities below this limit. For the Her X-1 system with its $2M_\odot$ companion this limit is higher than for an LMXB with a $1M_\odot$ companion, such that the latter one may not be able to reach a $|z|$ distance as large as that of Her X-1.

Accretion-induced collapse (AIC). This model was suggested already long ago for the origin of LMXBs (cf. Whelan and Iben[357]; van den Heuvel[306]; see the review by Canal et al.[32] and references therein), and has been worked out especially by Canal and co-workers, Taam and Fryxell[267] and Nomoto and co-workers. For a recent review see Nomoto and Kondo[182], Nomoto and Yamaoka[185], Woosley et al.[368], Woosley and Baron[367], Timmes and Woosley[281].

Since LMXBs have many characteristics in common with CV binaries, and since mass is being transferred to the white dwarfs in CVs, an evolutionary link between LMXBs and CVs by AIC seems quite plausible.

Still it appears that for AIC to work and to produce a neutron star many boundary conditions have to be fulfilled concerning:

a. the mass and chemical composition of the white dwarf;

b. the mass transfer rate;

c. possibly: the age of the white dwarf which may lead to a sinking of the heavier elements (O, Ne, Mg) to its center, which may facilitate e-capture collapse.

A comprehensive summary of the requirements was given by Taam and Fryxell[267].

Required Composition. As to condition (a): O-Ne-Mg white dwarfs present the most favourable conditions for AIC. The required mass transfer rate is at least $10^{-8} M_{\odot}/yr$ (see figure 73a). Also massive CO white dwarfs may, under certain special conditions, be triggered to undergo an e-capture collapse but here very much fine-tuning is required since under most circumstances these stars will C-deflagrate completely and no remnant will be left. This is most probably what occurs in type Ia supernovae (see section 4). For a review: see Nomoto and Yamaoka[185].

Only if element separation has occurred (i.e. in very old CO white dwarfs) perhaps AIC to a neutron star may occur, in a certain range of $\dot{M}$-values (see

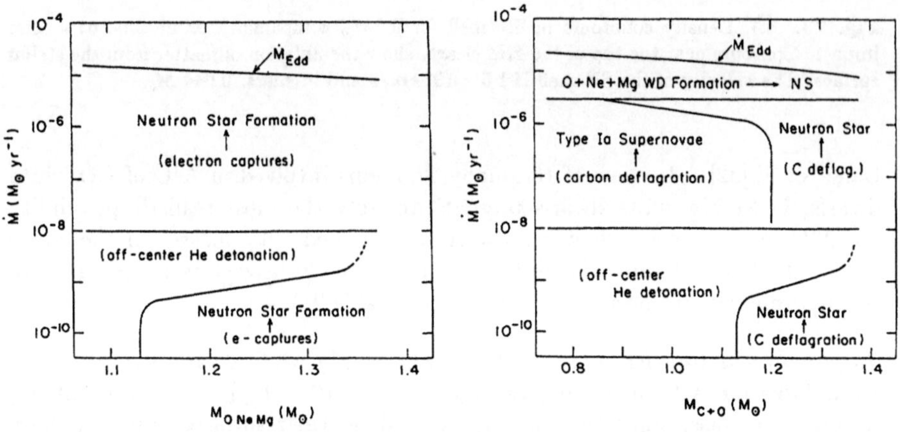

Fig. 73. **a.) Final fate of accreting O+Ne+Mg white dwarfs (left) and C+O white dwarfs (right) as a function of accretion rate and initial white dwarf mass. In O+Ne+Mg white dwarfs collapse to a neutron star is triggered by *e*-capture on ^{24}Mg and ^{20}Ne. For C+O white dwarfs, in two regions carbon deflagration is ignited at central density as high as $10^{10}\, g\, cm^{-3}$ and very likely leads to neutron star formation (after Nomoto and Kondo [182]).**

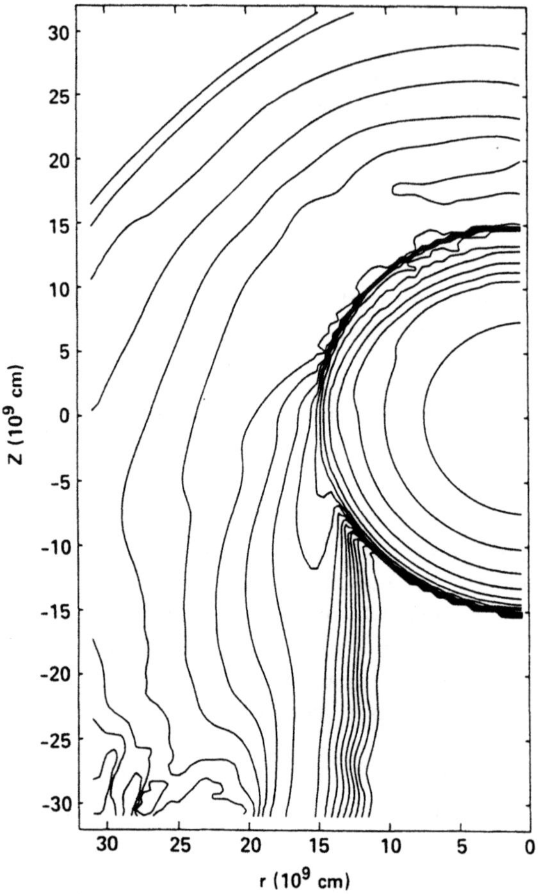

Fig. 73. b.) Density countours in SN shell for $0.2 M_{\odot}$ companion star at time 61 s after impact. Contours near the top of the grid clearly show the ablation of matter from the stellar surface. The velocity of the SN-shell is 1.5×10^4 *km/s* and its mass: $0.044\ M_{\odot}$.

Canal et al.[32]). In view of the many problems involved in AIC of CO-white dwarfs, O-Ne-Mg white dwarfs present probably the most realistic possibility for AIC in nature. Such white dwarfs are formed with masses in the range $\gtrsim 1.0 - 1.2 M_{\odot}$ and can only be formed in binaries in a very resticted range of primary masses and periods (see fig. 68 and table 21).

Required mass transfer rates. Although O-Ne-Mg white dwarfs seem the best candidates for AIC, in reality, as long as $\dot{M} < 10^{-7} M_{\odot}/yr$, they are unlikely to grow in mass, since the matter accreted on their surfaces will then burn intermittenly in strong nova-like thermo-nuclear outbursts.

The amount of matter ejected in such outbursts is probably larger than the amount that was accreted (see Livio, this volume), hence the white dwarf will not experience a net growth in mass and therefore will never reach collapse.

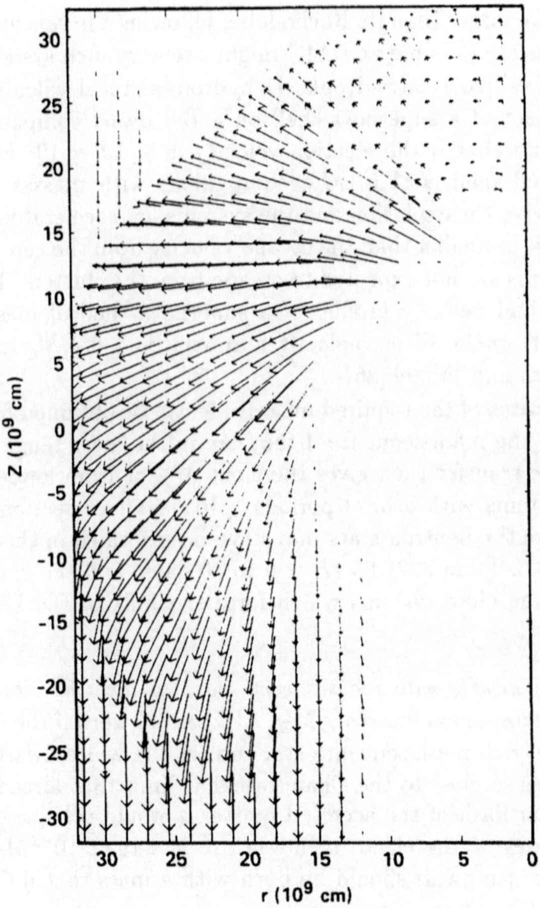

Fig. 73. c.) Velocity field for the SN shell of figure 74b at time 61 s, showing fan-shaped reverse ablation flow (after Taam and Fryxell[267]).

Only if $\dot{M} > 10^{-7} M_{\odot}/yr$ the white dwarf may go into more steady nuclear burning in its surface layers and net growth may occur. However, if $\dot{M} > (3 - 4) \times 10^{-7} M_{\odot}/yr$ (somewhat dependent on the white dwarf mass) a red giant envelope will form around it (see Iben[119], van den Heuvel et al.[318]).

As there is no place for such an envelope in a close binary, this envelope will be expelled and large mass loss from the system will ensue.

So, only in the range $\dot{M} = (1 - 4) \times 10^{-7} M_{\odot}/yr$ may net growth of the white dwarf, and AIC, occur. For details, see van den Heuvel et al.[318].

Such mass-transfer rates are practically never found in CV binaries, which typically have $\dot{M} \simeq 10^{-10}$ to $10^{-8} M_{\odot}/yr$. So, CVs turn out to be unfit for producing LMXBs by AIC in this way. An exception to this rule may be the case of LMXBs in globular clusters. Here a massive white dwarf may collide (physically or tidally) with a main-sequence star and form a binary. As the star, a red dwarf, may be larger than its Roche-lobe, following the capture event, high mass transfer rates are possible and AIC might occur in such systems.

Taam and Fryxell[267] carried out 2D hydrodynamical calculations of the effects of the impact of a supernova-shell on a red dwarf companion in a CV binary. They found that if the ejection velocity is $\leq 1.5 \times 10^4\ km/s$ and the mass of the ejected shell is $\leq 0.1 M_\odot$, companions with masses in the range 0.2 to 1.0 $M_\odot$ survive the explosion and the systems are accelerated to velocities $\leq 18\ km/s$. As this is smaller than the escape velocity from the core of a globular cluster, such systems are not expected to escape from the clusters. Figures 73b,c show the density and velocity profiles of a supernova shell of mass 0.044 $M_\odot$, with $v = 1.5 \times 10^4\ km/s$, 61 seconds after impact on a 0.2 $M_\odot$ companion as calculated by Taam and Fryxell[267].

Mass-transfer rates of the required magnitude can be obtained from companions in or close to the main-sequence if they are in the mass range 1.4 to 2 $M_\odot$ (thermal timescale transfer then gives the right $\dot{M}$), or from lower mass giants (1-1.2 $M_\odot$) in systems with orbital periods $\gtrsim 10$ days (see section 7). So, the only LMXBs where the neutron stars may have been formed in this way are the wide systems Cyg X-2 and 0921-63 ($P \simeq 9$ to 10 days) and the globular cluster sources, but not the close systems which form the bulk of the LMXBs in the galactic disk.

Origin from white dwarfs with masses very close to the Chandrasekhar limit. For systems with companion masses $\lesssim M_\odot$, AIC cannot work if the companion is a normal hydrogen-rich main-sequence star- unless the white dwarf would have formed with a mass so close to the Chandrasekhar limit that already before the first thermo-nuclear flash of the accreted matter it would collapse to a neutron star. As the masses of the ejecta in novae are less than $10^{-3} M_\odot$ (see Livio, this volume) the white dwarf should be born with a mass that differs less than $10^{-3} M_\odot$ from the Chandrasekhar mass.

One can make an estimate of the probability for this to happen, as follows. The white dwarfs with masses close to the Chandrasekhar limit will originate from stars with masses very close to the lower mass limit for neutron star formation, which we assume to be at $M = 8 M_\odot$ (see section 4). As the mass of a white dwarf originating from an $M = 8 M_\odot$ star is about 0.15 M, the mass range from which white dwarfs originate that differ by $< 10^{-3} M_\odot$ from the Chandrasekhar limit will be $\Delta M \leq \frac{1}{0.15} 10^{-3} M_\odot \simeq 7.10^{-3} M_\odot$.

Assuming the initial mass function to be a Salpeter function:

$$\psi(M) = C/M^{2.5} \tag{109}$$

one will have, for a steady state, and a constant star formation rate, that the Galactic neutron star formation rate is

$$r_n = \int_{8M_\odot}^{M_u} \psi(M) dM \tag{110}$$

where M_u is the upper mass limit for neutron star formation. As $M_u > 40 M_\odot$ (see next section) one has, in good approximation:

$$r_n = C/8^{1.5} \tag{111}$$

On the other hand, the formation rate of white dwarfs in the mass range ΔM near $M = 8M_\odot$ will be

$$r_{wdc} = \psi(M).\Delta M \leq 7.10^{-3}C/8^{2.5} \tag{112}$$

Hence, the ratio of the formation rates of these near-Chandrasekhar limit white dwarfs and of neutron stars will be $r_{wdc}/r_n \lesssim 10^{-3}$.

Since between 5 and 10 per cent of the nuclei of planetary nebulae are close binaries resembling CVs, one expects the formation rate of white dwarfs with masses within $10^{-3}M_\odot$ from the Chandrasekhar limit, that are members of close pre-CV binaries, to be less than $(0.5-1)\times 10^{-4}$ times the neutron star formation rate. As these pre-CVs will be the only ones in which the white dwarf can be driven over the Chandrasekhar limit by accretion from a close $\lesssim 1M_\odot$ companion, we expect that less than between 1 and 2 out of 2.10^4 neutron stars can have been formed in an LMXB this way.

It should be kept in mind here that we assumed that $\Delta M \leq 10^{-3}$ can be accreted before the first flash occurs. This is, however, the upper limit to the amount of ejecta observed in novae. Hence, the formation rate of LMXBs from AIC of white dwarfs with masses within $10^{-3}M_\odot$ from the Chandrasekhar limit is the upper limit to the expected LMXB formation rate.

As this upper limit is close to the real LMXB formation rate, which we estimated in the introduction of section 6.8, it seems quite possible that a sizeable fraction of all LMXBs -if not all- originated from AIC of such white dwarfs.

Triple star models -the origin of pulsar planets. These models were suggested by Eggleton and Verbunt[72]. The models make use of the fact that the later evolution of a close HMXB leads to the formation of a Thorne-Zytkov Object (see figure 70). This object has the radius of a red super giant. So, if the close HMXB had a low-mass ($\lesssim M_\odot$) companion, for example at a distance of a few AU, this star will be engulfed by the TZO's envelope and will spiral down into this envelope, as depicted in figure 74. If the mass-loss rate from the envelope of the TZO during spiral-in is such that the envelope is lost just by the time that the orbital period of the low-mass star has decreased to less than a day, a close system will remain, consisting of the neutron star and the low mass companion.

It is clear that this model requires much fine-tuning since, like in the model for the origin of Hercules X-1 (figure 71), a companion of mass $< M_\odot$ is not able to expell the envelope of the TZO by just the drop in orbital energy. For, this the same conditions hold as depicted in figure 72. The wind mass-loss rate from the TZO must therefore be fine-tuned such that precisely when the companion has reached an orbital period $\lesssim 1$ day the envelope has disappeared completely.

Only a tiny fraction of the systems of the type depicted in figure 74 will fulfill these conditions, and in the large majority of them the low-mass companion will spiral in too deeply and coalesces with the neutron star such that no binary is left.

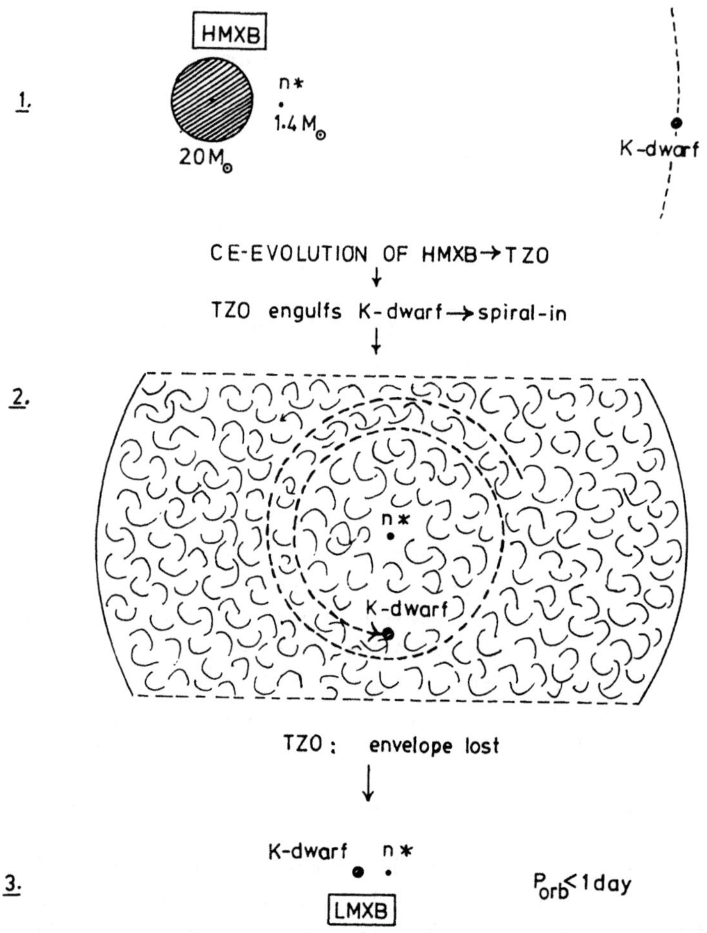

Fig. 74. Triple star model, for the formation of a Low-Mass X-ray binary, as suggested by Eggleton and Verbunt [72].

One therefore expects the triple star model of figure 74 to lead only very rarely to the formation of an LMXB. Since such triple stars are probably already rare to begin with, this model like the above described AIC models, does fulfill the condition that it gives a very low galatic formation rate of LMXBs. Unfortunately, it is not possible at this time to make a more quantitative statement about this.

If one replaces the low mass star in the model of figure 74 by one or more planets that spiral inwards, this model can similarly lead to the formation of a pulsar with several planets, in circular orbits. This was suggested by Wijers et al.[361] as a possible way to explain the presence of two planets in circular orbits around the 6 millisecond radio pulsar PSR 1257+12 discovered by Wolszczan and Frail[365].

Origins of LMXBs that contain a black hole.

a. Possible models. Three of these systems, consisting of a G or K-type main-sequence star and a black hole with a mass between ~ 3 and $15\ M_\odot$ are now known within about 3 kpc distance from the Sun, as described in section 3.5.D. These systems are all X-ray transients, and their total number in the Galaxy is estimated to be of order $10^2 - 10^3$ (Tanaka[269]; see also section 3.5.D). Also LMC X-3 is a system of this type, though here the companion to the black hole has a mass of $\sim 5 - 8M_\odot$. For the origins of these LMXBs the AIC model is not possible, as in this way never the mass of the black hole can be attained. Also, evolution from LMXBs that contain a neutron star -by accretion of matter onto the neutron star- is not possible since the companions of the neutron stars in LMXBs have not enough mass available to turn a $1.4M_\odot$ neutron star into a black hole with a mass $> 3M_\odot$, as observed in these systems.

The only remaining formation mechanisms are, therefore:

1. Origin from a wide massive binary with a very extreme mass ratio, similar to the model for the origin of Her X-1 depicted in figure 71, but with a much more massive primary star;

2. The Eggleton-Verbunt triple star model described in the last paragraph, as depicted in figure 74, with an HMXB that contains a black hole (i.e.: similar to Cyg X-1).

In both these models the black hole is formed by the direct core collapse of a star with a mass larger than the lower mass limit for black hole formation, M_B.

b. The lower mass limit for black hole formation. M_B is known to be $\geq 40M_\odot$ for the following reason (van den Heuvel and Habets[320]). The HMXB 1223-62 (GX 301-4) is an X-ray pulsar that is orbiting around a B1.5 Ia super giant with a mass $> 40M_\odot$. Model calculations for the formation of this system, taking stellar wind mass loss during the evolution into account, show that its neutron star must have started out as a primary star with a mass $> 40M_\odot$ (van den Heuvel and Habets[320]). From this it follows that the lower limit for black hole formation M_B is $> 40M_\odot$.

On the other hand this limit cannot be larger than 50-60 $M_\odot$, since otherwise an abundance distribution of heavy elements in the Galaxy would result that differs significantly from the observed one. This is because in the formation of a neutron star, all elements produced outside the $\sim 1.4M_\odot$ core are ejected from the star and enrich the interstellar medium. If a black hole of $\sim 10M_\odot$ forms, the elements formed in the inner $10M_\odot$ disappear into the hole and will not enrich the ISM. Calculations of explosive nucleosynthesis show that if stars $\geq 50 - 60M_\odot$ would produce neutron stars, an abundance distribution significantly different from the observed one would result (Nomoto and Hashimoto[181] and Nomoto, private communication).

We therefore conclude that $M_B \sim 50 \pm 10M_\odot$.

c. The preferred model. We favour the "Her X-1 like" model for the formation LMXB-BH systems, because we are not sure that indeed TZOs form as a later evolutionary phase of HMXBs that contain a black hole, for the following reasons. The mass of the black hole in such a system (like Cyg X-1) tends to be $\sim 10 M_{\odot}$, i.e. much higher than that of the neutron stars in HMXBs. Furthermore the masses of the companions in such systems are expected to be systematically larger than those of HMXBs that contain neutron stars, simply since much more massive systems are required to produce a black hole than to produce a neutron star. Like in Cyg X-1, the companions in such systems are, therefore, expected to be more massive than 20-25 $M_{\odot}$.

Such stars later in life develop very strong winds with mass loss rates $\gtrsim 10^{-6} M_{\odot}/yr$. Hence, before filling their Roche lobes they will have lost a large fraction of their envelope mass. For a 30 $M_{\odot}$ companion this may easily be $10 M_{\odot}$ or more, while the helium core in such a star has a mass of $10 M_{\odot}$ as well. With a $10 M_{\odot}$ black hole companion this results, according to eq. (107), into an orbital shrinking during the CE phase by about a factor 8. As the present separation in Cyg X-1 is about 48 $R_{\odot}$, the orbital radius after the spiral-in will be $\sim 6 R_{\odot}$, leaving a Roche-lobe radius of $\sim 2.5 R_{\odot}$ for the 10 $M_{\odot}$ helium star. Since such a star has a radius of about $R_{\odot}$, the system will not spiral-in further, and a TZO will not form. We thus do, in general, not expect that HMXBs that contain a black hole will form a TZO later in life.

For this reason we consider Her X-1 like models (like that of figure 71) the most likely ones for the formation of LMXB-BH systems. A very important difference with the model depicted in figure 71 is, that, in order to form a black hole, the primary star must have started out with a much larger mass, namely $\gtrsim 40 M_{\odot}$. This makes a profound difference with systems with primary stars $< 20 M_{\odot}$, since stars $\gtrsim 40 M_{\odot}$ spontaneously lose a large fraction of their envelope by the very strong winds ($\dot{M} \simeq$ several times $10^{-6} M_{\odot}/yr$ already on the main sequence and still larger later on). So, contrary to the case of the progenitor of Her X-1, the drop in orbital binding energy of the low mass companion during spiral-in is not needed for driving off the envelope of the massive star. In fact, single stars in this mass range do lose their entire hydrogen-rich envelopes and turn into Wolf-Rayet stars. The absence or presence of a companion makes here no difference (see Chiosi & Maeder[40]. Nevertheless, due to the large friction which it undergoes in the strong wind -and later on- possibly due to a CE phase in a possible remnant of the massive star's envelope- the low-mass star will spiral inwards and may end up in a close orbit around the remaining helium core of the massive star. Since, however, its drop in orbit energy is not required to drive off the envelope, there is no lower limit to its mass such as that for the Her X-1 progenitor as diagrammed in figure 72.

This important aspect for the origin of LMXB-BH systems -due to the large spontaneous stellar wind mass loss in very massive stars- was first realized by De Kool et al.[59] in constructing a model for the origin of A 0620-00, and, independently, discovered by Romani[233]. The model by De Kool et al.[59]is depicted in figure 75. It is an extension of an earlier model, for the origin of the LMC X-3 system, proposed by van den Heuvel and Habets[320].

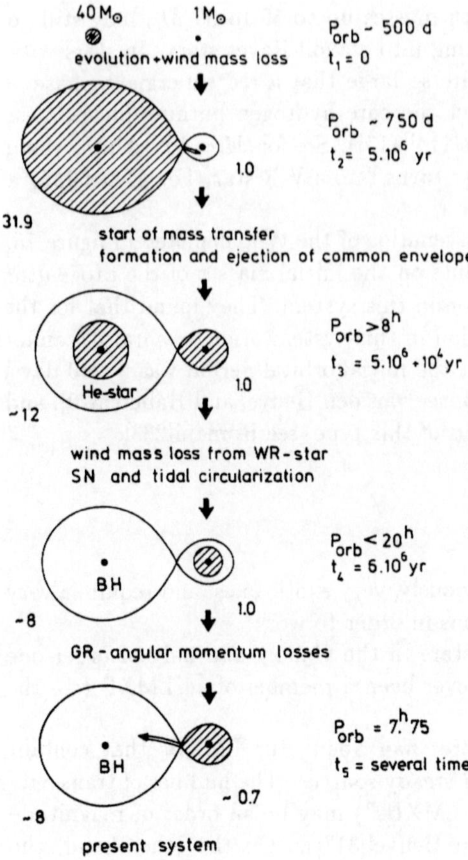

Fig. 75. A possible model for the formation of a Low-Mass X-ray Binary such as A0620-00, in which the compact star is a black hole (after de Kool et al.[59]).

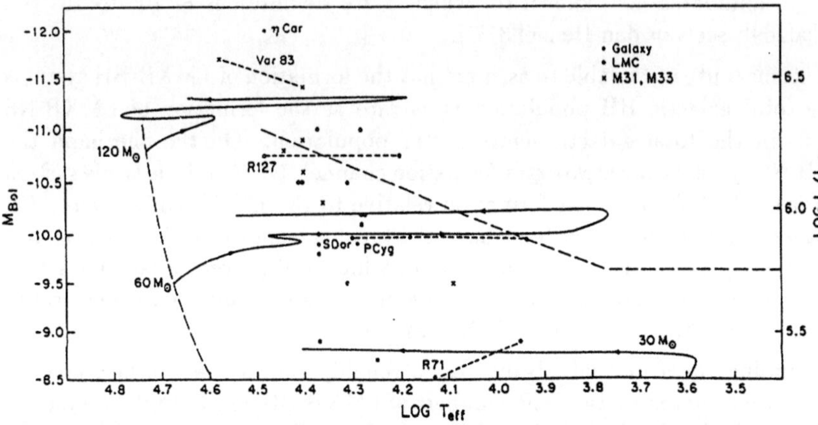

Fig. 76. The schematic HR diagram ($M_{bol} \leq -8.5$ mag) showing the empirical upper luminosity boundary and location of the luminous blue variables. The dashed lines show the shifts in the apparent photospheric temperature as they vary. Post main sequence stars $\geq 50 M_\odot$ never become red supergiants. They lose their envelopes and turn into Wolf-Rayet stars. The evolutionary tracks are from Maeder (see Chiosi and Maeder [40], after Humphreys [114]).

As depicted in figure 76, stars with masses up to 50 to 60 $M_\odot$ may still go to a red supergiant phase before turning into a Wolf-Rayet stars. In stars with M> $60M_\odot$ the wind mass loss rates are so large that a red supergiant phase is absent. These stars will, directly after the core hydrogen burning phase, turn into Wolf-Rayet stars (see Humphreys[114][115]). So, for $M \sim 40 - 60M_\odot$ there may still be a CE phase before the star turns into a WR star. For $M > 60M_\odot$ a CE phase will be absent.

De Kool et al.[59], by considering scenarios of the type depicted in figure 75, were able to place important constraints on the initial masses of the progenitor of the black hole and of the companion in this system. They found that for the original mass of the K-dwarf companion in this system prior to spiral-in cannot have been larger than 2 $M_\odot$, and that the initial orbital period was > 240 days.

For a similar scenario for LMCX-3, see van den Heuvel and Habets[320], and for a more detailed disussion of models of this type, see Romani[233].

6.9 Discussion.

The LMXB formation models are, obviously, very exotic ones, and require a very special combination of initial conditions in order to work.

There are some 10^9 old neutron stars in the Galaxy and only of order one out of several times 10^4 of these has ever been a member of an LMXB (see the introduction of section 6.8).

At present, there are are not more than about 10^2 LMXBs that contain neutron stars in the Galaxy. These are steady sources. The number of transients that contain a neutron star ("hidden LMXBs") may be an order of magnitude larger (see the discussion by van den Heuvel[317]). On the other hand, the number of potential LMXB-BH transients in the Galaxy, estimated from the about 7 candidates now known (of which 3 are definite), is $\sim 10^2$ to 10^3 (cf. Tanaka[269]; see van den Heuvel[317]).

It seems quite reasonable to assume that the formation of LMXB-BH systems in the total galactic BH population is as rare as the formation of LMXB-NS systems in the total galactic neutron star population. On the one hand the LMXB-NS systems have two extra formation channels (AIC and the triple system scenario), which favour their formation relative to the LMXB-BH systems. On the other hand the latter systems have the great advantage of the spontaneous loss of their progenitor's envelope by wind, which will prevent a larger fraction of their progenitor systems (with very large mass ratios) from complete spiral-in and coalescence than in the LMXB-NS case.

Assuming these mutual advantages to roughly cancel one another, one expects, in first approximation, the numbers of LMXB-BH and LMXB-NS systems relative to the total galactic BH and NS populations, respectively, to be roughly similar.

The relative numbers of the two types of systems in the Galaxy will then give us the relative numbers of old neutron stars and (old) stellar mass black

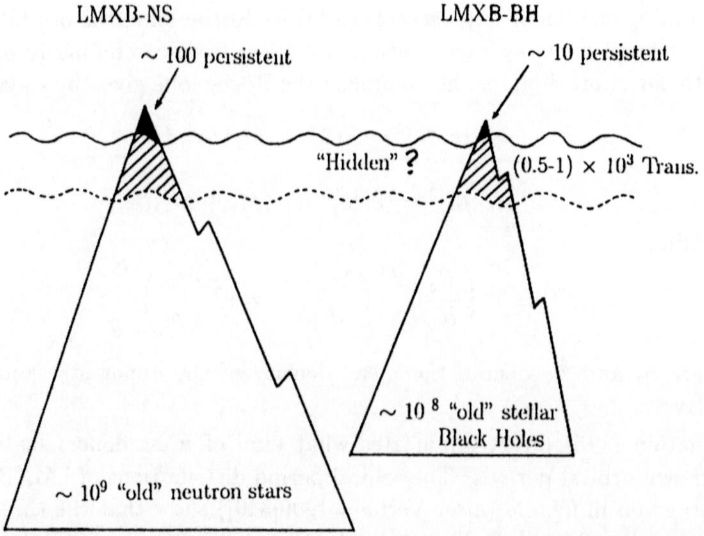

Fig. 77. The observed LMXB-NS and LMXB-BH systems are the 'tips' of enormous 'icebergs' of old neutron stars and stellar black holes in the Galaxy. The probability for a neutron star or a black hole to retain a low-mass companion in a close orbit is extremely low. It is thanks to enormous numbers of old neutron stars and stellar black holes in the Galaxy that we are able to observe a few such systems around us.

holes in the Galaxy, as depicted in figure 77. We see to our surprise that the number of stellar black holes in the Galaxy is likely to be at least $\sim 10^8$ (van den Heuvel[317]). Such a number is indeed expected with a Salpeter-like IMF if all stars between 50 $\pm 10 M_\odot$ and 120 $M_\odot$ that ever formed during the history of the Galaxy have left black holes.

7. Mechanisms driving the mass transfer in low-mass X-ray binaries and the formation of low-mass binary radio pulsars.

7.1 Introduction.

As the winds from low-mass stars are never strong (except perhaps when the stars reach the asymptotic red-giant branch), in general mass transfer by Roche-lobe overflow is required to power these sources.

The companion is here in general less massive than the neutron star, so the mass transfer will be stable as it leads to an increase of the orbital separation and Roche-lobe radius. The only exception is Her X-1, where the companion has $M \simeq 2M_\odot$; here beginning atmospheric Roche-lobe overflow can power the source for hundreds of thousands of years before it reaches the Eddington limit (Savonije 1983a[241]).

Assuming the donor star (star 2) to fill its Roche lobe, the orbital period is directly related to its average density, as follows from combining Kepler's third law with the expression for the volume of the Roche lobe given by equation (92):

$$(2\pi/\mathcal{P})^2 = G(M_1 + M_2)/a^3, \tag{113}$$

$$(R_L/a)^3 = (0.46)^3 M_2/(M_1 + M_2), \tag{114}$$

yielding:

$$\mathcal{P} \simeq 9^h \left(\frac{R_2}{R_\odot}\right)^{3/2} \left(\frac{M_\odot}{M_2}\right)^{1/2} = 9^h \left(\frac{\bar{\rho}_\odot}{\bar{\rho}_2}\right)^{1/2}, \tag{115}$$

where $\bar{\rho}_2$ and $\bar{\rho}_\odot$ denote the mean density of the donor star and the sun, respectively.

Equation (115) directly indicates what kind of mass donors fit in systems with known orbital periods. The orbital period distributions of LMXBs and CV binaries given in fig. 26 (after Verbunt 1990[340]) show that the three types of donor stars listed in table 30 and graphically represented in fig.78a are possible in short- period ($< 10^h$) systems. The table and the figure show that systems with orbital periods less than 80 minutes most probably contain helium stars or degenerate stars.

Table 30. Approximate mass–radius relations for Roche-lobe-filling low-mass ($\lesssim$ 1 $M_\odot$) stars in thermal equilibrium (during the mass transfer the thermal equilibrium may gradually be disturbed such that deviations from these relations become possible, after Verbunt 1990[340]).

Main sequence	$R_2/R_\odot = M/M_\odot$	$\mathcal{P} = 9^h\ (M_2/M_\odot)$
Helium main sequence	$R_2/R_\odot = 0.2\ M_2/M_\odot$	$\mathcal{P} = 0.^h9(M_2/M_\odot)$
Degenerate star	$R_2/R_\odot = 0.013(1+X)^{5/3}(M_\odot/M_2)^{1/3}$	$\mathcal{P} = 48^s(1+X)^{5/2} M_\odot/M_2$ [a]

[a]Here X is the fractional hydrogen abundance.

Binaries with orbital periods in the range 80 minutes to 10 hours are expected to contain normal (hydrogen-rich) main sequence stars (i.e. unevolved solar-type stars). Low-mass X-ray binaries with orbital periods longer than 10^h cannot contain Roche-lobe-filling unevolved (= main sequence) stars. If the donor is a low mass star—as is clearly the case in systems such as Sco X-1 ($\mathcal{P} = 0.78^d$), Cygnus X-2 ($\mathcal{P} = 9.8^d$), X0921−63 ($\mathcal{P} = 9^d$)—it must be an evolved star, i.e. a (sub-)giant. One usually assumes a neutron-star mass of ~ 1.3 to $1.4 M_\odot$. The X-ray luminosities of the LMXBs are mostly in the range $10^{36} - 10^{38}$ erg s^{-1}, corresponding to an accretion rate of 10^{-10} to $10^{-8} M_\odot$ yr^{-1}. With a donor mass of $\sim 1 M_\odot$, an X-ray lifetime of 10^8 to 10^{10} yr is possible —if most of the donor star were to be consumed. This is some 3 to 5 orders of magnitude longer than the expected lifetime of the standard HMXBs. As the numbers of LMXBs and standard HMXBs in the galaxy are similar to one another, this implies that the formation rate of LMXBs is some 3 to 5 orders of magnitude lower than that of

HMXBs,as we already mentioned in section 6.8. We will now examine how the observed mass-transfer rates can be obtained.

7.2 Mass transfer driven by losses of orbital angular momentum.

We will largely follow here the analysis given by Verbunt[340]. For simpicity we will tentatively assume that the total mass of the binary is conserved, i.e., the mass lost by the donor is accreted onto the compact companion, such that $\dot{M}_1 = -\dot{M}_2$. (For a straightforward generalization to the case of mass loss from the system see Rappaport, Joss and Webbink[222]; the outcome in most cases is not much different.) From eq. (5) one then easily derives:

$$\frac{\dot{a}}{a} = 2\frac{\dot{J}_{orb}}{J_{orb}} - 2\left(1 - \frac{M_2}{M_1}\right)\frac{\dot{M}_2}{M_2}. \tag{116}$$

Since we designated the donor as star 2, $\dot{M}_2 < 0$. As already mentioned in section 5, a necessary condition for stable mass transfer is that $M_2 < M_1$, since in this case the orbital radius (and thus the Roche-lobe radius) of the donor star increases when mass is being transferred. In order to examine whether the transfer is indeed stable—i.e., whether the condition is also sufficient—one has to compare the change in radius of the donor star with that of its Roche lobe, when mass is being transferred.

Substituting the logarithmic derivative of eq. (114) into eq. (116) one obtains:

$$\frac{\dot{R}_L}{R_L} = 2\frac{\dot{J}_{orb}}{J_{orb}} - 2\left(1 - \frac{M_2}{M_1}\right)\frac{\dot{M}_2}{M_2} + \frac{1}{3}\frac{\dot{M}_2}{M_2}, \tag{117}$$

which gives the change of the Roche-lobe radius in terms of the angular momentum loss rate from the system. We will now consider various ways in which angular momentum can be lost from the system.

If one writes the mass–radius relation of the donor star as $R_2 \propto M_2^n$ one will have

$$\frac{\dot{R}_2}{R_2} = n\frac{\dot{M}_2}{M_2} \tag{118}$$

In order to have stable mass transfer R_2 should remain equal to R_L, which implies $\dot{R}_2 = \dot{R}_L$, such that eq. (117) and eq. (118) together yield

$$\frac{\dot{J}_{orb}}{J_{orb}} = \left(\frac{5}{6} + \frac{n}{2} - \frac{M_2}{M_1}\right)\frac{\dot{M}_2}{M_2} \tag{119}$$

For stable mass transfer both sides of eq. (119) should be negative, which implies that

$$\frac{M_2}{M_1} < \frac{5}{6} + \frac{n}{2} \tag{120}$$

For a given value of n, this sets a limit to the mass ratio above which mass transfer is unstable. For example, for a (non-relativistically) degenerate donor star one has that $n = -1/3$, and $M_2/M_1 < 2/3$ is required for stable mass transfer. On the other hand, for a main-sequence donor $n = 1$, and the mass transfer is stable for $M_2/M_1 < 4/3$.

If the mechanism of angular momentum loss is prescribed, the evolution of the rate of mass transfer can be calculated with eq. (119).

The two mechanisms that are generally accepted to drive the mass transfer in close ($\mathcal{P} < 10^h$) systems are: angular momentum loss by (i) gravitational radiation [see eq. (74)] and (ii) "magnetic braking".

The latter process takes place when the Roche-lobe-filling donor is a dwarf (solar-like) star with a convective envelope. Such stars are observed to have hot coronae and winds, similar to the solar wind, as well as surface magnetic fields. Studies of solar-type stars of different ages show that their rotation is slowed down in the course of time. This is generally attributed to the fact that angular momentum is carried away by the stellar wind, which out to a certain distance (5 to 10 stellar radii or more) is forced by the magnetic fields to co-rotate with the star. This so-called "magnetically coupled stellar wind" involves negligible mass loss ($\sim 10^{-13} M_\odot$ yr^{-1}), but provides an important drain on the angular momentum of the star.

In a close binary, the angular momentum loss of the contact star by a magnetically coupled wind causes its rotation to slow down. On the other hand, the tidal forces in the system will continuously act to spin the star up back into co-rotation with the orbital revolution. This spin-up happens at the expense of the orbital angular momentum, and thus the magnetically coupled wind indirectly produces a drain on the orbital angular momentum of the system.

Using the observed "magnetic braking" laws for solar-type stars given by Skumanich[252] and Smith[255] one obtains numerically for the orbital angular momentum loss rate (see also eq. (131) of Livio, this volume):

$$\dot{J}_{mb} = -0.5 \times 10^{-28} f^{-2} I_d R_d^2 \omega^3 \tag{121}$$

where I_d and R_d are the moment of inertia and radius of the star, ω is the (orbital) angular velocity and f is a constant of order unity (Skumanich[252]; Smith[255]. I_d is given by

$$I_d = k^2 M_d R_d^2, \tag{122}$$

where k is the "radius of gyration" of the star and M_d its mass.

A more generalized braking law for binaries with solar-type stars (Rappaport, Verbunt and Joss[224]) yields:

$$\frac{\dot{J}_{mb}}{J_{orb}} = -3.8 \times 10^{-30} R_0^{4-\gamma}(M_1 + M_2) R_2^{\gamma} \frac{\omega^2}{M_1 a^2}, \tag{123}$$

where γ is a parameter which expresses the dependence of the braking on the radius of the donor. This law assumes that the observed rotational slow-down of young G stars involves the entire star and not just its convective envelope. In the case of angular momentum losses by gravitational radiation one has:

$$\dot{J}_{orb}/J_{orb} = \frac{32G^3}{5c^5} \frac{M_1 M_2 (M_1 + M_2)}{a^4} s^{-1} \tag{124}$$

7.3 Results for main sequence donors.

Equating eqs. (119) and (124) and eliminating a by using (114) together with the mass–radius relation $R_2 = \alpha M_2^n$, one obtains the mass-transfer rate $\dot{M}_2$ as a function of M_1 and M_2, for the case that mass transfer is driven by gravitational-radiation losses.

For a main-sequence donor one has $n = 1, \alpha = 1$. Assuming $M_1 = 1.4M_\odot$ (neutron star) one then obtains for M_2 in the range 0.1 $M_\odot$ to $1M_\odot$ that the mass-transfer rate $-\dot{M}_2 \approx 10^{-10} M_\odot$ yr^{-1} as depicted in fig. 78b; this will yield an X-ray luminosity $\sim 10^{36}$ erg s^{-1}.

A sizeable number of LMXBs has higher X-ray luminosities in the range $10^{37} - 10^{38}$ erg s^{-1}; to explain these, additional angular momentum loss mechanisms should be considered, of which magnetic braking is the most plausible one. Inclusion of this mechanism yields

$$\dot{J}_{orb} = \dot{J}_{GR} + \dot{J}_{mb}. \tag{125}$$

Using here $\dot{J}_{mb}$ from eq.(123), one observes that the variation of $\dot{M}$ with $\mathcal{P}$ depends on the parameter γ. For $\gamma = 4$, $\dot{M}_2$ is an order of magnitude lower at $M_2 = 0.3$ than at $M_2 = 1.0$. There is, however, an effect which counteracts the decrease of $|\dot{M}_2|$ in a system with decreasing M_2, as follows. The central entropy in thermal equilibrium is smaller in less massive main-sequence stars. Hence, a mass-losing star has to radiate away extra energy in order to remain in thermal equilibrium. The timescale for this is the thermal one given by eq. (85):

$$\tau_{th} = GM_2^2/R_2L_2 = 3.10^7(M_\odot/M)^2 \tag{126}$$

Thus, for decreasing M_2, the timescale for readjusting to thermal equilibrium of the donor star may become longer than the mass-transfer timescale—upon which this star will deviate more and more from thermal equilibrium. As its thermal energy content becomes larger than the equilibrium value, its radius will become larger than its thermal equilibrium radius. Although this effect will enhance the mass-transfer rate, the resulting changes in $\dot{M}$ are much smaller than the uncertainties in $\dot{M}$ due to the uncertainties in $\dot{J}_{mb}$. Figure 78b (after Verbunt[340]) shows the mass-transfer rates resulting from gravitational radiation losses in LMXBs and CV binaries. The figure shows that a binary with an original donor of mass $M_2 = 1M_\odot$ will evolve from $\mathcal{P} = 9^h$ to $\mathcal{P} < 3^h$ when M_2 becomes $< 0.3M_\odot$, and that the mass-transfer rate produced by gravitational-radiation losses alone is always $\lesssim 10^{-10} M_\odot$ yr^{-1}. Figure 78c shows the evolution of a system where "magnetic braking" was included (Rappaport, Verbunt and Joss[224]); here at the onset the mass-transfer rate is very high ($> 10^{-9} M_\odot$ yr^{-1}), but towards the end it tapers off to a low rate for a very long period of time.

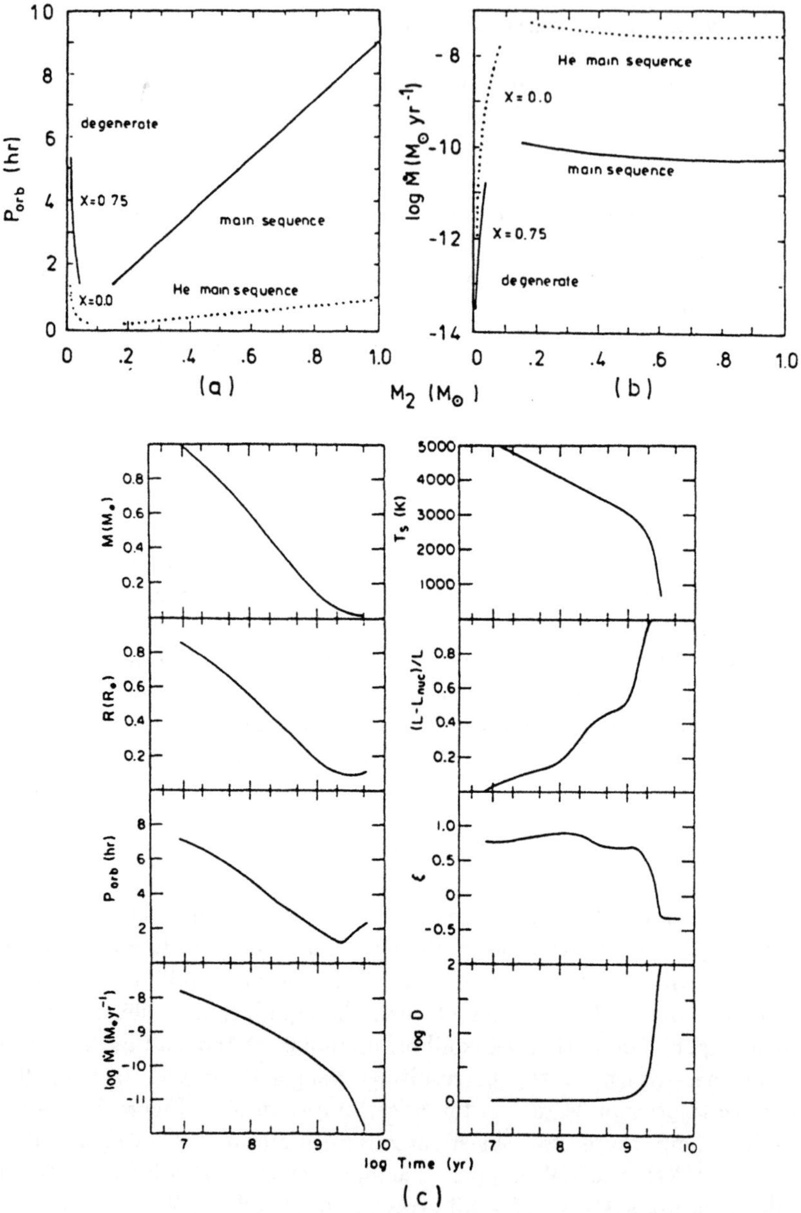

Fig. 78. (a) Orbital period P_{orb} and (b) mass-transfer rate $\dot{M}$ in low-mass X-ray binaries that evolve by means of angular momentum loss by gravitational radiation, as a function of the mass of the donor star, M_2. Donors are assumed to be in thermal equilibrium. Helium stars have shorter orbital periods and higher $\dot{M}$ than hydrogen-rich stars (after Verbunt [340]). (c) Evolution of a close binary with angular momentum losses by gravitational radiation and magnetic braking as calculated by Rappaport et al. [224]. The initial system consisted of a $1M_\odot$ main-sequence star and a $1.2M_\odot$ compact star with an orbital period of 7.2 hours. Plotted are: the evolution with time of the stellar mass, radius, mass-transfer rate, orbital period as well as other parameters characterizing the system.

7.4 The period gap for CV binaries.

For cataclysmic variables there are practically no systems with $\mathcal{P}$ in the range 2–3 hr (the "period gap"), whereas many systems are found with $\mathcal{P} > 3$ hr. A plausible theoretical explanation for this period gap, involving the fact that as a result of its mass-loss history the 0.3 $M_\odot$ secondary deviates strongly from thermal equilibrium and has an inflated radius, was proposed by Spruit and Ritter[256] and Verbunt[339]. At $M_2 \approx 0.3 M_\odot$ this star becomes completely convective, upon which magnetic braking switches off and the mass-transfer rate drops by an order of magnitude. This drop in the mass-transfer rate allows the donor to shrink back to thermal equilibrium, causing it to detach itself from the Roche lobe, upon which the mass transfer stops entirely. Further decay of the orbit due to gravitational radiation losses causes the Roche lobes to shrink so far that the secondary with its reduced (= thermal equilibrium) radius once again overflows its Roche lobe. This happens when the orbital period is ~ 2 hr. The Spruit–Ritter/Verbunt theory thus qualitatively explains why there is a period gap for the CV binaries. For more details, see Livio (this volume), section 11.

7.5 The occurrence of degeneracy, and the period minimum for CV binaries.

The absence of systems below $\mathcal{P} = 80$ minutes is due to the reversal of the mass–radius relation around $\mathcal{P} \sim 80$ min, at which point a hydrogen-rich star, with its reduced mass, becomes degenerate, and begins to follow a mass–radius relation of the form $R_2 \propto M_2^{-1/3}$. For hydrogen-rich donors in close mass-transfer binaries the minimum possible binary period is therefore ~ 80 minutes (Paczyński and Sienkewicz[204]); this minimum period can be reached only if the secondary is not driven far out of thermal equilibrium by the the preceding mass loss. If this is the case, the minimum period becomes longer. It turns out that in order to reach the observed period minimum of around 80 minutes, $\dot{J}_{mb}$ should be negligible in comparison to $\dot{J}_{GR}$ in systems with secondaries $< 0.3 M_\odot$ (Paczyński and Sienkewicz[204], Rappaport, Verbunt and Joss[224])

After passing the period minimum the mass-loss rate in the out-spiralling systems drops to negligibly low values (see fig. 78b), and their evolutionary timescale becomes longer than a Hubble time, such that these systems will no longer be observable as CV binaries.

We thus see that systems with normal hydrogen-rich donor stars can never evolve to periods < 80 minutes. The systems with much shorter periods, of which there are a few, both among the CV binaries and the LMXBs, therefore cannot contain hydrogen-rich donors.

The donor stars in these very short-period systems can either be helium-burning stars or degenerate stars of lower than normal hydrogen content. Figure 78 shows the mass-transfer rates and orbital periods of systems with helium-burning star donors. (Some CV binaries with ultra-short periods indeed appear to have helium-rich companions (Robinson[230]). These systems have much higher mass-transfer rates than those with hydrogen-rich companions and reach a minimum orbital period of $\sim$ 10 minutes (see figs. 78a,b; see also Van den Heuvel[309]). Detailed evolutionary calculations to explain ultra-short-period systems such as X2259+59, X1626−67 and X1916−05 have, for example, been carried out by Savonije et al.[243].

7.6 Absence of the period gap for LMXBs.

The above described evolution of CV binaries which start out with low-mass ($\lesssim 1 M_\odot$) hydrogen-rich donor stars appears not to be entirely applicable to the low-mass X-ray binaries. The period distributions of the CV binaries and LMXBs with periods above the period gap (> 3 hr) are quite similar, suggesting that in this period range both types of systems follow a similar evolutionary pattern. However, the virtual absence of LMXBs with periods between 80 minutes and 2 hours is striking, as CV binaries in this range are numerous. In section 8 we will discuss possible explanations for this striking difference, in terms of "evaporation" of companion stars by the pulsar action of the neutron stars.

7.7 Mass transfer driven by the internal evolution of the low-mass companion: (sub-)giant companion stars.

In low-mass systems with $\mathcal{P} \gtrsim 10$ hr, the Roche-lobe-filling donor star has too large a radius to be unevolved: the hydrogen in its center must be near or beyond exhaustion. In evolved low-mass stars ($< 2.8 M_\odot$) the helium core becomes degenerate and the luminosity is generated by hydrogen burning in a shell around this core. For a given metal content Z, and mass M the outer radius of such stars is uniquely determined by the core mass M_c. For a heavy element fraction $Z = 0.02$ and $0.16 \leq M_c/M_\odot \leq 0.45$ the radii and luminosities of such stars following from detailed numerical calculations can be fitted by simple polynomial relations in $y = \ln(M_c/0.25 M_\odot)$ (Webbink, Rappaport and Savonije 1983):

$$\begin{aligned} \ln(R_2/R_\odot) &= a_0 + a_1 y + a_2 y^2 + a_3 y^3, \\ &\quad (a_0,\ a_1,\ a_2,\ a_3) = (2.53,\ 5.10,\ -0.05,\ -1.71), \qquad (127) \\ \ln(L_2/L_\odot) &= b_0 + b_1 y + b_2 y^2 + b_3 y^3, \\ &\quad (b_0,\ b_1,\ b_2,\ b_3) = (3.50,\ 8.11,\ -0.61,\ -2.13). \qquad (128) \end{aligned}$$

For $Z = 10^{-4}$ and $0.20 \leq M_c/M_\odot \leq 0.37$ these constants have somewhat different values (see, for example, Verbunt[340]).

The change in radius of the giant is related to the change in core mass by:

$$\dot{R}_2/R_2 = (a_1 + 2a_2 y + 3a_3 y^2)\dot{M}_c/M_c. \qquad (129)$$

The growth in core mass is directly related to the luminosity, as this luminosity is entirely generated by hydrogen shell burning:

$$\dot{M}_c = 1.37 \times 10^{-11}(L/L_\odot)\ M_\odot yr^{-1}. \tag{130}$$

The calculation of the evolution of binaries with a low-mass giant is a simple matter: one equates $\dot{R}_L/R_L$ (eq. 117) and $\dot{R}_2/R_2$ (eq. 129). In this case, a priori no angular momentum losses from the system are expected, hence $\dot{J}_{orb} = 0$. The mass-transfer rate is therefore determined by the rate of change of the giant radius, which itself is determined by the growth rate of the core mass M_c, through eqs. (130) and (128). The core mass at which the mass transfer starts is uniquely determined by the radius of the Roche lobe of the donor at the moment at which this star begins to fill its Roche lobe, which itself is determined by the initial orbital separation (or orbital period). The wider the initial system, the larger the radius and luminosity of the donor at the onset of the mass transfer. Figures 79 and 80 (after Verbunt (unpublished), see also Bitzaraki[21]) show the evolution of the mass-transfer rates and orbital periods, as a function of the initial core mass, for donor stars with an initial mass 1.0 $M_\odot$ and $Z = 0.02$, and a 1.4 $M_\odot$ neutron star and a 7 $M_\odot$ black hole. It turns out that for the neutron star case, during the evolution of these systems the mass-transfer rate slightly decreases from an initial maximum value and then approaches an almost constant value during some 80% of the mass-transfer stage. In the case of a black hole, the mass transfer rate increases i the course of time. Figure 81 represents, as a characteristic example, the evolution of a binary with an initial orbital period 12.5 days, an initial giant mass 1 $M_\odot$ and a neutron star mass 1.3 $M_\odot$, calculated by Joss and Rappaport[126]. Such a system more or less resembles systems like Cygnus X-2 and X0921−63 (which have orbital periods $\sim 9^d$). After a brief initial episode with a mass-transfer rate of several times the Eddington limit ($\dot{M}_{Edd} \approx 1.5 \times 10^{-8} M_\odot$ yr^{-1} for hydrogen-rich matter) the rate settles at a value $\sim 10^{-8} M_\odot$ yr^{-1} for some 60 million years. During this time the orbital period gradually increases to $\sim$ 120 days. In the end only the helium core remains, as a $0.31 M_\odot$ helium white dwarf (while at the onset of the mass transfer M_c was 0.24 $M_\odot$). It turns out that for systems of this type the average mass-transfer rate depends (for a companion of mass 1 $M_\odot$, $Z = 0.02$, and a 1.4 $M_\odot$ neutron star) in first approximation only on the initial orbital period $\mathcal{P}_0$, roughly as:

$$\langle \dot{M} \rangle = -6 \times 10^{-10}\,(\mathcal{P}_0/1\,day)\quad M_\odot yr^{-1}. \tag{131}$$

The mass-transfer lifetime of these systems is roughly:

$$\tau \approx 0.75 M_\odot/\langle \dot{M} \rangle \approx 1.25 \times 10^9 (1\ day/\mathcal{P}_0) yr. \tag{132}$$

For other companion masses and chemical compositions, the relation becomes slightly different (Webbink, Rappaport and Savonije[353]). Relations similar to eqs. (12)7 and (128) can also be derived for low mass stars with degenerate CO cores. These stars are on the asymptotic giant branch. They will occur as mass donors only in systems with initial orbital periods of a year or longer. In these systems very high mass-transfer rates ($\sim 10^{-7} M_\odot$ yr^{-1}) are expected (De Kool, van den Heuvel and Rappaport[60]).

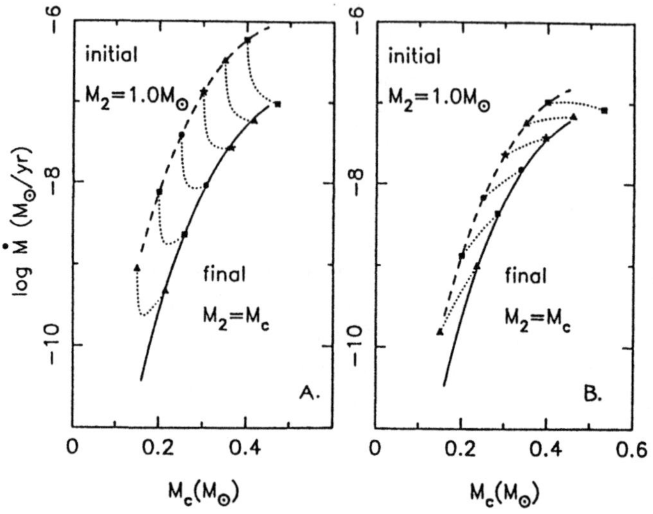

Fig. 79. Mass-transfer rate $\dot{M}$ for low-mass X-ray binaries driven by the evolutionary expansion of the mass donor (low-mass giant), as a function of its helium-core mass M_c. The curves are for metallicity $Z = 0.02$. The dashed line indicates the rates at the onset of the transfer, for a giant mass of 1.0 $M_\odot$. The full line indicates the end of the mass transfer, and dotted lines indicate the evolution of $\dot{M}$ during the transfer. Initial compact star masses of 1.4 and 7 $M_\odot$ (frame B) (black hole) were adopted, and the mass transfer is assumed to be conservative (after Verbunt (unpublished), see also Bitzaraki [21]). Figure 80 indicates the evolution of the orbital periods for the systems depicted here.

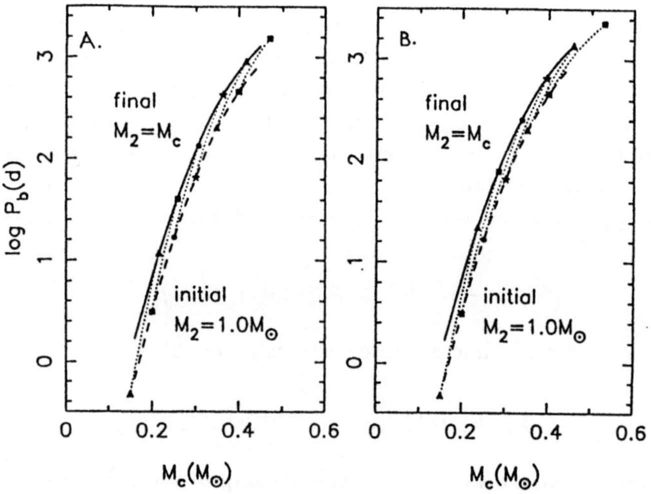

Fig. 80. Orbital periods of low-mass X-ray binaries in which the mass transfer is driven by the evolutionary expansion of an evolved companion (low-mass giant), as a function of helium-core mass. The systems are the same ones for which the evolution of $\dot{M}$ is depicted in fig.79. The meaning of the dashed, dotted and fully drawn curves is the same as in that figure. B corresponds to the 7 $M_\odot$ black hole case.

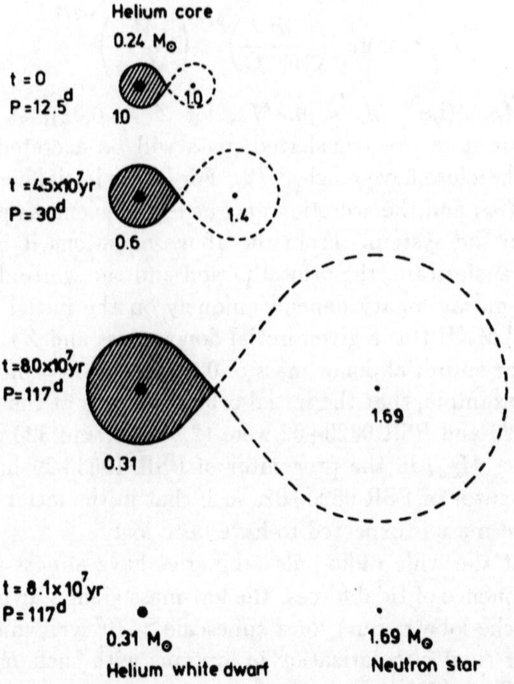

Fig. 81. Evolution of a wide low-mass X-ray binary such as Cygnus X-2 into a wide radio-pulsar binary with a circular orbit and a low-mass helium white-dwarf companion, such as PSR 1953+29. At the onset of the mass transfer, the low-mass companion is a (sub-)giant with a degenerate helium core of $0.24 M_\odot$. Its light is generated by hydrogen fusion in a shell around the core. The mass transfer from the giant to the neutron star is due to the slow expansion of the giant, driven by this hydrogen-shell burning. During the mass transfer the orbit gradually expands (due to angular-momentum conservation) and after 8×10^8 yr the system terminates as a wide radio-pulsar binary (after Joss and Rappaport [126]).

7.8 Origin of the wide binary radio pulsars with (nearly) circular orbits and low companion masses.

The orbital characteristics and companion masses of the wide radio-pulsar binaries show a striking resemblance to those of the systems that result from the evolution of wide low-mass X-ray binaries considered in the foregoing section, as was noticed by several authors (Joss and Rappaport[126], Savonije[242], Paczyński[203]).

As we argued in section 3 the companions of these pulsars must themselves already be dead stars (i.e. white dwarfs), as otherwise the radio pulses would be smothered, and therefore be unobservable. There can therefore be little doubt that the wide binary radio-pulsar systems as depicted in fig. 37 are the final products of the evolution of the wide LMXBs such as Cygnus X-2, X0921−02 and GX1+4. If the neutron star has accreted enough matter, $\gtrsim 0.1 M_\odot$, and its magnetic field is sufficiently weak ($\lesssim 10^9$ G) it will have reached a very short rotation period, in the range of milliseconds (see eq. 63):

$$P_{eq} \simeq 2 \text{ ms} \left(\frac{B}{10^9 \text{ G}}\right)^{6/7} \left(\frac{\dot{M}}{\dot{M}_{Edd}}\right)^{-3/7} . \tag{133}$$

For $\dot{M} < \dot{M}_{Edd}$ (i.e. $M_c < 0.3M_\odot$, for $Z = 0.02$), as was the case in PSR 1953+29, most of the transferred mass will be accreted by the neutron star which will therefore have reached P_{eq}. For larger initial core masses, $\dot{M}$ will be larger than $\dot{M}_{Edd}$ and the accretion may be less efficient, as part of the matter will be lost from the system. From the above equations it is easily observed that the mass-transfer rate, the orbital period and the white dwarf mass in the final wide radio-pulsar binary depend uniquely on the initial orbital period of the predecessor LMXB (for a given initial donor mass and Z). Figure 80 shows these relations for an initial donor mass of 0.9 $M_\odot$ and $Z = 0.02$. From this one easily finds, for example, that the initial orbital periods of the binary radio pulsars PSR 1953+29 and PSR 0820+02 were 12.5 days, and 320 days respectively, resulting in $\dot{M} < \dot{M}_{Edd}$ in the progenitor of PSR 1953+29 but $\dot{M} \sim 10^{-7} M_\odot$ yr^{-1} in the progenitor of PSR 0820+02, such that in the latter system over 90% of the transferred mass is expected to have been lost.

The fact that the wide radio-pulsar binaries have almost perfectly circular orbits is a consequence of tidal forces: the low-mass giant star had a large radius (equal to its Roche-lobe radius) for a timescale $> 10^7$ yr, which is longer than the timescale for tidal circularization in systems with such R/a values (cf.van den Heuvel and Taam[325]).

7.9 Evolution of LMXBs with orbital periods between 10^h and one day.

The above-mentioned equations for the evolution of wide low-mass X-ray binaries in which the donor star is a (sub-)giant with a degenerate helium core, and the mass transfer is driven entirely by the interior evolution of this giant, hold only for core masses $M_c > 0.16M_\odot$, which implies initial orbits $\mathcal{P} > 1$ day. For $\mathcal{P} < 10^h$, interior evolution of the companion plays a negligible role and the evolution of the system is driven by angular-momentum losses by magnetic braking and gravitational radiation. In the intermediate period range, between 10^h and about one day, both angular-momentum losses by magnetic braking and the radius expansion due to the interior evolution of the sub-giant play a role. The evolution of the systems due to the combination of these effects was calculated for a variety of initial system parameters by Pylyser and Savonije[218][219] and Pylyser[217]. They considered initial donor masses in the range 1.0 $M_\odot$ to 2.0 $M_\odot$ and initial orbital periods $\mathcal{P}_i$ ranging from 0.5 to 1.9 days. They found that for each donor mass and composition there is a "bifurcation period" $\mathcal{P}_b$ below which magnetic braking wins and the systems evolve to shorter periods, whereas above $\mathcal{P}_b$ nuclear evolution wins and the systems evolve towards longer periods. It will be clear that for $\mathcal{P}_i$ close to $\mathcal{P}_b$, the orbital period of the system may not change much during the entire evolution, as the two counteracting effects on the orbital period almost cancel each other. These results have an important bearing on the evolutionary history of radio pulsar binaries with circular orbits, (very) low companion masses and orbital periods in the range of 10 hours to 10 days. These systems must have evolved from LMXBs in which both the interior evolution of the companion and angular-momentum losses by magnetic braking were important. The remaining masses of the helium white dwarf in these systems will be smaller than those in the wider systems, i.e., less than 0.25 $M_{\odot}$. Based on the results of Pylyser and Savonije we shall now briefly discuss the history of observed systems of this type.

PSR 1831−00. This system has $\mathcal{P} = 1.81^d$ and a very low mass function $f(M) = 1.2 \times 10^{-4} M_{\odot}$ (Dewey, Maguire, Rawley Stokes and Taylor[68]). 1986). The initial orbital period must have been close to the bifurcation period. It turns out that such systems form the white dwarfs with the lowest masses. The calculations predict white-dwarf masses in the range 0.18–0.20 $M_{\odot}$. For example, sequence B 20 of Pylyser and Savonije starts out with $M_{donor} = M_{neutron\ star} = M_{\odot}$, $\mathcal{P}_i = 0.69$ day and terminates with $M_{white\ dwarf} = 0.20 M_{\odot}$, $\mathcal{P} = 1.72$ days.

PSR 1855+09. This is a 5.4 ms binary radio pulsar with an orbital period of 12.33 days and a mass function $5.2 \times 10^{-3} M_{\odot}$, which indicates a most likely companion mass 0.2–0.4 $M_{\odot}$. The lack of an optical identification indicates that its companion is a very cool white dwarf, T $\lesssim$ 4000 K (Callanan, Charles, Hassal, Machin, Mason, Naylor, Smale and van Paradijs 1990[31]), with a cooling age of at least 10^9 yr.

This system must have originated clearly above the bifurcation period. The calculations by Pylyser and Savonije[218] indicate that the initial orbital period of this system was around 1 day if the initial donor mass was about 1 $M_{\odot}$. In this case the white-dwarf mass is expected to be $\sim 0.23 M_{\odot}$, and during most of the mass transfer the mass- transfer rate was $\sim 5 \times 10^{-9} M_{\odot}\ \mathrm{yr}^{-1}$. If the initial donor mass was 1.5 $M_{\odot}$, the initial orbital period was also $\sim$ 1 day and the present white dwarf mass is expected to be $\sim 0.25 M_{\odot}$; but in this case the mass-transfer rate would have been $\sim 5 \times 10^{-8} M_{\odot}\ \mathrm{yr}^{-1}$, i.e., somewhat higher than the Eddington rate. Finally, it is even possible that the initial donor mass was as high as 2.0 $M_{\odot}$. If so, the initial orbital period was again $\sim$ 1 day, the white-dwarf mass is predicted to be between 0.25 and 0.50 $M_{\odot}$, and the mass-transfer rate was $\sim 3 \times 10^{-7} M_{\odot}\ \mathrm{yr}^{-1}$. In the latter case some 95% of the transferred mass must have been lost from the system and only some 0.1 $M_{\odot}$ will have been accreted by the neutron star—which is still sufficient to have spun up its rotation to a period of $\sim$ 2 ms.

8. From X-ray binaries to binary pulsars and single recycled pulsars.

8.1 Introduction.

In the foregoing sections we have seen that, in so far as orbital characteristics and companion masses are concerned, the formation of the two observed classes of binary radio pulsars depicted in fig. 37 can be explained as the result of the evolution of relatively wide High-Mass X-ray Binaries and wide Low-Mass X-ray Binaries, respectively. The evolutionary schemes leading to these two types of binary pulsars are depicted in fig. 70 and fig. 81, respectively.

The neutron-star binaries resulting from HMXBs are predicted to have relatively narrow orbits as a result of the preceding spiral-in phase. The companions of the recycled neutron stars in these systems can either be neutron stars or massive white dwarfs. In the first case the orbital eccentricity of the final system is expected to be high, as is the case in PSR 1913+16 and PSR 2303+46. In the second case the orbit is expected to be circular, as is the case in PSR 0655+64.

On the other hand, the recycled neutron stars resulting from wide LMXBs are predicted to have low-mass helium white-dwarf companions, in the mass range 0.2–0.4 $M_\odot$, exactly as is observed in the second class of binary radio pulsars depicted in fig. 37. In view of the long-lasting ($> 10^7$ yr) preceding evolutionary phase, in which the companion was a red giant, the orbits of these systems are expected to have been circularized by tidal friction in the red-giant envelope. This explains the circularity of the orbits of the systems in this class (van den Heuvel and Taam[325]).

A question which now naturally arises is: what are the final products of the close LMXB systems, with binary periods less than 12 hours, in which the companions are unevolved main-sequence dwarfs, and of the HMXB systems with orbital periods less than about one year?

We shall consider the possible final evolutionary stages of these two types of systems in the following sections.

8.2 The peculiar pulse periods and magnetic field strengths of binary pulsars: recycling.

Introduction. As noticed in section 3.6, the binary and millisecond pulsars as a group differ from the bulk of the normal single radio pulsars by having on average much faster spin and much weaker surface dipole magnetic fields than the ordinary single pulsars. This can be clearly seen in figure 36b: most binary pulsars and all of the millisecond pulsars fall outside of the "island" of ordinary pulsars in the magnetic field strength (B_s) versus period (P) diagram of pulsars. Less than 2 per cent of all pulsars in the galactic disk are found in binaries; however, of the millisecond pulsars some 50 per cent are found in binaries. Since, moreover, the millisecond pulsars also share the combination of very fast spin and weak magnetic field with the binary pulsars it seems most likely that the binary and millisecond pulsars form, in fact, one group with a common evolutionary origin.

Evolution of a neutron star born in a binary: You only live twice. One may wonder how a neutron star that is born in a close binary will evolve through the B_s vs P diagram. Figure 82 depicts this schematically: first, like in all normal pulsars, the spin of the neutron star will slow down and the star will move towards the right along a horizontal track, cross the deathline and disappear into the graveyard.

If it were a single neutron star, this would be the end of its life: it would be dead forever.

However, if the neutron star is born as a member of a binary with a normal companion star, its life is not over yet. As long as the companion remains unevolved and inside its own Roche lobe, little happens (if the companion is a Be star it may from time to time eject some matter causing the neutron star to become observable as a transient X-ray source for a few weeks). But when the companion evolves away from the main sequence and overflows its Roche lobe, it will begin to transfer matter to the neutron star which becomes an X-ray source.

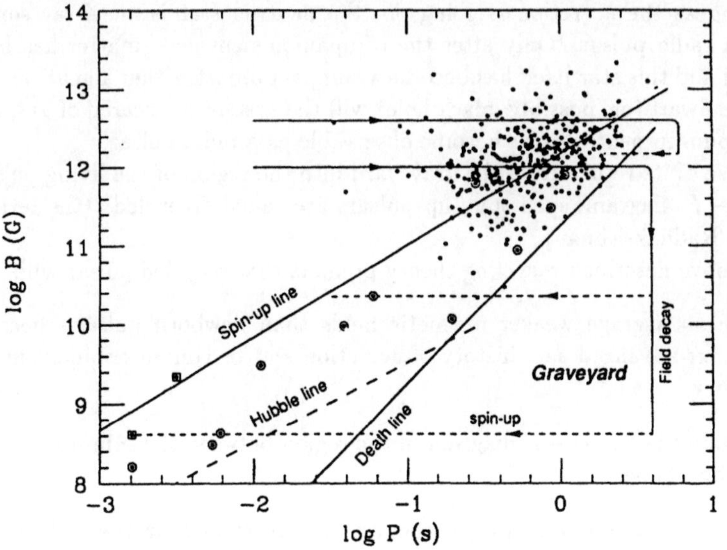

Fig. 82. Same diagram as Figure 36b with possible evolutionary tracks of pulsars in binaries indicated. Pulsars are born in the left upper part of the diagram and -if no field decay occurs-move towards the right along horizontal tracks (fully drawn). In the graveyard the field of a single pulsar probably does not decay (see text), but that of a binary neutron star does decay, presumably due to external circumstances (accretion, spin down). Furthermore, the accretion of matter with angular momentum causes the neutron star to be 'spun up' towards the left in the diagram along the dashed line until it reaches the Spin-up line. After the companion has itself become a compact star (neutron star or white dwarf) or has disappeared, the spun-up neutron star becomes observable as a radio pulsar. It will then slowly spin-down, i.e. move towards the right again in the diagram.

The accreted matter carries angular momentum as it is flowing in through a disk: this causes the rotation of the neutron star to be accelerated: it is spun up to shorter and shorter periods, as is observed in the pulsating X-ray sources with disk accretion (see figure 30). Before or after the onset of the accretion the magnetic field of the neutron star may have decayed to a lower value (either spontaneously, or due to its rotational evolution or due to accretion; cf. Bhattacharya and van den Heuvel[18]). This is indicated in figure 82 by the vertical part of the evolutionary tracks. During the spin up the neutron star moves to the left in the $B - P$ diagram along a (more or less) horizontal track and crosses now the deathline from the left, as schematically indicated.

The shortest spin period that it can reach is given by equation (133) for $\dot{M} = \dot{M}_{Edd}$ (the largest possible accretion rate). This equation shows that this shortest possible period is given by

$$P_{min} = (2ms)B_9^{6/7} \tag{134}$$

where B_9 is the dipole magnetic field strength in units of 10^9 G. This is the so-called "spin-up line" in the B_s *vs* P diagram of figure 82. The importance of the spin-up line was first realized by Radhakrishnan and Srinivasan[220], [221], Srinivasan and van den Heuvel[260] and by Alpar et al.[5].

As long as the accretion is going on, the neutron star is an X-ray source and not a radio pulsar. Only after the companion's envelope matter has been exhausted and this star itself has become a compact object, either a neutron star or a white dwarf (or, possibly, black hole) will the system be cleared of gas, and will the spun-up neutron star become observable as a radio pulsar.

Because of its return from the graveyard into the region of the living pulsars in the $B - P$ diagram, such spun-up pulsars are called "recycled" (the term is due to V. Radhakrishnan).

The above described recycling theory predicts that recycled pulsar will:

a. have on average weaker magnetic fields than newborn pulsars- because of their advanced age, history of accretion and/or complex spin-evolution history;

b. be found in the $B - P$ diagram in the region between the spin-up line and the deathline;

c. have companion stars which, themselves, are stars near the endpoint of their evolution, i.e. neutron stars or white dwarfs.

A further prediction is that one expects recycled radio pulsars to be most abundant in globular clusters, since in globular clusters X-ray binaries are a thousand times more abundant than in the Galaxy as a whole. All of these predictions are indeed fulfilled: figure 36b and figure 82 show that the binary pulsars are found in the predicted region of the $B - P$ diagram and figure 37 shows that their companions tend to be neutron stars or white dwarfs. [The only two exceptions are the two eclipsing radio-pulsar binaries PSR 1957+20, and PSR 1744-24A in the globular cluster Ter 5. We consider these in the next paragraph]. Also, the very large incidence of millisecond and binary pulsars in globular clusters exactly fits the prediction of the recycling model. It thus seems most likely that binary radio pulsars are indeed recycled. This includes the millisecond pulsars. Since single and binary millisecond pulsars are further completely indistinguishable, it seems most likely then that also the single millisecond pulsars have been recycled but have subsequently lost their companions in some way. In a globular cluster a companion can rather easily be lost by an encounter with a passing star.

The single millisecond pulsars in the galactic disk can, however, not have formed in this way.

There is still one type of LMXB whose final evolution we have not yet considered so far, namely those systems that were born with binary periods $\lesssim 12$ hours, and in which, therefore, the mass transfer is driven not by the interior evolution of the companion star but by angular momentum losses by gravitational radiation and/or magnetic braking. As we will show in the next section, they may well be the progenitors of the single millisecond pulsars.

8.3 The final evolution of close LMXBs: evaporation of companion stars.

Introduction. Ruderman, Shaham and Tavani[236], Ruderman et al.[237] and Kluźniak et al.[136] [135] pointed out that in low-mass X-ray binaries the interaction of the magnetic field of the neutron star with the accretion disk may produce a gamma-ray flux that is able to drive a wind from the companion star by heating the outer layers of this star. In this way the accreting neutron star may evaporate its companion. This evaporation may even cause the accretion to stop completely, after which the spun-up neutron star may appear as a millisecond pulsar. The gamma-ray flux from this pulsar may then continue the evaporation of the companion. If the companion is evaporated completely, a single rapidly rotating "recycled" pulsar will remain.

The model by Ruderman and co-workers was proposed primarily to explain the existence of the single 1.55 millisecond radio pulsar PSR 1937+21. The discovery of the 1.6 millisecond eclipsing binary radio pulsar PSR 1957+20, in which the companion star has a very low mass (about 0.02 $M_{\odot}$) and is clearly evaporating, appeared to provide a striking confirmation of this theoretical prediction. A further confirmation has come from the discovery of the eclipsing 11 ms pulsar PSR 1744−24A with a 1.9^h orbital period in the globular cluster Terzan 5 (Lyne et al. 1990). Here too the companion, a star with a mass $\sim 0.1 M_{\odot}$ appears to be evaporating, as the eclipse lasts for at least about half the orbital period. As to the actual evaporation mechanism there are still various views. Not all authors agree that already during the accreting X-ray emitting phase sufficient heating of the companion star may occur to cause evaporation. The only observational evidence for evaporation comes from systems in which the accretion has terminated and the neutron star has become a rapid radio pulsar, which with its energy flux causes the outer layers of the companion to be heated so much that evaporative wind is driven from these layers.

While Ruderman and co-workers see the gamma-ray flux of the pulsar as the main cause of atmospheric heating of the companion, Krolik and Sincell[141] have argued that the electron–positron flux of the pulsar is the main cause of this heating. In the model of Ruderman et al., the gamma rays are a by-product of the electron–positron flux, produced by pair creation from the curvature radiation generated by electrons and positrons accelerated in the electromagnetic fields generated by the rotating neutron star (or by the interaction of the star with a surrounding disk). In the model of Krolik and Sincell[141] the impinging of this "pulsar wind" on the companion heats its outer layers due to the annihilation of positrons. Since the bulk of the energy generated by the rotating neutron star is emitted in the form of this pulsar wind, the heating efficiency achieved by the annihilation is almost 100%. Part of the heat is radiated away, but Krolik and Sincell estimate that some 10% of the liberated energy is used to drive the evaporative wind from the companion. As the companion star is the only "conductor" close to the pulsar, a situation can be envisaged where practically the entire pulsar wind is conducted electromagnetically towards the companion along magnetic field lines, resulting in a very high evaporation efficiency.

A simplified evaporation model. In order to examine whether a pulsar has, in principle, enough energy available to evaporate its companion one should

i. compare the pulsar's rotational energy

$$E_{\rm rot} = \frac{1}{2} I \Omega^2 \simeq 2 \times 10^{52} P_{ms}^{-2} \; erg, \tag{135}$$

where P_{ms} is the rotation period of the pulsar in milliseconds, with the gravitational binding energy of its companion:

$$E_b \simeq \frac{GM^2}{R} \simeq 10^{48} \left(\frac{M}{M_\odot}\right)^2 \left(\frac{R}{R_\odot}\right)^{-1} erg; \tag{136}$$

ii. know the efficiency g of the evaporation process, defined as the fraction of the pulsar's rotation energy that is actually used for the evaporation of the companion. The condition under which the pulsar is able to completely evaporate its companion is:

$$g E_{rot} / E_b \geq 1. \tag{137}$$

Using eqs. (135) and (136) one observes that a pulsar with $P \simeq 15$ ms (as the Crab pulsar is thought to have had at its birth) would in principle be able to evaporate a 1 $M_\odot$ companion if the efficiency $g \gtrsim 0.01$. Similarly, a pulsar with a period of 1.55 ms will be able to evaporate a 1 $M_\odot$ companion if $g \gtrsim 0.0001$. Notice that g is the product of a geometric factor $(R/2a)^2$ and the "real" efficiency f of the evaporation process, i.e., the fraction of the energy flux falling on the companion's surface that is used to drive the wind. Here R is the radius of the companion and a its distance to the pulsar.

We will now make a simple estimate of the conditions under which evaporation may be expected, and of the timescales involved, following Van den Heuvel and Van Paradijs[326]. The pulsar's total energy emission ("luminosity") is

$$L_p = \frac{dE_{rot}}{dt} = \left(\frac{2R_n^6}{3c^3}\right) B_s^2 \left(\frac{2\pi}{P}\right)^4 , \tag{138}$$

where R_n is the radius of the neutron star, B_s its surface dipole magnetic field strength and c the velocity of light. Assuming that the energy is emitted in a spherically symmetric way, the amount of energy available to drive the wind from the companion is:

$$\frac{1}{2}\dot{M}v^2 = f\left(\frac{R}{2a}\right)^2 \left(\frac{2R_n^6}{3c^3}\right) B_s^2 \left(\frac{2\pi}{P}\right)^4 . \tag{139}$$

Assuming that, as usual for stellar winds, the wind velocity v is of the order of the escape velocity $v_e = (2GM/R)^{1/2}$ from the surface of the companion, the evaporation timescale τ_{evap} becomes

$$\tau_{evap} = \frac{M}{\dot{M}} = \left(\frac{3c^3}{2R_n^6}\right)\left(\frac{2GM^2}{R^3}\right)\left(\frac{a^2}{8\pi^4 f}\right)\frac{P^4}{B_s^2}. \tag{140}$$

On the other hand, the pulsar spin-down timescale is

$$\tau_{sd} = \frac{P}{2\dot{P}} = \left(\frac{P^2}{B_s^2}\right)\left(\frac{3c^3}{16\pi^2 R_n^6}\right) I. \tag{141}$$

Hence

$$\frac{\tau_{evap}}{\tau_{sd}} = \frac{1}{f(R/2a)^2}\frac{E_b}{E_{rot}}. \tag{142}$$

Evaporation of the companion can be achieved only if $\tau_{evap}/\tau_{sd} < 1$. This condition is equivalent to condition (137).

Equation (140) shows that the evaporation timescale is proportional to P^4/B_s^2. This implies that the evaporation timescale for a weak-field pulsar, such as the millisecond pulsars ($B_s \sim 4 \times 10^8$ G), is some 10^8 times longer than a strong-field pulsar ($B_s \sim 4 \times 10^{12}$ G), such as the Crab pulsar, for the same assumed rotation period. For $M \lesssim M_\odot$, the mass–radius relation of main-sequence companions is

$$R/R_\odot = M/M_\odot \tag{143}$$

and τ_{evap} scales as M^{-1}. For low values of M the star is degenerate and

$$R/R_\odot = 0.013(1+X)^{5/3}\left(\frac{M}{M_\odot}\right)^{-1/3} , \tag{144}$$

where X is the hydrogen mass fraction. In this case τ_{evap} scales as M^3. This shows that for very low companion masses the evaporation timescale becomes very short. For $X = 0.75$ relations (143) and (144) match one another at $M = 0.075 M_\odot$ Numerically, eq. (142) becomes

$$\frac{\tau_{evap}}{\tau_{sd}} = 10^{-2}\left(\frac{0.1}{f}\right)\left(\frac{a}{2R}\right)^2\left(\frac{P}{1.55\ ms}\right)^2\left(\frac{M}{M_\odot}\right)^2\left(\frac{R_\odot}{R}\right). \qquad (145)$$

This equation shows that a companion star like the sun, at a distance $a = 3R$, will be completely evaporated for $f = 0.1$ if the pulse period is less than 9 ms. If the companion mass is 0.3 $M_\odot$, as for systems at the upper boundary of the "period gap" (see section 7.2), even a 16 ms pulsar will evaporate it for $f = 0.1$ and a 3 ms pulsar will do so for $f = 0.01$. For this reason it is quite well possible that the 11 ms pulsar in Terzan 5 will be able to completely evaporate its ~ 0.1 $M_\odot$ companion.

For the case of the 1.6 ms binary pulsar PSR 1957+20 with $M = 0.02 M_\odot$, $a = 2.5 R_\odot$ and $R = 0.166 R_\odot$ [according to eq. 144], one obtains $\tau_{evap}/\tau_{sd} = 4 \times 10^{-3}$ for $f = 0.1$. Since this pulsar has a spin-down timescale of $\sim 10^9$ yr, one expects its $\lesssim 0.02$ $M_\odot$ companion to disappear in the next $\lesssim 10^7$ yr.

Evaporation and the absence of a "period gap" for low-mass X-ray binaries: relevance of the binary millisecond pulsar in Terzan 5. The evaporation model may be relevant for explaining the difference in the orbital period distributions of LMXBs and Cataclysmic Variable binaries (CVs), as depicted in fig. 26. Since the extensive X-ray observations of LMXBs with EXOSAT there is no obvious selection effect against the detection of orbital periods of LMXBs that would operate in the 1 to 2 hour range but not in the 3 to 4 hour range. Therefore the virtual absence of systems in the 1 to 2 hr range is real, and not due to observational selection. Contrary to the CV binaries, an LMXB contains an important source of power, in the form of a rapidly spinning neutron star. When the LMXB system arrives at the upper end of the period gap at $\mathcal{P} \sim 3^h$, the mass transfer stops and the rapidly spinning neutron star will become a radio pulsar. Thus, once an LMXB enters the period gap, one expects that the neutron star, provided its spin period is $\lesssim 15$ ms, will be able to destroy its companion completely, and very few LMXB systems with hydrogen-rich companions would be expected to survive below the period gap. This seems to provide a simultaneous explanation for the virtual absence of LMXBs in this period range and the existence of single millisecond pulsars.

In this respect the eclipsing binary millisecond radio pulsar in the globular cluster Terzan 5 is a most interesting case, as its orbital period is 1.9 hours, i.e. just below the period gap, and its companion has a mass of $\sim 0.1 M_\odot$. Its eclipses last for at least half the orbital period, and sometimes for the full orbital period, which indicates that the system is surrounded by a thick wind emanating from the companion star (Lyne et al.[159]). This system seems to be the perfect outcome of the evolution of an LMXB as described above: when it entered the period gap with $\mathcal{P} = 3^h$ and $M_{comp} = 0.3 M_\odot$, evaporation and shrinking of the orbit due to gravitational-radiation losses will have reduced $\mathcal{P}$ to 1.9^h and M_{comp} to $\sim 0.1 M_\odot$. In principle the 11 ms pulsar in the system has enough power to completely evaporate the 0.1 $M_\odot$ companion, provided that the evaporation efficiency factor f is $\gtrsim 0.01$. The fact that the system is largely embedded in a cocoon of matter ejected from the companion may imply that a very large fraction of the pulsar energy flux is unable to escape from this cocoon. It may therefore well be that a much larger fraction of the pulsar energy flux than given by the geometrical factor is impinging on the companion and assists in driving the wind. As the eclipses sometimes last for almost the full orbital period, it seems quite conceivable to us that evaporation efficiencies as large as 0.1 or more can be reached in this system.

Such a qualitative explanation for the virtual absence of a period gap in LMXBs was quantitatively worked out by Fedorova and Ergma[73] [75]). These authors computed full evolutionary sequences for a system originally consisting of a 1 $M_\odot$ neutron star and a 0.95 $M_\odot$ companion. They computed the evolution of the system using loss of orbital angular momentum by magnetic braking. They find that the period gap is situated between $\mathcal{P} = 3^h\!.33$ and $2^h\!.19$. At a time just before the system enters the period gap the companion has $M = 0.25 M_\odot$, and the mass transfer rate is 1.67×10^{-9} $M_\odot$ yr^{-1}. The spin period of the neutron star lies, due to the preceding mass accretion, close to its equilibrium value as given, for example, by eq. (133). For $B_s = 5 \times 10^8$ G the resulting spin period is $P = 1.8$ ms when the accretion stops. From here on the binary evolution is driven by the evaporation in combination with gravitational-radiation losses. Assuming that the specific angular momentum of the escaping wind matter is the specific orbital angular momentum of the secondary, one has (Tutukov and Yungelson[288]) that the change in a is given by (see equation (50):

$$\left(\frac{da}{dt}\right)_{evap} = -a\frac{\dot{M}_2}{M_1 + M_2} \quad (\dot{M}_2 < 0). \tag{146}$$

The change in a due to gravitational-radiation losses is given by:

$$\left(\frac{da}{dt}\right)_{GR} = -1.65 \times 10^{-9}\frac{M_1 M_2 (M_1 + M_2)}{a^3}. \tag{147}$$

These expressions have opposite signs. Together they determine the rate of change of a.

Ergma and Fedorova carried out calculations for f in the range 10^{-3} to 10^{-1}. They found that for $f > 0.03$ the stellar-wind mass-loss rate is so large ($\sim 10^{-8}$ $M_\odot$ yr $^{-1}$) that the secondary departs from thermal equilibrium and expands to overflow its Roche lobe. For example, for $f = 0.03$ they find $\dot{M}_{evap} = 4.1 \times 10^{-9} M_\odot$ yr $^{-1}$, which is 2.5 times larger than the mass-transfer rate before entering the period gap. As they did not know how to proceed further after the secondary overflowed its Roche lobe, they terminated the calculations here. It should be noted that even though the companion overflows its Roche lobe, accretion is unlikely to resume here, as the overflowing matter is outside the light cylinder of the pulsar, and is expected to be blown away by the pulsar wind. Consequently, the "Roche-lobe overflow" will continue, and the star may undergo an accelerated mass loss and expansion leading to its rapid dissipation.

Kluzńiak[135] also considered such a case, where the bloated companion could not overflow at the L_1-point due to the presence of the pulsar radiation flux, but instead overflowed from the L_2 point, leading to outflow of matter from the system and, subsequently, disruption of the companion. Ergma and Fedorova also carried out calculations for other combinations of parameter values, again for a 1 $M_\odot$ neutron star and a 0.95 $M_\odot$ companion. They find that for $0.005 \leq f < 0.03$ (and $B_s = 5 \times 10^8$ G, $P = 1.8$ ms) the evaporation takes place without complications when the system is in the period gap, and only the single millisecond pulsar remains at the end. For example, for $f = 0.01$ they find that after 4.8×10^8 yr M_2 has decreased to 0.014 $M_\odot$, and $\mathcal{P}$ increased to $3^h.93$. At the latter period the secondary again fills its Roche lobe, but the mass transfer rate due to this is so low (10^{-14} $M_\odot$ yr^{-1}) that the overflowing matter is blown away by the pulsar wind. Some 8×10^6 yr later the secondary is completely evaporated. For $B_s = 5 \times 10^8$ G and $f < 0.005$ the evaporation is so inefficient that the system survives the crossing of the period gap. For example, for $f = 0.001$ the lower boundary of the period gap is reached at $\mathcal{P} = 1^h.9$ and $M_2 = 0.20 M_\odot$. For $B_s = 10^9$ G and 5×10^9 G, the rotation periods at the moment of entering the period gap are 3.6 and 15 ms, respectively. In these cases complete evaporation can still be realized for $0.02 < f < 0.12$ and for $0.24 < f < 1$, respectively.

As mentioned above, the Terzan 5 binary pulsar suggests that evaporation efficiencies as high as f=0.1 are conceivable. It therefore seems well possible from the above that if $B_s < 5 \times 10^9$ G, systems are unlikely to survive the crossing of the period gap, and may leave behind a single pulsar at the end.

8.4 Alternative formation mechanisms for single millisecond pulsars; formation of planets around pulsars.

An alternative way to form a single millisecond pulsar is: from a close binary radio pulsar system in which the companion is a massive white dwarf ($M > 0.6 M_\odot$), and the orbital period is $\lesssim 15^h$. PSR 0655+64 is a system of this type, although its orbital period is about twice too long (see figure 37). From the models depicted in figure 70 one does, however, also expect that similar system with $P < 15^h$ will be produced in Nature. The orbits of such systems will decay by gravitational radiation losses on a timescale shorter than the Hubble time. When the orbit has shrunk so much that the white dwarf begins to overflow its Roche lobe, and mass transfer ensues, this star will, if its mass is $> 0.6 M_\odot$, become dynamically unstable, and will be disrupted on its own dynamical timescale, which is a few seconds (Bonsema and van den Heuvel[28]) The result will be a neutron star surrounded by a massive disk. If a small fraction of this disk matter ($\sim 0.05 M_\odot$) is accreted onto the neutron star, this star can be spun up to a rotational period of the order of milliseconds. Furthermore, due to accretion and outward transport of angular momentum the disk will spread outwards. It contains enough angular momentum to produce several planets (consisting of elements heavier than helium) of a few Earth masses and with orbital periods of several months. It has been suggested that the system of PSR 1257+12, which is a 6 ms pulsar with two planets of a few Earth mases and with orbital periods of 66 and 99 days -and nearly circular orbits (Wolzsczan and Frail[365])- has originated in this way (van den Heuvel[327]).

An alternative way to form such a system from a massive X-ray binary was described in section 6.8 (cf. Wijers et al.[361]). On the other hand Tavani and Brookshaw (1992). have put forward the idea that a millisecond pulsar with a planetary system may have resulted from a Low-mass X-ray binary in which the companion was finally evaporated in a way like described in section 8.4. If the evaporated matter flows out at low velocities it may form a disk around the system as described in section 2 (see figure 6, and Banit and Shaham[10][11]. From this disk planets may have condensed.

9. Some statistical considerations about the incidence of products of binary evolution in the Galaxy.

9.1 Introduction.

Abt and Levy[2][3] estimate that about half of the main-sequence stars of spectral type B2 to K have stellar companions with orbital periods $\lesssim$ 10 yrs. At some stage in their evolution the components of such systems will interact with one another (see chapter 5). It is clear, therefore, that remnants of close binary evolution will be produced in large quantities in the galaxy. Also, a large fraction of all supernovae is expected to take place in mass exchange binaries. As was pointed out long ago[297] the effects of conservative mass transfer will cause the post-mass-transfer (new) primary star of a close binary to often be more massive than the initial primary star of the system.

In view of the steepness of the Initial Mass Function (IMF), stars of lower mass are much more abundant than stars of higher mass. Consequently, since close binaries with primaries of lower mass -which are abundant- will produce new primaries of higher mass, i.e. among stars that are less abundant, close binary evolution will make a relatively large contribution to the formation rate of stars at this higher mass.

Due to the same effect, close binary mass transfer may cause a close binary consisting of components that both have too low a mass to explode as a supernova, to be transformed into a system in which the new primary star is massive enough to explode as a supernova. Again, since stars of higher mass, have a higher birthrate than stars of lower mass close binary evolution may result in a considerable increase of the supernova (SN) rate and the neutron star formation rate. The extra neutron stars produced by this effect of close binary evolution have the special property that *all* of them are produced in binaries. In this way, the formation of neutron stars will be (favourably) biased towards membership in a binary, and the same will be true for supernovae. Thus, one expects the fraction of neutron stars formed in binaries to be considerably larger than the -already high- fraction of potentially interacting binaries among stars in general. In this way, with an incidence of $\sim$ 40 to 50 % of interacting binaries one may achieve that $\sim$ 70% of or more of neutron stars are formed in relatively close binaries.

This was pointed out for example by van den Heuvel[311] [312], Meurs and van den Heuvel[171] and van Beveren[294].

If one wishes to make reliable estimates of the production rate of remnants of close binaries of various masses and types one needs as basic input ingredients:

i. the distribution at birth of the masses of primary stars;

ii. the distribution $f(a)$ of the orbital separations a of the components of binaries;

iii. the distribution $\varphi(q)$ mass ratios ($q = M_2/M_1$) of (potentially) interacting binaries, and the total fraction B of such binaries in the general stellar population.

We will discuss these input distributions in section 9.2. In section 9.3 we consider the evolution of the Hertzsprung-Russell (HRDs) of stellar groups that contain a realistic fraction of interacting binaries. In sections 9.4 to 9.8 we use the results of foregoing sections to make estimates of the formation rates of HMXBs and of binary and runaway pulsars, and we calculate the rate of coalescence of double neutron star systems in the Galaxy.

9.2 Input distributions.

A. The distribution at birth of stellar masses. For this we assume the Miller and Scalo IMF (see also Scalo[244]), denoted as $\psi(M)$. The star formation rate as a function of stellar mass M is the given by

$$\frac{dN(M)}{dt} = \Psi(M).f(t), \tag{148}$$

where $f(t)$ is the so-called rate function which describes how the galactic star formation rate changed in the course of time. The observations show that over the past 5.10^9yrs, $f(t)$ has been constant, therefore we will assume that for $M > 1.2\ M_\odot$ holds: $f(t) = 1$. (A star of $1.2\ M_\odot$ lives for about 5.10^9 yrs). As we are interested mostly in massive stars here, we will assume $f(t) = 1$ throughout this chapter.

In good approximation, for $M \geq 1.2 M_\odot$ one has

$$\Psi(M) \sim C.M^{-2.5}, \tag{149}$$

where C is a constant. The observed distribution of stellar masses $\Phi(M)$, for $M \geq 1.2 M_\odot$, is simply obtained by integrating equation (148) (with $f(t) = 1$ between $t = 1 - T(M)$) and $t = 1$, where $T(M)$ is the lifetime of a star mass M, in units of the age of the Galaxy. Thus for $M \geq 1.2 M_\odot$ one obtains:

$$N(M) = \Phi(M) = \Psi(M).T(M) \tag{150}$$

The lifetime T(M) of a star is given by

$$T(M) = \frac{M/M_\odot}{L/L_\odot}.10^{10}\ yrs \tag{151}$$

Using the mass-luminosity relation $L \sim M^{3.5}$, one finds:

$$T(M)(:)M^{-2.5} \tag{152}$$

Hence, eq. (150) yields:

$$\Phi(M)(:)M^{-5} \tag{153}$$

This is in good approximation, the distribution of stellar masses for $M > 1.2\ M_\odot$.

B. The distribution of orbital separations a. For simplicity we will assume circular orbits. The first attempt in obtaining a complete sample of orbital properties of early-type binaries was by van Albada[293]. He found that the frequency distributions of ($log\ a$) for close (spectroscopic) and wide (visual) binaries in OB associations are roughly equally high. For intermediate separations, i.e. for systems that are too wide to be detected as spectroscopic binaries, but too close to be seen as visual binaries, i.e. with separations between 0.5 AU and $\sim$ 40 AU, there was little information available. However,the presence of red supergiant eclipsing binaries such as 31 and 32 Cygni, ζ and ε Aurigae, VV Cephei, etc. (see chapter 3) shows that binaries with such separations must be quite common. Indeed, with new techniques such as Griffin's radial-velocity photometer and speckle interferometry, many binaries of these intermediate separations have recently been discovered (see e.g. Griffin[95] [96]). Even though this material is not yet free from selection effects there is no reason to doubt that the observations are consistent with an approximately flat distribution of ($\log\ a$) all the way from a few solar radii to one thousand astronomical units or more (see also Abt[1]).

This implies:

$$f(a).da = K da/a \tag{154}$$

or

$$\mathcal{F}(\log a).d\log a = K.d\log a \tag{155}$$

where K is a constant.

C. The distribution of mass ratios of unevolved close binaries $\varphi(q)$ and the interacting frequency B. Like the a-distribution, the distribution of mass ratios is very much affected by selection effects. The first investigations of $\varphi(q)$ were those of Kuiper[144] and Trimble[284].

Abt and Levy[2] [3] and Abt[1] studied samples that were expected to be complete for mass ratios $\gtrsim 0.3$. As mentioned in section 9.1 those studies showed that some half of the main-sequence B2-K stars are in binaries with $P < 10$ yrs. Roughly fifty per cent of these stars are in binaries with $P \lesssim 50^d$. In early investigations from catalogues it was often concluded (Trimble[284], Lucy and Ricco[158]) that $\varphi(q)$ has a peak near $q = 1$ and a second one near $q \simeq 0.2$ to 0.3 (cf. Kraicheva et al.[139], Staniucha[261]). However, the peak near $q = 1$ is almost certainly due to selection effects strongly favouring the discovery of spectroscopic binaries with q near unity, as follows (cf. Scarfe[245], Trimble[286]).

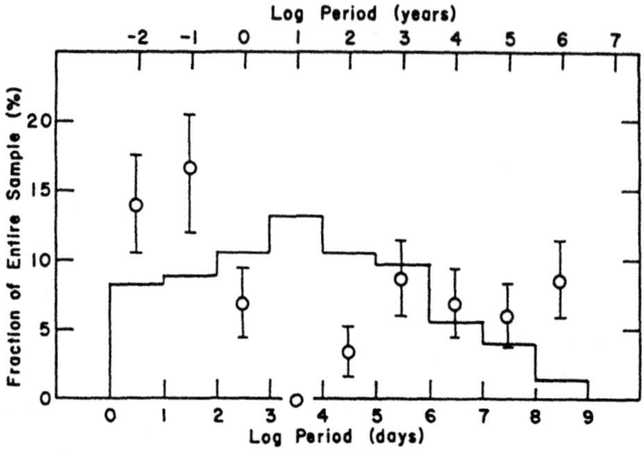

Fig. 83. The distribution of binary periods in a sample of 123 F3-G2 dwarfs (solid line) and 114 B2-B5 dwarfs (circles and error bars), according to Abt[1]. The *a*- and *P*-distributions are in good approximation represented by flat distributions of *log a* and *log P*.

A system with $q \gtrsim 0.7$ will be a double-lined spectroscopic binary (SB2). Because of the very strong mass-dependence of the mass-luminosity relation, systems with $q \lesssim 0.6 - 0.7$ will tend to be single-lined spectroscopic binaries (SB1), as the secondary star is often too faint to show itself in the spectrum.

Now an SB2 is discovered immediately if one takes one spectrum of the star: one observes the lines to be double, and one knows that it is a spectroscopic binary. So, in surveying stars for spectral classification -for which only one spectrum is sufficient- the SB2 are recognized, and will subsequently be put on observing programs of observatories, to have their orbits determined. Thus they will enter into catalogues of orbits of spectroscopic binaries.

On the other hand, in order to recognize an SB1, one has to take a number of spectra, in order to notice that the radial velocity of the star is variable. Since there is no a prior reason to do this in surveying stellar spectra (i.e. for classification purposes) most SB1 will remain unnoticed as spectroscopic binaries.

It is estimated that this simple selection effect favours the discovery of SB2 over SB1 by a factor of at least five and possibly as high as ten (Scarfe[245]; Hogeveen[111] [112]). SB2 will be overrepresented in catalogues of SB-orbits (such as Batten et al.[13]) by factor of this order, with respect to SB1. If one takes this into account the peak near $q = 1$ in the q-distribution of SBs, derived from catalogues, disappears (see Hogeveen[111] [112]; Tout[283]).

Recent careful discussions of the selection feffects affecting the q-distribution of SBs derived from catalogues, and of the works of earlier authors, were given by Halbwachs[102], Hogeveen[111] [112]) and Trimble[286]). They conclude that, after taking selection effects properly into account, the distribution $\varphi(q)$ increases towards the smaller q-values. For $q > 0.4$ a distribution of the form $\varphi(q)(:)q^{-2}$ fits the observations well (see figure 84). Trimble concludes to a distribution (:) q^{-1}.

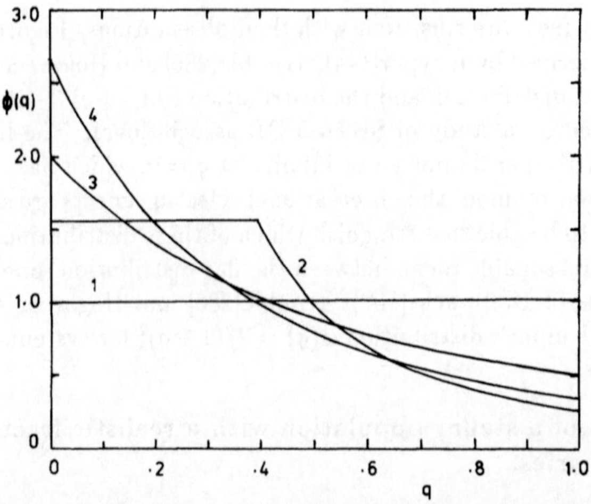

Fig. 84. Four possible q-distributions of spectroscopic binaries: (1) flat; (2) Hogeveen's (1991, 1992) distribution; (3) $\Phi(q) = 2/(1+q)^2$; (4) $\Phi(q) = 8/3(1+q)^3$. All distributions normalized to unit area.

If one assumes for $q < 0.4$ these distributions to be flat, one expects with Hogeveen's results some 60 and with Trimble's some 50 per cent of all unevolved systems to be born with $q < 0.4$. Upon mass transfer in a close binary these systems will enter a Common Envelope phase. From a completely different side. by studying the incidence of close binaries among the nuclei of planetary nebulae (which is at least 10-20 %), and their distribution of companion masses and periods, de Kool[56] [57] concluded that these observations require a q-distribution of unevolved close binaries that increases towards the lower q values.

Assuming indeed for $q < 0.4$ the distribution to be flat and half of all stars to be born as primaries in (potentially) interacting binaries, one sees that about 25 per cent of the stars are primaries in systems with $q > 0.4$, and will evolve more or less conservatively, whereas another 25 per cent will go into CE-evolution. [Of course, these are very rough estimates, as one also has to take into account the separations of the systems, which determine where the stars become largely convective and thus may go into CE-evolution even for $q > 0.4$]. Jaschek and Ferrer[124] came to the conclusion that roughly of order 30 per cent of all stars are primaries of spectroscopic binaries with $q \gtrsim 0.4$. Mermillod concludes et al.[170], contrary to the above, that $\varphi(q)$ is practically flat. Abt[1] also favours a rather flat distribution for the B-type stars. For O-type stars, Garmany et al.[86] conclude to a q-distribution that peaks near $q = 1$. They detected 31% of the O-type stars in their sample to be SB. As their lower detection limit for radial velocity variations is about 20 km/s (due to the great width of the spectral lines and the occurrence of erratic radial variations due to atmosheric motions), the real close binary percentage will be higher -and can easily be of order 50 per cent. Also, because of bias against detection of small velocity variations for $q =< 0.33$ the sample will be very incomplete. Taking this into account, a flat

q-distribution seems quite cinsistent with their observations. Figure 84 shows the q-distributions derived by Kuyper[144], Trimble[286] and Hogeveen[111][112], together with a flat distribution and the distribution $8/3(1+q)^{-3}$ (Blaauw and van Albada[24], based on a study of SB in 5 OB-associations). The latter distribution and that of Kuiper do not go to infinity at $q = 0$, which has computational advantages. In our opinion, the observational selection effects are still not known sufficiently well to be able to distinguish which of these distributions is the closest to reality. As a reasonable mean between the flat distribution (one extreme) and the distributions of Halbwachs[102], Trimble[286] and Hogeveen [111] [112] we will adopt here Kuiper's distribution $\varphi(q) = 2/(1+q)^2$ for systems of all masses.

9.3 Evolution of a stellar population with a realistic fraction of interacting binaries.

Using the input distributions described in section 9.2 one can construct a stellar population that contains a realistic fraction of potentially interacting binary systems.

A good example of such a population is formed by a star cluster in which all stars were formed at the same time by the fragmentation of a molecular cloud. Open clusters and associations are such coevally formed stellar systems.

One may construct a synthetic open cluster of N stars by the following procedure:

- pick N stars at random from a large collection of stars that have a smooth mass distribution according to the IMF.
- Pick at random a fraction B out of these N stars which are labelled "interacting binaries" ($P < 10$ yr).
- Assign to each binary an orbital separation a by picking a-values at random from a smooth a-distribution given by equation (155).
- Assign to each binary a mass ratio q by picking q-values at random from a smooth $\varphi(q)$ distribution such as one of those discussed above.

After this procedure one has a realistic open star cluster at age zero. The next step is to construct a stellar evolution program that is able to calculate the evolution of all single stars and binaries in the cluster.

This implies that the programme must be able to handle the conservative evolution according to the cases A, B (with its various possible branches: low-mass, intermediate mass and massive) and case C. This it will do if q is sufficiently large ($q > 0.4$) and the envelope of the Roche-lobe filling star is not convective. For systems where the lobe-filling star has a convective envelope, or which have $q < 0.4$, one expects that CE-evolution will occur. One thus has to build into the programme a procedure to follow CE evolution and to determine its outcome, taking large losses of mass and angular momentum into account. Also, one will have to allow for the cases BB, BC, CB, etc. and for the occurrence of reverse mass transfer, again with all its possible variations. Also, for stars that are exploding as a supernova, one has to build into the programme a subroutine to calculate the induced orbital changes, runaway velocities, etc. (with and without kicks due to asymmetries).

Of course, building such a programme and making it run is a large and complex project.

In order to make the programme not too time-consuming, one can simplify the calculation of the evolutionary tracks. This can be done by interpolating (logaritmically) between a limited number of tracks. This procedure works very well (see van den Heuvel[298]). The same can be done for calculating the evolution with mass exchange of close binaries.

For this Giannone et al.[87] were the first to publish grids of equilibrium stellar models with different ratios of helium core mass to total stellar mass (for case B) and with different central hydrogen content X_c and different ratios core mass to total stellar mass (for case A). Figures 85 and 86 show the positions in the HR-diagram of these grids of remnants of case B and case A mass transfer, respectively. Case C is expected to produce roughly the same remnants as case B, the only difference being that a part of the envelope mass may be lost from the systems during the exchange.

The first attempt to calculate the evolution of a synthetic star cluster with binaries was made in 1969 by the author (Lick Observatory) by using the above-mentioned grids of models by Giannone et al.[87] and using single star evolutionary tracks kindly put to his disposal by I. Iben. The HR diagrams of two clusters of stars were constructed.

Both were assigned 33% interacting binaries ($a < 10$ AU), and 33% visual binaries, an a-distribution according to eq. (155) and a q-distribution as given by Kuiper[144]:

$$\varphi(q) = \frac{2}{(1+q)^2} \tag{156}$$

For case C three possibilities were considered:

i. all of the envelope matter was transferred to the companion;

ii. one half of the envelope mass was transferred to the companion, the other half was lost from the system, and

iii. one quarter of the envelope mass was transferred to the companion, three quarters being lost from the system.

No allowance for CE evolution was made, as this was not yet known at the time, and cases A, B and C were calculated conservatively. Also, no cases BB, BC, CB, etc., or supernova effects were included. Figure 87 shows the HR diagram of cluster I at several ages. The figure shows that "blue stragglers" were produced in this cluster in a natural way by the mass-exchange processes in the cluster binaries. At every age one or more blue stragglers, practically always

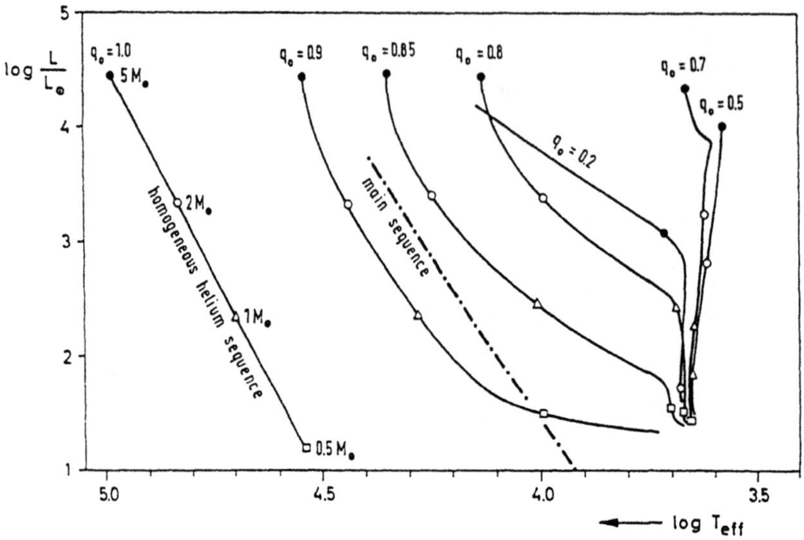

Fig. 85. HR diagram with sequences of inhomogenous equilibrium models with beginning central helium buring ('generalized main sequences' for case B). These models approximate the possible remnants of original primary components, for initial masses $\geq 2.8M_\odot$ at the end of a mass loss starting after central hydrogen burning (so-called case B; see also figure 60). q_0 gives the M_r/M value of the chemical discontinuity between the central helium core and the hydrogen rich envelope. Along the sequences, some mass values are indicated by different marks (• = $5M_\odot$ ∘ = $2M_\odot$; $\Delta = 1M_\odot$; □ = $0.5M_\odot$; the upper end of the sequence $q_0 = 0.2$ corresponds to $M \approx 10M_\odot$. The zero age main sequence is indicated (- •-) (after Giannone et al. [87]).

resulting from case B or C systems, are present (Case A turned out to be very rare).

In view of the many uncertainties involved (what to do with systems with low q? CE-evolution still had to be "invented" by Paczynski and Ostriker), these results were never published (although they were presented at colloquia in Brussels Paris and Geneva; van den Heuvel[300][301][302]). A considerably advanced set of calculations of similar type were made by Klumper[134] in his M. Sc. thesis at the University of Amsterdam, but also not published. A subsequent step forward was made by Collier and Jenkins[44] who calculated the evolution of star clusters with case B binaries, again with the primary aim to explain the existence of blue stragglers in star clusters. The most extensive programme known to us so far was constucted by Pols[212] and Pols and Marinus[214]. The programme uses the single star evolution tracks calculated by Maeder and Meynet[160] and takes full account of all the various possible types and of close binary evolution described above, including CE-evolution, reverse mass transfer, supernova effects, etc. etc. Also, the calculated luminosities and effective temperatures are converted into U, B and V magnitudes and (U-B), (B-V) colours, using the best available transformations (cf. Coté[45]). This is necessary if one wishes to compare the obtained HR-diagrams with the observed colour magnitude diagrams of star clusters.

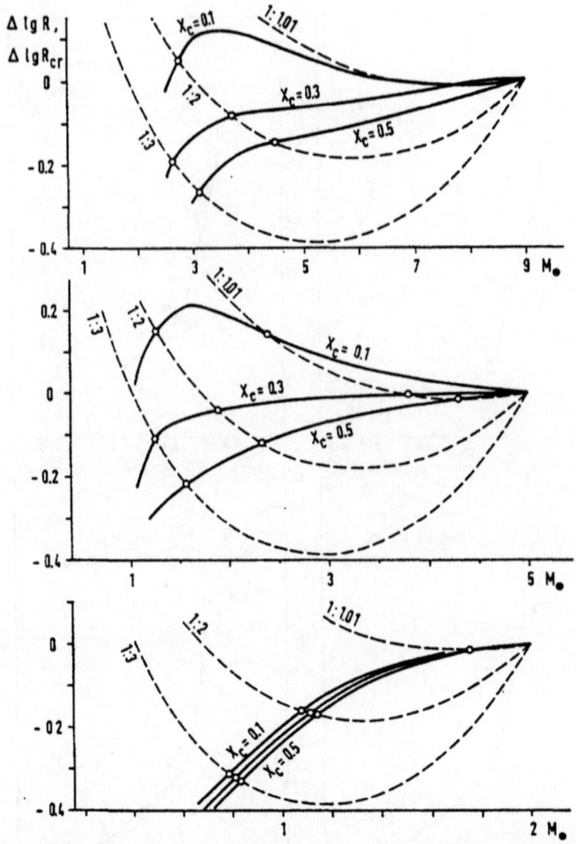

Fig. 86. End of rapid mass loss in case A, for three initial masses (9, 5 and 2 $M_\odot$)of the orginally more massive component, for three initial mass ratios:(1:1.01, 1:2 and 1:3) and for three initial separations of the components (corresponding to mass loss starting when the central hydrogen content is already down to X_c=0.5, 0.3 and 0.1, respectively). Solid curves ($\Delta lg R$): Plotted over M are the radii R of possible thermally adjusted remnants of the initially more massive components, with core mass q_0 = *const* and X_c=const. Dashed curves ($\Delta lg R_{cr}$): plotted over M (mass of the originally more massive component) are the radii R_{cr} of the critical Roche volume of this star. Conservation of total mass and total angular momentum of the system is assumed here. The zero points of lg R and of lgR_{cr} are chosen arbitrarily such that for all X_c values $R = R_{cr}$ at $M = M^0$ (beginning of rapid mass loss). The rapid mass loss stops for a given case at the intersection of the corresponding solid and dashed curves (open circles). Figures of this type can be used to calculate the end-result of case A mass transfer for systems with a large variety of initial parameters (after Giannone et al. [87]).

Figures 88a-f show some of the results of the calculations of Pols and Marinus for a cluster of 200 stars with 50% binaries. As input q-distribution the authors used Kuiper's distribution.

In a number of CE cases the two stars coalesced, and their evolution was then continued as that of a single star.

A number of results of these calculations merit special attention, as follows.

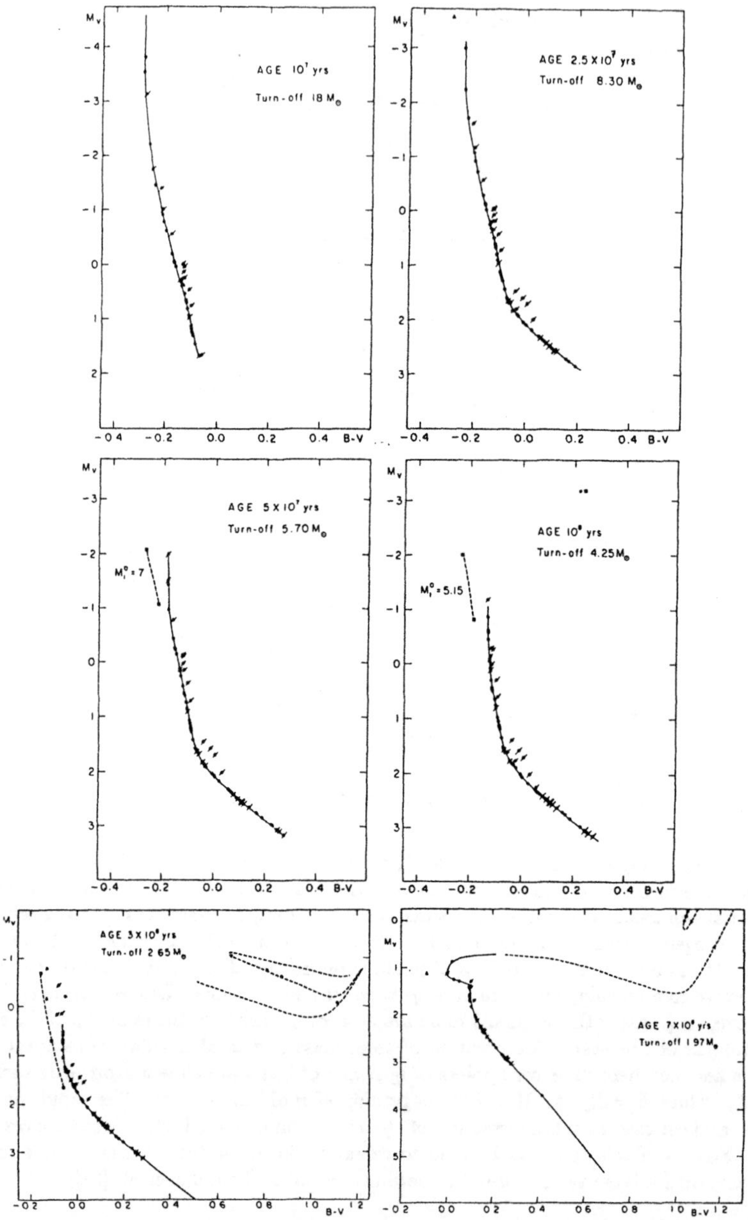

Fig. 87. Evolution of the HR diagram of a star cluster that contains 33% spectroscopic (=interacting) binaries and 33% visual binaries calculated by the author in 1969 (unpublished). The ways in which the cluster was constructed are described in the text. Unevolved binaries are indicated by a filled circle with a slash, single stars by filled circles, post-mass transfer case B systems by filled triangles (these are the new primaries), and post mass-transfer case C systems by filled squares. For case C three possible outcomes are given: all or half or one quarter of the envelope mass transferred; the remainder being lost from the systems. These three possible outcomes for the same system are connected with a dashed line. Common-envelope evolution was not taken into account, as this was not known at the time.

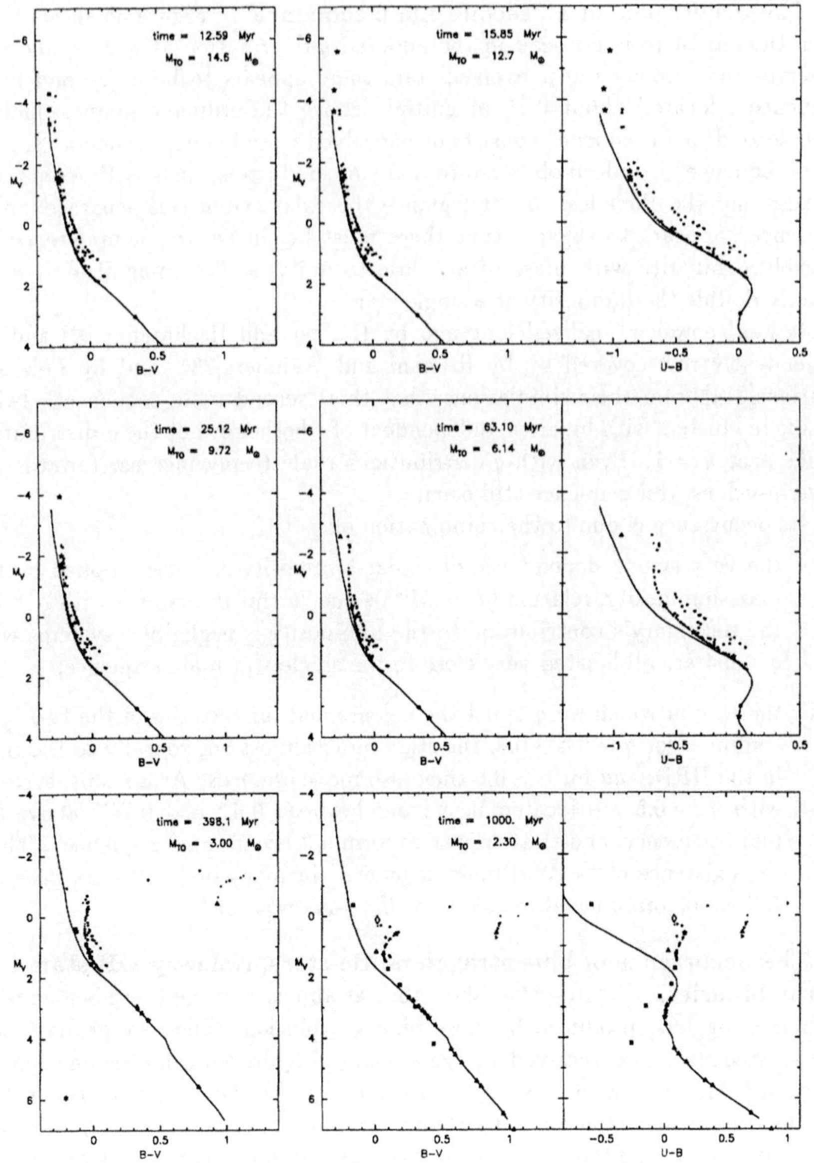

Fig. 88. Evolution of the HR diagram of a star cluster with a realistic fraction (50 per cent) of interacting binaries using a realistic distribution of primary masses, mass ratios and orbital separations. HR diagrams are given for 6 ages; for 3 of these also the M_v *vs.*$(U-B)$ diagram is shown. In these calculations all present knowledge of close binary evolution (including Common-Envelope Evolution) was properly taken into account. Calculations by Pols [212, 213] and Pols and Marinus [214]. The meaning of the symbols is :* =neutron star companion; □ = helium-star companion; △ = white dwarf companion; ◇ coalesced system. Notice that apart from blue stragglers even systems in the gap between the main-sequence and the giant branch are produced (as often observed in open clusters e.g. in Coma and Praesepe), while with single star isochrones, stars in this region are extremely rare. Notice that also subdwarfs below the main-sequence are produced, as are found for example in the Hyades.

a. The occurrence of a "second main sequence". Figure 88 shows that, even though there is no peak in the adopted q-distribution at $q = 1$, one still observes that among the unevolved stars there appears to be a "second main sequence", located about 0.75 magnitude above the ordinary main sequence. This second main sequence consists of unevolved close binaries. Such a "second main sequence" is indeed observed in many open clusters, such as Praesepe, the Hyades and the Pleiades. At first glance the existence of this separate "main sequence" appears to suggest that there must be, in nature, a preference for unevolved binaries with mass ratios close to unity, as 0.75 magnitude means: exactly double the lumnosity of a single star.

It was, however, realized long ago by Haffner and Heckmann[101] and independently rediscovered by by Romani and Weinberg[234] and by Pols and Marinus[214] from their simulations, that this "second main sequence" always occurs in clusters with binaries, independent of whether or not the q-distribution peaks near $q = 1$. Even with q-distributions that steeply increase towards the lower q-values, this sequence still occurs.

Its occurrence is due to the combination of:

i. the very strong dependence of stellar luminosity on mass implied by the mass-luminosity, relation ($L \sim M^{3.5}$); due to this in systems with $q < 0.4$ the secondary's contribution to the luminosity is negligible: systems with $q < 0.4$ are all located very close to the single-star main sequence;

ii. the way in which for $q > 0.4$ the colours and luminosities of the two stars combine: for $q = 0.4 - 0.6$, the stars move almost horizontally to the right in the HRD, and for $q > 0.6$ they also move upwards. As a result, systems with $q > 0.5$ are located in a band between 0.6^m and 0.75^m above the main-sequence, and thus appear to form a "second main sequence". Thus, the existence of "second main sequence" in many open clusters *does not tell us anything about the shape of the q-distribution!*

b. The occurrence of blue stragglers, Be-stars, runaway OB stars and X-ray binaries. Figures 88a-f show that at almost any age there is a number of blue stragglers, produced by close binary evolution. The new primary is a rejuvinated star: as it received a large amount of hydrogen from the envelope of its companion, it grew in mass and, consequently, also the mass of its convective core increased. Since the convective boundary moved outward into a region where no hydrogen-burning took place, fresh hydrogen from this region was mixed into the core, lowering the helium abundance of the core and causing the star to be rejuvinated: after the mass transfer it had moved close to the Zero Age Main-Sequence (ZAMS) whereas the most massive single stars in the cluster are close to core-hydrogen exhaustion and thus are located to the right of the ZAMS. Therefore, the new primary is in any case bluer and, if q was sufficiently large (such that the mass transfer was conservative), also sometimes brighter than the stars near the main-sequence turn-off.

One notices that at age $\sim 10^7$ *yr* there are also a number of blue stragglers that have neutron star companions. Such systems will have runaway velocities of

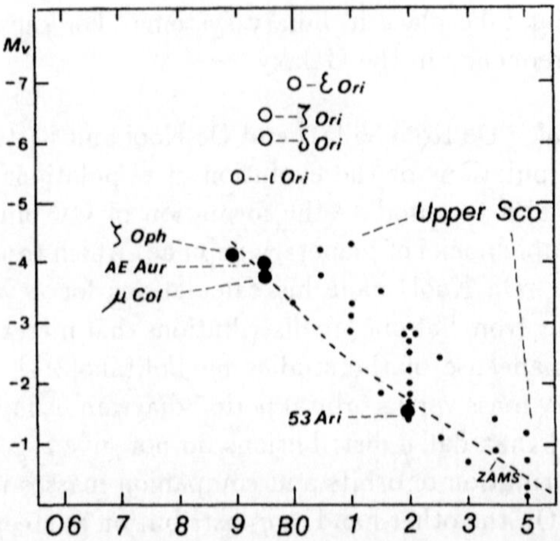

Fig. 89. Position in the HR diagram for the runaways Zeta Oph, AE Aur, Mu Col and 53 Ari for which accurate absolute magnitudes could be determined independent of luminosity classification, showing blue straggler character. Also shown: the zero age main sequence, stars in Upper Sco and nonrunaways of Ori OB1 (Zeta Oph is a runaway from Upper Sco; the other three runaways come from Ori OB1; after Blaauw [23]).

several tens of km/s as indicated in the figure. Blue stragglers observed in young open clusters often are B-emission stars. This is quite understandable, since the preceding mass transfer will have proceeded through an accretion disk around the mass-receiving component (see section 5): this star is therefore expected to have become a very rapid rotator. As mentioned in section 5 it was noticed by Blaauw[23] that 6 out of 7 runaway stars that have sufficiently well-determined space velocities to trace back the OB associations from which they originated are indeed (a) blue stragglers in these associations (figure 89), (b) are extremely rapid rotators and (c) have abnormally high helium abundances. None of them is observed to be spectroscopic binary. However, a neutron-star companion in an $\sim 10^d$ orbit will induce a radial velocity variation of $< 15km/s$ in a 20 $M_\odot$ OB-type companion. This is in practice impossible to measure in the spectrum of a star that has very large rotational line broadening. So, the absence of known radial velocity variations in these runaway stars, is not at all surprising.

The work of Tutukov & Yungelson. These authors and their associates have for many years studied the a- and q-distributions of unevolved binaries and have made important studies of conservative and non-conservative binary evolution. See, for example, Tutukov and Yungelson[289], Iben and Tutukov[120][121][122][123]. Tutukov and Yungelson have constructed a binary evolution programme that reportedly can follow many types of binary evolution (see e.g. Tutukov, Yungelson and Iben[290]). The precise set-up of the programme is not known to the author, but several interesting results have been published (Tutukov, Yungleson and Iben [290]). According to their calculations there may be as seventeen different types of SN-explosions that take place in binary systems. For each of these systems they estimate the frequency in the Galaxy.

The work of De Kool. De Kool[56][57] and De Kool and Ritter[58] have carried out Monte-Carlo simulations of the evolution of populations of close binaries, with especially the aim to simulate the formation of CV-binaries and CV-like systems (such as double nuclei of planetary nebulae) which result from Common-Envelope evolution. De Kool made his calculations for a variety of input q-distributions, ranging from flat ones to distributions that increase steeply towards low q-values. He made use of the studies by Politano[211] and identified the region in the primary mass versus orbital period diagram from which CV binaries originate. He found that flat q-distributions do not give a good representation of the observed distribution of orbits and companion masses in double nuclei of planetary nebulae. On the other hand, a q-distribution as steep as q^{-2} yields the correct observed distributions.

The work Meurs and van den Heuvel (1989). These authors calculated the fraction Y_{cbr} of primary stars of evolved close binaries among the main-sequence stars, as a function of stellar mass. Their results represent a summary over all orbital separations.

They tried several q-distributions, ranging from flat ones to ones increasing steeply to low q such as

$$\varphi(q) = \frac{8}{3}(1+q)^{-3}. \tag{157}$$

They assumed that a fraction f of the exchanged matter was lost from the systems during the exchange. Figure 90 shows the resulting Y_{cbr} values as a function of stellar mass, for f-values ranging from 0.0 to 0.9.

Stellar wind mass loss during the evolution of stars more massive than 15 $M_\odot$ was taken into account: this resulted into the dashed curves in figure 90. The figure shows that for $f = 0.3$ (which seems a reasonable estimate, certainly for systems with $q > 0.4$) one expects that for $M \lesssim 35 M_\odot$, $Y_{cbr} \simeq 0.75\,B$ where B is the fraction of (potentially) interacting binaries among unevolved main-sequence stars. The reasons for the high fraction of evolved close binaries among main-sequence stars were explained already in section 9.1.

[In the calculations CE-evolution was not included; this is not expected to affect the results very much, since CE-evolution takes place in systems with $q \leq 0.3 - 0.4$, which do not make an important contribution to Y_{cbr} for $f \geq 0.3$, since here the new primaries are always less massive than the old ones. Due to the IMF-steepness effect explained above, these systems will not contribute much to the total population of main-sequence stars of the same mass.] Figure 90 shows that, if one assumes, that $B \sim 0.5$, and $f = 0.3$, some ≥ 35 per cent of all main-sequence stars have an evolved binary companion. This companion can be: a helium star, a white dwarf, a neutron star or a black hole.

Meurs and van den Heuvel also calculated the total number of systems with a neutron star or black hole companion in the Galaxy. To calculate this number,

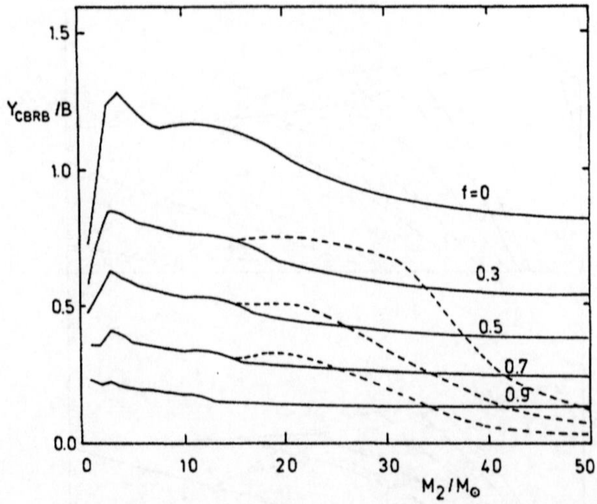

Fig. 90. The scaled fraction of case B CBRs, Y_{cbr}/B, among the main-sequence stars of mass M_2, for a number of values of *f*. Solid curves were calculated without corrections for stellar-wind mass loss among stars $\geq 15M_\odot$. Dashed curves (for f =0.3 to 0.7) include stellar-wind mass loss corrections (after Meurs and van den Heuvel [171]; f = fraction of envelope mass lost from system during exchange).

one has to calculate the number of systems resulting from those in which the initial primary star had a mass $\gtrsim$ 10-12 $M_\odot$ (the lower mass limit for core collapse in a binary system). Figures 91a,b show Y_{cbr}/B for systems resulting from those in which the primary star had $M \geq M_L$, with M_L ranging from 10 $M_\odot$ to 80 $M_\odot$ ($M_L \simeq 40 - 80M_\odot$ is the range in which the lower mass-limit for black hole formation is located: see section 6). The figures are for $f = 0.3$ and 0.5, respectively.

The figures show that for the mass range $M_2 \sim 15 - 35M_\odot$, for $f \simeq 0.3$, Y_{cbr} is of order $(0.7 - 0.8)B$.

Since for $M \geq 15M_\odot$ the helium star resulting from the original primary lives for less than 30% of the main-sequence lifetime of the new primary, one expects in most sytems the evolved companion to be a neutron star or a black hole.

The number of early-type main-sequence stars with a neutron or black hole companion. Most systems will not have been disrupted in the supernova explosion, as the exploding star was less massive than its companion at the time of the explosion. Even if, due to a asymmetries in the SN-mass ejection, the neutron star received a random kick-velocity of $\sim 100 - 150km/s$ at birth, still at least half of the systems will not have been disrupted, since -statistically- half of the kicks will have been retrograde, and thus will have caused the system become more strongly bound. This may have been the case in many Be systems. This argument will no longer hold if the systems are wide ($P \geq 100^d$) and $V_k \geq 200\,km/s$. If no systems were disrupted, Meurs and van den Heuvel calculate that there are some $(1 - 2).10^4$ unevolved early type stars with a neutron star or a black hole

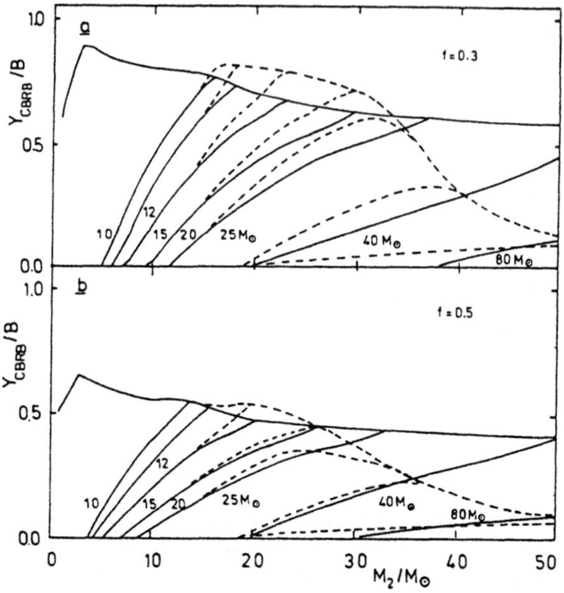

Fig. 91. As figure 90 with curves for systems with original primary mass larger than $M_L^0 = 10, 12, 15, 20, 25, 40$ and 80 $M_\odot$, for **a** :$f = 0.3$ and **b** :f =0.5. See text. Solid curves and dashed curves are again uncorrected and corrected for stellar-wind mass loss, respectively (after Meurs and van den Heuvel [171]).

companion in the Galaxy (this number is quite similar to earlier estimates made with cruder assumptions, cf. van den Heuvel[304][306][308]).

The estimated numbers of Be/X-ray binaries in the Galaxy derived from observations is of order 10^3 to 2.10^4 (cf. Meurs and van den Heuvel[171]; van den Heuvel and Rappaport[323]), so many -if not most- of the "quiet" early-type systems with a neutron-star or black hole component might be Be-like systems in their quiet phase.

The duration of the X-ray phase in "standard" HMXBs. Clearly, the standard HMXBs, with companion masses $15-25M_\odot$ form only a tiny fraction of the early type stars with a neutron star or a black hole component. This is due to the fact that the mass-transfer rate or wind-accretion rate must be very fine-tuned to a value in the range 10^{-11} to $10^{-8} M_\odot/yr$, to have an X-ray luminosity in the observed range 10^{35-38} $ergs/s$.

Only under very special circumstances -outlined in section 3- is an early-type companion able to offer such a limited mass transfer rate. One therefore expects this HMXB phase to be of very short duration. Some 4500 systems with $M > 15M_\odot$ and a compact companion are predicted for the entire galaxy (for $f = 0.3$) from Meurs & van den Heuvel's calculations. With in total some 50 HMXB-systems of this type in the Galaxy (there are 4 within 3.5 kpc from the Sun) and a total lifetime of the primary star of $\sim 5.10^6$ yrs, one finds that the HMXB X-ray lifetime is of order 5.10^4 yrs. This is indeed very short, but fits with the estimates given in section 3.

9.4 The fraction of supernova explosions that take place in stripped binary components: supernovae of type Ib, c.

Supernovae are, broadly, classed as type I and II, according to to the absence or presence of hydrogen in their spectra. All type I supernovae have lightcurves that are very similar in shape. In recent years it has become clear that there are at least three varieties of type I supernovae, designated types Ia, b and c, and two varieties of type II supernovae, the so-called "linear" and "plateau" types, according to the characteristic shapes of their lightcurves.

Table 31 lists the distinguishing properties of the five SN types (after Branch et al. [30]; Nomoto et al.[84]).

Table 31. The distinguishing properties of the five main classes of super novae (after Branch et al.[30]).

	H	He	Subtype	Late Spectrum	Suggested Progenitor Type
Type I	No	No	Ia	Fe	CO white dwarf (exploding)
	No	Strong	Ib	} O, CO	Helium Star
	No (perhaps weak)	Weak	Ic		($\geq 2.5 M_\odot$, probably in binary)
II	Yes (strong)		Linear (Bright) Plateau (faint, like 1987 A)		Single $\geq$ 8 $M_\odot$

Type Ib,c are charaterized by the presence of helium in their spectra, while in Ia's this element is absent. In Ic sometimes hydrogen is weakly present but the lightcurves is still very characteristis for type I.

Type II supernovae are always associated with early type stars and young stellar populations in galaxies: they clearly must be the core-collapse events of massive stars. While it was thought originally that type I supernovae were associated only with an old stellar population, it has become clear in recent years that this holds only for type Ia. Type Ib,c are found like type II supernovae in regions of early type stars and young stellar populations in galaxies. Therefore, they also must be the result of the evolution of massive stars. The absence (or weakness, in the case of Ic) of hydrogen in their spectra indicates that we must be dealing here with SN explosions of stars that lost most of their hydrogen envelopes (cf. Wheeler and Harkness[356]). This is exactly what is expected for the evolution of components of close binaries: these stars will, at the time of the explosion, be bare helium cores. Calculations of the lightcurves produced by exploding helium stars of 3.3 to 4 $M_\odot$ show a very good fit to the observed light-curves of Ib supernovae (figure 92, after Nomoto et al.[180]); they predict that some 0.25 $M_\odot$ of ^{56}Ni is produced in these explosions.

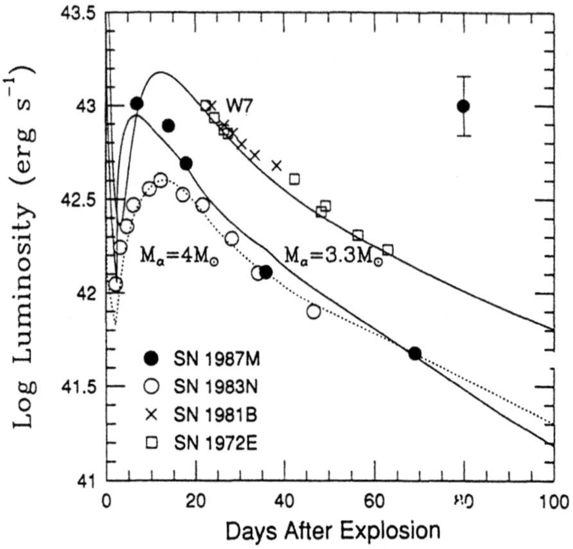

Fig. 92. **Quasi-bolometric (λλ3280-9100) light curve of type I SN 1987M, together with the bolometric light curves of SNe Ia 1981B and 1972E and of SN Ib 1983N. The predicted curves of the 3.3 $M_\odot$ helium star model for SN 1987M, the 4 $M_\odot$ helium star model for SN Ib, and a theoretical model for Ia with more than 1 $M_\odot$ in ^{56}Ni produced are indicated by solid and dotted lines (after Nomoto et al [180]). The figure shows that 1987M and 1983N fit very well with light curves expected for exploding helium stars.**

On the other hand, Ia supernovae must be due to another mechanism. Close scrutiny has shown that they are more than one magnitude brighter than types Ib,c (which implies that there is a selection effect favouring the discovery of Ia, and that the amount of ^{56}Ni is more than a solar mass.). The mechanism for type Ia now accepted by most workers, is that of the carbon-deflagration of a CO-white dwarf (rapid total conversion of CO into Fe), triggered by accretion. The absence of hydrogen and helium in the spectra shows that the accreting material must be CO (or heavier elements). The current idea is therefore that type Ia SNe are due to the merging of a double CO-white dwarf. Such close white dwarf binaries are expected to be produced in abundance by the evolution of intermediate mass binaries, through two CE-phases. Extensive studies by Iben and Tutukov[120][121] and Webbink[350] give support for this idea.

9.5 The expected incidence of bare-core supernova explosions among massive stars.

We will assume that for a binary component the lower mass limit for exploding as SN is 10 $(\pm 1) M_\odot$ (see section 5). With $f = 0.3$ one then has for $M_1 = 10 M_\odot$ that for $q \gtrsim 0.45$ the new primary also has $M > 10 M_\odot$, so the system will produce two neutron stars. For larger M_1 this will happen also for smaller q-values (i.e.: for $M_1 = 15 M_\odot$ it already happens for $q > 0.2$; however, systems with $q = 0.2 - 0.35$ will go into CE evolution).

Summarizing over all systems with $M_1 > 10 M_\odot$ one can say that roughly, for $q > 0.35$ the systems produce two neutron stars, whereas for $q < 0.35$ they will go into a CE phase and will produce only one neutron star (as the companion has too low a mass to explode as a supernova). The latter systems, if they are close, will coalesce completely (the secondary is absorbed in the primary) and thus will produce a single star explosion. The wider ones will produce a close binary consisting of a helium star and the low-mass companion. These systems which resemble the progenitor of Her X-1 (see figure 71) will often be disrupted in the SN of the helium star, since very few Her X-1-like systems are observed in nature. With $f(a)$ from eq. (155) and $\varphi(q)$ from eq. (156) one expects about half of the latter systems to coalesce and the other half to follow the Her X-1-like path, but be disrupted in the SN. Using the above $\varphi(q)$-distributions we expect some 0.4 of all systems with $M_1 > 10\, M_\odot$ to produce two neutron stars, some 0.3 of all systems to coalesce and from a single (hydrogen rich) star, and some 0.3 to produce a helium-star in a close binary with a low-mass companion. So, if B is the fraction of the potentially interacting binaries among unevolved stars with $M > 10 M_\odot$, one expects 0.4B to produce two bare-core explosions and two neutron stars; we expect 0.3B to produce one bare-core explosion and one neutron star in close systems like the Her X-1 progenitor (a low-mass companion, $M \sim 1 - 4\, M_\odot$), and 0.3B to produce a normal H-rich (type II) supernova.

Furthermore systems with $M_1 > 6.5 M_\odot$ and $q \gtrsim 0.5$ will produce wide binaries consisting of a white dwarf and a star of $M \geq 8 - 9\, M_\odot$. In a wide binary the latter ones will have helium cores sufficiently massive to explode as a SN, which they will do after a CE-phase, in which their envelope is removed, and a close system, consisting of a helium core plus a white dwarf results.

(The interesting point about the latter "bare-core" model is that it makes supernovae out of stars in the mass range 6.5-8 $M_\odot$, which as single stars would never be able to produce a SN and a neutron star; in the mass range 8-10 $M_\odot$ they would, in a binary, never make the primary star explode; they do, however, make the secondary explode). Single stars or components of wide (non-interacting) binaries will explode as SN if they have $M \geq 8 M_\odot$. So, the single star (plus wide binary) supernova rate will be

$$F_s = (1 - B) \int_8^\infty \psi(M) dM + 0.3 B \int_{10}^\infty \psi(M) dM \tag{158}$$

(the second term is due to the coalescing binary systems with $q < 0.35$ see above). The bare-core SN-rate will be:

$$F_{bc} = (2 \times 0.4) B \int_{10}^\infty \psi(M) d(M) + 0.3 B \int_{10}^\infty \psi dM + 0.5 B \int_{6.5}^{10} \psi(M) d(M) \tag{159}$$

Assuming $\psi(M) \sim M^{-2.50}$, one thus obtains that the ratio of bare-core explosions to explosions at normal single hydrogen-rich stars is:

$$F_{bc}/F_s = \frac{B}{0.83 - 0.66 B} \tag{160}$$

The observed rate of type Ib+c relative to type II is 1.37 (Muller et al. [175]) which fits with eq. (160) for $B = 0.60$. Thus, if the total fraction of potentially interacting unevolved binaries (summarized over all mass ratios and orbital separations) is 60 per cent, one expects that the ratio of bare-core explosions to single (hydrogen-rich) star explosions among early-type stars is 1.37, as observed.

An interacting binary percentage $B \sim 60\%$ does not disagree with the findings of Abt and Levy mentioned in section 9.1, if one accounts for selection effects. Thus some 60 per cent of all (1.37/2.37) neutron stars may be formed in bare-core explosions.

9.6 The origin of pulsar velocities.

It is important to notice that the neutron stars formed in bare-core explosions, when they are single, have been released from binaries that were very close: they either are released in the second supernova explosions in a very close system, or in the first SN in a close system in which the other star either is a normal low-mass star (Her X-1-like progenitor, figure 71) or a massive white dwarf. In all these systems the orbital velocities of the exploding stars are $\geq 100 km/s$, which is the velocity which they receive when the system breaks up. One thus expects -if most systems are disrupted in the explosions (which seems likely, since few binary pulsars or Her X-1 systems are observed) that some 60 per cent of all neutron stars is born with a space velocity $> 100\, km/s$.

This is in good agreement with the observations of the single pulsar population which show that a large fraction, of order two third of the (single) pulsars are runaway objects with velocities $\sim 100 - 200 km/s$ (cf. Harrison et al.[103]). Notice that to obtain this result, no asymmetries in SN mass ejection are required! However, as argued in section 5 (see especially the arguments given by Dewey and Cordes[67] Bailes[9] and Dewey[66]), it is very likely that asymmetries do occur. If one includes a kick of $\sim 150 - 200\, km/s$, about half of the potential double neutron star systems may be disrupted, and most of the Her X-1 like systems will be destroyed, due to the fact that prograde kicks will disrupt them and retrograde kicks will make the neutron star collide with the companion, causing its disruption (see Wijers et al.[361]). The formation events of the latter systems (together with their disruption) may well be the Ic supernovae (Nomoto, private communication).

9.7 Incidence and formation rate of LMXBs.

We have dealt already in section 6.8 with the formation mechanisms and formation rates of LMXB that contain neutron stars or black holes. The galactic formation rate of LMXB-NS is clearly very low ($\sim 10^{-6} yr^{-1}$); that of the LMXB-BH systems may be (somewhat) higher, as the systems are more likely to survive the SN-collapse of the massive star. This "advantage", however, may be offset by a much smaller resevoir of progenitor systems, leading to presumably roughly similar formation rates of LMXB-NS and LMXB-BH systems in the Galaxy. We refer to section 6.8 for details.

9.8 The formation rate of close double neutron star systems in the Galaxy.

This is a problem of considerable interest for fundamental physics as the orbits of close double neutron stars decay by the emission of gravitational waves on a timescale that can be much shorter than the Hubble time. The coalescence of a double neutron star is expected to produce a burst of gravitational waves that can be detected with future gravitational wave detectors on Earth (such as LIGO, see Whitcomb[358]) out to a distance of $\sim 200\ Mpc$. Also the train of very strong waves generated by the system during its last few thousand orbital revolutions before coalescence -when the orbital period is of the order of milliseconds- can be recorded on Earth with such detectors (see e.g.: Schutz[247]).

It is therefore important to know how often per year such coalescence events of double neutron stars occur in the volume of the Universe out to 200 Mpc distance. Our only information comes from the systems in our Galaxy. Thus, if one is able to estimate the formation (and coalescence) rate of these systems in our Galaxy, one can obtain the rate in the above mentioned volume by multiplying the galactic rate with the number of galaxies in a sphere with radius 200 Mpc.

A first estimate of the formation rate of close double neutron stars like PSR 1913+16 was made by Clark, van den Heuvel and Sutanyo[42]. This estimate was based on stellar-evolution considerations, together with statistical arguments. Since only one binary pulsar system was known at the time, the obtained formation rate of such systems -of order 10^{-4} to $10^{-5}\ yr^{-1}$ was, of course, rather crude. Now that three systems of this type are known (one of them ten times closer to us than PSR 1913+16, which -statistically- might imply an $\sim 10^2$ larger galactic formation rate) one can use these three systems to obtain an, however rough, statistical estimate of their formation rates. This was done, independently, by Narayan et al.[177] and Phinney[207]. We follow a slightly different approach which leads to about the same result as obtained by these authors. We notice that the presently known number of radio pulsars is about 600, while the total population of active pulsars (i.e. those that have not yet turned-off) is estimated to be about 3.10^5 (Taylor and Stinebring[277]). This means that for every detected pulsar there are 500 undetected pulsars of the same type in the Galaxy.

Of the three double neutron star systems in the galactic disk that we know, two are very close ones (PSR 1913+16 and 1534+12) and have very weak magnetic fields ($2.10^{10}\ G$ and $10^{10}\ G_1$, respectively), and consequently very long spin-down timescales: $\tau_{sd} = 10^8\ yr$ for PSR 1913+16 and $2.5 \times 10^8\ yr$ for 1534+12. The third binary pulsar is the much wider system PSR 2303+46 (see fig. 37) which has a much stronger magnetic field (almost $10^{12}\ G$) and $\tau_{sd} = 3.10^7\ yr$. Still all these three pulsars seems to be recycled (see figure 82). The average B-fields of the normal single pulsar population are (2-4) $\times 10^{12}\ G$, implying a maximum lifetime till turn-off of $\sim 5.10^6\ yr$. We will assume this to be the lifetime of a normal newborn radio pulsar.

If we assume that τ_{sd} is about the time that a pulsar remains observable (its 'true' lifetime; this seems reasonable in view of the mounting evidence that B-fields of neutron stars do not decay spontaneously; see section 8) we can calculate for each of the three above binary pulsars the formation rate of such pulsars in the Galaxy, by assuming that indeed of each of these 3 there are 500 times more of similar nature in the Galaxy. Table 31 lists the thus obtained formation rates for each of these systems.

The table shows that the combined formation rate of the two close systems (with weak B-fields) is $0.7 \times 10^{-5}\, yr^{-1}$, while that for 2303+46-like systems is $1.7 \times 10^{-5}\, yr^{-1}$.

These are what one might call the 'observed' formation rates of these systems which, taken together, give a total double-neutron star formation rate of $2.4 \times 10^{-5}\, yr^{-1}$ in the Galaxy. The table also lists the total number of "turned-off" binary pulsars in the Galaxy, calculated as follows. We assumed that a pulsar turns off (as a radio pulsar) after a time τ_{sd}. It then still lives on to coalescence by GW-losses, as a turned-off pulsar. For PSR 1913+16 $\tau_{GW} = 3.10^8$ yr, for 1534+12: 1.5×10^9 yr and for the wider systems: infinitely long. For the wide systems we therefore obtained the number of turned-off ones by multiplying the birth rate with the age of the Galaxy (1.5×10^{10} yr), for the close systems by multiplying it by τ_{GW}, and subtracting the number of "living" pulsars.

Let us now compare the above obtained "observed" formation rates of double neutron stars with those theoretically expected from binary evolution.

An estimate of the double-neutron star formation rate based on binary evolution statistics. In order to avoid the uncertainties involved in the first stage of mass transfer and in the effects of the first SN-explosion on the system, one can skip these first steps and use the observed characteristics of the HMXB-population as a starting point for the predictions of the formation rate of double neutron stars in the Galaxy. This approach was followed by van der Heuvel[327].

He noticed that the most likely progenitors of the double neutron stars are the B-emission X-ray binaries, as they have wide orbits and are the most numerous category of massive stars with a neutron-star companion known in the Galaxy, with more than 22 known members, and a total estimated number in the Galaxy of order 10^3 to 10^4, see above. The orbits of 9 out of 22 systems are known; 4 of these have $P > 100^d$, and only one has $P > 1\, yr$.

As shown in section 6.5 (example 2), we expect that for a CE-efficiency factor $\alpha_{ce} = 1$, that the orbit of a typical Be/X-ray binary will shrink during CE-evolution by a factor of order 250. This will turn a system with an initial orbital period of *one year* into one with $P_{orb} \simeq 4$ hours. In the latter system the remaining helium core practically fills its Roche lobe. Since α_{ce} is probably for Be-systems never larger than unity (as only the more massive stars, $\geq 20-25 M_\odot$ may expell part of their envelope spontaneously by wind), we thus expect that only Be/X-ray systems with an orbital periods $\gtrsim 1$ year will survive as (very close) binaries after spiral-in. Closer ones will spiral in completely, forming a Thorne-Zytkow Object.

Since only *one* out of nine Be/X-ray binaries with known orbital periods (X Per) has $P_{orb} > 1$ yr, we expect that no more than about 10 percent of the Be/X-ray binaries will survive as a (very close) binary after spiral in (resembling Cyg X-3) and may later on evolve into a close double neutron star. The other 90 per cent of the Be/X-ray binaries will then *not* contribute to the formation of double neutron stars.

According to the above mentioned calculations of Meurs and van den Heuvel[171] one expects, if not more than 50 per cent of the systems are disrupted in the first SN-explosion, that there are $\sim (1-2) \times 10^4$ potential Be/X-ray binaries in the Galaxy. As these systems are expected to live for $\sim 10^7$ yrs, one expects, if 10 per cent of them produce close binary pulsars, that the Galactic formation rate of close binary pulsars is $\sim (1-2) \times 10^{-4}\, yr^{-1}$.

This 'theoretical' rate is still a factor of 15 to 30 larger than the 'observed' birthrate of close double neutron stars of $0.7 \times 10^{-5}\, yr^{-1}$ (table 31).

In order to possibly remove this rather large difference between theory and observation one might think of one of the following four possiblities:

1. The possibilty that in most close double neutron stars the neutron star is born with a strong magnetic field and turns off within 5.10^6 yrs after its birth.

   Since we have not (yet) observed any close double neutron star with a strong B-field pulsar, the total number of such systems in the Galaxy cannot be larger than 500. Hence, an upper limit to the birthrate of such systems is $10^{-4}\, yr^{-1}$.

   We have listed this upper limit in table 31, and also one for wide systems (like 2303+46) with a strong field and $\tau_{sd} = 5.10^6$ yrs. For both these hypothetical types of systems we have also listed the maximum possible number of turned-off ones in the Galaxy.

2. The possibility that there are about ten times fewer Be/X-ray binaries in the Galaxy than predicted by Meurs and van den Heuvel[171], i.e. $(1-2)\times 10^3$ instead of $(1-2)\times 10^4$ (the statistical properties of the about two dozen Be/X-ray binaries are not well enough known to distinguish observationally between those possibilities).

   A ten times lower number of Be/X-ray binaries may result if 90-95 per cent of the progenitor systems are disrupted in the first SN-explosion.

3. The possibility that 90 per cent of the post-spiral-in systems spiral-in very deeply in a second (case BB) phase of mass transfer, during helium-shell burning. The orbits of the then resulting systems (a CO-core plus a neutron star) are so close that the double neutron stars resulting after the second SN-collapse have orbital periods so short ($\lesssim 1\text{-}2^h$) that they coalesce by GW-emission within $\lesssim 5.10^6$ yr.

   In this case PSR 1913+16 and 1534+12 would form the long-period tail of a distribution of orbital periods of double neutron stars at birth that peaks below $P \simeq 2^h$[316].

4. The possibility that 99 per cent (instead of 90 per cent) of all Be/X-ray binaries spiral-in to complete coalescence. This would reduce the formation rate of double neutron stars by a factor 100 instead of 10.

We now briefly discuss the merits of each of these possible explanations.

How realistic possibility (1) may be is hard to judge, as no close double pulsar with a strong magnetic field has (yet) been found. For the moment we it quite unlikely, although it cannot be discarded. Possibilty (2) is not at all implausible randomly directed kick-velocities imparted to the neutron star in the collapse may, if they are $\geq 200 - 250\, km/s$, disrupt a large fraction of these systems in the first SN-explosion. (Kick velocities $\leq 100 - 150\, km/s$ which Meurs and van den Heuvel[171] assumed, will not disrupt more than 50 per cent of the systems). In view of the mounting evidence that the kick velocities at birth are large, possibly $\geq 200 - 250\, km/s$ (cf. Harrison et al.[103]), disruption of a large fraction of the systems in the first SN seems a quite realistic possibility, which would at once solve the discrepancy between theory and observation, and would predict a formation rate of close double neutron stars of $\sim (1-2) \times 10^{-5}\, yr^{-1}$ in the Galaxy.

Possibilty (3) could be a very interesting one. If this possibilty is true, the absence of observed systems with $P \leq 1 \div 2^h$, indicates again an upper limit to their birthrate of $\sim 10^{-4} yr^{-1}$ which, like possibilty (1) would be just consistent with a Be/X-ray binary formation rate of (1-2) $\times 10^{-4} yr^{-1}$.

This possibility, though equally speculative as possibilty (1), would have the interesting consequence that the merger rate of double neutron stars in a Galaxy like ours would be as high as $10^{-4} yr^{-1}$. However, in view of the absence of observed systems of this type the only thing one can say at this moment is that such a rate is the absolute upper limit to the merging rate of double neutron stars in a Galaxy like ours.

Table 32. Birthrates of observed binary pulsars (double neutron star) systems and of hypothetical systems that may result from common envelope evolution of Be/X-ray binaries (explanation in the text).

	PSR name	τ_{sd} (yr)	τ_{gw} (yr)	Gal. birthrate (yr^{-1})	Number in galaxy: active pulsars	Number in galaxy: turned-off pulsars
Observed Systems	-1913+16	10^8	3.10^8	5.10^{-6}	500	1000
	-1534+12	2.5×10^8	1.5×10^9	2.10^{-6}	500	2500
	-2303+46	3×10^7	∞	1.7×10^{-5}	500	3.10^5
Hypothetical Systems	-strong-field Close Binary Pulsar	5.10^6	5.10^8	$< 10^{-4}$	< 500	$< 5.10^4$
	-strong-field wide systems	5.10^6	∞	$< 10^{-4}$	< 500	$< 1.5 \times 10^6$
	-ultra-close systems	5.10^6 (?)	5.10^6 (?)	$< 10^{-4}$	< 500	—

Finally, possibilty (4) can not at all be discarded, because if α_{ce} is somewhat smaller than unity, indeed probably not more than a few per cent of all Be/X-ray binaries will survive as binaries after CE evolution.

In summary, to us possibilities (2) and (4), or a combination of these two, seem very plausible ways to solve the discrepancy between the "observed" and theoretically predicted birthrates of close double neutron stars in the Galaxy. From this we conclude that most probably the Galactic formation rate of such systems is of order $10^{-5}\,yr^{-1}$, with an absolute upper limit of $10^{-4}\,yr^{-1}$.

Acknowledgements

I would like especially to thank Mr. Steven Bosch for his help and patience in preparing this manuscript, and Mr. E. Faverey for his help with the illustrations.

In this review figures were used from publications of the following authors, to whom the author expresses his gratitude:

H.A. Abt (fig. 83), W.D. Arnett (fig.47), A.H. Batten (fig. 5), A. Blaauw (fig. 89), C. Chiosi and A. Maeder (fig. 29), C.J. Forman (fig. 19b), W.H. Fowler (fig. 46), F.M. Habets (fig. 48,49,50,66), K. Haramundanis (fig. 8 and 38), R. Hjellming (fig. 55 and 56), R. Humphreys (fig. 76), R. Kippenhahn (fig. 58 and 59), W.H.G. Lewin (fig. 19a), Th.J. van der Linden (fig. 61 and 63), B. Margon (fig. 32 and 33), J.E. McClintock (fig. 21 and 22), F. Nagase (fig. 17 and 30), M. Niehues, A. Bruch and W. Duerbeck (fig. 12), K. Nomoto (fig. 73a and 92), J.P. Ostriker and K. Davidsen (fig. 15 and 28), J. van Paradijs (fig. 13 and 19a), O. Pols (fig. 88), S.A. Rappaport (fig. 78 c), H. Reeves (fig. 42 and 43), R.E. Taam and B. Fryxell (fig. 73 b,c), M. Tavani and L. Brookshaw (fig. 6), J.H. Taylor (fig. 39 and 40), J. Tomkin and D.L. Lambert (fig. 62), E. Schreier (fig. 14 a,b and figure 31), F. Verbunt, S.A. Rappaport and P.C. Joss (fig. 78 a,b, 19 and 80), A. Weigert (fig. 58,59 and 60).

References

[1] Abt, H.A. 1983, Annual Rev. Astron. Astrophys. *21*, 343-372

[2] Abt, H.A. and Levy, S.G. 1976, Astrophys. J. Suppl. *30*, 273

[3] Abt, H.A. and Levy, S.G. 1978, Astrophys. J. Suppl. *36*, 241

[4] Abbott, D.C. and Conti, P.S. 1987, Ann. Rev. Astron. Astrophys. *25*, 113-150

[5] Alpar, M.A., Cheng, A.F., Rudermann, M.A. and Shaham, J. 1982, Nature *300*, 728

[6] Arnett, W.D. 1978, in "Physics and Astrophysics of Neutron Stars and Black Holes" (eds. R. Giacconi and R. Ruffini), North Holl. Publ. Comp., Amsterdam. pp 356-436

[7] Arnett, W.D. 1991, in: Proc. Texas ESO-CERN Symposium on Relativistic Astrophysics etc., Annals N.Y. Acad. Sci. *647*, p. 1-10

[8] Backer, D.C. and Hellings, R.W. 1986, Ann. Rev. Astron. Astrophys. *24*, 537

[9] Bailes, M. 1989, Astrophys.J. *342*, 917

[10] Banit, M. and Shaham, J. 1992a, Astrophys.J. *388*, L19-22

[11] Banit, M. and Shaham, J. 1992b, in: "X-ray Binaries and Recycled Pulsars" (E.P.J. van den Heuvel and S.A. Rappaport eds.), Kluwer Acad. Publ. Dordrecht, 401-412

[12] Batten, A.H. 1973, "Binary and Multiple Systems of stars", Pergamon Press, Oxford, 278 pp.

[13] Batten, A.H., Fletcher, J.M. and Mc Carthy, D.G. 1989, "Eighth Catalogue of the Orbital Elements of Spectroscopic Binary Stars", Domin. Astrophys. Obs., Victoria, B.C., Canada

[14] Baym, G. 1991, in "Neutron Stars, Theory and Observation" (eds. J. Ventura and D. Pines), Kluwer Acad. Publ., Dordrecht, p. 37

[15] Benson, R.S. 1970, Ph.D. Thesis, Univ. of California, Berkeley (unpublished)

[16] Bertelli, G. Bressan, A.G. and Chiosi, C. 1985, Astron. Astrophys. *150*, 33

[17] Beskin, V.S., Gurevich, A.B. and Isomin, Ya. N. 1984, Astrophys. Sp. Sci., *102*, 301

[18] Bhattacharya, D. and van den Heuvel, E.P.J. 1991, Phys. Rep. *203*, 1-124

[19] Bhattacharya, D., Wijers, R.A.M.J., Hartman, J.W. and Verbunt, F. 1992, Astron. Astrophys. *254*, 198-212

[20] Biehle, G.T. 1991, Astrophys. J. *380*, 167

[21] Bitzaraki, O. 1993, Astron. Astrophys. (in prep.)

[22] Blaauw, A. 1961, Bull. Astron. Inst. Netherlands, *15*, 265-291

[23] Blaauw, A. 1992, in: "Massive Stars: Their Lives in The Interstellar Medium", ASP Conf. Proceedings 35 (eds. J. Cassinelli and E. Chruchwell), Astron. Soc. of the Pacific, San Francisco, p. 207.

[24] Blaauw, A. and van Albada, T.S. 1968, (unpublished -referred to in: "Birth & Evolution of Massive Stars and Stellar Groups", eds. W. Boland and H. van Woerden), Reidel Publ. Comp., Dordrecht 1985.

[25] Bodenheimer, P. and Taam, R.E. 1984, Astrophys. J. 280, 771

[26] Boersma, J. 1961, Bull Astron. Inst. Netherlands, *15*, 291-301

[27] Bonnet-Bidaud, J.M. and van der Klis, M, 1981, Astron. Astrophys. *101*, 299-304

[28] Bonsema, P.F. J., and van den Heuvel, E.P.J. 1985, Astron. Astrophys. *146*, L3

[29] Branch, D. 1990, in: "Neutron Stars and Their Birth Events" (ed. W. Kundt), Kluwer Acad. Publ., Dordrecht, p. 281

[30] Branch, D., Nomoto, K., Fillipenko, V. 1981, Comments on Astrophysics *15*, 221-237

[31] Callanan, P.J., Charles, P.A., Hassal, B.J.M., Machin, G., Mason, K.O., Naylor, T., Smale, A.P. and van Paradijs, J.A. 1990, Mon. Not.. R.A.S. *238*, 25 p.

[32] Canal R., Isern, J. and Labay, J. 1990, Ann. Rev. Astron. Astrophys. *28*, 183-214

[33] Cannon, R., Eggleton, P.P. Zytkow, A.N. and Podsiadlowski, P. 1992, Astrophys. J. *386*, 206

[34] Casares, J., Charles, P.A. and Naylor, T. 1992, Nature, *355*, 614

[35] Castellani, V. Chieffi, A., Pulone, L. and Tornambe, A. 1985, Astrophys.J. *294*, L31

[36] Chanmugam, G. 1992, Ann. Rev. Astron. Astrophys. *30*, 143-184

[37] Cherepaschuk, A.M. 1981, Mon. Not. R.A.S., *194*, 761

[38] Chevalier, R. 1981, Astrophys. J. *246*, 267

[39] Chiosi, C., Bertelli, G. and Bressan, A. 1992, Ann. Rev. Astron. Astrophys. *30*, 235-286

[40] Chiosi, C. and Maeder, A. 1986, Ann. Rev. Astron. Astrophys. *24*, 329-375

[41] Chiu, H.J. 1968, "Stellar Physics", Blaisdell, London & Toronto

[42] Clark, J.P.A., van den Heuvel, E.P.J. and Sutantyo, W. 1979, Astron. Astrophys. *72*, 120

[43] Clayton, D.D. 1968, "Principles of Stellar Evolution and Nucleosynthesis", McGraw Hill, N.Y. 612 pp.

[44] Collier, A.C. and Jenkins, C.R. 1984, Mon. Not. R.A.S. *211*, 391-419

[45] Coté, J. 1993, Ph. D. Thesis, Univ. of Amsterdam

[46] Counselman, C.C. 1973, Astrophys. J. *180*, 307

[47] Cowley, A.P. 1992, Ann. Rev. Astron. Astrophys. *30*, 287-310

[48] Cox, J.P. and Giuli, R.T. 1968, Principles of Stellar Structure I & II, Gordon and Breach, 1327 pp.

[49] Crampton, D., Cowley, A.P., and Hutchings, J.B. 1980, Astrophys.J. *235*, L131

[50] Crawford, J.A. 1955, Astrophys. J. *121*, 71

[51] Darwin, G.H. 1908, "Scientific Papers", vol. 2, Cambridge Univ. Press (original pub. 1879)

[52] De Greve, J.P. 1986, Space. Sci. Rev. *43*, 139-210

[53] De Greve, J.P., de Loore, C. and Sutantyo, W. 1975, Astrophys. Spa. Sci. *38* , 301

[54] De Jager, C. 1980, "The Brightest Stars", Dordrecht, Reidel Publ. Comp., Dordrecht, 458 pp.

[55] De Kool, M. 1987, "Models of Interacting Binary Stars", Ph. D. Thesis, Univ. of Amsterdam

[56] De Kool, M. 1990, Astrophys. J. *358*, 189

[57] De Kool, M. 1992, Astrophys. J., (in the press)

[58] De Kool, M. and Ritter, H. 1993, Astron. Astrophys. (in the press)

[59] De Kool, M., van den Heuvel, E.P.J. and Pylyser, E. 1987, Astron. Astrophys. *183*, 47

[60] De Kool, M., van den Heuvel, E.P.J. and Rappaport, S.A. 1986, Astron. Astrophys. *164*, 73-76

[61] Delgado, A.J. and Thomas, H.-C., 1981, Astron. Astrophys., *96*, 142

[62] De Loore, C.W.H., 1980, Space Sci. Rev., *26* 133

[63] De Loore, C.W.H., De Greve, J.P., van den Heuvel, E.P.J. and De Cuyper, J.P. 1975, Mem. Soc. Astron. Italiana (Proc. 2nd IAU Reg. Meeting, Trieste), *45*, 893

[64] De Loore, C.W.H. and De Greve, 1976 in: "Structure and Evolution of Close Binary Systems" (eds. P. Eggleton, S. Mitton and J. Whelan), Reidel Publ. Com., Dordrecht, pp. 27-34

[65] De Loore, C.W.H. and Doom, C. 1992, "Structure and Evolution of Single and Binary Stars", Kluwer Acad. Publ., Dordrecht, 458 pp.

[66] Dewey, R.J. 1992, in: "X-ray Binaries and Recycled Pulsars" (eds. E.P.J. van den Heuvel and S.a Rappaport), Kluwer Acad. Publ., pp. 453-463

[67] Dewey, R.J. and Cordes, J.M. 1987, Astrophys.J. *321*, 780

[68] Dewey, R.J., Maguire, C.M., Rawley, L.A., Stokes, G.H. and Taylor, J.F. 1986, Nature, *322*, 712

[69] Dolan, J.F. 1992, Astrophys. J. *384*, 249

[70] Dolan, J.F. and Tapia, S. 1989, Astrophys. J., *344*, 830

[71] Eggleton, P.P. and Pringle, J.E. 1985, "Interacting Binaries" Reidel, Dordrecht, 410 pp

[72] Eggleton, P.P. and Verbunt, F. 1986, Mon. Not.R.A.S., *220*, 13p.

[73] Ergma, E.V. 1991, Comments Astrophys., *15*, 239-254

[74] Fabian, A.C. and Rees, M.J. 1979, M.N.R.A.S. *187*, 13p

[75] Federova, A.V. and Ergma, E.V. 1991, Soviet. Astron., *35*, 640-645

[76] Ferluga, S. and Hack, M. 1985, Astron. Astrophys. *144* 395

[77] Flannery, B.P. 1977, Annals N.Y. Acad. Sci. *362*, 36-46

[78] Flannery, B.P. and Ulrich, R.K. 1977, Astrophys. J. *212*, 533

[79] Flannery, B.P. and van den Heuvel, E.P.J. 1975, Astron. Astrophys. *39*, 61

[80] Flowers, E. and Ruderman, M. 1977, Astrophys. J. *215*, 302

[81] Forman, C.J. 1974, Ph.D. Thesis, Harvard University

[82] Fowler and Hoyle, F. 1964, Astrophys. J. Suppl. *9*, 201

[83] Friedjung, M. and Viotti, R. 1982, "The Nature of Symbiotic Stars", IAU Coll.no. 70, Reidel Publ. Comp., Dordrecht

[84] Fruchter, A.S., Stinebring D.R. and Taylor, J.H. 1988, Nature *333*, 237

[85] Fryxell, B.A. and Arnett, W.D. 1981, Astrophys. J. *243*, 994

[86] Garmany, C.D., Conti, P.S. and Massey, P. 1980, Astrophys.J. *242*, 1063

[87] Giannone, P., Kohl, K. and Weigert, A. 1968, Zeitschr. für Astrophysik, *68*, 107

[88] Gies, D. 1987, Astrophys. J. Suppl., *63*, 545

[89] Gies, D. and Bolton, C.T. 1986, Astrophys. J. Suppl., *61* 419

[90] Gies, D. and Bolton, C.T. 1986, Astrophys. J. *304*, 371

[91] Glendenning, N.K. 1991, Nucl. Phys. B., Proc. Suppl., *24B*, 110-118

[92] Gold, T. 1968, Nature, *218*, 731

[93] Gold, T., Nature, *221*, 25

[94] Goldreich, P. and Julian, W.H. 1969, Astrophys. J., *157*, 869

[95] Griffin, R.F. 1985, in: "Interacting Binaries" (eds. P.P. Eggleton and J.E. Pringle), Reidel Publ. Comp., Dordrecht, pp.1-12

[96] Griffin, R.F. 1991, The Observatory, *111*, 291-298

[97] Gunn, J.E. and Ostriker, J.P. 1970, Astrophys. J. *160*, 979

[98] Gursky, H. 1973, Lecture presented at NATO Adv. Study Institute on "Physics of Compact Objects", Cambridge, United Kingdom (unpublished)

[99] Habets, G.M.H.J. 1985, "Advanced Evolution of Helium Stars and Massive Close Binaries", Ph.D. Thesis, University of Amsterdam

[100] Habets, G.M.H.J. 1986, Astron. Astrophys., *187*, 209

[101] Haffner, H. and Heckman, O. 1937, Veröff. Univ. Sternwarte Göttingen, *4 (nr. 55)*, 77

[102] Halbwachs, J.L. 1987, Astron. Astrophys. *183*, 234

[103] Harrison, P.A., Lyne, A.G. and Anderson, B. 1992, in: "X-ray Binaries and Recycled Pulsars" (eds. E.P.J. van der Heuvel and S.A. Rappaport), Kluwer Acad. Publ., Dordrecht, pp. 155-160

[104] Helfand, D.J. and Huang, J.H. (eds.) 1987, "The Origin and Evolution of Neutron Stars", Reidel Publ. Company, Dordrecht, 572 pp.

[105] Heintz, W.D. 1978, "Double Stars", Reidel Publ. Comp., Dordrecht, 174 pp.

[106] Hewish, A., Bell., S.J., Pilkington, J.D.H., Scott, P.F. and Collins, R.A. 1968, Nature *217*, 709

[107] Hillebrandt, W. 1987, in :"High Energy Penomena around Cpllapsed Stars" (ed. F. Pacini), Reidel Publ, Comp. Dordrecht., p. 73

[108] Hillebrandt, W. 1991, in: "Neutron Stars, Theory and Observation (eds. J. Ventura and D. Pines), Kluwer Acad. Publ., Dordrecht, pp. 125-141

[109] Hjellming, R.M. 1973, Science *182*, 1089-1095 (reprinted in 1977 in: "The New Astronomy and Space Science Reader" eds. J.C. Brandt and S.P. Maran, Freeman, San Francisco, pp. 306-317)

[110] Hjellming, M.S. 1989, Ph.D. Thesis, Univ. of Illinois, Urbana, Ill.

[111] Hogeveen, S. 1991, "The Mass Ratio Distribution of Binary Stars", Ph.D. Thesis, University of Amsterdam (132 pp).

[112] Hogeveen, S.J. 1992, Astrophys. Sp. Sci., *196*, 299

[113] Huang, S.S. 1963, Astrophys. J. *138*, 342

[114] Humphreys, R.M. 1987, Pub. Astron. Soc. Pacific, *99*, 5-14

[115] Humphreys, R.M. 1992, in: "Instabilities in Evolved Super- and Hypergiants" (eds. C. de Jager and H. Nieuwenhuyzen), North Holland, Amsterdam, pp. 13-17

[116] Hut, P. and van den Heuvel E.P.J. 1981, Astron. Astrophys., *94*, 327

[117] Hut, P. 1992, in "X-ray Binaries and Recycled Pulsars"(eds. E.P.J. van den Heuvel and S.A. Rappaport), Kluwer Acad. Pub., Dordrecht, 317-348

[118] Iben, I. 1967, Annual Rev. Astron. Ap. *5*, 571

[119] Iben, I. 1982, Astrophys J., *259*, 244-266

[120] Iben, I. and Tutukov, A.V. 1984, Astrophys. J. Suppl. *54*, 335

[121] Iben, I. and Tutukov, A.V. 1985, Astrophys. J. Suppl. *58*, 661.

[122] Iben, I. and Tutukov, A.V. 1989, Astrophys. J. *342*, 430

[123] Iben, I. and Tutukov, A.V. 1991, Astrophys. J. *370*, 615

[124] Jaschek, C. and Ferrer, O. 1972, Pub. Astron. Soc, Pacific. *84*, 292

[125] Johnston, J. Manchester, R.N. Lyne, A.G. Bailes, M., Kaspi, V.M., Guojun, Q. and D'Amico, N. 1992, Astrophys.J., *387*, L37

[126] Joss, P.C. and Rappaport, S.A. 1983, Nature *304*, 419

[127] Kenyon, S.J. 1985, in: "Interacting Binaries" (eds. P.P. Eggleton and J.E. Pringle), Reidel Publ. Comp., Dordrecht, pp. 179-203

[128] Kenyon, S.J. and Garcia, M.R. 1986, Astron. J. *91*, 125.

[129] Kenyon, S.J. and Garcia, M.R. 1989, Astron. J. *97*, 194.

[130] Kenyon, S.J. and Webbink, R.F. 1984,Astrophys. J. *279*, 252

[131] Kippenhahn, R., Kohl, K. and Weigert, A. 1968, Zeit. für Astrophysik *66*, 58

[132] Kippenhahn, R. and Weigert, A. 1967, Zeit. für Astrophysik *65*, 251

[133] Kippenhahn, R. and Weigert A. 1990, "Stellar Structure and Evolution", Springer, Heidelberg 468 pp.

[134] Klumper, A. 1981, "Enkele Aspecten van de Evolutie van Sterrenhopen", M. Sci. Thesis, Univ. of Amsterdam (unpub.)

[135] Kluzniak, W. 1992, in: "X-ray Binaries andRecycled Pulsars" (eds. E.P.J. van den Heuvel and S.A. Rappport,Kluwer Acad. Publ. pp. 413-424

[136] Kluzniak, W., Ruderman, M., Shaham, J. and Tavani, M. 1988, Nature *334*, 225

[137] Kluzniak, W., Ruderman, M., Shaham, J. and Tavani, M. 1989, in: "Timing Neutron Stars" (eds. H. Ögelman and E.P.J. van den Heuvel), Kluwer, Acad. Publ. 641

[138] Koyama, K. and Takeuchi, Y. 1989, in: "X-ray Binaries" Proc. 23rd ESLAB Symp. (eds. I.J. Hunt and B. Battrick), ESA, Noordwijk, The Netherlands, pp. 483-484

[139] Kraicheva, Z.I. Popova, E.I., Tutukov, A.V. and Yungelson, L.R. 1978, Soviet Astron. *22*, 670

[140] Kreiner, J.M. 1981, in: "Non-stationary Evolution of Close Binaries" (ed. A. Zytkow), Polish Scientif. Publishers, Warszawa, 133-135

[141] Krolik, J.H. and Sincell, M.W. 1990, Astrophys. J. *357*, 208

[142] Kruszewski, A. 1966, Adv. Astron. Astrophys. *4*, 233.

[143] Kudritzki, B.P. and Reimers,D. 1978, Astron. Astrophys. *70*, 227

[144] Kuiper, G.P. 1935, Publ. Astron. Soc. Pacific. *47*, 15

[145] Kuiper, G.P., 1941, Astrophys. J. *93*, 133

[146] Kulkarni, S.R. 1986, Astrophys. J. *306*, L85.

[147] Kulkarni, S.R. and Hester, J.J. 1988, Nature *335*, 801-803.

[148] Kundt, W. (ed.) 1990, "Neutron Stars and Their Birth Events", Kluwer Acad. Pub., Dordrecht, 572 pp.

[149] Lambert, D.L. and Sawyer,S.R. 1986, Publ. Astron. Soc. Pacific. *98*, 398-402.

[150] Law, W.-Y. and Ritter, H. 1983, Astron. Astrophys. *123*, 33

[151] Lewin, W.H.G., and Joss, P.C. 1983, in "Accretion Driven Stellar X-ray Sources" (eds. W.H.G. Lewin and E.P.J. van den Heuvel), Cambridge Univ. Press, p. 41

[152] Lewin, W. and van den Heuvel, E.P.J. 1983 (eds.), "Accretion-Drven Stellar X-ray Sources", Cambridge Univ. Press, 450 pp.

[153] Lewin, W.H.G. and van Paradijs, J.A. 1992 (private communication)

[154] Leung, K.-C. 1980, in: "Close Binary Stars: Observations and Interpretations" (IAU Symp. no 88, eds. M.J. Plavec et al.), Reidel Publ. comp. Dordecht, pp. 527-531

[155] Leung, K.-C. 1991, in: "Wolf-Rayet Stars" (eds. K.A. van der Hucht and B. Hidayat), Kluwer Acad. Publ., Dordrecht, pp. 207-212

[156] Liebert, J., Bergeron, P., and Saffer, R.A. 1991, in: "White Dwarfs" (eds. G. Vauclair and E. Sion), Kluwer Acad. Pub., Dordrecht, pp. 409-416

[157] Long, K.S. and van Speybroeck, L.P. 1983, in "Accretion Diven Stellar X-ray Sources" (eds W.H.G. Lewin and E.P.J. van den Heuvel) , Cambridge Univ. Press, 117-146.

[158] Lucy, L.B. and Ricco, E. 1979, Astron. J. *84*, 401

[159] Lyne, A.G. Manchester, R.N., D'Amico, N. Stavely,-Smith, L., Johnston, S., Lim, J., Fruchter, A.S., Goss,, W.M. and Frail, D. 1990, Nature *347*, 650

[160] Maeder, A. and Meynet, G. 1989, Astron. Astrophys. *210*, 155

[161] Maraschi, L. , Treves, A. and van den Heuvel, E.P.J. 1976, Nature, *259*, 292

[162] Maraschi, L., Treves, A. and van den Heuvel, E.P.J. 1977, Astrophys. J., *216*, 819

[163] Margon, B. 1980, "Proc. Ninth Texas Symp. on Relativ. Astrophysics", Annals N.Y,. Acad. Sci., *375*, 403

[164] Margon, B. 1983, in "Accretion Driven Stellar X-ray Sources" (eds. W.H.G. Lewin and E.P.J. van den Heuvel), Cambridge Univ. Press, pp. 287-301

[165] Margon, B. 1984, Ann. Rev. Astron. Astrophys. *22*, 507-536

[166] Massey, P. 1985, Publ. Astron. Soc. Pacific *97*, 5-24

[167] Mazurek, T.J. and Wheeler, J.C. 1980, Fundamental Cosm. Phys. *5*, 193

[168] Mc Clintock, J.E. and Remillard, R.A. 1986, Astrophys. J., *308*, 110

[169] Mc Clintock, J.E. 1992, in "X-ray Binaries and Recycled Pulsars" (eds. E.P.J. van den Heuvel and S.A. Rappaport), Kluwer Acad. Publ., Dordrecht, p. 27

[170] Mermillod, J.C., Rosvick, J.M. Duquennoy,A. and Mayor, M. 1992, Astron. Astrophys. *265*, 513-526

[171] Meurs, E. and van den Heuvel, E.P.J. 1989, Astron. Astrophys., *226*, 88

[172] Meyer, F. and Meyer-Hoffmeister, E. 1978, Astron. Astrophys. *78*, 167

[173] Milgrom, M. 1979, Astron. Astrophys., *76*, L3

[174] Morton, D.C. 1960, Astrophys. J., *132*, 146

[175] Muller, R.A., Newberg, H.J.M., Pennypacker, C.R., Perlemutter, S., Sasseen, T.P. and Smith, T.K. 1992, Astrophys. J. 384, L5.

[176] Nagase, F. 1989, Publ. Astron. Soc. Japan *41*, 1

[177] Narayan, R., Piran,T. and Shemi, A. 1991, Astrophys. J. *379*, L17-20

[178] Nauenberg, M. and Chapline, G. 1973, Astrophys. J., *179*, 277

[179] Niehues, M., Bruch, A. and Duerbeck, H.W. 1992, ESO Messenger Nr. 67 (March), p.38

[180] Nomoto, K. Filippenko, A.V. and Shigeyama, T. 1990, Astron Astrophys. 240, L1-L4

[181] Nomoto, and Hashimoto, M. 1988, Physics Reports *163*, 13-36

[182] Nomoto, K. and Kondo, Y. 1991, Astrophys. J. *367*, L19-L22

[183] Nomoto, K. Thieleman, F.K., Yokoi, K. 1984, Astrophys. J. *286*, 644

[184] Nomoto, K., Yamaoka, H., Shigeyama, T., Kumagai, S. and Tsujimoto, T. 1991a, in: "Neutron Stars Theory and Observation" (eds. J. Ventura and D. Pines) Kluwer Acad. Publ., Dordrecht, pp. 143-170

[185] Nomoto, K. and Yamaoko, H. 1992, in: "X-ray Binaries and Recycled Pulsars" (E.P.J. van den Heuvel and S.A. Rappaport, eds.), Kluwer Acad. Publ. Dordrecht, 189-206

[186] Ögelman. H.A 1989, in:"Timing Neutron Stars", (eds. H.A. Ögelman and E.P.J. van den Heuvel), Kluwer Acad. Publ., Dodrecht, p. 619

[187] Ögelman, H.A. and van den Heuvel, E.P.J. (eds) 1989, "Timing Neutron Stars", Kluwer Acad. Publishers", 774 pp.

[188] Ostriker, J.P. 1973, (private comm. to B. Paczynski, mentioned in Paczynski 1976).

[189] Ostriker, J.P. and Gunn, J.E. 1969, Astrophys. J. *157*, 1395

[190] Ostriker, J.P. and Davidson, K. 1973, in "X- and Gamma Ray Astronomy" (IAU Symp. nr 55), (eds. H. Bradt and R. Giacconi), Reidel Publ. Comp., Dodrecht, pp. 143-154

[191] Pacini, F. 1967, Nature, *216*, 567

[192] Pacini, F. (ed.) 1987, "High Energy Phenomena Around Collapsed Stars", Reidel Publ. Company, Dordrecht, 368 pp.

[193] Paczyński, B. 1966, Acta Astronomica *16*, 231

[194] Paczyński, B. 1967a, Commun. Obs. R. Belgique, Uccle, B17, 111

[195] Paczyński, B. 1967b, Acta Astronomica *17*, 1

[196] Paczyński, B. 1967c, Acta Astronomica *17*, 193

[197] Paczyński, B. 1967d, Acta Astronomica *17*, 287

[198] Paczyński, B. 1967e, Acta Astronomica *17*, 355

[199] Paczyński, B. 1970, in: "Mass Loss and Evolution of Close Bianries" (ed. K. Gyldenkerne and R. West), Copenhagen Univ. Pub. Funds, Copenhagen, p. 142

[200] Paczyński, B. 1971, Annual Rev. Astron. Ap. *9*, 183

[201] Paczyński, B. 1971b, Acta Astron. *21*, 1

[202] Paczyński, B. 1976, in: "Structure and Evolution of Close Binary Systems" (eds. P. Eggleton, S. Mitton and J. Whelan), Reidel Publ. Comp., Dordrecht, pp. 75-80

[203] Paczyński, B. 1983, Nature, *304*, 421

[204] Paczyński, B. and Sienkiwicz, R. 1981, Astrophys. J. *248*, L27

[205] Parmar, A. and White, N.E. 1988, in: "X-ray Astronomy with EXOSAT" (eds. N.E. White and R. Pallavacini), Mem. Astron. Soc. Ital. *59*, 147

[206] Payne-Gaposchkin, C. and Haramundanis, K. 1970, "Introduction to Astronomy", Englewood Cliffs, Prentice-Hall, 610 pp.

[207] Phinney E.S. 1991, Astrophys. J. *380*, L17-22

[208] Pines, D. 1991, in: "Neutron Stars, Theory and Observation" (eds. J. Ventura and D. Pines)Kluwer Acad. Publ., pp. 57-70

[209] Plavec, M.J. 1967, Commun. Obs. R. Belgique, Uccle, B17, 83

[210] Plavec, M.J., Popper., D.M. and Ulrich, R.K. (eds., 1980) "Close Binary Stars: Observations and Interpretation" (I.A.U. Symp. no. 88), Reidel Publ. Comp., Dordrecht, 598 pp.

[211] Politano, M. 1988, Ph.D. Thesis, Univ. of Illinois, Urbana

[212] Pols, O. 1992, Talk presented at Workshop on Blue Stragglers, S.T. Sci.I., Baltimore (October 12-14, 1992)

[213] Pols, O. 1993a, Ph. D. Thesis, University of Amsterdam (in preparation)

[214] Pols, O. and Marinus, M. 1993b, Astron. Astrophys. (in prep.)

[215] Popper, D.M. 1980, Ann. Rev. of Astron. Astrophys. *18*, 115-164

[216] Popper, D.M. and Tomkin, J. 1984, Astrophys.J. *285*, 208

[217] Pylyser, E. 1988, "The Evolution of Low Mass Binary Systems with a Compact Component", Ph.D. Thesis, Univ. of Amsterdam

[218] Pylyser, E. and Savonije, G.J. 1988, Astron. Astrophys. *191*, 57

[219] Pylyser, E. and Savonije, G.J. 1989, Astron. Astrophys., *208*, 52

[220] Radhakrishnan, V. and Srinivasan, G. 1981, Paper presented at 2nd Asia-Pacific Regional Meeting of the IAU, Bandung, Indonesia

[221] Radhakrishnan, V. and Srinivasan, G. 1984, Proc. 2nd Asia-Pacific Regional Meeting of the IAU (eds. B. Hidayat and M.W. Feast), Jakarta, Tira Pustaka, p. 423

[222] Rappaport, S.A. , Joss, P.C. and Webbink, R. 1982, Astrophys. J. *254*, 616

[223] Rappaport, S.A. and van den Heuvel, E.P.J. 1982, in: "Be Stars" (IAU Symp. 98) (eds. M. Jaschek and G. Groth), Reidel Publ. Comp., Dordrecht, p. 327

[224] Rappaport, S.A., Verbunt, F. and Joss, P.C. 1984, Astrophys. J. *275*, 713

[225] Reeves, H. 1968, "Stellar Evolution and Nucleosynthesis", Gordon & Breach, New York, 99 pp.

[226] Remillard, R.A., Mc Clintock, J.E. and Baylin, C.D. 1992, Astrophys.J. *399*, L145-150

[227] Reimers, D. and Koester, D. 1988, ESO Messenger *54*, 47

[228] Rhoades, C.E. and Ruffini, R. 1974, Phys. Rev. Lett. *32*, 324

[229] Ritter, H. 1990, Astron. Astrophys. Suppl. *85*, 1179

[230] Robinson, E.L. 1987, Bull. Am. Astron. Soc. *19*, 1057

[231] Robinson, E.L. and Shafter, A.W. 1987, Astrophys. J. *322*, 296-301

[232] Romani, R.W. 1990, Nature, *347*, 741

[233] Romani, R.W. 1992, Astrophys. J., *399*, 621

[234] Romani, R.W. and Weinberg, M.D. 1991, Astrophys. J., *372*, 487

[235] Romanishin, W. and Angel, J.R.P. 1980, Astrophys. J. *235*, 992

[236] Ruderman, M.A., Shaham, J. and Tavani, M. 1989a, Astrophys. J. *336*, 507

[237] Ruderman, M.A., Shaham, J., Tavani, M. and Eichler, D. 1989b, Astrophys. J. *343*, 292

[238] Saffer, R.A., Liebert, J. and Olszewski, E.W. 1988, Astrophys. J. *334*, 947-957

[239] Salpeter, E.E. 1964, Astrophys. J. *140*, 796

[240] Sandage, A.H. 1986, Ann. Rev. Astron. Astrophys., *24*, 241

[241] Savonije, G.J. 1983a, in "Accretion-driven Stellar X-ray Sources" (eds. W.H.G. Lewin and E.P.J. van den Heuvel), Cambridge Univ. Press, pp. 343-366

[242] Savonije, G.J. 1983b, Nature *304*, 422

[243] Savonije, G.J., De Kool, M. and van den Heuvel, E.P.J. 1986, Astron. Astrophys., *155*, 51-57

[244] Scalo, J.M. 1986, Fundamentals of Cosmic Physics, 11, 1

[245] Scarfe, C.D. 1986, J. Roy. Astron. Soc. Can., *80*, 257

[246] Schreier, E., Levinson, R., Gursky, H., Kellogg, E., Tananbaum, H. and Giacconi, R. 1972, Astrophys.J., *172*, L79

[247] Schutz, B.F. 1989, Ann. New York Acad. Sci. *571*, pp. 27-43

[248] Shapiro, S.L. and Teukolsky, S.A. 1983, "Black Holes, White Dwarfs and Neutron Stars" John Wiley & Sons, New York/Chichester, 645 pp.

[249] Shibazaki, N., Murakami, T., Shaham, J. and Nomoto, K. 1989, Nature, *342*, 656

[250] Shklovskii, I. 1979, Pis'ma Astron. Zh., *5*, 644.

[251] Shu, F.H. and Lubow, S.H. 1981, Ann. Rev. Astron. Astrophys., *19*, 277-293

[252] Skumanich, A. 1972, Astrophys. J. *171*, 565

[253] Slettebak, A. 1988, Pub. Astron. Soc. Pacific, *100*, 770

[254] Slettebak, A. and Snow, T.P. (eds.) 1987, "Physics of Be Stars", Cambridge Univ. Press, 557 pp.

[255] Smith, M.A. 1979, Publ. Astron. Soc. Pacific. *91*, 737

[256] Spruit, H.C. and Ritter, H. 1983, Astron. Astrophys. *124*, 267

[257] Srinivasan, G. 1991, Astron. Astrophys. Rev. *1*, 209

[258] Srinivasan, G., Bhattacharya, D. Muslimov, A.G. and Tsygan, A.I. 1984, Current Sci. *59*, 31

[259] Srinivasan, G. and Radhakrishnan, V. 1985, "Supernovae, Their Progenitors and Remnants", Indian Acad. Sci., Bangalore, 178 pp.

[260] Srinivasan, G. and van den Heuvel, E.P.J. 1982, Astron. Astrophys. *108*, 143

[261] Staniucha, M. 1979, Acta Astron. *29*, 587

[262] Stencel, R.E. (ed.) 1985, "1982-1984 Eclipse of Epsilon Aurigae", NASA Conf. Publ. *2384*

[263] Stone, R.C., 1991. Astron. J. *102*, 333-349

[264] Sutantyo, W. 1975, Astron. Astrophys. *41*, 47

[265] Sutantyo, W. 1992, in: "X-ray Binaries and Recycled Pulsars" (eds. E.P.J. van den Heuvel and S.A. Rappaport), Kluwer Acad. Publ., Dordrecht, pp. 293-310

[266] Taam, R.E., Bodenheimer, P. and Ostriker, J.P. 1978, Astrophys. J. *222*, 269

[267] Taam, R.E. and Fryxell, B.A. 1984, Astrophys. J., *279*, 166-176

[268] Taam, R.E. and van den Heuvel, E.P.J. 1986, Astrophys. J. *305*, 235

[269] Tanaka, Y. 1992, in "X-ray Binaries and Recycled Pulsars" (eds. E.P.J. van den Heuvel and S.A. Rappaport), Kluwer Acad. Publ., Dordrecht, p. 37

[270] Tanaka, Y. et al. 1983, IAU Circular, *3891*

[271] Tassoul, J.L. 1978, "Theory of Rotating Stars", Princeton Univ. Press, 506 pp.

[272] Tassoul, M. and Tassoul, J.L. 1992, Astrophys.J. *395*, 604

[273] Tavani, M. and Brookshaw, L. 1992, Nature *356*, 320-322

[274] Taylor, J.H. 1989, in: "Timing Neutron Stars" (eds. H. Ögelman and E.P.J. van den Heuvel), Kluwer Acad. Publ., Dordrecht p. 17

[275] Taylor, J.H. 1987, Proc. 13th. Texas Symp. on Relativ. Astrophysics (ed. M.P. Ulmer), World Scientific, Singapore, pp. 467-477

[276] Taylor, J.H. 1992, in: "X-ray Binaries and Recycled Pulsars" (E.P.J. van den Heuvel and S.A. Rappaport, eds.), Kluwer Acad. Publ. Dordrecht, pp. 87-92

[277] Taylor, J.H. and Stinebring, D.R. 1986, Ann. Rev. Astron. Ap. *24*, 285-327

[278] Taylor, J.H. and Weisberg, J.M. 1989, Astrophys. J. *345*, 434-450

[279] Terrell, J. and Priedhorsky, W.C. 1984, Astrophys. J., *285*, L15

[280] Thorne, K.S. and Zytkow, A.N. 1977, Astrophys.J. *212*, 832

[281] Timmes, F.X. and Woosley, S.E. 1992, Astrophys.J. *39*, 649

[282] Tomkin, J.. and Lambert, D.L. 1987, Sky and Telescope *vol. 74 (4)*, 354-357

[283] Tout, C.A. 1991, Mon. Not. R.A.S. *250*, 701-706

[284] Trimble, V. 1974, Astron. J. *79*, 967

[285] Trimble, V. 1982, Rev. Mod. Physics, *54*, 1183

[286] Trimble, V. 1990, Mon. Not.. R.A.S. *242*, 79

[287] Tsesevitch, V.P. (ed.) 1973, "Eclipsing Variable Stars", IPST/John Wiley and Sons Ltd, New York/Chichester, 310 pp.

[288] Tutukov, A.V. and Yungelson, L.R. 1971, Nauchnie Informatsie Astrosoveta ANSSSR *20*, 86

[289] Tutukov, A.V and Yungelson, L.R. 1973, Nauchnie Informatsie, *27*, 58

[290] , Tutukov, A.V. Yungelson, L.R. and Iben, I. 1992, Astrophys. J. *386*, 197-205

[291] Ulrich, R.K. and Burgher, H. 1976, Astrophys. J., *206*, 509

[292] Underhill, A.B. 1966, "The Early Type Stars", Reidel Publ. Comp., Dordrecht

[293] van Albada, T.S. 1968, Bull. Astron. Inst. Netherlands, *20*, 47

[294] van Beveren, D. 1988, Astrophys. Sp. Sci *149*, 1

[295] van Beveren, D. 1988, Astron. Astrophys. *189*, 109

[296] van Beveren, D. 1991, Space Sc. Rev. *56*, 249-311

[297] van den Heuvel, E.P.J. 1969a, Astron. J. *74*, 1095

[298] van den Heuvel, E.P.J. 1969b, Pub. Astron. Soc. Pacific. *81*, 815

[299] van den Heuvel, E.P.J. 1970a, in: "Stellar Rotation" (ed. A. Slettebak), New York, Gordon & Breach, pp. 78-86

[300] van den Heuvel, E.P.J. 1970b, Colloquium given at the Institut d'Astrophysique, Paris, France

[301] van den Heuvel, E.P.J. 1971a, Colloquium given at the Observatoire Royal, Uccle, Belgium (January)

[302] van den Heuvel, E.P.J. 1971b, Colloquium given at the Observatoire de Genève, Geneva, October 27, 1971

[303] van den Heuvel 1973, Nature Phys. Sci., *242*, 71

[304] van den Heuvel, E.P.J. 1974, in: Proc. 16th Solvay Conference on Physics (Univ. of Brussels Press, Brussels), p.119

[305] van den Heuvel, E.P.J. 1975, Astrophys. J. *198*, L109

[306] van den Heuvel, E.P.J. 1976, in: "Structure and Evolution of Close Binary Systems" (eds. P.P. Eggleton, B. Mitton and J. Whelan), Reidel Publ. Comp., Dordrecht, p. 35

[307] van den Heuvel, E.P.J. 1977, Proc. 7th Texas Symp. Rel. Astrophysics, Annals N.Y. Acad. Sci. *302*, 14

[308] van den Heuvel, E.P.J. 1978, in: "Physics and Astrophysics of Neutron Stars and Black Holes" (eds. R. Giacconi and R. Ruffini), North Holl. Pub. Comp., Amsterdam, pp. 828-871

[309] van den Heuvel, E.P.J. 1980, in "X-ray Astronomy" (eds. R. Giacconi and G. Setti), Reidel, Dordrecht, pp. 115-128

[310] van den Heuvel, E.P.J. 1981a, Vistas in Astronomy, *25*, 95

[311] van den Heuvel, E.P.J. 1981b, in: "Pulsars" (eds. W. Sieber & R. Wielinski), Reidel Publ. Comp., Dordrecht, p. 379

[312] van den Heuvel, E.P.J. 1981c, in: "Fundamental Problems in the Theory of Stellar Evolution" (ed. D. Sugimoto et al.), Reidel Publ. Comp., Dordrecht, p. 155

[313] van den Heuvel, 1984, J. Astrophys. and Astron., *5*, 209

[314] van den Heuvel, E.P.J. 1985, in: "Birth and Evolution of Massive Stars and Stellar Groups" (eds. W. Boland and H. van Woerden), Reidel. Publ. Comp., Dordrecht, p. 107

[315] van den Heuvel, E.P.J. 1992, Nature, *356*, 668

[316] van den Heuvel 1992a, in: "X-ray Binaries and Recycled Pulsars" (E.P.J. van den Heuvel and S.A. Rappaport, eds.), Kluwer Acad. Publ. Dordrecht, pp. 233-256.

[317] van den Heuvel, E.P.J. 1992b, in: "Proc. Internat. Space Year Confer. (Satellite Symp. No. 3), ESA ISY-3 (July 1992), pp. 29-36

[318] van den Heuvel, E.P.J. Bhattacharya, D. Nomoto, K. and Rappaport, S.A. 1992, Astron. Astrophys. *262*, 97-105

[319] van den Heuvel, E.P.J. and De Loore, C. 1973, Astron. Astrophys. *25*, 387

[320] van den Heuvel, E.P.J. and Habets, G.M.H.J. 1984, Nature *309*, 698

[321] van den Heuvel E.P.J. and Heise, J. 1972 Nature Phys. Sci. *239*, 67

[322] van den Heuvel, E.P.J. Ostriker, J.P. and Petterson, J.A. 1980, Astron. Astrophys. *81*, L7

[323] van den Heuvel, E.P.J. and Rappaport, S.A. 1987, in "Physics of Be Stars" (eds. A. Slettebak and T.P. Snow), Cambridge Univ. Press, p. 291

[324] van den Heuvel and Rappaport, S.A. (eds.) 1992, "X-ray Binaries and Recycled Pulsars" Kluwer Acad. Publ. Dordrecht, 566 pp.

[325] van den Heuvel, E.P.J. and Taam, R.E. 1984, Nature *309*, 235

[326] van den Heuvel, E.P.J. and van Paradijs, J.A. 1988, Nature *334*, 227

[327] van der Hucht, K.A. 1992, Astron. Astrophys. Rev. *4*, 123-159

[328] van Kerkwijk, M.H., Charles, P.A., Geballe, T.R., King, D.L., Miley, G.K., Molnar, L.A. and van den Heuvel, E.P.J. 1992, Nature, *355*, 703-705

[329] van Kerkwijk, M.H. 1993, Ph. D. Thesis Univ. of Amsterdam (in prep.)

[330] van der Klis, M., Hasinger, G. Dotani, T., Mitsuda, K., Verbunt, F., Murphy, B.W., van Paradijs, J.A., Bellow, T., Makashima, K., Morgan E. and Lewin, W.H.G. 1992, M.N.R.A.S. (in press)

[331] van der Linden, Th.J. 1982, "Close Binaries and Stellar Evolution", Ph.D. Thesis, Univ. of Amsterdam

[332] van der Linden, Th.J. 1985, Astron. Astrophys.

[333] van Oyen, J. 1989, Astron. Astrophys., *217*, 115

[334] Ventura, J. and Pines, D. (eds.) 1991, "Neutron Stars, Theory and Observation", Kluwer Acad. Publ. Dordrecht, 592 pp.

[335] van Paradijs, J.A. 1991, in "Neutron Stars, Theory and Observation" (eds. J. Ventura and D. Pines), Kluwer Acad. Publ. Dordrecht, 289-317

[336] van Paradijs, J.A. 1993 (preprint)

[337] van Paradijs, J.A., Allington-Smith, J., Callanan, P. Charles, P.A., Hassall, B.J.M., Machin, G. Mason, K.O., Naylor, T. and Smale, A.P. 1988, Nature *334*, 684-686

[338] van Paradijs, J.A. and Lewin, W.A.G. 1990, in "Neutron Stars and Birth Events" (ed. W. Kundt), Kluwer Acad. Publ., Dordrecht

[339] Verbunt, F. 1984, Mon. Not. R.A.S., *209*, 227-240

[340] Verbunt, F. 1990, in "Neutron Stars and Their Birth Events" (ed. W. Kundt), Kluwer, Dordrecht, pp. 179-218.

[341] Verbunt, F. 1993, Ann. Rev. Astro. Astrophys. (in press).

[342] Verbunt and van den Heuvel, E.P.J. 1993 in: "X-ray Binaries" (eds. W.H.G. Lewin et al.), Cambridge University Press (in prep.)

[343] Verbunt, F., Wijers, R.A.M.J. and Burm, H. 1990, Astron. Astrophys. *234*, 195

[344] Vermeulen, R. 1989, Ph.D. Thesis, Leyden University, The Netherlands

[345] Vogel, M. 1990, "Modelle für Symbiotische Sterne", Ph.D. Thesis, Eidgen. Tech. Hochschule, Zürich (Diss. ETH Nr. 9089).

[346] Wakatsuki, S. Hikita, A. Sato, N. and Itoh, N. 1992, Astrophys J. *392*, 628

[347] Weaver, T.A., Axelrod, T.S. and Woosley, S.E 1980, in: "Type I Supernovae", Univ. of Texas, Austin, p. 113

[348] Weaver, T.A. and Woosley, S.E. 1980, in: "Supernova Spectra, AIP Conf. Proc. No. 63, (eds. R.E. Meyerott and G.H. Gillespie), N.Y. Acad.Sci., New York, p. 15

[349] Weaver, T.A., Zimmerman, G.B. and Woosley, S.E. 1978, Astrophys. J. *225*, 1021

[350] Webbink, R.F. 1984, Astrophys. J. *277*, 355

[351] Webbink, R.F. 1991, "Highly Evolved Close Binary Systems with Known or Suspected Periods", Dept. of Astron., Univ. of Illinois, Urbana, Ill.

[352] , Webbink, R.F. 1992, in: "X-ray Binaries and Recycled Pulsars (eds.E.P.J. van den Heuvel & S.A Rappaport) Kluwer Acad Publ., Dordrecht, pp. 269-280

[353] Webbink, R.F., Rappaport, S.A. and Savonije, G.J. 1983, Astrophys. J. *270*, 678

[354] Weber, F. and Glendenning, N.K. 1991, Phys. Lett.B., *265*, 1-5

[355] Weidemann, V. 1990, Ann. Rev. Astron. Astrophys. *28*, 103-137

[356] Wheeler, J.C. and Harkness, R.P. 1990, Report Progr. Phys. *53*, 467

[357] Whelan, J. and Iben, I. 1973, Astrophys. J., *186*, 1007

[358] Whitcomb, S. 1993, "Classical & Quantum Gravity", (Proc. Journéés Relativ. 1992, held in Amsterdam) (in press).

[359] Wijers, R.A.M.J. 1991, "On the Properites and Evolution of Binary and Single Neutron Stars", Ph.D. Thesis, Univ. of Amsterdam

[360] Wijers, R.A.M.J., van den Heuvel, E.P.J. van Kerkwijk, M.H., and Bhattacharya, D. 1992, Nature, *355*, 593

[361] Wijers, R.A.M.J., J.A. van Paradijs and van den Heuvel, E.P.J. 1992, Astron. Astrophys. *261*, 145-153

[362] Wilson, O.C. 1960, in: "Stellar Atmospheres" (ed. J.L. Greenstein), University of Chicago Press, pp. 436-465

[363] Wolszczan, A. 1991, Nature, *350*, 688

[364] Wolszczan, A. 1992, in: "X-ray Binaries and Recycled Pulsars" (E.P.J. van den Heuvel and S.A. Rappaport eds.) Kluwer Acad. Publ. Dordrecht, 93

[365] Wolszczan, A. and Frail, D.A. 1992, Nature *355*, 145

[366] Woosley, S.E. and Weaver, T.A. 1986, Annual. Rev. Astron. Ap. *24*, 205-253

[367] Woosley, S.E. and Baron, E. 1992, Astrophys. J., *391*, 228

[368] Woosley, S.E., Timmes, F.X. and Baron, E. 1992, in: "X-ray Binaries and Recycled Pulsars" (eds. E.P.J. van den Heuvel & S.A. Rappaport), KLuwer Acad. Publ. Dordrecht, pp. 167–187

[369] Zeldovitch, Ya.B. 1964 , Sov. Phys. Dokl. *9*, 246

[370] Zeldovitch, Ya.B. and Guseinov, O.H. 1965, Astrophys. J., *144*, 840

Index

Springer-Verlag and the Environment

GPSR Compliance

The European Union's (EU) General Product Safety Regulation (GPSR) is a set of rules that requires consumer products to be safe and our obligations to ensure this.

If you have any concerns about our products, you can contact us on ProductSafety@springernature.com

In case Publisher is established outside the EU, the EU authorized representative is:

Springer Nature Customer Service Center GmbH
Europaplatz 3
69115 Heidelberg, Germany

Batch number: 10704030

Printed by Printforce, the Netherlands